I0885532

Deterministic and Stochastic Approaches in Computer Modeling and Simulation

Radi Petrov Romansky
Technical University of Sofia, Bulgaria

Nikolay Lyuboslavov Hinov
Technical University of Sofia, Bulgaria

IGI Global
PUBLISHER of TIMELY KNOWLEDGE

A volume in the Advances in Computational Intelligence and Robotics (ACIR) Book Series

Published in the United States of America by
IGI Global
Engineering Science Reference (an imprint of IGI Global)
701 E. Chocolate Avenue
Hershey PA, USA 17033
Tel: 717-533-8845
Fax: 717-533-8661
E-mail: cust@igi-global.com
Web site: http://www.igi-global.com

Library of Congress Cataloging-in-Publication Data

Names: Romansky, Radi, author. | Hinov, Nikolay, 1970- author.
Title: Deterministic and stochastic approaches in computer modeling and simulation / authored by Radi Romansky and Nikolay Hinov.
Description: Hershey, PA : Engineering Science Reference, [2023] | Includes bibliographical references and index. | Summary: "The purpose of this book is to make a summary of the possibilities of modeling research, mainly in the computer field, by discussing the areas of problems in mathematical formalization and abstract description, discrete and probabilistic modeling approaches, computer simulation and the empirical approach of statistical modeling"-- Provided by publisher.
Identifiers: LCCN 2023024055 (print) | LCCN 2023024056 (ebook) | ISBN 9781668489475 (hardcover) | ISBN 9781668489482 (paperback) | ISBN 9781668489499 (ebook)
Subjects: LCSH: Computer simulation. | Stochastic processes--Computer simulation. | Deterministic chaos--Computer simulation.
Classification: LCC QA76.9.C65 R65 2023 (print) | LCC QA76.9.C65 (ebook) | DDC 003/.76--dc23/eng/20231005
LC record available at https://lccn.loc.gov/2023024055
LC ebook record available at https://lccn.loc.gov/2023024056

This book is published in the IGI Global book series Advances in Computational Intelligence and Robotics (ACIR) (ISSN: 2327-0411; eISSN: 2327-042X)

British Cataloguing in Publication Data
A Cataloguing in Publication record for this book is available from the British Library.

For electronic access to this publication, please contact: eresources@igi-global.com.

Advances in Computational Intelligence and Robotics (ACIR) Book Series

Ivan Giannoccaro
University of Salento, Italy

ISSN:2327-0411
EISSN:2327-042X

Mission

While intelligence is traditionally a term applied to humans and human cognition, technology has progressed in such a way to allow for the development of intelligent systems able to simulate many human traits. With this new era of simulated and artificial intelligence, much research is needed in order to continue to advance the field and also to evaluate the ethical and societal concerns of the existence of artificial life and machine learning.

The **Advances in Computational Intelligence and Robotics (ACIR) Book Series** encourages scholarly discourse on all topics pertaining to evolutionary computing, artificial life, computational intelligence, machine learning, and robotics. ACIR presents the latest research being conducted on diverse topics in intelligence technologies with the goal of advancing knowledge and applications in this rapidly evolving field.

Coverage

- Artificial Intelligence
- Machine Learning
- Agent technologies
- Neural Networks
- Evolutionary Computing
- Intelligent Control
- Natural Language Processing
- Pattern Recognition
- Automated Reasoning
- Computer Vision

IGI Global is currently accepting manuscripts for publication within this series. To submit a proposal for a volume in this series, please contact our Acquisition Editors at acquisitions@igi-global.com or visit: https://www.igi-global.com/publish/.

The Advances in Computational Intelligence and Robotics (ACIR) Book Series (ISSN 2327-0411) is published by IGI Global, 701 E. Chocolate Avenue, Hershey, PA 17033-1240, USA, www.igi-global.com. This series is composed of titles available for purchase individually; each title is edited to be contextually exclusive from any other title within the series. For pricing and ordering information please visit https://www.igi-global.com/book-series/advances-computational-intelligence-robotics/73674. Postmaster: Send all address changes to above address.

Titles in this Series

For a list of additional titles in this series, please visit: https://www.igi-global.com/book-series/advances-computational-intelligence-robotics/73674

Advances in Artificial and Human Intelligence in the Modern Era
S. Suman Rajest (Dhaanish Ahmed College of Engineering, India) Bhopendra Singh (Amity University, Dubai, UAE) Ahmed J. Obaid (University of Kufa, Iraq) R. Regin (SRM Institute of Science and Technology, Ramapuram, India) and Karthikeyan Chinnusamy (Veritas, USA)
Engineering Science Reference • © 2023 • 420pp • H/C (ISBN: 9798369313015) • US $300.00

Handbook of Research on Advancements in AI and IoT Convergence Technologies
Jingyuan Zhao (University of Toronto, Canada) V. Vinoth Kumar (Jain University, India) Rajesh Natarajan (University of Applied Science and Technology, Shinas, Oman) and T.R. Mahesh (Jain University, India)
Engineering Science Reference • © 2023 • 372pp • H/C (ISBN: 9781668469712) • US $380.00

Scalable and Distributed Machine Learning and Deep Learning Patterns
J. Joshua Thomas (UOW Malaysia KDU Penang University College, Malaysia) S. Harini (Vellore Institute of Technology, India) and V. Pattabiraman (Vellore Institute of Technology, India)
Engineering Science Reference • © 2023 • 286pp • H/C (ISBN: 9781668498040) • US $270.00

Handbook of Research on Thrust Technologies' Effect on Image Processing
Binay Kumar Pandey (Department of Information Technology, College of Technology, Govind Ballabh Pant University of Agriculture and Technology, India) Digvijay Pandey (Department of Technical Education, Government of Uttar Pradesh, India) Rohit Anand (G.B. Pant DSEU Okhla-1 Campus, India & Government of NCT of Delhi, New Delhi, India) Deepak S. Mane (Performance Engineering Lab, Tata Research, Development, and Design Center, Australia) and Vinay Kumar Nassa (Rajarambapu Institute of Technology, India)
Engineering Science Reference • © 2023 • 542pp • H/C (ISBN: 9781668486184) • US $350.00

Multi-Disciplinary Applications of Fog Computing Responsiveness in Real-Time
Debi Prasanna Acharjya (Vellore Institute of Technology, India) and Kauser Ahmed P. (Vellore Institute of Technology, India)
Engineering Science Reference • © 2023 • 280pp • H/C (ISBN: 9781668444665) • US $270.00

Global Perspectives on Robotics and Autonomous Systems Development and Applications
Maki K. Habib (The American University in Cairo, Egypt)
Engineering Science Reference • © 2023 • 405pp • H/C (ISBN: 9781668477915) • US $360.00

IGI Global
PUBLISHER of TIMELY KNOWLEDGE

701 East Chocolate Avenue, Hershey, PA 17033, USA
Tel: 717-533-8845 x100 • Fax: 717-533-8661
E-Mail: cust@igi-global.com • www.igi-global.com

Table of Contents

Preface

The digital society is characterized by the increasing impact of information and communication technologies (ICT) in the immediate activities of people and communications between them. In this reason, is mass "informatization" of society is observed, which is a comprehensive process for the transformation of various types of information and the formation of new knowledge with the aim of continuously increasing the quality of life and status of society in social, economic, technological, and scientific-technical directions (Romansky, 2021a).

The term "*information*" is subject to different definitions depending on the field of use. In a broad sense, information means any data about the features, type and properties of objects and phenomena in the environment, which reduces the existing incompleteness of knowledge. This also applies to the ongoing processes in living and non-living nature. In the contemporary digital age, information is the main object in our society, which is perceived as knowledge that can be used in society to generate new knowledge. Hence, information is understood as the content (meaning) that is assigned to specific data and is maintained and processed by a corresponding information environment, most often realized through computer means. Thus, the main goal of specialists in this field is to create technological and organizational conditions and mechanisms for automated collection and search processing of information, its systematization, storage, and distribution. Information, as a concept, makes sense when forming a given decision as a result of processing source data or when communicating (transmitting) through messages between two parties (from a source to a receiver of messages). This also defines two main sides related to information – information processing and information communication.

Information processing is the process of converting the source set of elements (data set) into another set, called the result set. From a practical point of view, it should be known that information processing (automatic or not) does not add new information, but only extracts interesting information contained in the set of source data. Adding new information and drawing logical conclusions is the task of knowledge (fact) processing in systems with elements of artificial intelligence. From a mathematical point of view, information processing can be considered as a functional transformation $f{:}D{\rightarrow}R$ from the set of input data D to the set of output results R. The technical implementation of this functional transformation uses appropriate means that form the two main subsystems – hardware and software. Therefore, the real implementation of computer processing can be represented as $f^{*}{:}X{\rightarrow}Y$, with components the physical realizations of f, D and R. To transition between the two levels, additional means are applied: encoding $\varphi_1{:}D{\rightarrow}X$ (transformation of input information to binary code) and decoding $\varphi_2{:}Y{\rightarrow}R$ (inverse conversion of binary code to output information).

Information communication is the process of transmitting messages from a source to a receiver using a certain type of transmission medium. It arises with the need to ensure effective and reliable communication links when the volume of information transmitted through communication systems (telephone, telegraph, etc.) increases. The importance of this process increases with the entry of computer equipment and technology into material production and public spheres. The development of multimedia technologies allowed communication between people to be put on another, different interactive basis. The global network (Internet) is no less important for information. The World Information System (www) covers the entire planet like a "web" and enables the exchange of information over considerable distances.

The organization of any computer processing uses the resources of the hardware and software components of the computer environment, which is built based on a certain architectural model. Computer architecture (Hennessy & Patterson, 2017) represents a set of quantitative and qualitative characteristics (system parameters) for the implementation of a certain type of computer processing model in a given application area, reflecting the functional requirements and the relationships between individual components. It sets the basic recommendations and determines the functionality based on defining specific parameters, such as the instruction set (Instruction Set Architecture), machine word size, supported data and instruction formats, basic system feature levels, memory access methods, interrupt rules, peripheral connection management, operating modes, etc.

The design of any such environment requires prior assurance of adequate implementation of the planned functionalities, which can be done by conducting research with a selected approach. Usually, any such research, regardless of the field, starts with an initial formalization of the processes to clearly formulate the main goal. The formalization is based on abstractions, the demonstration of their different levels is presented in (Hanson & Walker, 2021). In this case, the goal of the presented computational hierarchy is to form a more general set of criteria when solving the problem posed by applying the capabilities of finite state machines. In principle, this theory is applicable to the study of deterministic processes (such as computing), because any process can be described by successive transitions between different states depending on the functional organization. Research in this field is presented in (Zozulia et al, 2022) where algorithmization of software testing as a stage of digital devices design is discussed. The formalization of information processing in each computer system (CS) can be represented by formal means, which include: ✓ a set of abstract elements (concepts); ✓ terms reflecting data and algorithms for their processing; ✓ procedures for the interaction between data and algorithms. The formal means include an alphabet for data presentation, methods and means of their coding, algorithms for data presentation, transformation and processing, input-output procedures, procedures for storing and searching data, procedures for management, etc. In a generalized aspect, the work of a CS can be represented by an ordered pair $\langle S, A \rangle$, where S reflects the functional organization and features of the computer resource for the implementation of information processes, and A represents the algorithmic and programmatic features of computer processing.

The classical organization of computer processing obeys the hierarchical model of two levels – macro & micro levels (Romansky, 2020a), formed by system resources characterized by their own parameters. The proper organization of the processes requires the determination of quantitative estimates of the actual levels of the parameters, which can be done by conducting research experiments and analyzing behavior. Any process, regardless of level, can be described by a functional algorithm that organizes the computations and manages the transitions between the states they fall into (Safyannikov et al, 2021). Processes can be represented as a flow of events, and the paper proposes a generalized functional diagram and presents timing diagrams of the functional organization of the flow conversion components. Such an approach allows to make a preliminary formalization in accordance with Turing Computational Theory (Ramos et al, 2022) by using suitable approach, for example using Graph theory (Munir et al, 2020) for formal description of computer process as a State Transition Network – STN (Romansky, 2020b). In addition, the study of temporal and structural parameters of computer processing, more of which are stochastic in nature, can be carried out (Romansky, 2021b; Romansky, 2022a). The general goal of any research is to determine a solution that provides adequate results at a reasonable cost. For this purpose, it is necessary to draw up a correct plan of the experiment based on predefined primary and secondary factors.

In scientific literature, there are different, although close, interpretations of the term "computer architecture". For example, in (Hennessy & Patterson, 2017) three aspects are defined: ✓ the capabilities of the set of supported instructions; ✓ the ways of organizing the processes; ✓the features of the hardware configuration. Another interpretation defines computer architecture as a multi-level hierarchy of hardware and software tools from which the computer system evolves. In this reason, each computer system is a concrete implementation of a selected architectural model by combining the system resource (hardware and software) to provide effective levels (values) for the set of system characteristics when implementing the applications. The two subsystems are characterized by the respective features presented below, the organization of which affects the overall performance and efficiency of computer processing.

1. Hardware
   1.1. Basic structural organization of the system.
   1.2. Organization of the system memory (primary, secondary, virtualization).
   1.3. Organization of the Input/Output (I/O) system.
   1.4. Processes managing in hardware aspect.
2. Software
   2.1. System software (Operation system, system programs, programs for system support, etc.)
   2.2. Program languages (universal and specialized) and program environments for program development and verification.
   2.3. Applied software (application software packages and application software systems).
   2.4. Processes managing in software aspect.

The structural organization of a computer system is a concrete implementation of a selected architectural model by uniting basic components that interact in support of purposeful computer processing: ✓ processing devices for the implementation of the selected model of computer processing (sequential, pipelined, parallel, etc.); ✓ different types and characteristics of memories for building the system memory; ✓ peripheral and external devices included in the composition of the I/O system; ✓ Interconnection network (interfaces and/or connection lines) for communication between individual components and connection to other systems. The functional realization of processing a given task in a computer environment is associated with transitions between a finite number of discrete states in which it can fall, which allows for the formalization of processes to use a directed graph with nodes representing the states of the process and directed arcs for the transitions between them.

The main goal in the organization of computer processes is to ensure the necessary level of performance, for which it is recommended to carry out various experiments to study the behavior of the environment and its system characteristics. The general definition of performance as the amount of work performed in a certain period of time is inapplicable in computer processing, and in this respect specific indexes (factors) of performance are defined, such as number of instructions executed per unit time (MIPS) or number of operations performed in a certain time (FLOPS). To determine the levels of the main parameters when designing or refining a computing environment, it is recommended to conduct an initial investigation of the prototype before proceeding to its concrete implementation (Romansky, 2017). For this purpose, different approaches can be applied in a deterministic and stochastic aspect, and the specific research must be planned and organized depending on the main task set. In this reason, the purpose of this book is to make a summary of the possibilities of modeling research, mainly in the computer field, by discussing the areas of problems in mathematical formalization and abstract description, discrete and probabilistic modeling approaches, computer simulation and the empirical approach of statistical modeling. A brief summary of the contents of this book is presented below.

Chapter 1 is an introduction to the field of system investigation with presenting basic directions used in the computing field such as benchmark, modelling and measuring (monitoring). A short discussion for each approach is presented to determine the place of modeling in the sphere of methods and tools for investigation and evaluation of computer parameters and processes. Some examples of carried out experiments for discussed approaches are presented as initial information. In general modeling theory, two main groups of models can be defined (physical and mathematical), and the place and features of computer modeling as a part of the mathematical models is discussed. Another topic discussed in this first chapter is the consideration of modeling as a method of scientific knowledge. In addition, the expediency of using computer modeling, as well as the main directions of its application are discussed.

Chapter 2 is devoted to computer systems and processes as an object of modeling. A brief review of the role of "information", "information processing" and "information technology" is made, to specify the place of the technological environment for implementation of information processing and to define its main functionalities. The two basic levels (micro- and macro-) of a traditional computer environment are discussed with an initial formalization of basic system resources. The main objects of analysis are the performance (with its indices) and the workload based on input workflow. In this respect, a formal description of workload and specifying the role of the process profile are presented. In addition, a formalization of information technology is presented.

Chapter 3 discusses technological aspects of computer modeling and approaches using mainly computer systems and processes investigation. For this purpose, a classification of the applied computer models was made, defining two groups - the homogeneous models and the heterogeneous combined models. A brief description of the main features of each of the models specified in the classification is presented, and relevant examples are also presented. A general procedure for conducting a model study is defined and a methodological scheme is presented, including separate independent phases, and an explanation is given for each of them.

Chapter 4 deals with technological organization of model research, discussing as a first part the requirement for adequacy of computer model and its efficiency. The second part is discussion of model time (synchronous and asynchronous), which should represent the real time in the studied processes. The last part discusses the technical and software tools used for developing computer models. The first group includes stochastic machines, multiprocessor and hybrid complexes, and distributed systems. The second group is presented by universal and specialized programming languages and simple examples are given.

Chapter 5 is dedicated to the possibilities of formalization of an investigated object and its abstract description with appropriate mathematical means. The need for decomposition of complex systems and application of formal theory for mathematical formalization and abstract description is discussed. A major place is devoted to the application of discrete and stochastic structures in the initial formalization of the studied systems and processes. The possibilities of formalization using sets, graphs, finite automata, as well as the application of elements of probability theory, the theory of Markov processes and the theory of queues are discussed.

Chapter 6 discusses the possibilities for functional modeling of systems and processes, presenting the features of the used tools. The first part deals with the application of functional modeling means for study of system resources and their presentation as a synchronous or asynchronous network of functional blocks. The second part presents selected technologies and tools for functional describing and investigation of algorithmic structures (logical scheme of an algorithm, Dijkstra D-cards, discrete and stochastic graph description). The third part deals with functional modeling of information processes and structures by IDEF technology, Data flow diagrams, Dialog transition networks, Generalized net and UML apparatus.

Chapter 7 discusses deterministic analytical modeling, and its features and possible application are presented. The object of the first section is the general organization of this approach and presents its main implementation in deterministic, stochastic, and combined versions. The second section deals with the use of mathematical approximations as an apparatus for discrete model investigation, presenting the analytical approximation with a mathematical description and the application of analytical modeling in the study of power electronic converter. The next two sections are devoted to the application of graphs and Petri nets for the realization of deterministic models and their investigation, with presenting selected examples for illustration. The last section discusses pseudo-stochastic analytical modeling, which is a deterministic variant of the description of processes that are stochastic in nature.

Chapter 8 deals with the features of stochastic modeling and its application in the investigation of probabilistic computer processing. A brief overview of the possibilities of probabilistic analytical modeling was made in the first part and the relationship of discrete and continuous random variables to stochastic modeling is discussed. The second part is devoted to the stochastic analytical approximation. The third part is directed to the discussion of the stochastic modeling based on the theory of Markov processes, concentrating on random processes with discrete states and discrete time (Markov Chain) and on analogous processes, but with continuous time. Various experiments are presented to illustrate the possibilities offered by this approach with an analysis of the obtained results.

Chapter 9 discusses the stochastic approach for model investigation by using queuing systems (QS) with waiting buffer and service unit. Single-channel QS is discussed in the first part with presentation of main characteristics such as workload of the resource, queuing length, total number of requests in system, waiting time and time for presence in the system. The second part deals with organization of analytical investigation of QS based on the theory of Markov Chain (with discrete and continuous time). The investigation of conditions for transient and steady state regimes is presented for three types of QS – with an infinite buffer, with a limited buffer, and QS with request rejections. In the third part, some basic parameters of QS are summarized, and priority QS and queuing networks are presented. Discussion of stochastic workload modeling is made in the last part.

Chapter 10 presents approaches and solutions in organizing the study of complex systems using queuing systems (QSs) and networks (QNs). In the first part, the object of the model study by using QS is a multicomputer system with a cube architecture. The second part is dedicated to priority QSs whit discussion of various applications of queuing modeling and presentation of concrete examples. The third part deals with QNs which are combinations of different QSs. The stochastic workload modeling is the object of the last part where the type of specification and the possibilities for analytical representation and modeling are considered.

Chapter 11 is devoted to the technological aspects of computer simulation and its application in solving specific problems. A comparison of the features of simulation and emulation is made, and the main advantages and disadvantages of computer simulation are discussed with presentation of the mine phases in simulation model developing and clarifying its adequacy. The two main types of simulation – discrete-event and continuous, and the mine structural components of a simulation model are discussed. Different examples for application of computer simulation are presented in the last section.

Chapter 12 introduces basic concepts of random sequence generation that can be used in the simulation modeling of random flows. Two main types of Random Number Generators (RNGs) are presented – True RNGs and Pseudo-RNGs and a brief overview of these types was made in the first part. The second part is devoted to Pseudo-RNGs, which are mainly used in simulation modeling and their implementation based on discrete and continuous probability distribution law is discussed. Several examples related to program realization of random sequences are presented.

Chapter 13 has a min goal of presenting basic principles for organization and realization of simulation experiment. The first part discusses the basic forms of simulation research (statistical simulation and functional simulation), and a summary of simulation languages and environments is presented. The second part is devoted to the planning of simulation experiments and the organization of implementation. This is related to determining the main factors and their levels, and optimal time for simulation. In addition, pre-processing of the experimental data is formulated. The last part presents examples for organization of simulation experiments.

Chapter 14 deals with principles of organization and application of statistical modeling. The first section discusses the essence of the statical approach and its application in scientific research. The second part is devoted to mathematical statistics, with a brief presentation of basic concepts (simple and statistical order, statistical estimates, numerical characteristics, and basic methods). Particular attention is paid to the method of least squares. In the third part, the characteristic surface is presented and the two main groups of statistical models – univariate and multivariate are discussed. Some statistical experiments are presented in the last part.

Chapter 15 discusses the possibilities for adequate processing of experimental data collected based on research experiments and the appropriate interpretation of results and assessments from its analysis. In principle, these data are numerical, and processing mainly is based on statistical analysis and graphical interpretation of the estimates is the last phase of processing. Three parts are included in the chapter. In the first part, an approach is proposed for primary processing of experimental data and the main activities are proposed including determination of individual criteria for evaluation, logical completeness of the estimations, and the need for preliminary filtering of the sample). The second part is dedicated to the basic processing of experimental data, where statistical approaches for this are discussed. In the last part, the basic methods are tabular and graphical interpretation of estimates from the analysis of experimental data.

Radi Romansky
Technical University of Sofia, Bulgaria

Nikolay Hinov
Technical University of Sofia, Bulgaria

REFERENCES

Hanson, J. R., & Walker, S. I. (2021). Formalizing falsification for theories of consciousness across computational hierarchies. Neuroscience of Consciousness.) doi:10.1093/nc/niab014

Hennessy, J., & Patterson, D. (2017). *Computer Architecture: A Quantitative Approach* (6th ed.). Morgan Kaufman / Elsevier Inc.

Munir, S., Jami, S. I., & Wasi, S. (2020). Knowledge graph based semantic modelling for profiling in Industry 4.0. *International Journal on Information Technologies and Security*, *12*(1), 37–50.

Ramos, T. M. F., Almeida, A. A., & Ayala-Rincón, M. (2022, January). Formalization of the computational theory of a Turing complete functional language model. *Journal of Automated Reasoning*, *30*(4), 1031–1063. doi:10.100710817-021-09615-x

Romansky, R. (2017). *Information servicing in distributed learning environments. Formalization and model investigation.* LAP LAMBERT Academic Publishing.

Romansky, R. (2020a). An approach for mathematical modelling and investigation of computer processes at a macro level. *Mathematics*, *8*(10), 1838. doi:10.3390/math8101838

Romansky, R. (2020b). Formalization and discrete modelling of communication in the digital age by using graph theory. In M. Pal, S. Samanta, & A. Pal (eds), Handbook of Research on Advanced Applications of Graph Theory in Modern Society. IGI Global. doi:10.4018/978-1-5225-9380-5.ch013

Romansky, R. (2021a). Informatization of society in the digital age. *Biomedical Journal of Scientific & Technical Research*, *33*(3), 25902–25910. doi:10.26717/BJSTR.2021.33.005418

Romansky, R. (2021b). Mathematical modelling and study of stochastic parameters of computer data processing. *Mathematics*, *9*(18), 2240. doi:10.3390/math9182240

Romansky, R. (2022a). Stochastic approach to investigate protected access to information resources in combined e-learning environment. *Mathematics*, *10*(16), 2909. doi:10.3390/math10162909

Safyannikov, N., Chepasov, A., & Bondarenko, P. (2021). *Functional organization of elements of stream converters with actualization of states*. 10th Mediterranean Conference on Embedded Computing (MECO), Budva, Montenegro. ()10.1109/MECO52532.2021.9460167

Zozulya, M.M., & Kravets, O.Ja., Atlasov, I.V., Aksenov, I.A., Bozhko, L.M. & Rahman, P.A. (2022). Algorithmization of the software testing system based on finite automata. *International Journal on Information Technologies and Security*, *14*(1), 77–86.

Acknowledgement

This study is financed by the European Union-NextGenerationEU, through the National Recovery and Resilience Plan of the Republic of Bulgaria, project № BG-RRP-2.004-0005.

Chapter 1
Modeling as a Research Approach in Computing

ABSTRACT

This chapter is an introduction in the field of system investigation with presenting basic directions used in the computing filed as benchmark, modelling and measuring (monitoring). A brief discussion of features and capabilities for each approach without going into depth is presented to determine the place of modeling in the sphere of methods and tools for investigation and evaluation of computer parameters and processes. Some examples of carried out experiments for discussed approaches are presented as initial information. In general modeling theory, two main groups of models can be defined (physical and mathematical), and the place and features of computer modeling as a part of the mathematical models is discussed. Another topic discussed in this first chapter is the consideration of modeling as a method of scientific knowledge. In addition, the expediency of using computer modeling, as well as the main directions of its application are discussed.

1. INTRODUCTION: METHODS AND MEANS FOR RESEARCH IN COMPUTER FIELD

The computer system (CS) is a set of hardware and software components that function together in solving a certain class of tasks, which requires the application of an appropriate approach to studying their functionality. Each study must be tailored to a set main goal and an appropriate approach must be chosen to achieve it. On the other hand, for the effective conduct of the research, it is necessary that it is based on a correctly prepared experiment plan based on predetermined primary and secondary factors. Primary factors are essential for conducting the research, because they reflect basic parameters of the object or process, important for achieving the set goal. Secondary factors are additional parameters and characteristics that somehow influence the primary factors and therefore cannot be ignored.

When conducting each research, the necessary information is accumulated, which is subjected to analysis (Wrigh & Ma, 2022) to obtain estimates for the selected performance indicators, allowing to make an adequate assessment of behavior (Romansky, 2022b). This is valide for conducting research in

DOI: 10.4018/978-1-6684-8947-5.ch001

all forms of computing, from traditional computing, e.g., in healthcare data volumes (Shukla, 2023a) to modern cloud computing technologies, Internet of Things, fog computing (Rathi et al., 2022), etc. An important requirement is to ensure reproducibility of research results, which is discussed in (Raghupathi et al., 2022), where research in computing was reviewed based on three factors – method, data, and experiment. The conclusion drawn is that appropriate methods must be chosen for each specific study and correct determination of factors to ensure reproducibility of experimental results.

One of the main requirements for the successful conduct of an effective experiment in the field of computer technology is the choice of the right method and means of research, and the general classification of the main directions (methods) is presented in Figure 1. Three main directions have been formed with the corresponding technological approaches and research methods, which are oriented towards experiments for the study of computer systems and processes. Each one of these directions has its own specifics and offered opportunities for conducting effective research, and it is possible to apply them together in combined experiments.

Figure 1. General classification of methods for investigation in computing

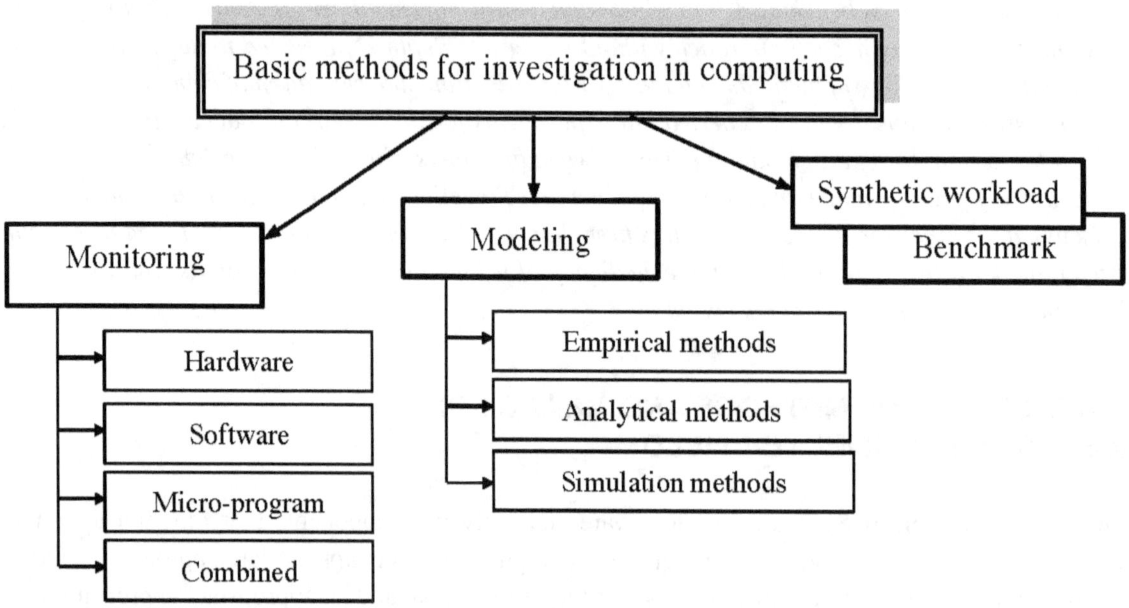

Synthetic Workload and Benchmark

Analytical dependencies (program mixtures) are one of the initial directions in the study of computer system performance parameters. They are implemented as software tools giving an approximate estimate of processor performance through usability weights for individual classes of operations when solving a given class of tasks. The program load on the processor is defined as a priori based on a statistical analysis of executed typical programs for various applications. So, for example, if the number of operation classes (in particular, all operations) is n, then the performance of the processor is calculated using the formula:

$$P = \frac{\sum_{i=1}^{n} a_i}{\sum_{i=1}^{n} a_i.t_i},$$

where: t_i – execution time of the i-th operation; a_i – coefficient (weight) of using the i-th operation in the calculation process. These analytical mixtures are particularly suitable for classical computer architecture. They were obtained as a result of statistical analysis of a large number of tasks, and the estimates obtained through them are relevant mainly for processors processing scientific and applied tasks. A basic requirement is to have a maximum match of the operations (instructions) supported by the processor under study with those of the used program mixture.

The development of the above idea is represented by synthetic workload means, which are techniques for evaluating processor performance based on sessions of interdependent requests with dynamically generated responses (Krishnamurthy et al., 2006). The article states that the requirement that dependencies between requests be reflected in synthetic workloads *"poses significant challenges for automating the construction of representative synthetic workloads and manipulating workload characteristics for sensitivity analyses"* and offers a tool to overcome them by applying a technique for automatically creating a synthetic workload. Another opportunity for applying synthetic workload technique in cloud computing is discussed in (Bahga & Madisetti, 2011), based on two different models – benchmark application and model of workload, and design and implementation of a synthetic workload generator is presented. The main idea is to combine *"the specifications of benchmark and workload models to create workload specifications that are used by a synthetic work-load generator to generate synthetic workloads for performance evaluation of cloud computing applications*".

Another key issue in processor and system design is energy efficiency research. In (Du Bois et al., 2011) the limitations of some existing tools offered by Embedded Microprocessor Benchmark Consortium (EEMBC), Standard Performance Evaluation Corporation (SPEC), etc. are discussed, and in this aspect the author's product SWEEP is proposed for generating synthetic workloads with specific behavioral characteristics. The tool allows changing the instruction mix while maintaining ILP (Instruction Level Parallelism), the memory access model and the intensity of I/O operations. The development of mobile communications and especially mobile cloud computing poses the need to create a new framework for conducting research, and in this direction in (To et al., 2016) a specialized synthetic workload generator SCAWG is proposed to create common data sets for conducting of the experiments, allowing for tractable research. In addition to the product, an *"open-source toolbox to demonstrate the feasibility and applicability of SCAWG"* has been developed with the possibility of expansion. The toolbox is designed to be generic and extensible, which would expedite experimental studies.

Benchmarking is the act of evaluating the relative performance of a computer system (processor), in which a computer program or set of programs is run to obtain a comparison of the characteristics of a given object with some benchmark. The benchmarks used provide a comparison of the performance of subsystems of different computer architectures, because a comparison based on clock frequency does not give an adequate assessment. A thorough discussion of modern benchmarks and their development is done in (Kounev et al., 2020). The presented material can be used as a manual for comparative analysis of systems and their constituent components in the field of modern information and communication technologies. This book includes specific applications and case studies based on data from consortia such as the Standard Performance Evaluation Corporation (SPEC) and the Transaction Processing Performance Council (TPC)

The performance of a CS can be evaluated by analyzing a group of factors, such as response time (time from the start of a task to its completion), the load on individual devices and the entire system, the throughput of connection lines, etc. It should give that the traditional MIPS and FLOPS estimates give a subjective result, because they directly depend on the instruction set and the operations supported by the processor (the specific machine level). A general systematization of basic benchmark means determines the following groups:

- Application-based tests which start real programs for execution time measuring.
- Playback tests used protocols for system calls which are suitable for evaluation of separate subsystems (hard disk, CD ROM, graphical sub-system, etc.).
- Synthetic test to imitate the activity of programs in individual subsystems of the computer and is most often used when testing the performance of the processor.
- Inspection tests which work directly with real components of the CS.

Standard Performance Evaluation Corporation (SPEC, 2022) is one of the companies that develop and offer benchmark tools for performance evaluation of various technologies – CPU, cloud, high performance computing, graphical processors, server machines, etc. SPECint is a computer benchmark specification for CPU integer processing power and the first SPEC test suite, CPU92, was announced in 1992. The latest version SPEC CPU2017 benchmark suite is widely used in industry and academic communities, but due to the large size and complexity in evaluation, its use is not easy, as shown by the analysis in (Panda et al., 2018). The analysis is based on experiments using a performance counter and statistical techniques. It also states that many of the SPEC CPU2006 benchmarks have been replaced by larger and more complex workloads in the SPEC CPU2017 suite. In addition, it is concluded that there is no evidence of significant differences between the CPU2006 and SPEC CPU2017 benchmark suites regarding performance or machine load requirements. An example of obtained result bus using some versions of SPECint is shown in Figure 2.

Figure 2. Evaluation of CPU performance in different time periods

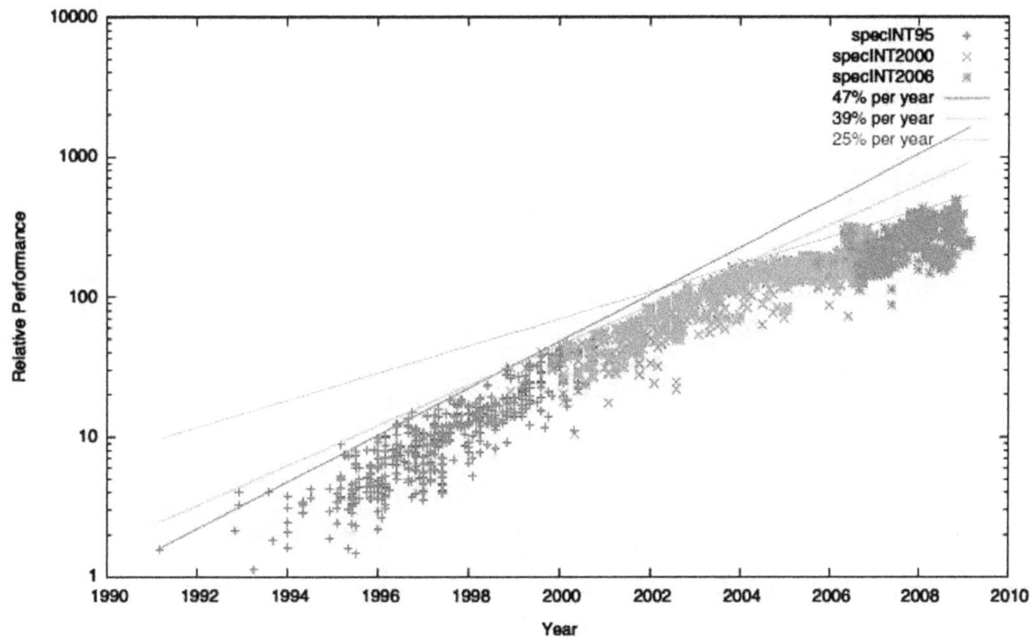

Monitoring of Computer Characteristics

When measuring computer characteristics, the information necessary to determine the ratings is obtained directly from the studied CS through various measuring means united under the name monitors. A basic requirement is the accessibility of the measured object to include probes to selected control points. The measuring device (monitor) registers the occurrence of certain events in the object of measurement and accumulates the information in an appropriate way, and the general structure of a monitor is shown in Figure 3.

Figure 3. Basic structure of a monitor

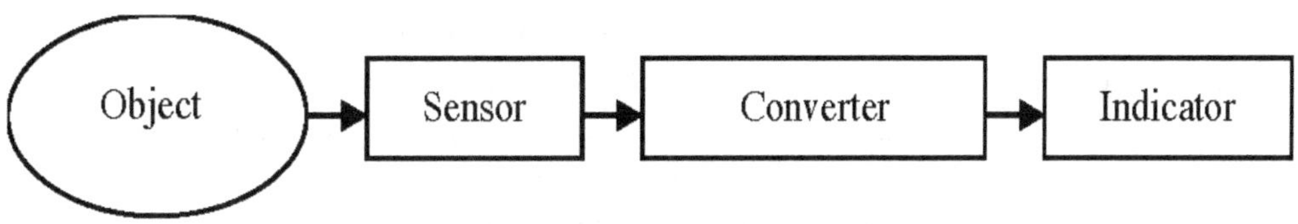

The following requirements are placed on the monitors: measurement accuracy; minimal impact on the object of measurement; high resolution; wide range of applicability; possibility of pre-processing the data. To measure values of system parameters of CS, the means presented below are applied.

Embedded system monitors. They are pre-included in the CS equipment by the manufacturer and represent registers, timers, counters that are programmatically accessible by the user or by the operating system (OS). They serve to continuously monitor a specific characteristic of the CS and can be read if desired. In addition to hardware-implemented built-in measuring means, specialized monitoring instructions included in the general system of instructions for a given processor can also be used (firmware monitoring).

Figure 4 shows the format of an example system instruction-monitor for monitoring the occurrence of an event, which is defined by an event indicator E. At each specific execution of the instruction and after deciphering the operation code (Code), the values of the system timers associated with the event and are stored together with the indicator E in the memory (RAM) at address $\alpha=(X)+(B)+D$, formed by the contents of the index register X, the base register B and the offset register D. After the registration the index register is incremented, i.e., $X=(X)+1$.

Figure 4. System instruction for registration of an event in computing

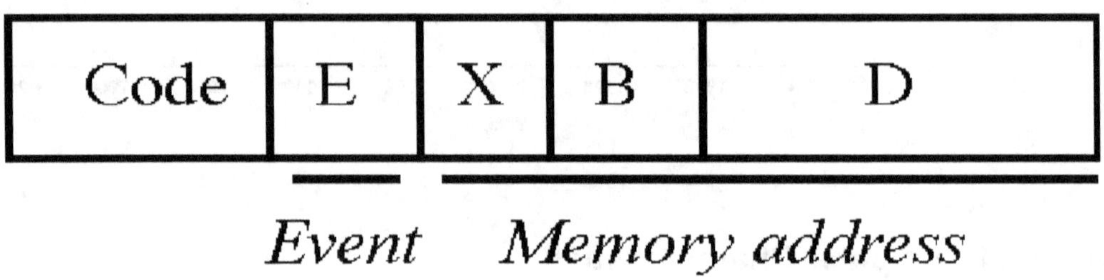

Microprocessors from the P6 (Pentium) family have built-in real-time logging and performance monitoring. For real-time marking, they have a 64-bit TSC (Time Stamp Counter), which is reset when a RESET signal is received and increments by 1 each machine cycle. It can be read by RDTSC instruction.

The built-in monitoring tools for the P6 processors allow monitoring various parameters during program execution - they fix the occurrence of certain events or their duration in time. Some of the main groups of monitoring events are as follows:

a) cache memory usability control - number of reversals for data selection or instruction selection; number of rows filled with data; number of rows with modified content; number of clocks for exchange with the RAM; downtime of the instruction decoder, etc.
b) control of access to the cache memory of the second level - number of fetched or written commands or data; number of rows filled; number of rows removed; number of turns, etc.
c) control of access to the system bus - total number of turns to the bus; number of clocks to refer to memory; number of cycles to access the periphery, etc.
d) control of the implementation of interruptions – number of received requests for hardware interruption, number of bars with prohibited service, etc.

To perform performance monitoring, two 40-bit counters PerfCtr0 and PerfCtr1 are built in, each with its own 32-bit control register PerfEvtSel0 and PerfEvtSel1 – Figure 5.

Figure 5. Format of register for monitoring realization in P6 microprocessor family

31 24	23	22	21	20	19	18	17	16	15 8	7 0
Counter Mask	INV	EN	–	INT	PC	E	OS	USR	Unit Mask	Event Select

The contents of the Event Select and Unit Mask fields determine the type of monitored event for the respective counter. The Counter Mask field, together with the INV bit, defines the switching mode of the counter. If INV=0 and the number of tracked events occurring within one cycle is greater than or equal to the Count Mask value, then the counter is toggled. Otherwise, the state of the counter does not change. With a value of INV=1, the counter is toggled only if the number of events within one cycle is less than or equal to the contents of the Count Mask field. In this way, several events (for example, execution of several instructions) can be monitored in one machine cycle. The remaining bits serve to refine the registrations when a monitored event occurs. For example, setting E=1 allows the collection of information on the number of transitions performed under the monitored conditions, which allows the calculation of the average dwell time in a given state (for example, average service waiting time).

Hardware monitors. To access the object of measurement, measuring probes are used, which are connected to selected points of the CS depending on the planned measurement experiment. They register and send the information to a digital filter for logical processing and dynamic selection of informative data. The registration itself is carried out by block counters, periodically their contents are transferred to a memory area for storage. There are two main types of hardware monitors.

Hardware fixed monitors (monitors with fixed algorithm) represent an external block of counters N_i for registering specific events and a general counter N for the duration of the measurement. The occurrence of a given event is registered by the corresponding probe S_i and through a logic 'gate' (scheme AND) the counter N_i is incremented by the number of clocks (generated by a common clock generator TG) during which the door is 'open' (Figure 6). After the measurement is completed, the activity of event i (probe S_i) is evaluated by $F(S_i)=N_i/N$.

Figure 6. Principle of hardware monitor with fixed program

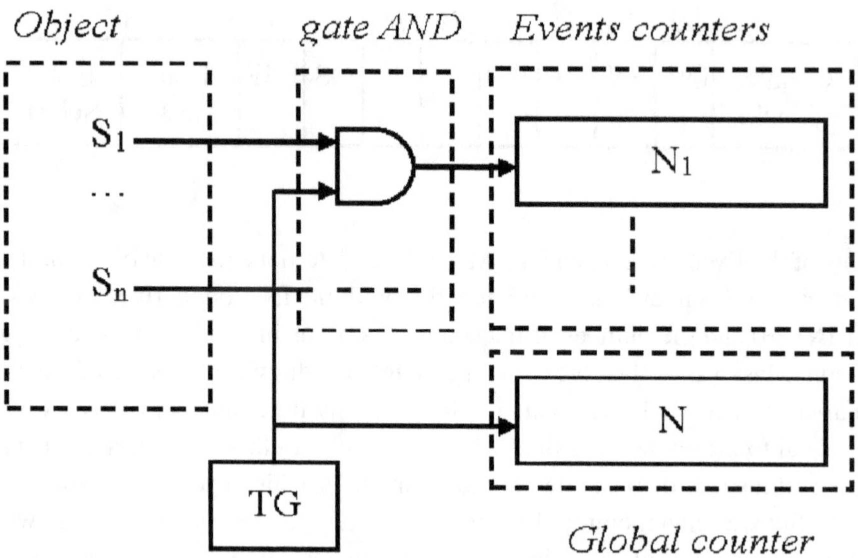

With the *hardware programmable monitors (program switchable monitors)* the measurement is controlled by a program stored in the memory of the measuring processor. In this way, it is possible to readjust the functions of the monitor. A control computer with its own peripheral is usually used, to which is added a digital filter (like that of Figure 6) connected to the interface outputs of the processor. The presence of external memory allows permanent storage of monitoring data and their subsequent processing based on the system software of the control computer.

Software monitors. They use software and firmware tools that are added to the system software of the measured CS. Thus, the registration itself takes place by executing a program that is located in the computer's memory. This program needs to be activated (initialized), which is done depending on the type of program monitor. During the execution of the program, the intended registration of events and accumulation of information is carried out. Thus, programmatic measurement introduces a time (due to execution of the monitor itself) and memory (the monitor is also located in memory) biases in the monitoring estimates. Despite this drawback, software monitors are much cheaper than hardware monitors and are easy to build.

Program monitors can be divided into two groups:

a) With internal control - identify internal events through control points (program probes) in the executed program, which call a subroutine for analyzing the event code and, depending on the result, the next subroutine is called, following this event.
b) With external control – information on the state of processes in the measurement object is collected at certain time intervals depending on external events – timer (measurement by time), counter (measurement by event). It is carried out by a program that, at certain time intervals, captures the contents of system variables or registers and stores them in a certain area of memory or in a supported monitor file. Thus, the system profile can be taken for the usability of CS blocks or certain

information objects, for example, access to files or a database (DB), measurement of the intensity of network traffic, measurement and analysis of packet transmission errors, etc.

Combined monitors. They are built based on combining some of the above types of measuring means to improve the effect of monitoring at a reduced value. They can be divided into two groups as folowing: *mixed* - combine similar measurement methods (for example, program and microprogram measurement); *hybrid* - apply different methods (most often a combination of hardware and software measurement).

Examples for Investigations by Using Monitoring

The growing importance of network communications and real-time operation is related to the application of modern technologies, as one example of the application of real-time monitoring and conducting predictive analytics is discussed in (Shukla, 2023b). One of the main applications in a network setting is to monitor and measure data traffic.Various products are available for conducting experiments related to network communications parameters, some of which are presented below.

ActiveXperts Network Monitor 2022 (https://www.activexperts.com/network-monitor/#introduction) is a software monitor of ActiveXperts available in two versions (64-bit and 32-bits). These are software solutions based on Windows with the possibility for graphical visualization of obtained measured data which are stored in database of MS SQL Server or MySQL.

Iris is an open-source network traffic management software product that provides a graphical user interface for monitoring incoming and outgoing network traffic. It supports modern containerization technology and presents a new model for assembling a sustainable and maintainable Internet metering architecture (Gouel et al., 2022). In the cited paper, it was used to create a system for distributed IP-level tracing of the routes that packets take over the IPv4 Internet, with the developed Zeph algorithm tracing the route between agents at multiple monitoring points. Iris functions as a complete management and monitoring system with capabilities to monitor network activity (Figure 7a), decrypt and recover captured data, assemble the session by tracking packets and their sequence in the session, filter connection attempts (Figure 7b), shows network statistics.

Traffic Analyzer (TA) is a user monitor program that successfully performs functions like LANAlyzer (Novell's continuous network traffic monitoring tool). Allows description of traffic categories through specially prepared templates or through created own rule files. So, it can filter by protocols and their characteristics, services and machines involved in the communication. Rule files are compiled by the protocol analyzer when it is started and arranged into internal structures that are convenient to access at the time of traffic analysis. Modification of the rules file is possible when monitoring conditions change.

Figure 7. Functionality of Iris

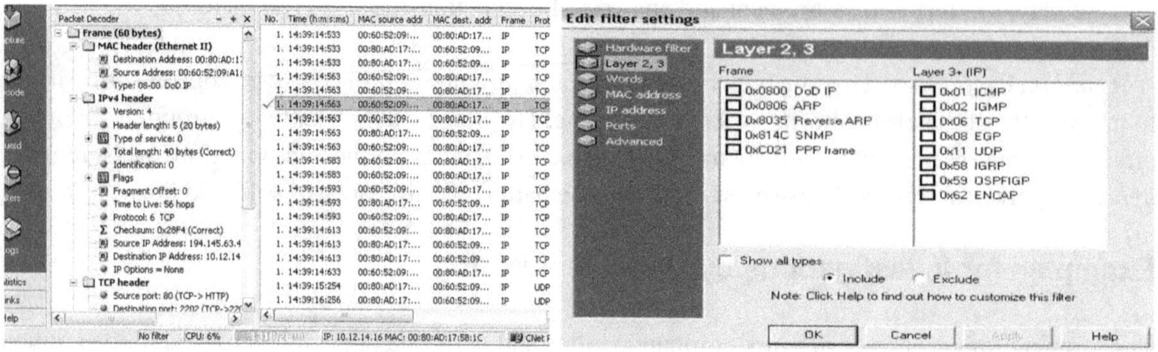

(a) Protocols and packets *(b) Edit filter settings*

Distinct Network Monitor 2022 (Distinct, 2022) of the Distings Corporation is a program that captures packets and analyzes network protocols on the monitored segment. It represents complex protocol interactions in natural language, pinpointing errors (Figure 8). It has a detailed statistics module allowing reporting of network performance in the environment under study. Supports high-speed and wireless networks 10/100/1000 Mbps Ethernet, wireless TCP/IP adapters and 4/16 Token Ring cards. It works over a serial line with a PPP connection. It allows diagnosing problems in the monitored network and discovering bottlenecks in it, as the integration of the statistical module Report Builder provides the following possibilities:

- Graphical interpretation of the collected data.
- Information about the system that uses the most traffic in the monitored network segment.
- Information about the established connection and the protocols used, including statistics by application protocols.
- Information about the most active MAC addresses and about the traffic divided by MAC address.
- Adapter statistics, number of passed packets, information about incoming and outgoing network traffic.

In addition, the application allows the use of filters to detect specific problems, discovers who owns unauthorized IP addresses and generates HTML reports for the analysis performed. It has several advantages, some of which are precise traffic capture, the ability to display a single TCP stream and select fields from the IP header, using filters, easily detect and mark bad packets, etc.

The products "Iris" and "Distinct Network Monitor" were used to form empirical data samples to conduct additional statistical analysis of network traffic characteristics (Romansky, 2022c; Romansky, 2022d). Their selection is based on the capabilities provided for precise traffic capture, single TCP flow display, remote IP monitoring with selection of fields from the IP header for visualization, easy detection and flagging of errant packets, etc. Subjects of monitoring are dependencies related to average length of packages, total number of packages, number of packages for different network protocols, number of packages for different IP protocols, incoming and outgoing traffic, etc. The first phase is carried out experiments for real behavior monitoring and graphical interpretation of obtained results for MAC and IP traffic are presented in Figure 9.

Figure 8. Basic screen form of distinct network monitor

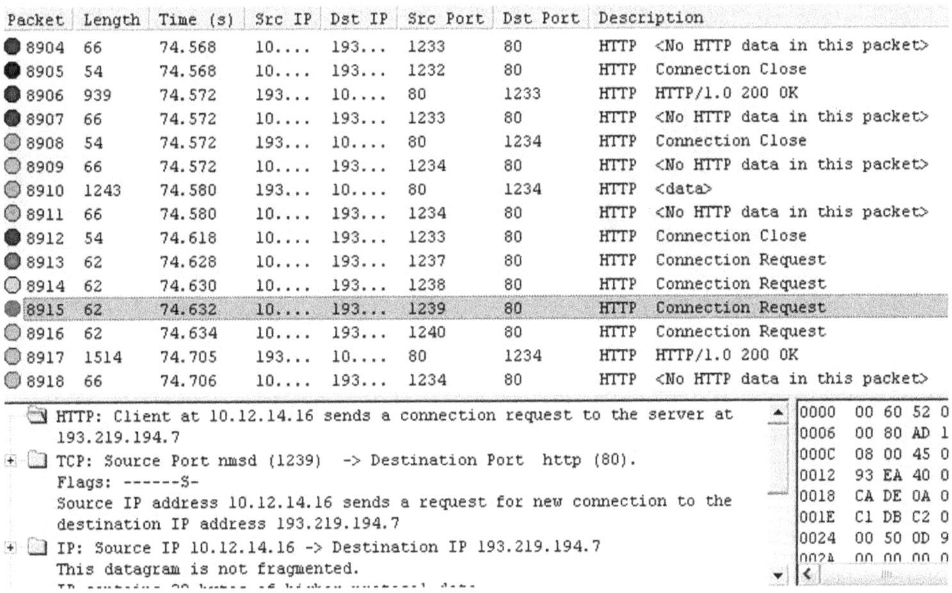

Figure 9. Graphical interpretation of monitored data

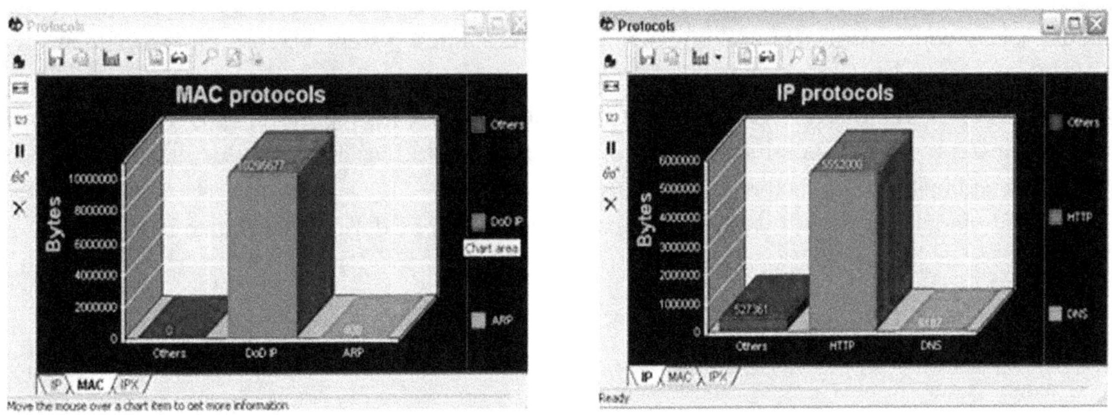

Regarding the distribution of packages by size, both monitors are applied, and one of the results is shown in Figure 10. The formed estimates show that the main traffic in a traditional information service in a distributed environment is formed by mostly short packets with a length of up to 64 B. Statistical analysis of accumulated empirical data by using author's program application is made and different statistical assessments for average values, correlation coefficient, regression, etc. are calculated.

Figure 10. Network traffic investigation: Distribution of packages by size

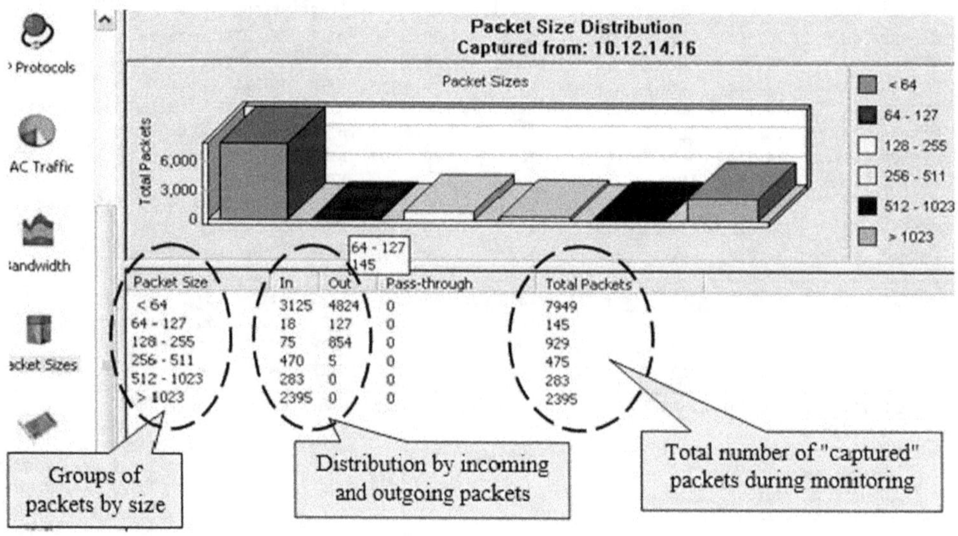

Modeling of Systems and Processes

Modeling is a method of scientific knowledge, allowing the collection of information in cases where the object of study does not exist or is difficult to access. The main task of modeling theory is to provide technology for creating and applying models in the study of systems and processes of a different nature. Modeling as a process consists of building a model of the studied system, performing experiments with the model, and interpreting the obtained results. It is appropriate when the real object does not allow its immediate study. It should be noted that the model cannot completely replace the physical object. In the process of designing the model, its adequacy is artificially distorted due to the idealization of several components and influencing factors.

In general modeling theory, two main groups of models can be defined, as shown in Figure 11.

Figure 11. Types of models

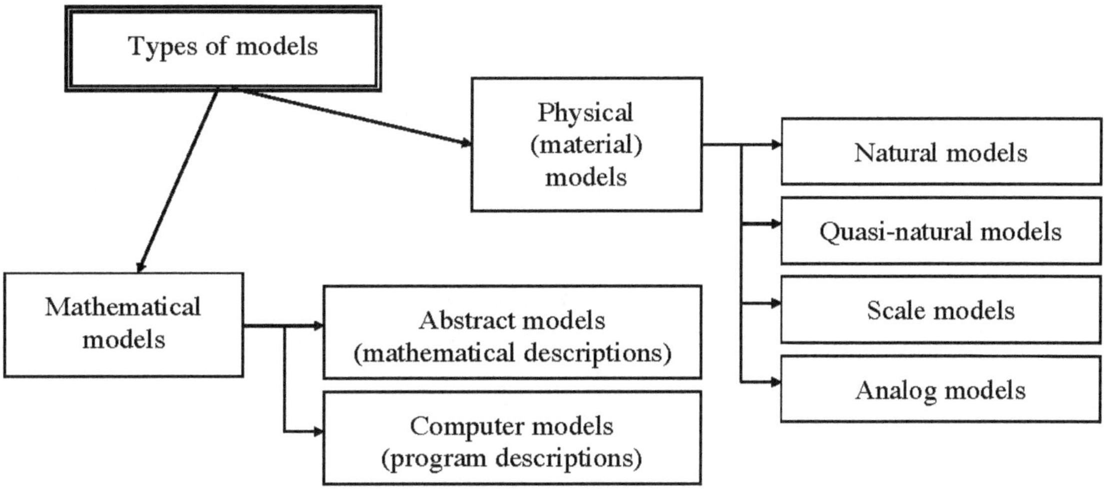

Physical models are usually equivalent or similar to the original or have a different nature but function like it. Natural (real systems, fully adequate to the original), quasi-natural (natural models supplemented with mathematical means), large-scale (the same physical nature, but with a different scale), analog (with a different nature, but analogous functional behavior), etc. are applied.

Mathematical models represent formalized descriptions of the studied object through an abstract language (mathematical formal system) or through mathematical relations describing the functional behavior. Various mathematical tools can be used for their creation (algebraic, differential, and integral calculus, theoretical systems, etc.).

Computer Modeling

Computer modeling is a form of mathematical modeling which uses computer space and tools for designing the model and its realization (execution) for carried out experiments by them. The means of computer technology play an essential role in it, together with the very diverse software provision - universal and specialized. The specific implementation of a computer model is based on the application of a selected modeling method, which can be referred to one of the groups presented below.

- Empirical methods - mainly involve quantitative methods for accumulating empirical data on the behavior of the real object, through which a suitable statistical model is derived.
- Analytical methods - use mathematical means for formalization and abstract description of the real system and the processes taking place in it, and numerical methods are most often used for the solution.
- Imitation (simulation) methods – the behavior of the main system is imitated by the program, which, during its execution in time, uses the main workload or model of the workload.

The main advantages of computer modeling are the following:

a) makes it possible to make an adequate decision already during the design of a given system based on a conducted model study.
b) offers opportunities to study a given system in a wide range of changes in its influencing factors, saving expensive laboratory equipment.
c) significant reduction of the experimental time for the studied processes compared to their real implementation.
d) allows studying the behavior under critical and risky input and external impacts, which could be fatal for a real system.
e) allows easy adaptation of the developed computer model for different situations of working conditions, as well as its easy modification for a specific implementation.
f) in some cases, is the only applicable method for studying systems and processes.

The tools used in the development and implementation of computer models are grouped into three main categories:

a) *graphic means* - allow an appropriate abstract presentation of the main idea for developing the computer model and presentation of the basic concept for defining the mathematical description.
b) *software tools* - specialized and universal language and operating environments for program description of the model, its adjustment and verification, as well as for the implementation of model experiments.
c) *technical means* - hardware environments with a specialized and universal application for creating the program description of the model, entering, storing, and outputting the experimental data, managing, and conducting the experiments, etc.

2. MODELING AS A METHOD OF SCIENTIFIC KNOWLEDGE

Modeling is a method of scientific knowledge, uniting the stages of creating a model of the research object, conducting experiments with it and transforming the experimental results into information about the research object itself (Romansky, 2017). In this sense, modeling is replacing an original object Ω_O with another model object Ω_M in order to study the properties or behavior in certain situations of the original by experimenting with the model.

The original object Ω_O is an arbitrary system or process that may not actually exist. Nevertheless, its system properties can be described by finite sets, such as:

S_o – system parameters characterizing the internal state of the real system, its structure and functioning.
Y_o – quantitative characteristics of the system parameters, describing mainly behavior-resulting signs, important in the interaction with other systems.
X_o – external influences influencing the behavior of system parameters.

When studying a given system, a certain subset $\{y_0\} \in Y_0$ is usually studied under specific external influences $\{x_0\} \in X_0$, and a given characteristic y_{oi} depends on some subset $\{s_0\} \in S_0$ (usually the influence of other parameters is neglected). The system parameters organized in this way define the so-called *field of the system* and characteristics are the data describing its organization. In this sense, both static real systems (field and characteristics do not depend on time) and dynamic real systems (field and/or characteristics depend on time) are the subject of research.

A model is a representation of an object, system, or concept in a form different from its actual existence. It reflects or reproduces the object of the study to a sufficient extent, allowing obtaining reliable information for its study.

The object-model Ω_M can be considered as an image of Ω_O, realized in a set-theoretic aspect based on the sets: $\{s_m\} \in S_m$ – model object parameters; $\{y_m\} \in Y_m$ – characteristics of the model object; $\{x_m\} \in X_m$ – external influences on the object-model.

The replacement of the real system or process Ω_O with some model Ω_M is admissible if the specific model characteristics $\{y_m\} \in Y_m$ sufficiently reflect the corresponding quantitative characteristics $\{y_o\} \in Y_o$, defined in the modeling process. In this sense, modeling is the replacement of a real functional dependence $\{y_o\} = \Phi[\{s_o\}, \{x_o\}, T_o]$, describing the behavior of the original object over time, with a suitable equivalent dependence $\{y_m\} = \Phi[\{s_m\}, \{x_m\}, T_m]$, where usually the model time T_m is related to the real time T_o by a scale factor. The design of a model is related to the basic requirement of finding such an analytical dependence that describes the behavior of the original with sufficient accuracy for the purposes of the study. This justifies both the requirement to correctly define the main model parameters (primary and secondary factors of the experiment) and the acceptable level of error obtained in this substitution. This is how an image $f\colon Y_o \rightarrow Y_m$ is defined, through which, with sufficient validity of the initial (a priori) information in the modeling, conclusions can be drawn (with some approximation) about the behavior of the system parameters $\{s_o\}$ of the original.

If there is not enough information about the behavior of some system parameters s_{oi} under certain external influences x_{oj}, a hypothesis is defined through which a model of the behavior of the real object is built. If the reliability of the image $f\colon Y_o \rightarrow Y_m$ is proven, the hypothesis made for $\{s_o\}$ is confirmed and $\{s_m\}$ allows the study of the original object.

The main task of modeling theory is to provide technology for creating and applying models in the study of systems and processes of a different nature. Modeling as a process consists in building a model of the studied system, performing experiments with the model and interpreting the obtained results. In the process of designing the model, its adequacy is artificially distorted due to the idealization of a number of components and influencing factors. This means that first of all, it brings a definite break in the gap between the model and the original. All this can lead to a significant difference of the model results compared to the real situation. In this sense, it is necessary to continuously evaluate the model's accuracy.

There are different approaches to model research, but each of them is based on a certain descriptive formalism and includes two phases: *modeling* (description of the object through the formal means of a chosen method) and *analysis* (experiments with the model to obtain quantitative estimates of the observed parameters) – Figure 12

Figure 12. Phases of the model investigation

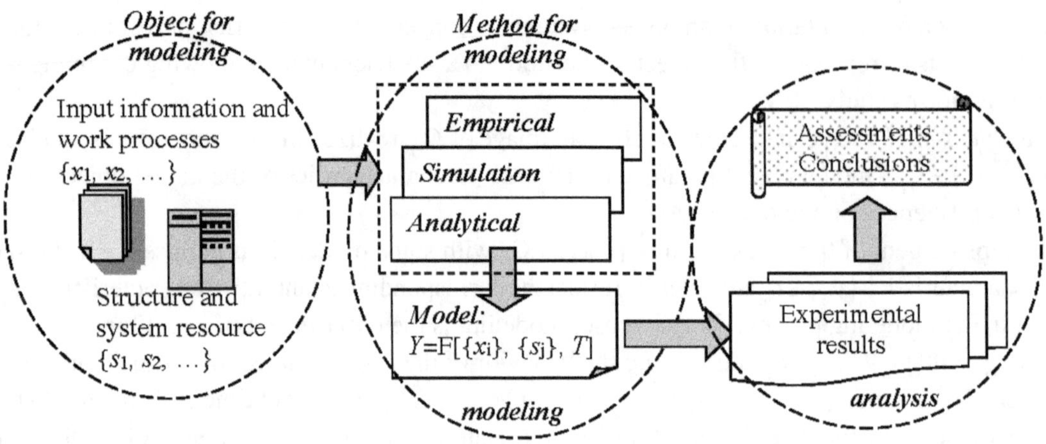

3. APPLICATION OF COMPUTER MODELING AND SIMULATION

Regardless of the significant advantages of computer modeling, its application in the study of systems and processes must be consistent with the goals of the experiment and ensure sufficient adequacy and efficiency. The feasibility of using computer modeling should be based on the following criteria:

- Impossibility of using empirical methods and means for direct measurement and study of the object.
- Sufficient conviction in the effectiveness of modeling in solving a specific problem, as well as in the possibilities of developing a sufficiently adequate model.
- The presence of a large volume of calculations during the analysis, for the performance of which the use of computer tools is imperative.
- Ineffectiveness or unsuitability of other research methods.
- Availability of sufficiently effective means (hardware and software) for implementing a computer model and conducting experiments with it.

Below are listed some main areas of application of computer modeling, without exhausting all possibilities and areas of applicability.

1. Design and functional research of devices and systems in various technical fields (electronics, computer technology, electrical engineering, mechanics, etc.) in the process of preliminary design, layout and confirmation of the workability and efficiency of the project.
2. Comparative analysis and evaluation of different options in decision-making in engineering projects, as well as in sociological, political, and economic analyses.
3. Study of traffic and the transfer of information in communication networks, evaluation of throughput and load of switches, analysis of routing algorithms and possibilities of failures in communication elements.

4. Solving transport problems when choosing an optimal route, planning the transport system, analyzing the mutual influence of different types of transport (land, river, sea, air), as well as the impact on the social and economic sphere.
5. Trend analysis and risk assessment in the economic sphere, the development of market relationships, resource management and reserve maintenance.
6. Study of problems and processes in the natural sciences, such as the development of biological populations, changes in the environment, analysis and quantitative assessment of physical phenomena, research of the influence of meteorological factors on people and the environment, etc.
7. Application of simulators and test methods in the field of education and retraining related to the development of new methods and means of training, training courses and training systems.
8. Military application for analysis and evaluation of military doctrines, strategies and tactics, resource reallocation, exercise plan optimization and societal impact analysis.
9. Conducting demographic studies related to the development and movement of the population, the features of zoning and the impact on public, social and economic infrastructure.
10. In the aviation field for simulation of flights (simulators) and training of pilots and astronauts, conducting space research, optimization of trajectories, evaluation of aircraft components, etc.

Some major applications of model research in computer science are presented below.

In the process of designing computer devices and systems. The design goes through several successive phases (schematic, functional, systemic), and at each of them model concepts and tools for solving specific tasks are applied.

In schematic design, automated environments (CAD – Computer Aided Design) are used for model description and study of the workability of the created prototype. Testing of behavior over time or testing of output responses to certain input impacts is most commonly performed.

Functional design of computer systems and devices is concerned with developing the functional algorithm for general behavior of the device. There are specialized language environments for describing the functional structure and studying the processes in it. An IEEE standard has been developed for representing the low-level logical structure through a description with a high level of abstraction - the language environment Very high-speed Hardware Description Language, VHDL (Lipsett et al., 2012; Roth et al., 2016). It is a high-level language offering appropriate language constructs for the functional description of standard and user-defined components. It offers rich and specific capabilities for description and functional simulation in device design.

When developing software systems, functional modeling is an important stage of conceptual design, ensuring the necessary efficiency and reliability of the designed software product. It is the basis of the so-called IDEF-technology (Muminov & Bekmurodov, 2020), which includes multiple positions - for functional modeling of business processes (IDEF0), for documentation of technological processes (IDEF3), for analysis of data flow (DFD – Data Flow Diagrams) and of the workflow (WorkFlow), etc. When developing object-oriented systems for defining, visualizing, and documenting the functional essence, the UML (Unified Modeling Language) standard was created. This unified modeling language is a suitable tool for standardizing the analysis process and the design of its components - semantic model, syntactic descriptions, diagrams.

At the stage of system design, tasks related to the choice of structure, management, interaction and communications between the individual components of the system are solved. In most cases, it is necessary to examine several options in order to choose the best one under specific working conditions. This is done by conducting multiple model experiments while varying the main parameters and comparative analysis of the results under defined criteria. Most often, the "performance-price" ratio is analyzed, which is a criterion for the level of efficiency in the organization of computer processing.

When evaluating processor performance. When evaluating various performance indices, CPU time is used as the basic unit, which is associated with the speed of the processor. Typically, the rate at which an event occurs per unit of time (second) is used to evaluate performance. Related to real time concepts, the object of research and evaluation are response time (Response Time), execution time (Execution Time), elapsed time (Elapsed Time). To these times, the time for communication with other devices (Communication Time) should be included, which leads to a delay in computer processing - access to external memory, input-output operation, loading of a system program, exchange of messages with other processors, etc. The study of processor work during program execution requires the analysis of two types of processes - application and system, and therefore with the corresponding implementation times. Usually the system time is ignored, which in cases of significant system code would cause the time estimate to be skewed.

In modern processors, the flow of processes is related to the internal interaction of functional devices, which directly affects the overall delay. One standard estimate of processor performance is "million instructions executed per second" (MIPS). When solving scientific and technical problems using arithmetic over floating-point numbers, MFLOPS (million floating-point operations performed in one second) is used. In this case, however, the problem is that different processors support different sets of floating-point operations, which before the benchmarking. It is also important whether the estimates are made in the "canonical" or "normalized" form of the numbers.

To form a general estimate of processor performance in different applications, emulating or tracer test suite packages are usually applied (see "Synthetic workload and benchmark" in sub-section 1.1), some of which are as follows:

- LINPACK – a package of programs for solving systems of linear algebraic equations. A comparative analysis of the performance of different computers is used. The test suite evaluates in MFLOPS the processing of a matrix of size 100x100, which can be located entirely in a cache memory of 1 MB. There are also parallel versions of the package for studying multiprocessor systems. For example, by using LINPACK is determined assessment of 93 PFLOPS for supercomputer Sunway TaihuLight which is installed in 2017 and in the top of ranking TOP500, but in June 2018 it is replaced at the top of the list by Summit, an Oak Ridge National Laboratory based supercomputer with a theoretical peak of 188 PFLOPS, and an achieved peak of 122 PFLOPS (Wright, 2019).
- SPECint test suits the Standard Performance Evaluation Corporation, based on real application programs for evaluating the performance of new generations of high-performance computers, allowing the import of code on different platforms (SPEC, 2022). They are used to emulate computer processing at the stage of developing new architectural solutions. They are available in two versions – for integer arithmetic and for processing floating-point numbers. Each individual test simulates a CPU load, giving an estimate of performance on the machine under study against a benchmark time.
- TPC-A, TPC-B, TPC-C – Transaction Processing Performance Council test suites for evaluating system transaction processing when working with databases and when transmitting various data, being oriented to the economic sphere. They are applied in the evaluation of information exchange

parameters and comparative analysis of the used computer systems and platforms. For example, TCP-C has been used for workload evaluation in the study of *"distributed database that supports replication, horizontal partitioning (sharing), a flexible document schema and ACID guarantees on the document level"* (Kamsky, 2019). The package was chosen based on the capabilities it provided, although some adaptation was necessary to ensure the necessary compliance with best practice.

In research and optimization of network traffic. When studying computer networks and network processes, the aim is to determine the necessary speeds for data transmission and the size of the buffers, as well as to determine the characteristics of the expected network traffic. This is usually achieved through model research, applying a probabilistic approach associated with the principles of mass service theory. The reason for this is that traffic is considered as a random process that can be described by the Poisson probability distribution. With the development of high-speed networks, it turned out that a larger part of network traffic is self-similar, which requires a more specific approach to its modeling. The concept of self-similarity is closely related to chaos theory and the concept of "fractal". At its core, self-similarity is the repetition of individual similar series over larger time intervals, also observing recursion (i.e., the presence of structures that repeat themselves). An investigation of self-similarity of network traffic is presented in (Tatanikova et al., 2021), where the *"properties were checked on different time scales obtained on the available daily traffic data"*. This was done for the purposes of a statistical analysis performed, with an estimate of tail weight obtained by constructing a regression line. A comparison of artificial network traffic generators was made by the criterion of the method of least squares for the approximation of the base values when developing a suitable model for simulating real network traffic.

Data transfer in high-speed networks requires a high level of network parallelism, which creates the need to conduct online optimization, as in (Arifuzzaman & Arslan, (2021) a product Falcon *"that combines a novel utility function with state-of-the- art online optimization algorithms ... that can maximize the throughput while keeping system overhead low and ensuring fairness among competing transfers"*. Research shows that the product can find a near-optimal solution in just 20 seconds and outperforms the existing transfer application by 2-6 times In principle, the design of high-speed networks is also directly related to the possibility of modeling and evaluating the amount of traffic and the value of its various characteristics, which can be divided into two groups: (a) characteristics related to throughput (average speed, maximum speed, unevenness, etc.); (b) characteristics related to delays (transmission delay, delay variance, etc.)

These statistical characteristics affect various aspects of the configuration, such as routing protocols, resource reservation protocols, queue disciplines in routers and ATM switches, size of buffers, etc. To make effective decisions, it is necessary to design a sufficiently accurate model of information traffic in the communication structure. In this sense, mass service theory (queuing theory) offers a simple and convenient means to study network traffic and solve network problems by developing an appropriate analytical model. The model with queues (Queuing Network - QN) allows to obtain an approximate estimate of the behavior of the system quickly and sufficiently accurately under different loads. Although the probabilistic theory of mass service and queuing analysis is generally quite complex, in many cases when evaluating network performance, the solution to a given problem turns out to be relatively simple. The study is reduced to knowledge of elementary statistical concepts, such as mathematical expectation, variance, and mean square deviation. This approach, however, is suitable for traffic that obeys the Poisson distribution. In the case of self-similar traffic, this is not the case, and the basic formulas of queuing theory cannot be applied. When modeling the series of sectors, it is necessary to introduce delays that cause a reduction in throughput.

4. BASIC APPROACHES OF MODEL INVESTIGATION IN COMPUTER FIELD

Computer modeling is one of the alternatives in the study of computer systems and processes and finds application in various fields. Each study must begin with a preliminary formalization of the object, which will define the main essence of the developed model. The organization of computer calculations covers the two hierarchical levels associated with high-level applications (macro-level) and their specific implementation at the internal micro-program level (micro-level). It is known that in a traditional computer environment, programs are implemented as a sequence of instructions (defining individual macro-level operations), and after selecting each one for its execution, a micro-program of micro-instructions (micro-level micro-operations) is activated. A formal description of this process as an example is presented in Figure 13.

In the modeling study of computer processes, the peculiarities of both levels should be considered.

- In the macro-level organization, each instruction from the control program activates execution in the processor of a certain operation, which defines a discrete state of the calculation process at discrete moments of time.
- When implementing micro-level computer processing, the activated operation is "broken" into separate elementary operations (micro-operations), performed in some discrete sequence depending on the transitions between them, which brings a probabilistic character to the micro-process (Romansky, 2021c).

Figure 13. Formal description of computation in the two hierarchical levels

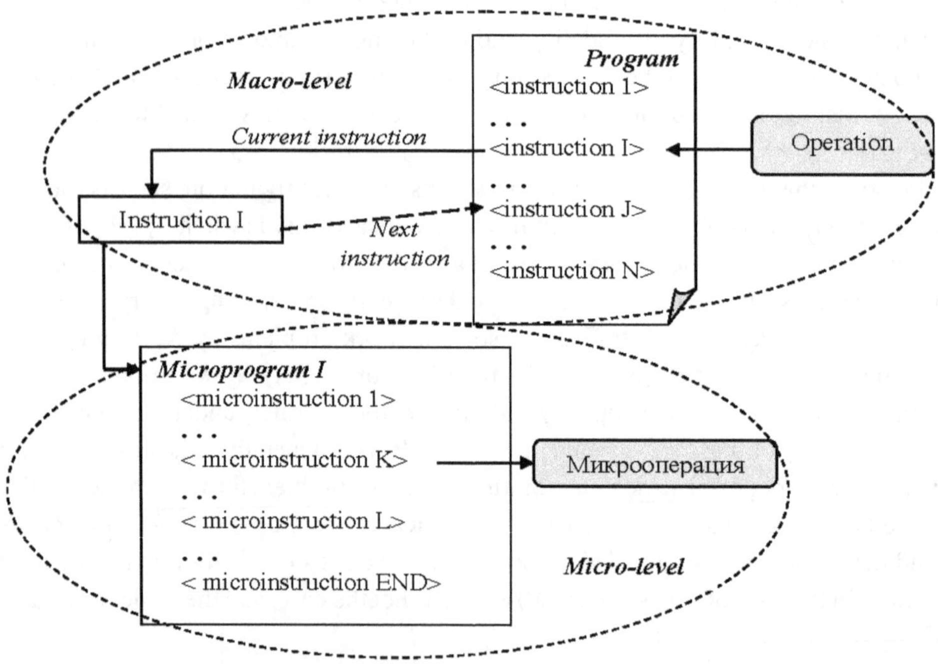

This setting of the processes determines the relevance of two basic approaches in model research in the computer field - deterministic and stochastic.

The application of the ***deterministic approach*** is justified by the discrete nature of traditional computing environments and the processes in them as transitions between discrete states. This allows the use of tools of discrete mathematics and discrete structures such as mathematical analytic dependencies, finite automata, graph theory, etc. (Romansky, 2020b).

A graph is a suitable consequence for modeling because it allows easy description and programmatic exploration, as well as convenient graphical visualization. Its application in the computer field is determined by the possibility to represent the functioning of computer objects as a finite set $S=\{s_i, s_2,..., s_n\}$ of states representing the nodes of a given graph. The graph is a discrete structure $G(V, A)$, where $V=\{V_1,..., V_n\}$ is a finite non-empty set ($V\neq\varnothing$) of the nodes, and $A=\{a_{ij}\}\subset V\times V$, $i=1\div n$ $j=1\div n$, is the finite set of arcs connecting different pairs of vertices (V_i, V_j). Depending on a defined direction of the arcs, the graphs are divided into unoriented if $a_{ij}=a_{ji}$ or oriented if $a_{ij}=<V_i\rightarrow V_j> \neq a_{ji}$ and depending on the associated value of the arcs – into binary ($a_{ij}\in\{0,1\}$) and weighted ($a_{ij}=k\geq 0$). They are defined by the adjacency matrix (AM) according to the following scheme (Figure 14):

Binary graph: $$AM[i,j]=\begin{cases}0; if \nexists a_{ij} for \forall i,j\\ 1; if \exists a_{ij} for \forall i,j\end{cases}$$

Weight graph: $$AM[i,j]=\begin{cases}0; for \forall i=j\\ k; if \exists a_{ij} for \forall i\neq j; k\in N^{+}\\ \infty; if \nexists a_{ij} for \forall i\neq j\end{cases}$$

Figure 14. Definition and graph presentation of binary and weight graphs

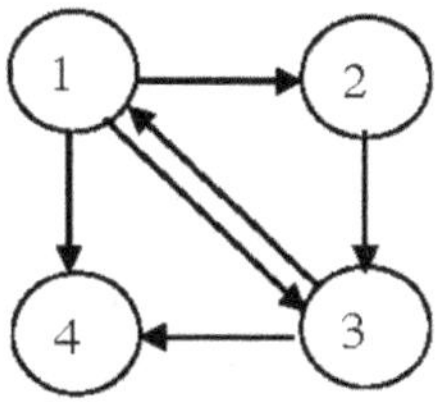

AM	1	2	3	4
1	0	1	1	1
2	0	0	1	0
3	1	0	0	1
4	0	0	0	0

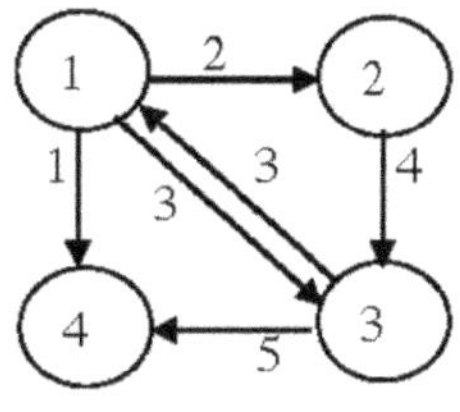

AM	1	2	3	4
1	0	2	3	1
2	∞	0	4	∞
3	3	∞	0	5
4	∞	∞	∞	0

In the general case, we consider the function f_G: $A\rightarrow V\times V$, which maps to each arc an ordered pair of vertices and is called a *finite directed multigraph* – denoted by $G(V, A, f_G)$, such that $f_G(a_{ij}) = (V_i, V_j)$. It follows that the structure "graph" is a concrete implementation of a multigraph. For graphs, the function f_G is unique, and each ordered pair (V_i, V_j) of $V\times V$ can participate at most once in the definition of a graph. In the multigraph, the function f_G: $A\rightarrow V\times V$ can generate for a given pair of nodes (V_i, V_j) more than one connecting arcs with the same direction. An example for graph presentation of computation based on communicating processes (programs/algorithms A) is shown in Figure 15.

Figure 15. Simple example of graph presentation of computation in macro-level

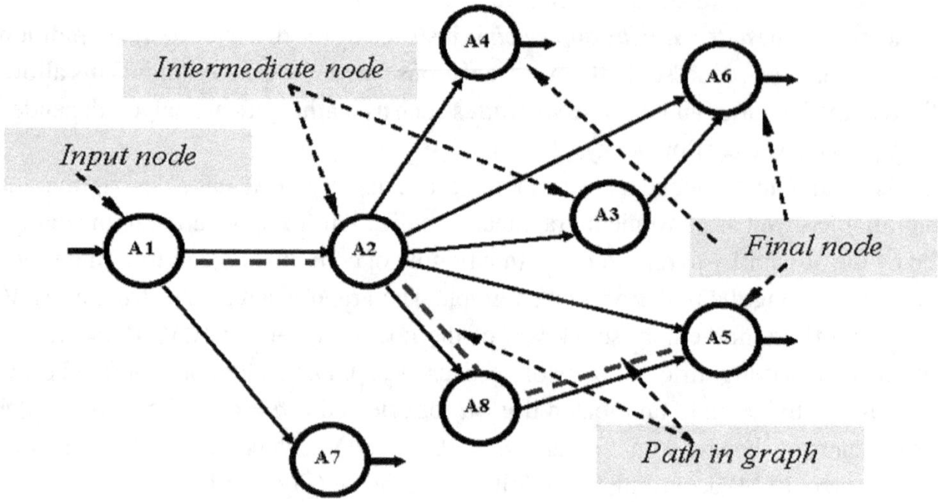

The stochastic approach is based on probability theory, queuing theory, theory of Markov and semi-Markov processes (Baker et al., 2022). This approach is adequate to the probabilistic nature of computer processes. The essence consists in defining probabilistic parameters to reflect the stochastic nature of processes (Romansky, 2021b). It also considers a set of states describing a given process or the behavior of a structure, but the transitions between these states obey probabilistic laws. Most often, these transitions are described by a matrix of transition probabilities, each transition having probability $p_{ij} = P[s_i(t) \rightarrow s_j(t+1)]$, where s_i and s_j are allowed states and t and $t+1$ are two successive moments or two successive steps in the process. Graphically, the behavior can be represented by a *state graph* – Figure 16

The theory of random processes is a branch of the theory of probabilities for studying regularities in random phenomena in the dynamics of their development. Basically, in nature, all processes are more or less random. For processes that are defined as deterministic, the random factor is so small that it can be neglected. In other processes, the probabilistic nature is clearly expressed, and it is necessary to apply a probabilistic approach to the description. A random process (RP) occurring in system S is a sequence of random transitions from one state of the system to another. Thus, the behavior of the process manifests itself in time and can be written as $X(t)$, and at a given time $t = t_i$ takes a specific value $X(t_i)$, representing a random variable. This allows the behavior of a CS to be represented as RP $X(t)$, taking a finite and countable set of states $S = \{S_1, S_2, \ldots, S_i, \ldots, S_n\}$ at discrete time moments $t_i \in T$. In the successive steps of its implementation, the process will fall into one of the permissible states $S_i \in S$.

Figure 16. An example for graph of states for processing in micro-level (microprogram)

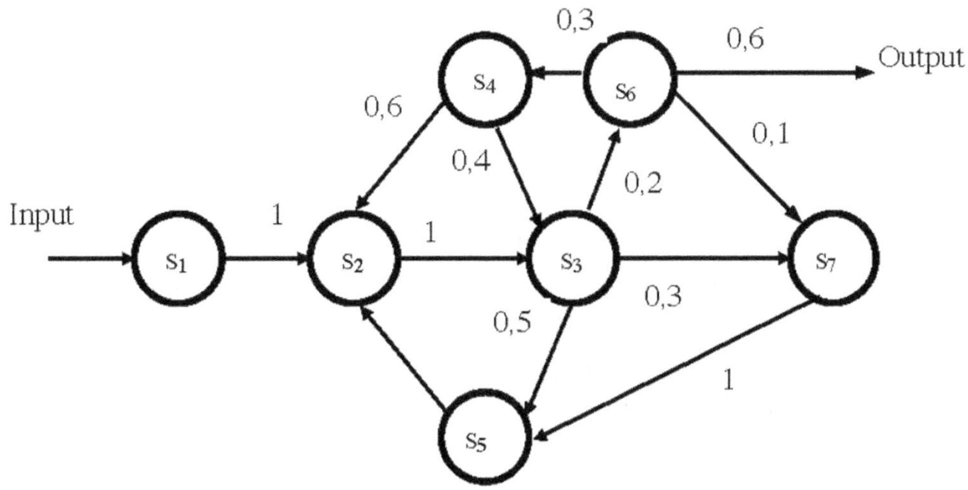

REFERENCES

Arifuzzaman, M., & Arslan, E. (2021). Online optimization of file transfers in high-speed networks. *Proceedings of the International Conference for High Performance Computing, Networking, Storage and Analysis*. IEEE. 10.1145/3458817.3476208

Bahga, A., & Madisetti, V. K. (2011). Synthetic workload generation for cloud computing applications. *Journal of Software Engineering and Applications*, *4*(07), 396–410. doi:10.4236/jsea.2011.47046

Baker, E., Barbillon, P., Fadikar, A., Gramacy, R. B., Herbei, R., Higdon, D., Huang, J., Johnson, L. R., Ma, P., Mondal, A., Pires, B., Sacks, J., & Sokolov, V. (2022). Analyzing stochastic computer models: A review with opportunities. *Statistical Science*, *37*(1), 64–89. doi:10.1214/21-STS822

Distinct Corporation. (2022). *Distinct Network Monitor*. District Corporation. https://www.distinct.com/monitor/monitor.htm (Visited on 5 Jan 2023)

Du Bois, K., Schaeps, T., Polfliet, S., Ryckbosch, F., & Eeckhout, L. (2011). SWEEP: Evaluating computer system energy efficiency using synthetic workloads. In *HiPEAC'11 Proceedings of the 6th International Conference on High Performance and Embedded Architectures and Compilers*, (pp. 159-166). IEEE.)10.1145/1944862.1944886

Gouel, M., Vermeulen, K., Mouchet, M., Rohrer, J. P., Fourmaux, O., & Friedman, T. (2022). Zeph & Iris map the internet: A resilient reinforcement learning approach to distributed IP route tracing. *Computer Communication Review*, *52*(1), 2–9. doi:10.1145/3523230.3523232

Kamsky, A. (2019). Adapting TPC-C benchmark to measure performance of multi-document transactions in MongoDB. *Proceedings of the VLDB Endowment International Conference on Very Large Data Bases*, *12*(12), 2254–2262. doi:10.14778/3352063.3352140

Kounev, S., Lange, K. D., & von Kistowski, J. (2020). *Systems benchmarking: for scientists and engineers* (Vol. 1). Springer International Publishing. doi:10.1007/978-3-030-41705-5

Krishnamurthy, D., Rolia, J. A., & Majumdar, S. (2006). A synthetic workload generation technique for stress testing session-based systems. *IEEE Transactions on Software Engineering*, *32*(11), 868–882. doi:10.1109/TSE.2006.106

Lipsett, R., Schaefer, C. F., & Ussery, C. (2012). *VHDL: Hardware description and design*. Springer Science & Business Media.

Muminov, B. B., & Bekmurodov, U. B. (2020, October). IDEF models and innovative system for search data in stochastic information environment. *2020 IEEE 14th International Conference on Application of Information and Communication Technologies (AICT),* (pp. 1-6). IEEE.

Panda, R., Song, S., Dean, J., & John, L. K. (2018, February). Wait of a decade: Did SPEC CPU 2017 broaden the performance horizon? In *2018 IEEE International Symposium on High Performance Computer Architecture (HPCA)* (pp. 271-282). IEEE. 10.1109/HPCA.2018.00032

Raghupathi, W., Raghupathi, V., & Ren, J. (2022). Reproducibility in computing research: An empirical study. *IEEE Access : Practical Innovations, Open Solutions*, *10*, 29207–29223. doi:10.1109/ACCESS.2022.3158675

Rathi, S., Nagpal, R., Mehrotra, D., & Srivastava, G. (2022). A metric focused performance assessment of fog computing environments: A critical review. *Computers & Electrical Engineering*, *103*(October), 108350. doi:10.1016/j.compeleceng.2022.108350

Romansky, R. (2017). *Information servicing in distributed learning environments. Formalization and model investigation*. LAP LAMBERT Academic Publishing.

Romansky, R. (2020b). Formalization and discrete modelling of communication in the digital age by using graph theory. In M. Pal, S. Samanta, & A. Pal (eds) Handbook of Research on Advanced Applications of Graph Theory in Modern Society. IGI Global. doi:10.4018/978-1-5225-9380-5.ch013

Romansky, R. (2021b). Mathematical modelling and study of stochastic parameters of computer data processing. *Mathematics*, *9*(18), 2240. doi:10.3390/math9182240

Romansky, R. (2021c). Program environment for investigation of micro-level computer processing. *International Journal on IT and Security*, *13*(1), 83–92.

Romansky, R. (2022b). Evaluation of experimental data from monitoring and simulation of network communication parameters. *International Journal on Information Technologies and Security*, *14*(2), 75–86.

Romansky, R. (2022c). Statistical analysis of empirical network traffic data from program monitoring. *International Journal on Information Technologies and Security*, *14*(3), 15–24.

Romansky, R. (2022d). Investigation of network communications by using statistical processing of monitored data. *2022 IEEE International Conference on Information Technologies (InfoTech-2022)*, (pp. 37-40). IEEE. 10.1109/InfoTech55606.2022.9897115

Roth, C. H. Jr, & John, L. K. (2016). *Digital systems design using VHDL*. Cengage Learning.

Shukla, S. (2023a). Unlocking the power of data: An introduction to data analysis in healthcare. *International Journal on Computer Science and Engineering*, *11*(3), 1–9. doi:10.26438/ijcse/v11i3.19

Shukla, S. (2023b). Real-time monitoring and predictive analytics in healthcare: Harnessing the power of data streaming. *International Journal of Computer Applications*, *185*(8), 32–37. doi:10.5120/ijca2023922738

SPEC. (2022). *SPEC's Benchmarks and Tools*. SPEC. https://www.spec.org/benchmarks.html

Tatarnikova, T., Sikarev, I., Karetnikov, V., & Butsanets, A. (2021). Statistical research and modeling network traffic. *E3S Web of Conferences, 244*, 07002.

To, H., Asghari, M., Deng, D., & Shahabi, C. (2016). SCAWG: A toolbox for generating synthetic workload for spatial crowdsourcing. In *2016 IEEE International Conference on Pervasive Computing and Communication Workshops (PerCom Workshops)*. IEEE. 10.1109/PERCOMW.2016.7457121

Wright, J., & Ma, Y. (2022). *High-dimensional data analysis with low-dimensional models: Principles, computation, and applications*. Cambridge University Press., doi:10.1017/9781108779302

Wright, S. A. (2019). Performance modeling, benchmarking and simulation of high-performance computing systems. *Future Generation Computer Systems*, *92*(March), 900–902. doi:10.1016/j.future.2018.11.020

Chapter 2
Computer Systems and Processes as an Object of Modeling

ABSTRACT

The chapter is devoted to the main objects that will be discussed in this book (structures, systems, processes) in the field of computing. A brief review has been made, starting from specifying the role of "information" and "information processing," in order to specify the place of the technological environment for the implementation of information processes and to define its main functionalities. Given that the technological environment is a computer implementation, the two basic levels are presented – high user level (macro-level) and low internal machine level (micro-level) with an initial formalization of basic system resources. The main object of analysis is the performance with its factors (performance indices) which are defined and which have an impact on the workload of the computing environment based on input workflow. In this respect, a formal description of workload and specifying the role of the process profile are presented. In addition, a formalization of information technology is presented.

1. ORGANIZATION OF INFORMATION PROCESSING

The term "information" is derived from the Latin word "informare" (to teach) and was originally associated with the process of transmission between people of knowledge and content. In the digital society, the meaning of the concept of information is significantly developing, but it is always associated with the processes and technologies for its processing, as a generalized formalism for this is presented in Figure 1. The goal is to represent real information from the surrounding world through a system of rules and means in coded form as a message with a certain format suitable for computer processing.

DOI: 10.4018/978-1-6684-8947-5.ch002

Figure 1. Principle of information processing

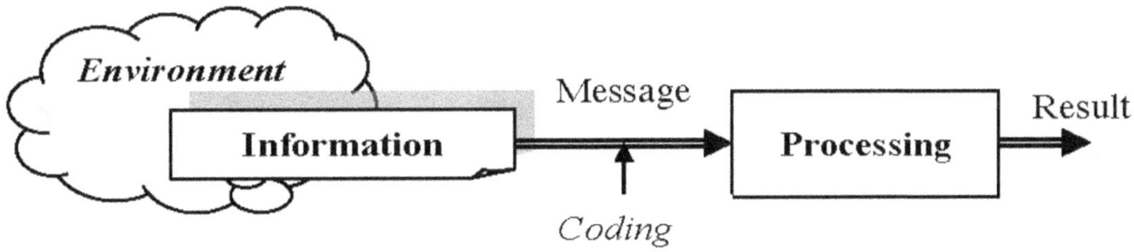

The process approach brings together technologies and methodologies for describing, analyzing, implementing and controlling real-world processes, allowing not only their proper management, but also regulating efforts to effectively obtain correct results. The benefit is increasing the transparency and quality of control, building an effective organizational structure, optimizing resources, increasing their quality and reducing risk. This is achieved through the interaction between the two levels – of real information and processes and their transfer into a computer environment (Figure 2.).

Figure 2. Relationship between reality and information processing

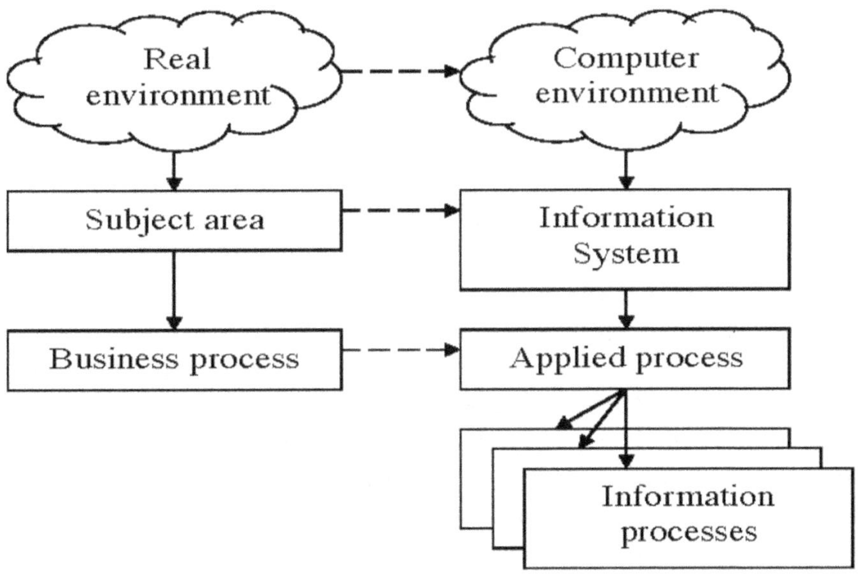

In a generalized sense, an organizational process is an organized set of one or more interconnected actions (procedures, operations, functions) that implements a certain task with the aim of creating a product or service with a specific purpose for the end customer. According to ISO 9000:2000, this set of interrelated activities transforms input impacts into output objects, effectively bringing together the flow of activities, functions, personnel and equipment (resources), decision-making information (knowledge), and rules for performance of these activities and functions.

The initial information determines the prerequisites necessary for making decisions on the management of the activity, reflecting: ✓ the requirements of the users; ✓ interaction with other processes; ✓ necessary resources; ✓ management methods; ✓ evaluation methods and improvement approaches. For its effective processing, it is necessary to implement the relevant technological environment in which to develop information processes based on information technology (Englander & Wong, 2021; Wickens & Carswell, 2021) – Figure 3.

Figure 3. Components of information processing

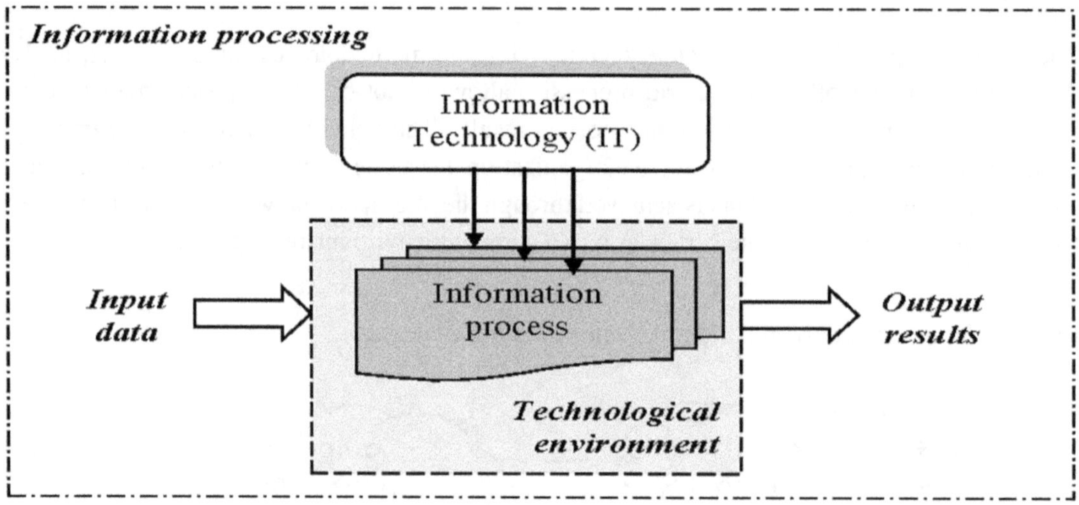

The information process is a set of sequential actions (operations) performed on the input data to obtain output results (the information necessary to realize a given goal). These processes take place in a certain type of system and are characterized by their peculiarities. The following main types of processes in information processing can be specified.

Gathering information – processes for searching, discovering and selecting information, carried out with a specific purpose, which determines the applied search methods and the procedures for selecting appropriate information. Search methods may include: direct observation; expert assessments; company literature; library archives; access to databases and knowledge bases, etc. When selecting information, procedures such as comparative analysis, parameter estimation, correlation, etc. can be applied.

Data entry – processes for receiving information from an external source for its subsequent processing, which can be implemented manually (by a person) or automatically (by specialized input devices), including appropriate conversion (encoding and decoding).

Data processing – processes for converting input data, leading to a change in their form and content. Usually in informatics it is assumed that this processing is done in a computer environment by the main processing node (processor). Processing can be "deferred" in time or "real-time" (providing the output results within a reasonable time interval after their input). Also, it can be sequential (performed by a single processor) or parallel (several processors work simultaneously).

Data storage – processes related to saving the entered data and the obtained intermediate results for the purpose of their repeated use. Depending on the level of the processes, different technical storage devices (memories) are applied.

Data output – processes for providing the results of information processing for further use. They are mainly related to appropriate visualization of the internal (for the processing system) data for easy perception by the user (speech, image, printed text, graphics, etc.).

To support the information processes, a technological environment with certain functionalities is required for the implementation of the global algorithm A_F:$\{D\} \rightarrow \{R\}$. System means (hardware and software) must be provided in the structure to support the operations of storing, converting and transmitting the data, as well as for the general management of the processes. All this defines the basic functionalities, which are presented below (Figure 4).

- "Information processing" – to implement the procedures for maintaining the global functional algorithm A_F:$\{D\} \rightarrow \{R\}$.
- "Storage" – providing the necessary information of a different type to maintain the above functionality without changing the content.
- "Control" – general organization of the behavior of the environment by coordinating the interaction of all functionalities.
- To carry out the processing, auxiliary procedures are supported for input (input conversion) and output (output conversion) of the information.

Figure 4. Basic functionalities of a typical technological environment

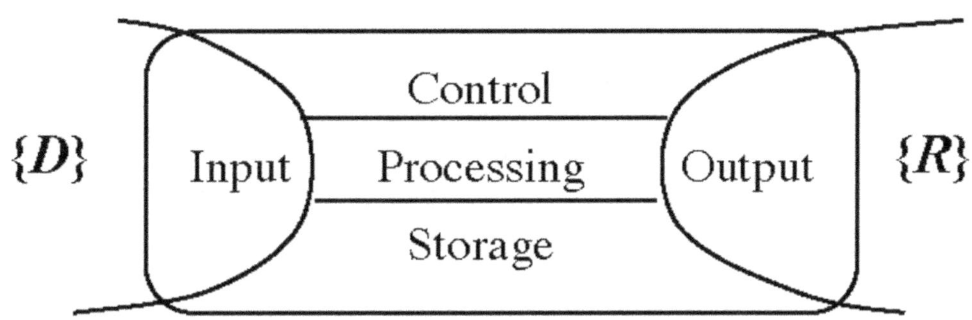

Each technological (computer) environment has a specific application and represents a specific implementation of a selected architectural model (computer architecture). The goal is to provide effective levels (values) for the set of system characteristics in the specific application based on the joint action of hardware and software components. The latter is subject to analysis and evaluation, with computer modeling as one of the possible approaches.

2. SYSTEM RESOURCES AND PERFORMANCE

Computer systems (CS) represents a complex hierarchical system of interacting functional subsystems. The only way to describe its structure and functioning is through a hierarchical decomposition. With it, a given system is presented as a limited set of subsystems, each with its own functional characteristics and possibly also with a hierarchical internal structure. This hierarchical approach can also be applied to individual subsystems (the so-called "top-down" approach), which defines levels in the hierarchical structure as shown in Figure 5.

Figure 5. Functional levels in organization of typical computer environment

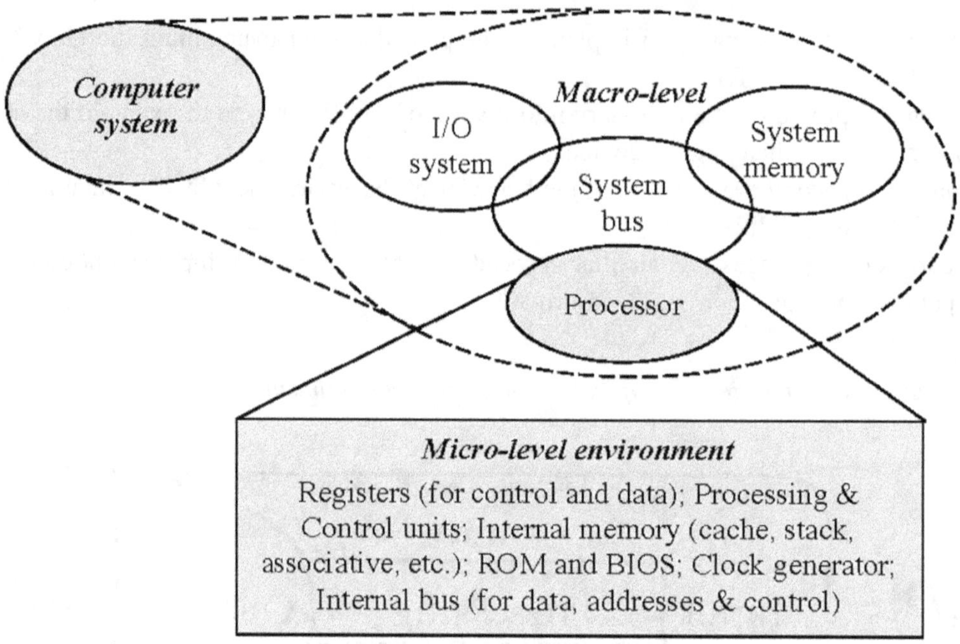

Components of the macro-level organization of traditional computer system are the following subsystems.

- Processor (CPU – Central Processing Unit) which manages the functioning of the entire system and performs the main functions of information processing based on an executed program.
- System memory for storing all the necessary information (programs and data) for the implementation of computer processing, representing a hierarchical organization of memories with different functions, capabilities, characteristics and place in the hierarchy.
- System bus unites a number of standardized connection lines that carry out internal communications based on a strategy (mechanism) for exchanging information between the other components;
- I/O system connects the processor to the environment based on various peripheral devices (keyboard, mouse, display, printer, scanner, etc.), as well as data exchange between internal memory and external memory (disk, optical, magnetic, etc.).

The features of these components are listed below.

- Processors characterize the computing power of the CS by the size of the processed data and the type of basic arithmetic-logical operations;
- Memories differ in the way of accessing the data and in their relation to the work of the processors;
- The connecting network provides a physical environment for the implementation of internal communications and connections with other CSs through I/O lines.

A basic definition of CS is as a set of hardware and software components that work together (Hennessy & Patterson, 2017). These components form the system resource and are characterized by their own attributes (system parameters). Their research and evaluation is important for determining the level of system performance and process efficiency. A need to investigate the performance of a CS arises in three situations.

1. When choosing alternative solutions, for the implementation of which it is necessary to obtain estimates for the essential factors (parameters) influencing the specific application.
2. When improving the equipment, related to modifying the system resource in order to improve the working parameters and productivity.
3. When designing systems and devices, as the preliminary research at the development stage ensures the necessary level of operability of the developed prototype and reduces the costs of the project activity.

When modeling computer objects, two levels of presentation and research are defined – system resource and workload.

The *system resource* unites all hardware and software components that participate in the organization of an environment for the implementation of computer processes. In their main part, these are the individual devices, the topology of the connecting network of relations, as well as management methods applied in the functional organization of the system. Thus, the set of system components $\{S_j/j=1\div n\}$ can be defined, building the environment for implementation and maintenance of the generated processes in the organization of computer processing. An example formal description of the behavior of basic structural components is given in Table. 1.

Table 1. Formal description of resources

Resource	*Formal description of the behavior*
Processor	Functional transformation $F=\{f_1, \ldots, f_m\}$ over a set of data (operands) $D=\{d_1, \ldots, d_k\}$ in order to obtain new data (results) $R=\{r_1, \ldots, r_q\}$
Memory	Stores the elements $d_i \in D$ in a specific order to access them
Controller	Generator of control sequences $C=\{c_1, \ldots, c_w\}$ for changing system functionality based on input data
Terminal	A passive device for generating (input) or destroying (output) requests (data) through which a connection with the environment is made
Communication medium	Topological means of defining the structural relationships between system components

Performance is a complex concept (Rathi et al., 2022). Each individual device is characterized by its own parameters, which are involved in determining the nominal performance of the system. Such are the workload of the device, the response time when solving a specific task, etc. In principle, a performance rating is determined by the work performed per unit of time.

$$P = \frac{Work}{Time}$$

In a computer environment, obtaining a direct quantitative assessment according to the classical formula is difficult due to the impossibility of precisely and clearly defining the concept of “amount of computational work”. This is the reason for determining grades for the so-called performance indices (factors). These are descriptors (parameters) by which aspects of system performance are studied and evaluated (e.g. number of instructions executed per unit time; speed of information exchange; memory cycle; channel throughput, etc.). Typically, the rate at which an event occurs per unit of time (second) is used to evaluate performance. One standard estimate of processor performance is “million instructions executed per second” (MIPS), defined by the formula:

$$\lambda = \frac{Number\_of\_Instructions}{Time}$$

Ideal performance depends on the match between the hardware capabilities and the software behavior of the computer system (CS). Ensuring coherence leads to good characteristics of the functional behavior of the CS. Main factors related to performance in classical computer processing are given in Table. 2.

Table 2. Sample performance factors (indexes)

Processor cycle (cycle time) [ns]	τ
Processor frequency (clock rate) [MHz]	$f=1/\tau$
Average number of instructions in a program (instruction count)	I_c
Number of cycles per instruction (cycles per instruction)	CPI
Number of processor cycles to decode an instruction	p
Number of memory accesses required	m
Ratio of memory cycle to CPU cycle	K
Total number of clock cycles	$C=I_c CPI$

A given computer architecture directly defines parameters I_C and p, and the application compilation technology used affects the set {I_C, p, m}. The management of the computer processes and their performance is related to the parameters:

- *Total_Processor_Time* = $p.\tau$ (for organization of processing);

- *Memory_Access_Latency* = $k.\tau$. (for organization of the system memory)

Some basic assessments for performance indexes are presented in Table 3.

Table 3. Basic assessment for performance indexes

Processor time for a program execution	$T = I_C.CPI.\tau = I_C.(p+mk).\tau = C.\tau = C/f$
Processor speed (MIPS)	$\tilde{A} = \dfrac{I_C}{T.10^6} = \tilde{A} = \dfrac{I_C.f}{C.10^6}$
Processor throughput	$W_p = \dfrac{f}{I_C.CPI} = \dfrac{f}{C} = \dfrac{1}{T}$

System characteristics reflect the organization of computer processing and CS behavior in performing user tasks, having a direct bearing on overall system performance. Some of the main features are presented below.

Unit and system workload. The occupancy of a device S_j during the intervals τ_{ij} ($i = 1, 2, ..., k$) for a time period T determines its pre-period workload. This is an individual (partial) assessment as for the specific device. When multitasking in CS, several processes develop that load different devices (parallel unit's workload), and then we speak of a system load R, determined on the basis of individual workloads:

$$R_j = \frac{1}{T}\sum_{i=1}^{k} \ddot{A}_{ij} \leq 1 \quad R = \sum_{j=1}^{n} R_j .$$

Average response time. The time u_i from the entry of the current task "*i*" for processing to the receipt of the final result of the calculation defines the response time for the particular case. A useful feature is the performance score for N tasks, defined as the average response time by the formula:

$$u_{av} = \frac{1}{N}\sum_{i=1}^{N} u_i .$$

Value (price) of the calculation – a summary estimate of the value of technical means and software when solving a task, where W_i is the volume of the *i*-th system resource used when solving the task, and k_i is a value (weight) coefficient for the corresponding resource.

$$C = \sum_{i=1}^{n} k_i.W_i$$

Speed of information exchange – the speed of information transfer through communication channels in the communication structure.

Memory cycle – the time to access information and perform a read or write operation.

Throughput capacity of a channel – volume of information passed per unit of time through the formed channel for information exchange.

Effectiveness of CS – how appropriate is the application of CS to solve a given class of tasks when assessing the value and costs of the organization of calculations. This includes the costs related to the operation of the CS, maintaining its operability, checking the technical indicators and the costs of troubleshooting.

Reliability and fault tolerance are characteristics which determine the ability of the CS to correctly perform its functions for a certain period of time. To evaluate these characteristics, the failure intensity (number of failures per unit of time) is analyzed, and for a given time-period τ, the probability of failure-free operation $p(\tau)$ at the density $f(t)$ and the probability of failure occurrence $\delta(\tau)$ are determined:

$$p(\tau)=\int_{\tau}^{\infty} f(t).dt \rightarrow p(\tau)\approx\left[n-N(\tau)/n\right] \text{ and } \delta(\tau)=\frac{N(t)-N(t+\tau)}{t.N(t)};N(t)=n-N(\tau),$$

where $N(\tau)$ is a number of components failed in time τ, and n is total number of observed components.

Multiple access to the resource – defines the capabilities of the CS to manage multiple processes that try to simultaneously occupy one or several resources (Liu et al., 2022). A manager is usually supported, which takes over the functions of system resource distribution between parallel processes and sets the rules for access and protection of common objects (data, programs, devices, etc.). Multiple access to resources is typical of network processes and especially in traffic management over mobile cellular networks. Improving network communications and increasing capacity in the 5G generation affects throughput, reliability, and low latency, which requires new and innovative solutions (Liang et al., 2022).

A common resource for multiple processes is an object (device, program module, individual program section or statement, variable or block of data, etc.) used jointly by them. Some types of shared resources allow different processes to refer to them at the same time and do not impose any requirements on the behavior of the processes. Others allow only strictly sequential access to them, requiring synchronization between processes. This is true for traditional multiple access (including for processes in network space) as well as in pipelined and parallel processing, which necessitates the development of pragmatic data channels (Shukla, 2022).

A critical resource is an object, together with multiple operations on it, which can only be "owned" by one user for the time that one operation is performed on it. When requesting such a resource from several processes, it must be provided to only one of them, and for a finite time. The latter is the responsibility of the process, which must minimize the time of possession of a critical resource. There are two main aspects of regulated multiple access to resources in the joint action of different processes presented below.

- Inter Process Communication, representing a system mechanism for ensuring the exchange of data and messages between different processes by regulating access and use (processing) to certain non-proprietary data (data sharing). It is clearly expressed in systems with distributed memory based on the exchange of messages managed by the operating system. Two types of communications are possible – between related processes that are initialized by a single process and between two or more unrelated processes.

- Inter Process Synchronization, which ensures the correct access order and parallel processing of common data (data protection). Generally, processes access shared objects to modify their contents, which requires entering their correct sequence. On the other hand, in Shared Memory Systems, two processes can communicate with each other by exchanging data through a shared object with synchronized access to it. This defines explicit synchronization while implicitly maintaining communication.

Level of parallelism – the ability of the CS to process information units of different volumes in parallel. Depending on this, two from can be defined: local parallelism (at the level of individual operations or data processing) and global parallelism (at the level of independent tasks or individual parts of a task). Another formulation is low and high parallelism in computing. The last classification refers to the volume of information processed, called a granule or grain. This defines fine-grained parallelism (with a low level of granularity) and coarse-grained parallelism (with a high level of granularity) respectively. An example for levels of granularity in parallel processing is presented in Figure 6. The main levels are as follows:

(1) Coarse-grained parallelism which has two sub-forms: (a) Large grain parallelism (Task Level) for parallel execution of separate independent programs; (b) Medium grain parallelism (Control Level) for parallel execution of independent parts (functions) of a program.
(2) Fine-grained parallelism, which determined based on: (c) parallel processing of data (Data Level) and (d) Very Fine Grain/ Multiple Instruction Issue.

Figure 6. Levels of granularity in parallel processing

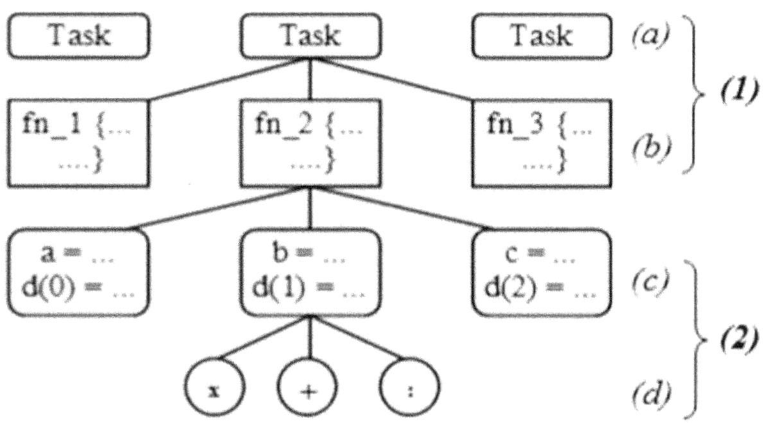

One index for evaluating parallelism is "Degree of Parallelism" (DOP), which determines how much the program parallelism covers the hardware-defined parallelism, i.e., number of processors used at a given time to execute the parallel program. The time evolution DOP(t) represents a discrete function which is called "Parallelism profile of the program" – Figure 7.

Figure 7. An example for a parallelism profile

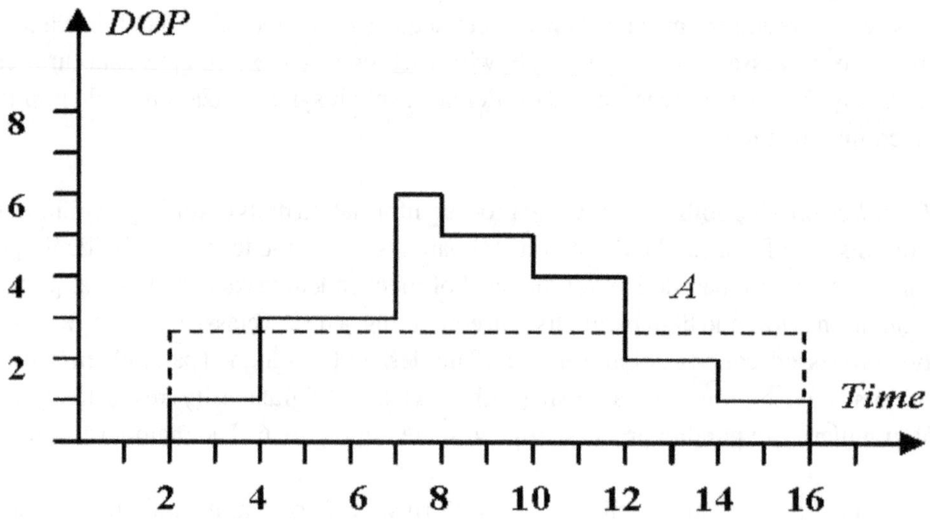

Based on a specific parallelism profile, the "Average Parallelism" (DOP_{AV} = A) can be determined – dotted line in Figure 7. Let's consider the case of a parallel CS built from n homogeneous processors, maintaining a maximum degree of parallelism $DOP_{MAX} = m$ (for the ideal case $n>m$) in a given profile. Each processor has a computational capacity Δ (analogous to computation time in MIPS or MFlops) and for a given period k_i processors are occupied ($DOP_i = k_i$). The total amount of work done will be determined by:

$$W = " \int_{t_1}^{t_2} DOP(t).dt \text{ , or in discrete form: } W = " \sum_{i=1}^{m} k_i.t_i \text{ ,}$$

where the different parameters t_i are the periods in which $DOP_i = k_i$. As a result, average parallelism A is:

$$A = \frac{1}{t_2 - t_1} \int_{t_1}^{t_2} DOP(t).dt \Rightarrow A = \frac{\sum_{i=1}^{m} k_i.t_i}{\sum_{i=1}^{m} t_i}$$

For the example presented in Figure 7 is determined:

$$A = \frac{1.2 + 3.3 + 6.1 + 5.2 + 4.2 + 2.2 + 1.2}{16 - 2} = \frac{41}{14} \approx 2{,}93$$

Additionally, the parameter "Available Parallelism" can be defined, representing an estimate of the potential parallelism in a given program. It was found that the highest degree can be reached with data parallelism (500-3500 arithmetic operations), while with instruction parallelism the limit is up to 5-7 parallel executable instructions. This parallelism is limited by the real allowable parallelism in a parallel environment, which is determined by the existing parallel nodes (processors) in it

Task scheduling strategy – a mechanism for dispatching independent processes in the computer, which introduces rules for servicing the incoming tasks for processing. An investigation of parallel processes dispatching is presented in (Romansky, 2022e). Scheduling is applied at three levels presented below.

- A low level of scheduling, where the dispatcher (Scheduler) monitors the release of a processor and activates the next process for execution.
- An intermediate level of scheduling as the dispatcher (Swapper) determines when a process should be transferred to disk memory and when to return to CS main memory again).
- High level of long-term planning of tasks (Job Scheduling) with determination of the order of their activation. Some of the planning strategies applied are as follows: FCFS (First-Come-First-Served); LCFS (Last-Come-First-Served); SJF (Shortest Job First); SRTF (Shortest Remaining Time First); HPF (High Priority First), etc.

An example of traditional dispatching strategies in time-sharing multiprogramming processing is presented in Figure 8. The scheduling of processes in these strategies must take into account the quantum of time q provided to each of the processes involved in the CPU processing. This determines the cyclical allocation of CPU time to each process with analysis of its completion or not. The two presented dispatch models have the following features.

RR (Round Robin) is a mode of cyclic processing of incoming tasks flow with intensity λ and with provision of a fixed time-quantum q. If the processing completes in the allotted time, the process exits the system. In case of incomplete processing (probability p), the task is placed at the end of the input queue (FIFO/FCFS) for further processing, shifting the processes by one position. With RR, short processes will be processed faster, giving long processes better opportunities to use the CPU. It is typical for systems with teleprocessing, time sharing and network architectures.

FB (Foreground-Background) supports hierarchical levels of queues for waiting processes when serving them with backlinks. This discipline is accepted as one of the best for process planning, being a modification of the RR algorithm. Each new process enters the first queue (level 1 queue). If after completion of the first allocated time quantum task Z_i has not completed, it is sent to a lower queue (next level). For each failure, the task (process) "falls" down one level. If the number of required time-quanta exceeds the number of levels N, the task loops in the lowest queue. Queues implement an FCFS strategy. The processing of a task of level (j+1) is moved only if all tasks of level (j) have been exhausted. A constant or variable quantum q is allowed. This is supported by the CPU busyness discipline, which ensures that a non-empty queue with the lowest number in the hierarchy is serviced. In many systems it is accepted that the time quantum increases as the number of the service queue increases.

Figure 8. Examples of dispatching strategies for time-sharing processing

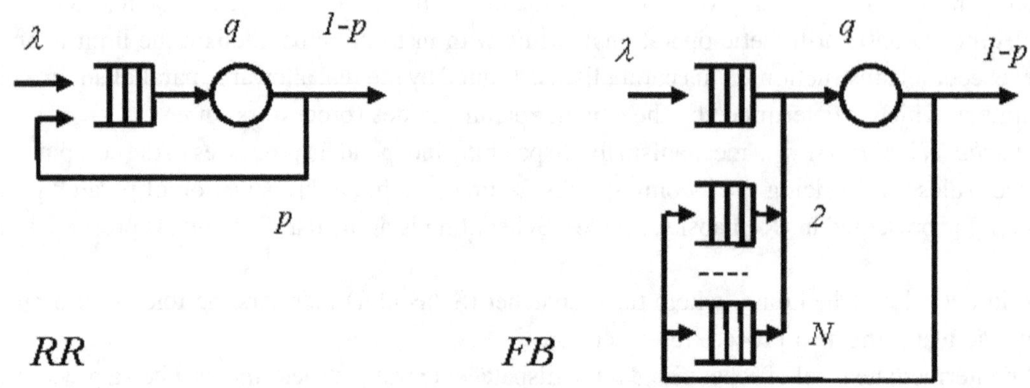

In the contemporary digital society, the main means of connection between different persons and objects is information communication, which is a process of transmitting messages from a source to a receiver using a certain type of transmission medium. Initially, it arose due to the need to ensure effective and reliable communication links when the volume of information transferred through communication systems, which were initially telephone, telegraph, etc., increased. The importance increased with the entry of computer equipment and technologies into material production and public areas. The development of multimedia technologies has allowed communication between people to be put on another, different, interactive basis. The global network (Internet) is no less important for information communications. The World Information System (www) covers the entire planet like a "web" and enables the exchange of information over considerable distances.

The above determines a requirement for continuous research of performance indicators (indexes) in the organization of communication processes by applying different approaches corresponding to the nature of the specific object. The point is that two relatively independent directions can be defined in the network space – processing and communication, each of which has its own specific features. In addition, this allowed the concept of "information" to be exposed in other concepts that reflect the modern world - informatization, information system, information resources, information security, information culture, information society, information industry, etc. All this led to the development of the modern strategy for the globalization of information resources and the construction of a common information space with possibilities for remote access, distribution of information and its use from different points of the planet.

3. WORKLOAD FORMALIZATION

Workflow (WF) is the full volume of information forming the set of processed tasks, received at the input of CS (input WF) and processed by its components (hardware and software resources). The elements of the WF can be represented as separate entities circulating in the internal structure of the system and requesting service from the resources S at different stages. Thus, a flow is formed that has a probabilistic (stochastic) nature, i.e. has stochastic characteristics and obeys probabilistic laws. This is true for all kinds of computing systems, including quantum computing, which has been developing in the last decade (Weder et al., 2020). In the article, besides the claim that quantum computing "*are promising to solve problems more efficiently than possible on classical computers*", issues with the organization and modeling of work processes are discussed and an extension to imperative model languages is proposed, allowing mapping and defining own constructions for workflow modeling.

Computer processing itself is an organized set of processes related to serving user tasks, which forms the general *workload (WL)* of the CS. A process can be described by the ordered triple I=<t, A, T>, where: t – starting moment of the process; A – defining attributes that define the source of the process (e.g. user, program, etc.); T – process path, characterizing the sequence of events in occupying elements of the system resource S. Workload investigations are conducted in various directions, for example high-performance computing, cloud computing, IoT, smart applications, etc. An analysis of WL and performance in several directions (traditional computations, machine learning, graphical analysis, etc.) is presented in (Akram et al., 2021). The goal is to investigate the possibilities of security of the conducted scientific calculations, observing the location of the data in the memory of large-scale parallel machines, the impact of virtualization on the productivity of the workload and the flexibility of the program management.

A given event e_j is characterized by a time of occurrence (t_j), a name of an occupied system resource (S_i) and parameters for the availability of the resource. This defines a computer profile as a sequence of discrete events to occupy selected resources, an example representation of which is given in Figure 9.

Figure 9. Profile of computational process

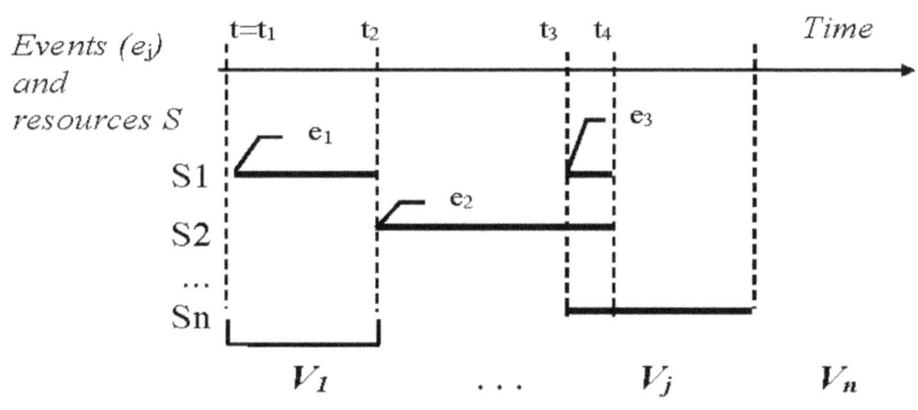

The events e_j can be interpreted as different actions, for example: read data; waiting in an input queue for scheduling; occupying a system resource; processor processing; work with external memory; data exchange etc.

A workload is most often represented as a stochastic flow of requests, which can be either uniform or non-uniform (the latter being a collection of uniform flows). Depending on whether the statistical characteristics of the flow are constant in time or not, it can be stationary or non-stationary. An example of WF as a stochastic process with significant non-stationarity of the mathematical expectation *M* of the intensity is shown in Figure 10.

Formalization of workload and processes for computation can be made by using mathematical formal means which include: ✓ a set of abstract elements (concepts); ✓ terms reflecting data and algorithms for their processing; ✓ procedures for the interaction between data and algorithms. The formal means include the alphabet for data representation, methods and means of their coding, algorithms for data representation, transformation and processing, input-output procedures, for storing and searching data, for management, etc.

Figure 10. Description of non-stationary WF

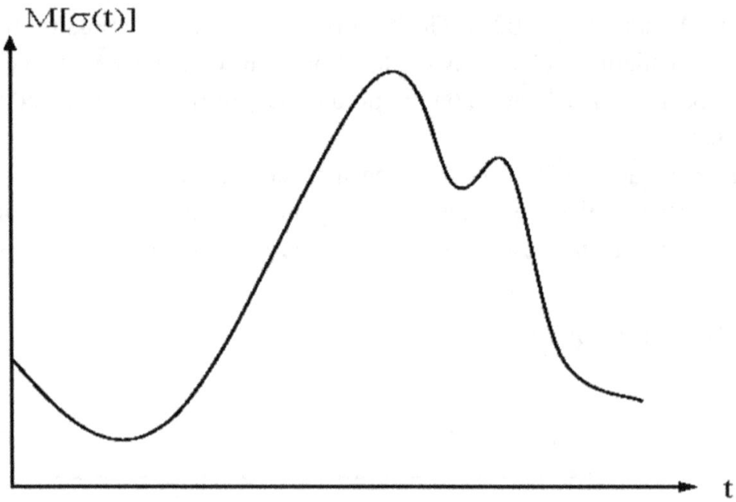

In a general, the work of a CS can be represented by an ordered pair <S, A>, where S reflects the functional organization and features of the computer resource for the implementation of information processes, and A represents the algorithmic and programmatic features of computer processing.

One of the traditional approaches to the formalization of computer processing of information is the graph representation, which is possible because, in a functional aspect, computer processing can be considered as a sequence of transitions between a finite set of discrete states. Transitions between states are assumed to be instantaneous, with the specific values for system parameters $\{s_i\}$ and characteristics $\{y_i\}$ being determined by the functional algorithm A_F and the set of input effects $\{x_i\}$. An example of formalization is the description of processes in computer processing under the control of the operating system as a network of transitions between states (State Transition Network - STN) – Figure 11. Six main states are defined with corresponding transitions between them:

(1) Input – a process of inputting a program and data for a given task through an input device and placing them in the main memory of the computer, which is performed under the condition of "Read" (input and storage).
(2) Memory – storing the information in the corresponding memory areas (for programs and for data).

(3) Readiness – a state in which the task is planned for execution and awaits provision of the necessary system resource. The order of execution of incoming tasks is determined by a "dispatcher" system program (part of the operating system OS), implementing a certain strategy for optimizing system characteristics during information processing, one of them being processor performance. The dispatcher forms a plan for the execution of the received tasks, initializing them for execution (Initiate) and, if possible, allocating them to the necessary system resource (Dispatch). Upon completion of information processing on a given task, it terminates its participation (Terminate) in the readiness process, and the result is stored in memory.

(4) Execution – active state of implementation of a given task in the processor by extracting and executing the instructions from the control program. The execution can be interrupted (temporarily suspended) by another task with a higher priority (Priority-interrupt) or when there is a need to perform an I/O operation (I/O_Start).

(5) Waiting – state of a task that is waiting for the realization of a certain event outside of processor work. For example, when an I/O operation (I/O_Start) is requested, the active program is temporarily interrupted, and control is transferred to the peripheral management system (the corresponding driver program is started). After completing an I/O operation, the execution of the task can be continued (I/O_End), and for this purpose it is sent to the "Ready" state.

(6) Output – last state of a given task, in which the obtained results can be output to an external device (Write) and the active presence of the application in the computer is terminated.

Figure 11. Formal description of basic states of traditional processing by using STN

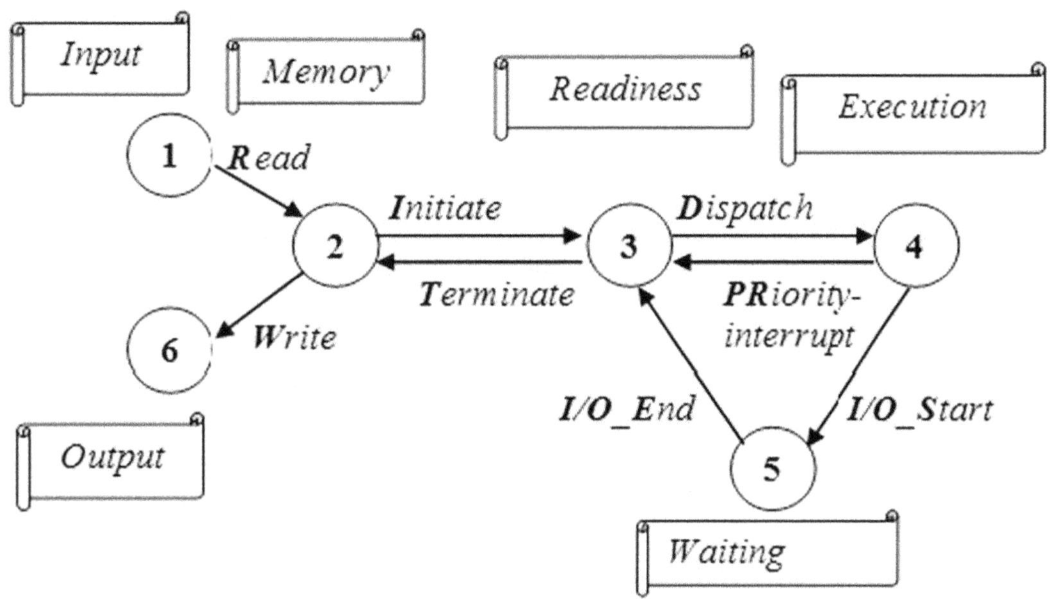

The presented formalization is related to strategies of multitasking / multiprogramming, which permit simultaneous execution of several independent tasks in the presence of one processor (Figure 12). A task (application) can be in one of the presented states, which is determined by the management strategy supported by the operation system.

Figure 12. Formalization in multitasking

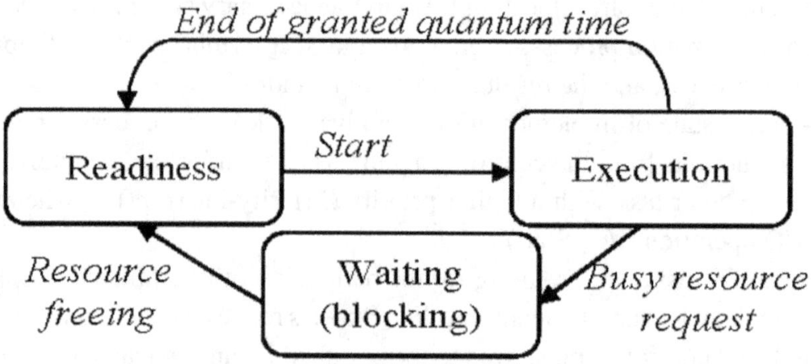

2.4. INFORMATION AND INFORMATION TECHNOLOGY

The Role of Information in the Society

Information, as a concept, has undergone development over the years depending on the existing realities. In the modern world, information is understood as "transmitted or received message, information, and knowledge about someone or something", which connects it to information communications for the transmission of messages between users through computer networks. Depending on the sphere in which the essence of the information is discussed, the interpretation of the concept has its specificity. This defines different aspects of the concept of "information".

- The philosophy considers information as a reflected variety that arose as a result of the interaction of objects – the characteristics of the transmitting object are reproduced (reflected) in the receiver (transmission of information). In this sense, information is the result of reflection and, together with matter and energy, is an attribute of matter.
- Sociology perceives information as information, data, concepts, reflected in human consciousness and changing people's ideas about the real world.
- Information theory (Claude Shannon, 1948) considers information only as information to eliminate a given uncertainty. Each signal appears with some probability, and the smaller this probability, the more significant the information it represents.
- In the science of cybernetics (management theory), founded by Norbert Wiener in 1948, information is defined as a carrier of content obtained from the outside world in the process of adapting a person to it and is such knowledge that is used in making decisions in the process of managing a given system.

- In the computer field, information is any message formed according to an established format (presentation standard) and represented as an ordered set of signals (discrete or analog), stored, transmitted or processed using technical devices.
- The public concept "information" includes the totality of information about the surrounding reality and ongoing processes that form human knowledge.

In a general sense, *information processes* carry out processing of a certain type of information by applying appropriate technology depending on the subject area. They are related to the information activity of the person and are implemented in a certain technological environment. The information process from a computer point of view is a set of sequential actions on the input data to obtain output results that provide the necessary information to realize a given goal – see Figure 3.

Informatization is a concept that, according to a UNESCO definition, is the development and application of methods and means for collecting, transforming, storing and disseminating information, enabling the formation of new knowledge and its application in society for its current management. In this sense, the informatization of society is a purposeful socio-economic and scientific-technical process for changing the social information environment by creating optimal conditions for satisfying information needs and realizing the rights of citizens, authorities and organizations (Romansky, 2021a). The term "informatization" was introduced into the public space at the end of the 1970s by two publications by Marc Porat (Porat, 1977) and by S. Nora & A. Mincs (Nora & Mincs, 1978) to define the role of information in industrial society. Later, G. Wang (1994) associated informatization with the processes of popularizing information and accelerating its dissemination to increase the economic, political, social and cultural status of society (Wang, 1994). At the end of the 20th and the beginning of the 21st century, the term acquired the meaning of applying modern means of information processing and the "World Wide Web" in social development. The goal is to build and develop the global communication infrastructure and increase efficiency in the use of territorially distributed information resources through the systematic computerization of society.

The main goal is to generate processes that create conditions for important changes in people's lives, which requires significant efforts by government, business and all ICT users to form an adequate information culture. Approaches to informatization can be defined in two directions:

- A technological approach related to the development of technical and technological means to increase the efficiency and productivity of production processes.
- Sociological approach, reflecting the role of informatization in human activity and specifically on their knowledge, skills, moral, economic and cultural interests, personal development, etc.

A connection is often made between two concepts in modern society - informatization and digitalization, which defines the features of cyberspace and network communications. However, these are two different approaches, assuming that informatization is the basis for the development of the digital age. In this regard, various concepts are defined that reflect the ways of processing and disseminating information in society – electronization, mediatization, computerization, and intellectualization. Some publications even suggest the introduction of the term internetization to reflect the role of the global network in society. In addition, it is suggested that the used terms "e-" / "electronic" or "digital" be replaced by more modern words starting with "intelligent", "intelligent", "Cyber-physical", etc.

Formalization of Information Technology

An information technology (IT) is a set of methods and means for processing and transmitting data in order to obtain information of a new quality (information product) about the state of the object, process or phenomenon (Wickens & Carswell, 2021). As a rule, computer and communication tools (hardware and software) are used for implementation. Nevertheless, in addition to the application of computer means, a given information technology usually also foresees elements of manual processing or collection of the information. An example of this is the information processes in various sociological surveys, the initial accumulation of information in modeling technologies, the formation of the initial sample in forecasting and statistical research of trends, etc.

The main purpose of an IT is the production of new information satisfying the informational needs of man. For this, it is necessary for IT to ensure separation of the information processing process into standardized and unified phases, operations and actions, as well as to carry out purposeful management of information processes.

The toolkit of an IT includes the following components:

- Hardware – computer and communication means for performing operations on information objects;
- Software – a set of system, application, methodical and informational provision;
- Orgware – human organizational procedures and actions when applying the technical and software tools for the implementation of information processes.

Depending on the specifics of a given IT, some or all components of the system will be involved in the implementation of the information process, which leads to their load to a certain degree (Englander & Wong, 2021). This is the user's understanding (from a high level) of an information process, and for its implementation, a sequence of low-level processes is generated. From a computer processing perspective, a process is the implementation of a program unit (executable code) in an environment that includes low-level hardware components (processor, registers, program counter, status flags, and part of RAM). This environment determines the current state of the process, and therefore each information process is an organized set of individual computer processes generated on the basis of specific IT.

Formalization of IT can be done on the basis of information conversion. One possibility is to define processes and procedures as basic functional elements. This allows us to build a *conceptual model* of basic information technology which is shown in Figure 13.

The conceptual space is divided into two parts – a field of processes and a field of procedures. Functional blocks are included in each, with those with predominantly manual processing represented by a dashed line. For the rest (solid line), automatic processing performed by computer means prevails. Two horizontal levels are defined. The upper level depicts the presentation of information in a form that is easily perceived by humans. The lower level reflects the machine (computer) representation of the data.

Figure 13. Conceptual formal model of IT

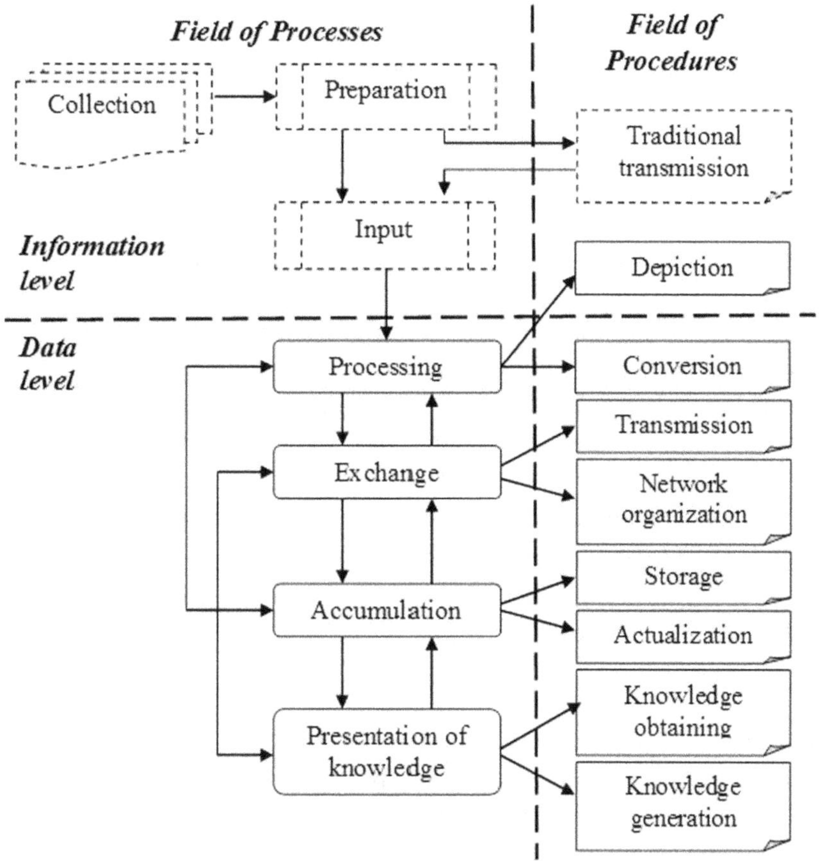

Information processing technology begins with the formation of an information resource, which is transformed into an information product through a sequence of processes, each associated with certain procedures listed below.

- Collection – collection of information to reflect the conceptual area (characteristics, parameters, state, etc.).
- Preparation – organization of information in an appropriate form and structure (table, array, etc.).
- Input – after transmission of the information (telephone, courier, mail, etc.), its input is performed in a corresponding machine form.
- Processing – machine processing of the entered data, which is related to appropriate visualization or conversion.
- Exchange – two ways of implementation are possible: transmission through connection channels (teleprocessing) or application of network procedures for information exchange (network organization).
- Accumulation – presentation of the information in such a data type as to ensure their long-term storage and use (editing, updating, deletion, etc.).

- Representation of knowledge – a finalized process, insofar as knowledge is the ultimate product of an IT. It combines the formalization (representation) of knowledge, its accumulation in a formalized form and formalisms for generating new knowledge. In this way, knowledge becomes an important strategic resource of society, determining the directions for its future development and related to artificial intelligence systems.

The *main characteristics* of an IT are the following:

- Expediency – increasing the efficiency of processes by using computer, information and communication tools.
- Integrity – IT is a complete system capable of solving tasks that are not characteristic of its components.
- Modularity – IT is a set of functional modules specialized according to the processes they provide.
- Dynamic development – IT is the fastest developing technologies that are constantly modified and include new components.
- Subject area – each IT is directly related to a specific subject area for which it was developed. Usually, the subject area is reflected through the development of a conceptual model (reflects the basic concept of the created applications in relation to the real world), and the implementation takes place through the so-called a logical model for defining the structure and relationships between application elements.

Every IT is related to carrying out research activity, which in a given organization is the task of the Research & Development team, as in (Vendrell-Herrero et al., 2021) a study of the relationship between IT processes and innovation efficiency is presented with the application of the model approach. A multi-indicator, multi-causal model is estimated based on a unique sample of multinational manufacturing enterprises to compare different variants of the IT – R&D team relationship. Another research presented in (Krak et al., 2022) focused on using a visual analytics system in machine learning to analyze the intellectual capabilities of humans in model building. The paper explores the forms and purposes of using the visual analysis workflow to form the final product to build a model as an information processor and decision-making mechanism. Model building concepts are explored and classification information technology is proposed.

Technological organization in the field of information transformation has been applied in various directions, such as in the optimization of processes for managing organizational systems (Lvovich et al., 2021). The article states that "*the introduction of the aggregation-balance direction of optimization modeling can be considered as an effective approach for coordinating the transition from personnel adaptation in the conditions of single-channel control to adaptation in the conditions of omnichannel control*". An optimization modeling of the technological management process was carried out and a block diagram for its adaptation during the transition from single-channel to multi-channel implementation was demonstrated. Similar research is presented in (Cheryshov et al., 2020), where the sequence of actions for forming an optimization model and an algorithm for intellectual support of the process of managing the distribution of resources in an organizational system is discussed. The technological aspect of the research allows establishing a sequence of specific management tasks that form a decision-making cycle, as well as "*extreme and boundary requirements can be formed on the basis of optimiza-*

tion model". The development of the algorithm for intellectual support uses the Games technology and block linear programming.

REFERENCES

Akram, A., Giannakou, A., Akella, V., Lowe-Power, J., & Peisert, S. (2021). Performance analysis of scientific computing workloads on general purpose TEEs. *2021 IEEE International Parallel and Distributed Processing Symposium (IPDPS)*, (pp. 1066-1076). IEEE. 10.1109/IPDPS49936.2021.00115

Cheryshov, A.B., Choporov, O.N., Preobrazhenskiy, A.P., & Kravets, O.Ja. (2020). The development of optimization model and algorithm for support of resources management in organizational system. *International Journal on Information Technologies and Security*, *12*(2), 25–36.

Englander, I., & Wong, W. (2021). *The architecture of computer hardware, systems software, and networking: An information technology approach*. John Wiley & Sons.

Hennessy, J., & Patterson, D. (2017). *Computer Architecture: A Quantitative Approach* (6th ed.). Morgan Kaufman / Elseviar Inc.

Krak, I., Barmak, O., & Manziuk, E. (2022). Using visual analytics to develop human and machine-centric models: A review of approaches and proposed information technology. *Computational Intelligence*, *38*(3), 921–946. doi:10.1111/coin.12289

Liang, B., Gregory, M. A., & Li, S. (2022). Multi-access Edge Computing fundamentals, services, enablers, and challenges: A complete survey. *Journal of Network and Computer Applications*, *199*, 103308. doi:10.1016/j.jnca.2021.103308

Liu, Y., Zhang, S., Mu, X., Ding, Z., Schober, R., Al-Dhahir, N., Hossain, E., & Shen, X. (2022). Evolution of NOMA toward next generation multiple access (NGMA) for 6G. *IEEE Journal on Selected Areas in Communications*, *40*(4), 1037–1071. doi:10.1109/JSAC.2022.3145234

Lvovich, K. I., Preobrazhenskii, A. P., Choporov, O. N., Aksenov, I. A., & Ivaschenko, A. V. (2021). Modelling of optimization process of personnel adaptation to digital management in organizational systems. *International Journal on Information Technologies and Security*, *13*(4), 71–82.

Nora, S. & Minc, A. (1978). *L'informatisation de la société* (*Vol. 11*). Paris: La documentation française.

Porat, M. U. (1977). The Information Economy: Sources and Methods for Measuring the Primary Information Sector (Detailed Industry Reports). *The Office*, *77*(12).

Rathi, S., Nagpal, R., Mehrotra, D., & Srivastava, G. (2022). A metric focused performance assessment of fog computing environments: A critical review. *Computers & Electrical Engineering*, *103*(October), 108350. doi:10.1016/j.compeleceng.2022.108350

Romansky, R. (2021a). Informatization of the society in the digital age. *Biomedical Journal of Scientific & Technical Research*, *33*(3), 25902–25910. doi:10.26717/BJSTR.2021.33.005418

Romansky, R. (2022e). Formalization and investigation of parallel processes dispatching. *2022 IEEE International Conference on Information Technologies (InfoTech-2022)*, (pp. 94-97). IEEE. 10.1109/InfoTech55606.2022.9897104

Shukla, S. (2022). Developing pragmatic data pipelines using Apache Airflow on Google Cloud Platform. *International Journal on Computer Science and Engineering*, *10*(8), 1–8. doi:10.26438/ijcse/v10i8.18

Vendrell-Herrero, F., Bustinza, O. F., & Opazo-Basaez, M. (2021). Information technologies and product-service innovation: The moderating role of service R&D team structure. *Journal of Business Research*, *128*, 673–687. doi:10.1016/j.jbusres.2020.01.047

Wang, G. (1994). *Treading different paths: informatization in Asian nations.*

Weder, B., Breitenbücher, U., Leymann, F., & Wild, K. (2020). Integrating quantum computing into workflow modeling and execution. *2020 IEEE/ACM 13th International Conference on Utility and Cloud Computing (UCC)*, (pp. 279-291). IEEE. 10.1109/UCC48980.2020.00046

Wickens, C. D., & Carswell, C. M. (2021). Information processing. Handbook of human factors and ergonomics, 114-158.

Chapter 3
Technology of Computer Modeling

ABSTRACT

Technological aspects of computer modeling in its application to the investigation in computer field are discussed. A detailed classification of the applied computer models is presented, with two relatively independent groups - homogeneous and the heterogeneous models. For each of the models an introductory description of the main features is given, and corresponding relatively simple examples are presented. As a next step, a general procedure for the overall organization of a model research is defined, and the importance of the correct planning of each model experiment is indicated. A methodological scheme as a sequence of relatively independent phases is defined and each of them are discussed to present the mine goal and details, specific features, and requirements. The difference and features between the three successive types of models when conducting research are defined - conceptual model (formalized description), mathematical model (analytical description), and program model (concrete implementation in a suitable program environment).

1. CLASIFICATION AND FEATURES OF COMPUTER MODELS

Modeling is an approach used in various fields, and the choice of a specific method is made based on a preliminary analysis of the proposed application possibilities when solving a given task. One example is the critical review of different approaches for photo-voltaic arrays modeling presented in (Jena & Ramana, 2015). Basic approaches based on various analytical methods, classical optimization techniques and soft computing techniques are discussed. A classification of modeling techniques for both uniform and non-uniform conditions is presented. Another application of modeling research approaches is shown in (Taifa et al., 2000), analyzing the processes of allocating, sharing or dividing order quantities between cooperating business entities. A discrete event simulation software package for many-to-many processes was used for the study in order to optimize the monitoring and control of manufacturing relationships.

An analysis of the impact of computerization on modern science is made in (Varenne & Turnbull, 2018), explaining the crucial interaction between technology and science, discussing particularities of

DOI: 10.4018/978-1-6684-8947-5.ch003

factors such as formalization, computation, data collection and visualization that led to the development of computer models and simulation in the time. The aim of the book is to explain the successive transition from mathematical models to computer simulations of the last decades.

A generalized classification of the main types of computer models used in the investigation in computer field is presented in Figure 1. Depending on the internal organization of the models, two main directions are defined – uniform and non-uniform models. In the case of the first, one of the basic modeling methods was applied and a homogeneous model of the original object was built. In the models of the second group, different approaches and/or means are used, which creates a heterogeneous structure of the developed model. A summary of special features of basic models with simple applications is presented below.

Figure 1. Generalized classification of computer models

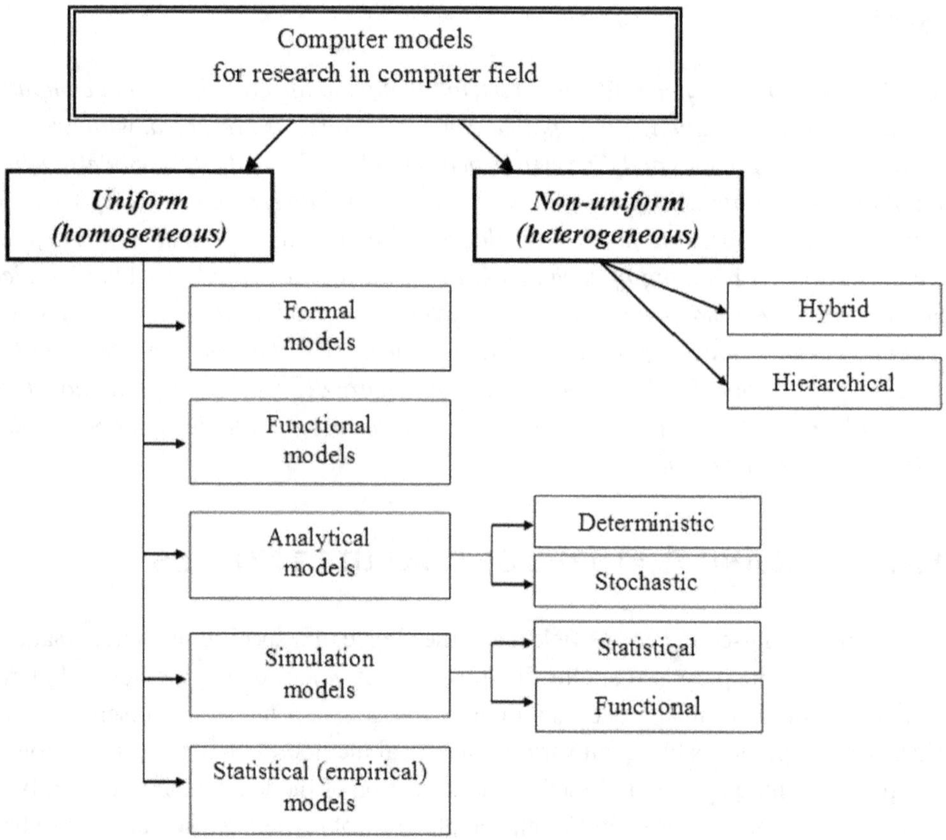

Formal and Abstract Models

Formal models are built on the basis of strictly defined rules and hypotheses and can use various deterministic and stochastic means of abstract representation of the formalized object. Formalization allows converting the researched object (system resource and workload) into a form convenient for creating a computer model using one of the main modeling methods. In this reason, the formal model is the first step in model research. For its development, an appropriate formal system is chosen (or defined), including an alphabet, grammar, and rules for defining primitives. The formalization is based on the perception of each discrete system as a creative object of interconnected but relatively independent subsystems. These subsystems are described by formal techniques and means, and then the relationships between them are defined. For example, in (Ramos et al., 2022) a formalization of the theory of computation is presented for a computational model given as a class of partial recursive functions and built on basic operators (constants, successor, projections, bijections of tuples) leading to a formalized complete Turing model. Formalizations of the recursion theorem, the Rice theorem, the fixed-point theorem, etc. have been made. Another example is the developed ontology suite for modeling in different fields of science, presented in (Fathalla et al., 2020), discussing the design considerations in its development, as well as applying the formalization in deploying applications in cloud environments, discussed in (Saatkamp et al., 2020). The reason for the latter work is that multi-cloud deployments increase the complexity of provisioning, and the paper formalizes and presents algorithms for automating the pattern separation and matching method. A formalization is used in (Alturki et al., 2022) to explore the computational complexity of processes and distinguish rules that are under the control of Cyber Physical Systems (CPS).

While the formal model is some reasoned description of the object and its behavior through mathematical hypotheses, terms and rules, the abstract model is its appropriate visualization through graphical means. Various means can be used to describe the formalized object and create an abstract model, such as block diagram, structural schemes, binary and weight graphs, Entity-Relationship Diagrams (ERD), Data Flow Diagrams (DFD), Dialog Transition Networks (DTN), Petri nets (PN), stochastic networks, etc. The special thing in this case is that these tools are applied only to the abstract presentation of the formalized description of the studied object, regardless of the fact that many of these techniques are used to develop deterministic or stochastic models. Some of these will be presented in subsequent chapters depending on their specific relevance to the modeling approach.

An example formal model of a hypothetical single-processor CS operating in multiprogramming mode with a constant multiprogramming factor N is given below, as in Figure 2 the abstract model based on a directed graph is presented. The following assumptions are made:

a) The workload is considered as a uniform flow of tasks of the same type, occurring at the same intensity equal to the average.
b) Each task requires time t for processor work and k turns to external memory (EM).
c) Requests to EM are submitted at constant intervals, each request resulting in a fixed and constant volume of data exchange.
d) A constant multiprogramming factor $N=N_1+N_2=const$ is maintained, where N_1 and N_2 are the number of tasks served in the processor and in the EM, respectively.

The defined formal description of computer processing includes one passive state S_0 and two active ones: task processing in the processor (S_1) and service in the EM (S_2). It is assumed that if a task goes into passive state, a new task is activated instantly ($N=const$). At equilibrium, the average input flux intensity must equal the average output flux intensity for both subsystems (processor, EM).

Figure 2. Abstract description of traditional processing in CS with classical architecture

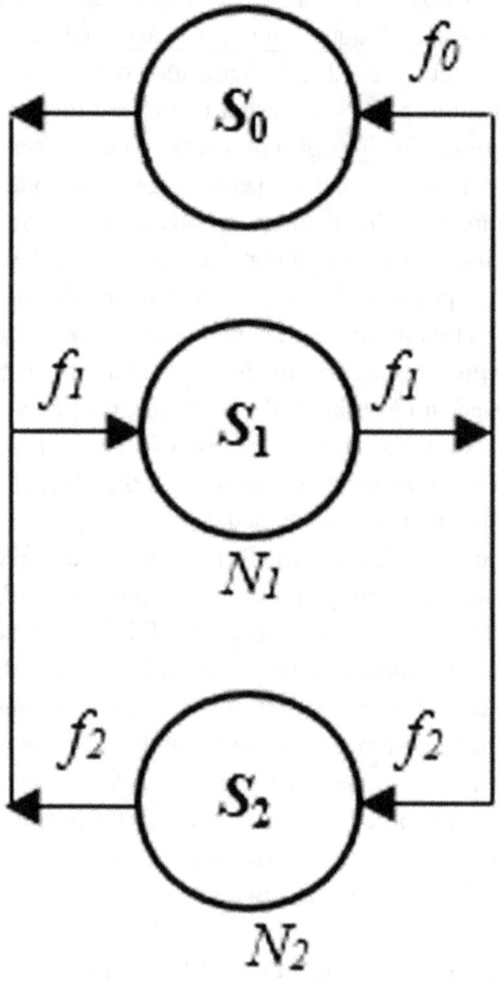

Functional Models

This group brings together models that describe the functionality of a given system without considering the available structural connections. They are aimed at researching the management of processes, information flows, algorithms, programs, etc., as well as at individual stages of the design of devices and systems. The functional model is designed to study the general behavior of a system and the reliability of the implementation of its functional algorithm. In most cases, it is sufficient to analyze individual events related to key situations in the system's operation. This approach is applied in (Williams, 2020) to define a functional modeling framework to provide a simple general mechanism for representing models of consciousness and cognition. Techniques are borrowed from complex systems and software engineering to define the full set of features in model development and to compare different models.

An important conclusion is that the functional modeling must be carried out before the actual development of the respective system, in order to be able to examine the expected functionalities of the prototype in advance. This thesis is also advocated in (McIntire et al., 2016), where the place and role of functional modeling as an initial stage of architectural design is discussed, when many of the functions are not yet specified or the possible problems are not clear. In order to analyze the effectiveness of the design of a given system, represented by functions and workflows, two different models have been developed – a behavioral model for simulating the behavior of system components and a functional model for mapping the relationships of components to functions. For each basic determined element of the functional flow, a behavior is defined in advance, and each flow is represented by an object with two variables – a variable for its state and a variable of the flow rate. The result is that users can easily describe functional models with consistent behavior as Python NetworkX graphical objects, which will allow modeling the functional behavior of the system architecture under development.

The applicability of functional modeling is in various fields and directions. One such possibility is discussed in (Yoshikawa et al., 2013) in the development of a new risk monitoring system, enabling both the prevention of accidents in daily work and the reduction of radiological hazard after a serious accident. A defense-in-depth risk monitoring model and a reliability monitor were developed with a preliminary study using the functional modeling approach. Especially for the applicability of functional modeling in the field of computer systems and processes, the following two directions can be considered.

- Functional study of the CS system resource as an interconnected set of hardware components.
- The functional study of computer processing, i.e., of algorithmic and program components for organization and management of data processing.

In essence, these are tasks of system analysis, considering the mutual influence of individual subsystems and devices. One possible solution is clustering, as such research whit application in healthcare is presented in (Shukla, 2023c). In the general case, to solve these tasks, typical structural modules are defined, or logical schemes are applied to describe individual work nodes and transitions between them.

An example of the application of functional modeling in the description and study of information processes and computer processing is given below. When developing a software system, a multilayer architecture (Figure 3) with three basic subsystems is recommended:

a) Processing (operational) subsystem – performs the main processing of the input data based on the functional algorithm for obtaining the corresponding output results.
b) Information subsystem – usually includes a database for servicing the main processing processes.
c) Management subsystem – for making logical decisions related to information processing.

Figure 3. An example for software multilevel architecture

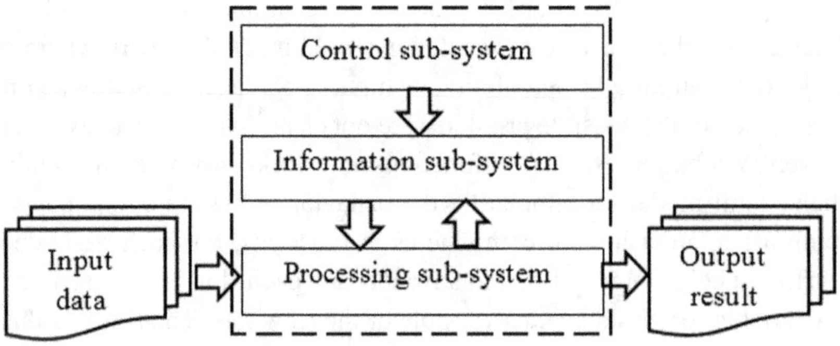

When developing the program components, it is recommended to describe the basic concept as a functional model. One suitable tool for this (oriented towards database design) is the Entity Relationship Diagram (ERD). It is a graphical means of abstract representation of the relations in the real object, considered as a set of entities with certain relations between them (Chapple, 2021). It is possible to define different sets of entities of the same type E_1, E_2, ..., E_n, through which the individual classes of objects are represented, as well as sets of relations of the same type R_1, R_2, ..., R_m for the formalization of relations (the relations) between these classes. The main elements of ERD are presented in Table 1, and two examples of ERD-segments are presented in Figure 4. Relationships can be "many-to-one" (N:1), "one-to-many" (1:M), "one-to-one" (1:1), and "many-to-many" (N:M).

Table 1. Basic graphical elements of traditional ERD

Entity	Entity – object or conception for presentation of class of objects in the designed application
Attribute	Basic (key) attribute of an entity, which is unique and important for it
◇	Defines the relationship between different classes of objects (entities) in diagram

Figure 4. Examples of ERD-segments

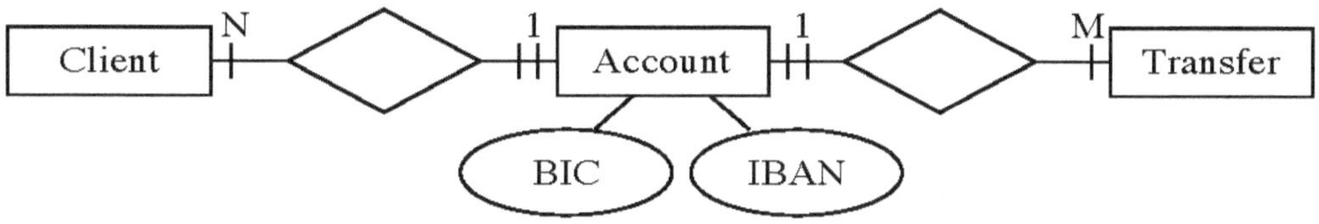

(a) Bank system

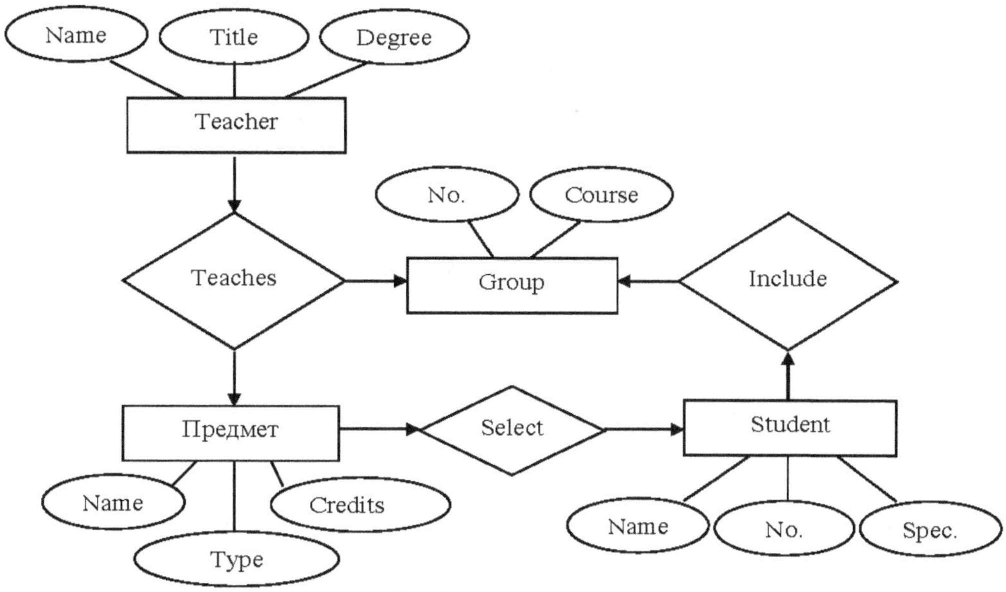

(b) University system

The ERD model has gained wide distribution due to the following more important features.

- Ease of building ERDs that well represent the subject area semantically.
- Ability of ERDs to reflect not only a specific (static) state of the subject area, but also the behavior of entities and relationships in a certain period of time, i.e. The ER-model can reflect the dynamics of the subject area.
- Methods for converting an ER-model into a relational model have been developed and applied, rules for natural language representation with ERD have been formulated, which optimizes database design processes and expands the possibilities of users.

Analytical Models

Analytical models are based on mathematical methods for describing the behavior and physical properties of the studied system as mathematical objects (deterministic or probabilistic) and relations between them, expressed through mathematical operations. This requires precision in formulating mathematical relations, assumptions, and approximations. In the general case, the mathematical model is a system of equations with defined initial and boundary conditions, which are passed through mathematical observations and numerical methods. Thus, the analytical model can be a description in explicit form $Y = F(X, S)$ or implicit form $\Phi(X, S, Y) = 0$ of the relationships between the quantitative characteristics, external influences, and system parameters of the studied object.

When applying analytical modeling, it should be considered that the studied object is generally a deterministic system and the input variables are probabilistic in nature. This defines different approaches to analytical description based on deterministic and stochastic mathematical means, which define the two main groups presented below.

Deterministic Analytical Models (DAMs) – they are characterized by a constant and definite structure and operate with finite quantities, with sub-directions being the types of automatic, graphical, network, pseudo-stochastic, etc. (Romansky, 2020a; Romansky, 2020b; Romansky & Noninska, 2022). DAMs find applications in a variety of fields, including non-conventional computing, bioinformatics, and neuromorphic artificial intelligence systems. An example of this is the article (Akther et al., 2021), which is dedicated to a deterministic model study of the processes of the recently developed "diffusive memristor" resembling the behavior of biological cells. These devices are shown to generate a vast array of interesting phenomena through the interplay of regular, chaotic and stochastic behavior, i.e. essentially have stochastic dynamics. However, the paper proposes a deterministic model, based on the Fokker-Planck description, to study the dynamics of the object, which allows to apply the bifurcation theory to describe the transition between dynamic regimes. Often, research is conducted in the combined application of deterministic modeling with other approaches, for example in (Lemiale et al., 2022) to the deterministic modeling with the application of finite elements, an additional statistical analysis was conducted to determine the most important factors in the model study. Another example is the seemingly problematic combination of deterministic modeling with stochastic modeling because the two approaches have different meanings (Atangana & Araz, 2021). The presented research is related to modeling of chaos problems by conducting analyzes of partial existence, uniqueness, and partial numerical scheme situations in cross-effect processes.

Stochastic Analytical Models (SAMs) use a stochastic apparatus (theories and tools) to create the model description, and when modeling the stochastic aspects of the researched object, they operate with random variables and processes based on probability theory, Markov chains or queuing theory for request servicing (Romansky, 2021b; Romansky, 2022a). SAMs are widely used in various fields and numerous examples can be found. One of them is the use of the stochastic approach in the study of energy stability of buildings, applied in (Carlucci et al., 2021), where 144 permutations of 15 different stochastic models were used for sensitivity analysis. To evaluate the relationship between the stochastic models, the method of generalized estimating equations was applied. Another example is the overview of existing stochastic models for predictive control (SMPC) in cyber-physical system (CPS) presented in (Chen & Shi, 2021), and based on an overview of existing control algorithms and probabilistic constraints, an extension of stochastic framework for resilience against malicious attacks. As a result, an architectural stochastic SMPC-based framework for sustainable CPS is proposed and we identify future research challenges.

An example for DAM is the following description of the formalized CS from Figure 2. It is assumed that the intensities f_1 and f_2 for the two active states depend directly on the number of tasks N_1 and N_2 processed in the two subsystems. The passive state is also characterized by a constant intensity f_0, which coincides with the throughput P of the system.

During processing, each task falls (k+1) times into state S_1, of which the first k times it makes a transition to state S_2 – assumption (b) from the conceptual model from Figure 2, and in the last – to S_0. This defines the following analytical dependencies:

Entity

To solve the mathematical model, it is necessary to determine the intensity $f_1(N_1)$, which is based on the following:

(a) There is zero intensity for the load on the processor when it does not process a task (N_1=0), for example due to an infinitely large time for servicing tasks in state S_2.
(b) Due to the only processor node in the system in the presence of $N_1 \geq 1$ service task in state S_1, the analytical dependence $f_1(N_1 \geq 1)=const$ will be saturated and limited by the maximum throughput $(k+1)/t$.
(c) To solve the situation $0<N_1<1$ it is necessary to know $f_2(N_2)$, which is usually a non-linear dependence.

As a result, the following analytical description can be presented:

Attribute

This DAM is easily programmable, requiring a pre-setting of the parameters t and k. Through successive implementations, a functional dependence of $f_1(N_1)$ and P on parameter N_1 can be constructed – illustration for N=5; t=10 and k={2, 3, 4, 5} is shown in Figure 5.

Figure 5. Graphical interpretation of result after DAM execution

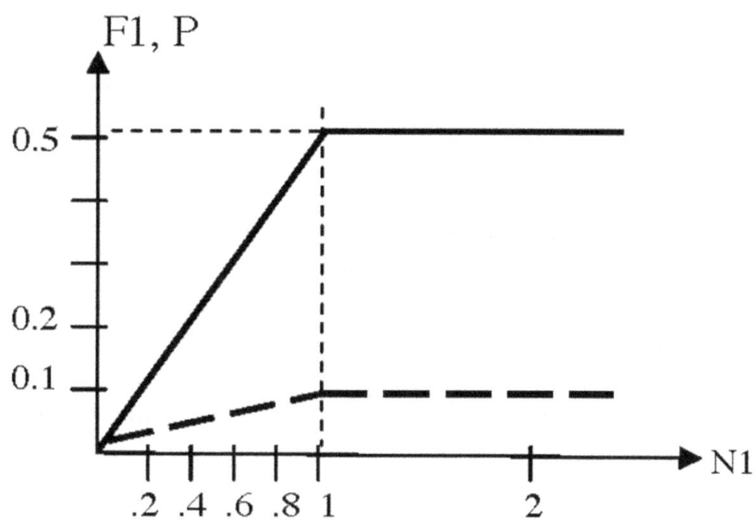

When constructing a SAM, it is most often assumed that a CS behavior can be described as a set of states $\{s_1, s_2, \ldots, s_n\}$ and transitions between them at successive time points $t_1, t_2, \ldots, t_m$. Such behavior of CS can be described as a discrete random process (RP). A similar approach is using the analytic queuing theory (AQT) which allows to present the input workflow and the servicing processes as random processes and probability functions. An example is shown in Figure 6, where the discussed CS (presented as a formal model in Figure 2) is described as a single channel queuing system (QS). For analytic definition of the SAM the two probability lows (for input workflow and for the servicing flow) must be determined.

Figure 6. Traditional single channel queue system

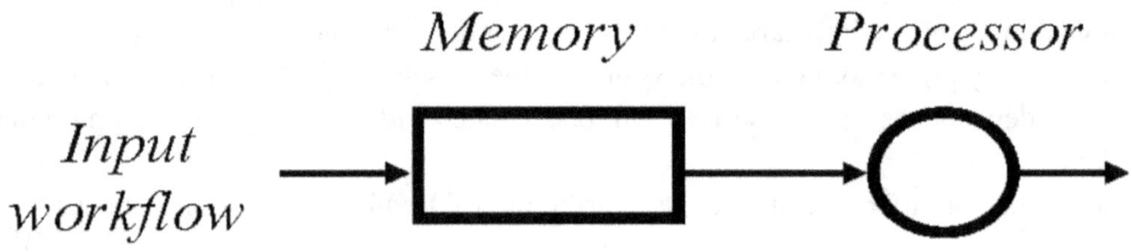

Simulation Models

The simulation model is an imitator of the behavior of the real object in time, where it registers the occurrence of certain events and summarizes the collected data about the ongoing processes. Computer simulation allows the study of systems of arbitrary complexity and provides an unlimited amount of behavioral data. For this purpose, appropriate mathematical models of the system and external influences are created, describing the set of blocks and the relationship between them, the rules for their behavior in relation to the processed elements and processes, the reaction of the system to external influences, etc. In this reason, the simulation model is a set of two interconnected parts – a system model (to represent the static structure of the system) and a workflow model (to reflect the dynamic features of the external influences forming the workload).

To develop a reliable and valid simulation model, it is necessary to correctly formulate the problem by determining correct assumptions, determining the important factors by applying preliminary analysis and comparison with an existing system, if any (Law, 2019). The increasing use of simulation modeling in the study of complex systems requires optimization of the technique to determine the best decision-making approach. One possibility in this direction is the application of metamodeling in the optimization, as in (do Amaral et al., 2022) a systematic review of the literature on simulation optimization based on metamodeling was made. The aim is to survey techniques to assist researchers by presenting the most common metamodeling contexts, applications, methods, tools, and metrics found in optimization through simulation problems.

Depending on the chosen simulation approach and the amount of input data, two types of simulation models are defined below.

(1) *Statistical simulation models* – simulating the behavior of the system in order to obtain statistical information about the occurrence of certain events and the main characteristics of the constituent components.
(2) *Functional simulation models* – complexes of simulating programs that preserve in detail the structural features and functional aspects of the individual hardware and software components of the system.

An example for program simulation model of single processor CS based on the description in Figure 6 by using GPSS language is presented in Figure 7.

The presented simulation model is a typical description of single channel QS with infinite buffer and strategy FIFO (First In, First Out), and non-priority service of tasks. Input workflow is modeled by random flow λ with a uniform distribution in the range [5, 10]. The same probability law of distribution is applied in modeling the intensity μ of serving the requests (tasks) in the processor (the serving device), i.e., the service time in the processor PROC is a random variable uniformly distributed in the range [2, 7]. In the simulation, a time scale is introduced, defining the relationship of the model time with the real time in the object.

Figure 7. Simulation GPSS model (UMT – unit of machine time)

```
GENERATE     10,5     -- task generation by time 5÷15 UMT
QUEUE        MEM      -- input in memory (queue type FCFS)
SEIZE        PROC     -- occupation of processor (if it is free)
DEPART       MEM      -- output from memory
ADVANCE      5,3      -- time for task processing
RELEASE      PROC     -- device release (processor)
TERMINATE             -- serviced task (destroy)
GENERATE     500      -- the global simulation time determining
TERMINATE    1
START        1        -- starts a single run
END
```

Statistical Models

This group includes models created based on accumulated empirical information about the object's behavior and the subsequent construction of a hypothesis-model that is subject to verification. Depending on the purpose, such a model is built for research based on certain univariate or multivariate statistical analyzes to confirm or reject the selected statistical model representing the interrelationships between the components of the object and the influence of the ongoing processes (Freedman, 2009).

It is applied when performing statistical analyzes to assess the relationship between individual factors based on accumulated observations through monitoring and other types of registrations (Huser & Wadsworth, 2022). The goal is to determine statistical estimates and verify hypotheses about the behavior of the object under study by applying traditional statistical analysis - descriptive, correlational, regression, dispersion, etc.

An example of a regression statistical model *Y(X)* formed based on a given sample is shown in Figure 8 below.

Figure 8. Exemplary source sample and regression statistical model

X	Y
2	6
4	12
8	24
12	32
20	36
32	38
40	39
45	39
50	39

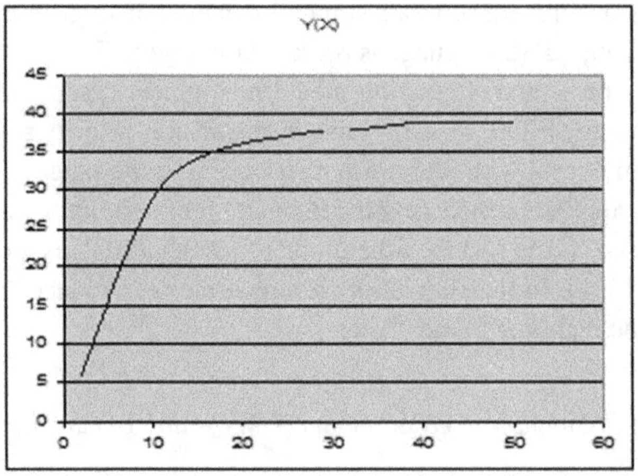

Heterogeneous Models

These are models that are built as a set of two or more constituent parts (sub-models), implemented as one of the uniform models presented above. Different modeling techniques are applied to the individual components. In this way, higher efficiency and reliability of the research is achieved, especially in complex complexes with a non-uniform structure and a mixed workload. In their implementation, these models can be built as *hierarchical* (separate relatively independent levels are distinguished in the modeling process, each applying a different technique) or *hybrid* (a relatively unified model is built, parts of which are implemented based on different methods).

2. GENERAL PROCEDURE AND METHODOLOGY FOR MODEL INVESTIGATION

Examining a discrete system or process to obtain estimates of system characteristics is a task requiring various experiments and analyses. When organizing computer modeling, it is necessary to define two groups of factors (performance indices): ✓ Primary factors (they are essential for the research being conducted); ✓ Secondary factors (cannot be ignored because they affect the primary indices in some way).

The planning of a model experiment for the investigation of computer systems and processes is based on the following main steps.

(1) Collection of empirical data on current values of the basic system parameters associated with the primary and secondary indices.

(2) Structuring and processing the empirical information and developing a functional scheme of the model.
(3) Determining the "a priori" information and definition areas for the operating parameters used in creating an appropriate mathematical model for the study of the original object.
(4) Conducting model experiments, collecting model information (modeling results) and then analyzing the results to obtain performance index estimates.

A general procedure for the formalization of the model research is given in Figure 9 in the form of a block diagram. The initial goal is determined by the need to investigate a real system or process (object-original), and the main stages are presented below.

Stage 1. Preliminary formulating conceptual model for the investigated object (system or process) by using decomposition to obtain separate functional sub-systems, defining some initial conditions and approximation for the general behavior of the original Ω_O. Realization of these activities can be made on the base of some actual empirical information about system parameters to determine the frame of the definition space for main factors.

Stage 2. Mathematical formalization of the structure and determining functional relationships between subsystems which are determined based on the decomposition of the investigated object-original (Ω_O). This is made on the base of determined conceptual mode and by suitable formal system.

Stage 3. Development of functional algorithm – the formal description should be transformed in the algorithmic sequence by using specific tools of the formal system.

Stage 4. Mathematical model building is based on preliminary choice of appropriate approach and method for modelling. The second part in this stage is adaptation of the formalization and functional algorithm to the selected method with the main goal of the stage – mathematical model Ω_M building in the formed working frame of mathematical description.

Stage 5. Mathematical model programming by using selected program environment and language (universal or specialized), and testing and verification to determine the levels of correctness and effectivity.

Stage 6. This is the last stage that completes the model experiment and includes conducting experiments based on a developed experiment plan, accumulating correct experimental results (with filtering of deviant values if necessary), their processing to obtain various estimates and conclusions.

Figure 9. A generalized formalized procedure for model research

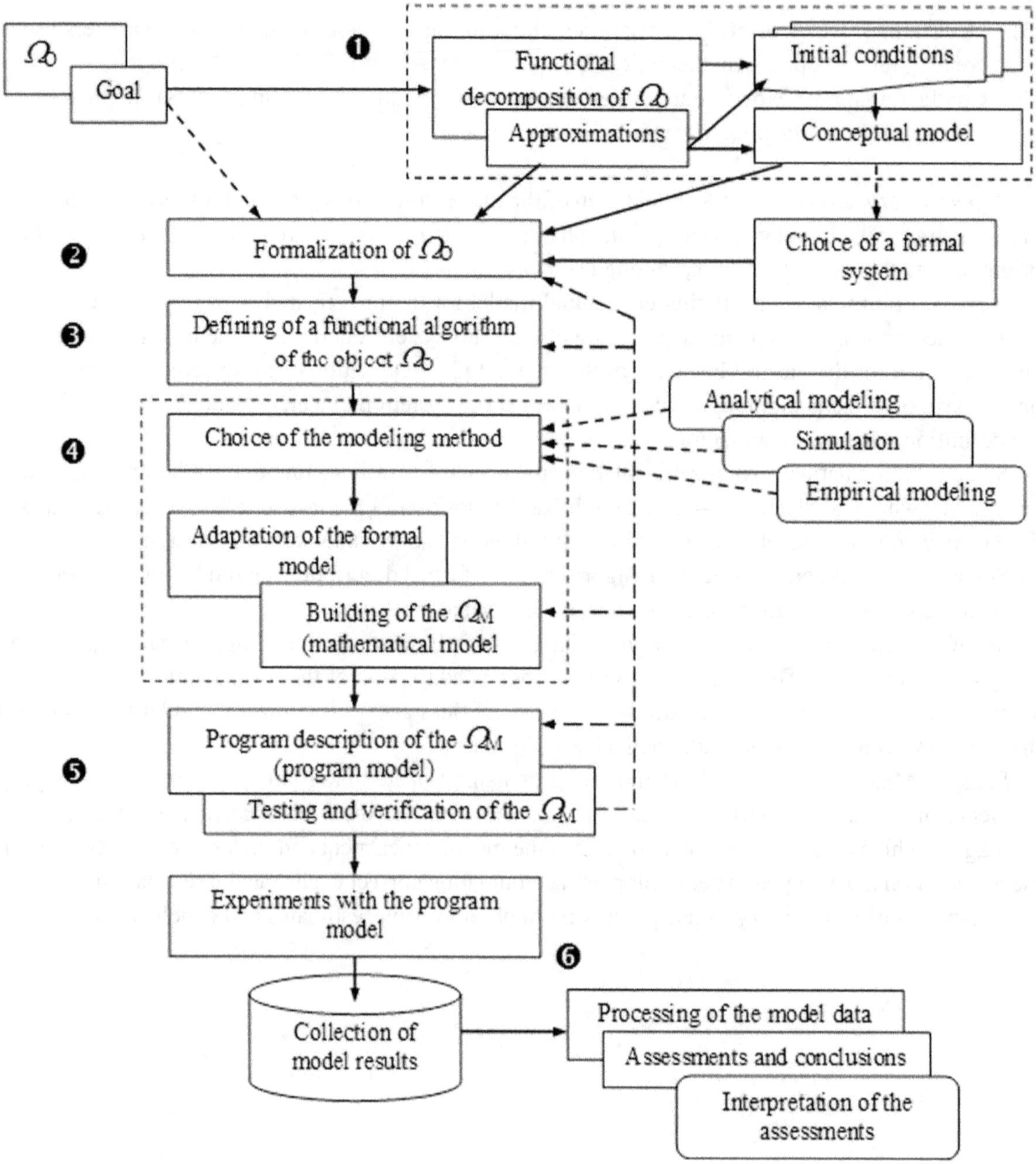

Based on this formalized procedure, a technological scheme for developing an effective computer model was created, which is presented in Figure 10. In it, several successive phases are distinguished, the implementation of which defines a methodology for model research.

In essence, the proposed organizational scheme is an iterative procedure to consistently reach the desired credibility of the developed computer model. This also includes the ability to develop the full model through modifications of successive complexity, starting from a simplified version. Such a modeling approach is recommended for objects that are very complex to describe or when there is insufficient information about them.

Figure 10. Technological organization of model investigation

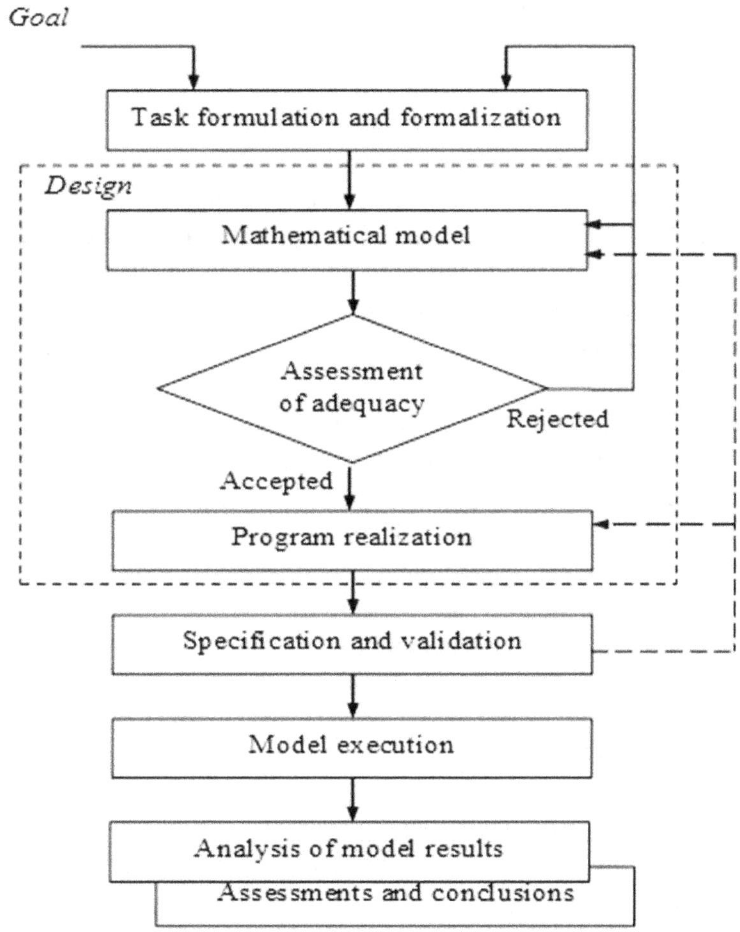

3. PHASES OF MODEL RESEARCH

Phase "Formulation and Formalization"

The first phase of computer modeling is related to specifying the task and objectives of the model investigation, as well as making decisions about the methods and tools used. At this stage, it is necessary to define the object of the modeling precisely and clearly, the initial conditions and the hypotheses of the research, as well as to determine the criteria for evaluating the model effectiveness. The result is the development of a *conceptual model* – describing and defining the problem in abstract terms and concepts.

The initial concept of the model is built in the mind of the developer, for which he must have deep enough knowledge of the studied object. It is important not only what to include in the model, but also which sides or elements can be neglected to obtain reliable results for a particular study. The formulation of the conceptual model determines the level of detail (decomposition of the system) when defining the static structure of the model, as well as determining the functional processes that most accurately reflect the dynamics in the studied system. The formulation phase involves the following sequential steps.

1. *Definition and analysis of the task.* It includes clearly defining the main task and planning the actions to solve it. The definition of the task aims to give a general idea of the solved problems, the volume of the actions carried out and the possible formation of subtasks. The latter is related to decomposition of the task, which can continue to the next level. The number of levels of decomposition is determined by the complexity and requirements placed on the conceptual model. In addition, the general formulation of the task should also include the procedure for solving it.
2. *Clarifying the requirements for the initial information.* This is information through which quantitative and qualitative output data are obtained for the developed model. It is required to specify the type of information required, how it can be obtained and by what methods it should be processed. Only after solving the mentioned questions can, one proceeds to the immediate accumulation of the necessary information.
3. *Acceptance of hypotheses and assumptions.* It is required in the absence of sufficient information about the realization of the model. Hypotheses refer to the type of possible outcomes or the environment in which the processes take place. Assumptions replace missing or incomplete data. This is usually associated with simplifications that do not significantly alter the modeled environment. When formulating the hypotheses and assumptions, the following factors must be considered: the amount of available information and its applicability to the specific modelling; the subtasks for which there is insufficient initial information; the resources for solving the task; expected model results. In the process of experimentation, hypotheses and assumptions can be confirmed or rejected.
4. *Defining the main content of the model.* It is related to the applied modeling method, considering the features of the real environment, the task at hand and the means to solve it. The real environment is the set of operations, means and relationships for the realization of the real processes and provides the source material for the modeling. It helps to find an answer to questions that arise when solving the task, related to the type of implemented functions and their approximation, the influence of the main factors on the operation of the system, the determinism of the processes, etc. The correct answers determine the credibility of the model, as well as the correct selection of the necessary technical and technological means.
5. *Determination of model parameters and selection of performance criteria.* Before creating the mathematical description of the model, it is necessary to determine the main and additional system parameters (primary and secondary factors), as well as the input impacts and output responses (model variables). In order to achieve a high accuracy of the mathematical model, it is necessary to determine the significant parameters, as well as the influence of the insignificant ones on the overall operation of the system. Each model parameter or variable must have a description that includes the chosen symbol, dimension, range of change, characteristics (single or multi-valued, controllable or observable, etc.), application in the model, source of the parameter or variable.
6. *Abstract description of the model.* The phase of the general formulation of the task ends with the creation of an abstract model, representing a description of the conceptual model in a selected system of abstract terms and concepts. It is necessary to specify precisely the deterministic and

probabilistic characteristics and the possibility of replacing some of them with average values. The abstract description should also include a validation of the conceptual model.

In conclusion, in the first phase a conceptual model is formulated, and some functional dependencies are approximated to build an abstract model used in designing the computer model in the next phase.

Phase "Design"

The phase unites the actions on the realization of a computer model based on the conceptual model and its abstract description, including the two levels – mathematical model and program model.

The *mathematical model* provides a formalized description of the static structure and functional processes of the studied system and provides an apparatus for its analytical study. In the general case, the mathematical model is a description through appropriate analytical or probabilistic equations of the abstract model and the accepted approximations of functional dependencies for the processes taking place in the real object based on the introduced assumptions and hypotheses. For this purpose, known formulas, dependencies or laws can be used, as well as creating equations for specific parameters through them.

For certain classes of systems, formalization tools and mathematical methods for functional description have been developed. Such are the automatic and set-theoretic description, transition graphs and diagrams, stochastic networks, and queuing systems (QSs), etc. Appropriate mathematical means have been developed for them, allowing an accurate and clear description of the specific structural elements of the abstract model and representation through clear mathematical dependencies of the ongoing processes.

Based on the criteria defined in the previous phase, the created mathematical model is evaluated for the level of descriptive credibility and the degree of implementation of the conceptual model, which can confirm or reject it.

The *program model* is an implementation of the mathematical description in a given modeling language and operating environment. For this purpose, a modeling method and appropriate technical and technological means are chosen for computer implementation. It is recommended that the following two stages be implemented, and their more detailed explanation is presented below.

(a) Specification of the structure of the model, related to the creation of a logical block diagram for a clear, unambiguous, and concrete implementation of the mathematical model.
(b) Programming of the model – presentation of the logical block diagram in the terms of the selected programming language.

Specifying the structural model involves decomposing the general function into sub functions, each of which can be described as a separate module in the program implementation. This approach allows the use of standard modules, easy description, modification and extension, as well as convenient verification of the operability of the program model.

When creating the logical block diagram of the model, two types of modules are used – main and auxiliary. The main modules represent the basic elements in the structure of the modeled object, the specific operations in the functioning and the logic of process management. The auxiliary modules usually have a universal purpose and are not related to the specifics of the particular model. Examples of such blocks are the various generators of random numbers and distributions, sensors for registering events, means for accumulating and processing statistical data, blocks for managing model time, etc.

The program implementation of the model is essentially a task for creating a program application, the solution of which is subject to the technological principles of programming. To a large extent, the first two phases of modeling cover the main stages of the software development life cycle (analysis, specification, overall structure design, module design, program implementation, module testing, and integration with generic testing). In this reason, model programming is a description of conceptual and mathematical models through the linguistic means of a selected programming environment designed for a specific computer system. When choosing the technical and technological means, the possibilities of the language for describing the specific model, the availability of computer means for the implementation of the experiments, the possibilities of quickly obtaining reliable results with minimal loss of time and funds, etc. are evaluated. The specific requirements regarding the computer model being created affect the choice of a universal programming language (procedural or object-oriented) or a specialized problem-oriented language designed to describe a certain type of model.

The transition from the structural model to the programmatic implementation includes development of the immediate algorithm and its programming, as well as verification of their reliability. The block diagram of the algorithm may not clearly repeat the logical structure, but it must accurately realize the processes taking place during the specific program execution. For this purpose, there must be complete correspondence between the operations in the program algorithm and the analogous actions in the structural model.

Appropriate tests can be used to verify the reliability of the program model, both for individual modules and for the entire program implementation. The tests should be relatively simple, but also sufficiently effective in checking the mathematical calculations and the logic of the general control of the processes.

Phase "Specification and Validation"

The purpose of the actions envisaged in this phase is a comprehensive check for correct behavior of the designed computer model and confirmation of its adequacy. Assessments of the current adequacy in the previous two phases, especially for the conceptual and mathematical models, play an essential role in their effectiveness. In this sense, greater attention is paid here to the reliability of the program implementation of the model, as the specific conclusions may lead to changes in the results of the previous phases. In general, the phase can be presented as an iterative procedure (Figure 11), allowing several modifications of the developed model. The goal is to achieve the stability of the model - reliable results for the entire set of values of the input parameters.

Figure 11. An iterative procedure for specifying a model

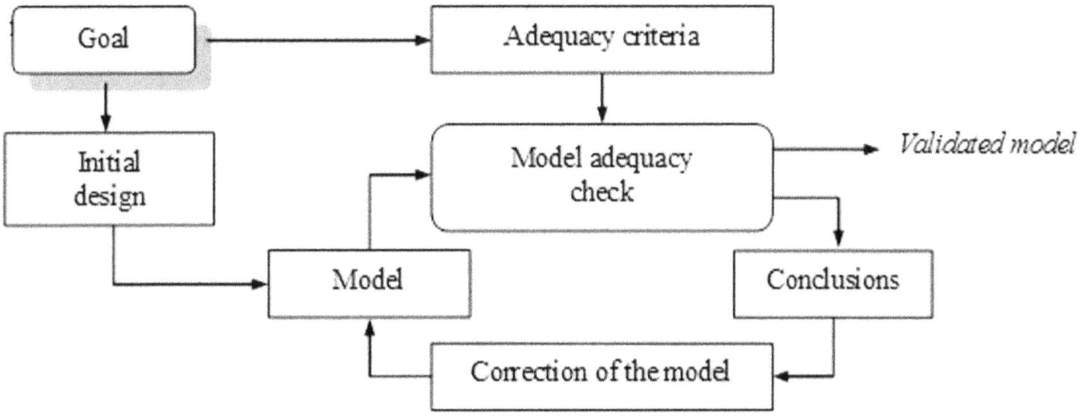

The main test for the correctness of the model is how accurately it reflects the real object and how it performs the registration of the necessary data. In general, a computer model is a collection of individual structural elements, mathematical equations, expressions, and dependencies. The use of elements of proven reliability (for example, confirmed equations for mathematical and natural laws) ensures the adequacy of the entire model. However, the inclusion of one unreliable element can have fatal results for the overall adequacy of the model.

The adequacy of the model can only be tested for certain sets of input data and there are no guarantees of the effect of all possible combinations. At this stage, it is not dangerous to get approximate results, but the bad thing is if they are contradictory (this can happen due to improper simplification). The test for the adequacy of the model must answer the following questions: 1. Does the model make it possible to do the given task on foot? 2. Does the structural model require the required transparency? 3. Is the created functional algorithm of the model correct? 4. Does the software implementation correspond to the sequence of the algorithm? 5. Are the obtained output results within acceptable parameters?

The positive answers to the posed questions testify to a reliable model, correctly reflecting the structure of the studied object and the dynamics of its functional behavior, as partial answers to some of them can be obtained already in the previous phases. For example, the formulation of the task, the created conceptual model and its abstract description are related to the first two questions. The third question is related to the algorithmic consistency of the model's behavior, as well as the correctness of the mathematical equations used to create the mathematical model. The fourth question relates to immediate programming in the chosen medium. The last question is related to execution of the program model for selected control examples, which include checking the output results for essential operations in the model.

Phase "Model Execution"

It represents the realization of the model through the organization of a model experiment (analytical, simulation, statistical). The goal is to obtain maximum information in a minimum time depending on the given input (initial) data. The implementation of the model experiment involves two main stages.

Planning a model experiment. It includes specifying the memory rectangles, input data, and technical combinations that must perform the calculations, or on which to run the model. One of the tasks is to define two groups of variables - observables (which will be recorded during the model experiments) and controllable (which will be assigned specific values in each experiment). The properly created plan of the experiment gives the possibility of obtaining the maximum volume of information with minimum expenditure of machine time. The main model variables are observable (accessible for measurement), and the input variables are controllable.

The choice of a particular experimental plan depends on what results are expected from the modelling. Statistical methods are usually used to obtain an economical and sufficiently effective plan. The statistical experimental plan includes: ✓ selection of values for the controllable factors when conducting the experiments; ✓ determination of an appropriate way to form statistical samples for the observable factors; ✓ specification of the necessary number of repetitions of the experiment.

In most experiments, the influence of multiple factors (inputs, parameters, variables) is of interest. In this case, two main approaches are possible:

- One-factor plan, allowing the values for one selected factor to vary with random combinations of the other factors assumed to be constant.
- Multifactorial design using the methods of factor analysis – full factorial design, random sampling, randomized design, systematic sampling, statistically independent design, random balance method, etc.

In a statistical aspect, the design of the model experiment should ensure the following: ✓ Obtaining unbiased estimates of the influence of different factors under different working conditions; ✓ Elimination of the possibilities of joint influence between the observable factors; ✓ Adequate estimate of the allowable experimental error; ✓ Minimum possible correlation between the remaining quantities, etc.

Implementation of the experiment. It foresees the preparation of the input data, computer implementation of the experiment and accumulation of model results. It is recommended that the model experiment be performed in two stages.

(1) Preliminary modeling which aims to verify the model's performance, to determine the sensitivity of the results to changes in the model's parameters and input data, and to provide an overview of the machine in question. This stage is another opportunity to check the reliability of the compiled program model. The goal is to verify that the intended operations and actions are performed correctly and that the output variables respond to the change in the input parameters. To check the sensitivity of the model, a partial random design can be applied with a choice of output parameters and analysis of their change with a choice of input effects.

(2) Work modeling is the real modeling realized on the base of the developed experimental plan and the results of the model execution are obtained. For this purpose, the input data must be correctly completed at each successive execution. It matters whether intermediate results will be stored, reset, or accumulated over successive experiments. Experimentation continues until the plan is fully implemented.

Phase "Analysis of Model Results"

Conducting the model experiments leads to the accumulation of a large amount of experimental data, which are subject to analysis and processing to obtain estimates for the studied system characteristics and model parameters. The most used methods for basic analysis of the results of model research based on statistical data processing: ✓ correlational analysis establishes the relationship between two or more random variables; ✓ the detailed analysis examines the causal relationship between quantitative facts; ✓ the dispersion analysis is applied to establish the relative influence of different facts among the four points of the external practices. During the analysis, various estimates are obtained, such as frequency histograms, regression dependencies, correlation coefficients, estimates of the original value, mathematical expectation, estimation of variance, etc.

The interpretation of the model results is associated with an appropriate presentation (tables, histograms, graphical dependencies, diagrams). Modern software systems for modeling or those used for mathematical processing of information have good graphical means for visualization of model results and appropriate graphical interpretation of the obtained estimates. Nevertheless, to reflect the specific nature or certain aspects of the object under study, a graphical representation may be additionally made by the experimenter. When choosing the appropriate graphic tools, the applied mathematical method for data analysis and illustrative capabilities are essential.

REFERENCES

Akther, A., Ushakov, Y., Balanov, A. G., & Savel'ev, S. E. (2021). Deterministic modeling of the diffusive memristor. *Chaos (Woodbury, N.Y.)*, *31*(7), 073111. doi:10.1063/5.0056239 PMID:34340321

Alturki, M. A., Ban Kirigin, T., Kanovich, M., Nigam, V., Scedrov, A., & Talcott, C. (2022). On the formalization and computational complexity of resilience problems for Cyber-Physical Systems. *International Colloquium on Theoretical Aspects of Computing* (pp. 96-113). Springer, Cham.

Atangana, A., & Araz, S. I. (2021). *Deterministic-Stochastic modeling: A new direction in modeling real world problems with crossover effect*. Hal Science. https://hal.science/hal-03201318

Carlucci, S., Causone, F., Biandrate, S., Ferrando, M., Moazami, A., & Erba, S. (2021). On the impact of stochastic modeling of occupant behavior on the energy use of office buildings. *Energy and Building*, *246*, 111049. doi:10.1016/j.enbuild.2021.111049

Chapple, M. (2021). *Entity-Relationship Diagram Definition- Use ER diagrams to illustrate relationships between database entities*. LifeWire. https://www.lifewire.com/entity-relationship-diagram-1019253

Chen, J., & Shi, Y. (2021). Stochastic model predictive control framework for resilient cyber-physical systems: Review and perspectives. *Philosophical Transactions - Royal Society. Mathematical, Physical, and Engineering Sciences*, *379*(2207), 20200371. doi:10.1098/rsta.2020.0371 PMID:34398650

do Amaral, J. V. S., Montevechi, J. A. B., de Carvalho Miranda, R., & de Sousa, W. T. Junior. (2022). Metamodel-based simulation optimization: A systematic literature review. *Simulation Modelling Practice and Theory*, *114*, 102403. doi:10.1016/j.simpat.2021.102403

Fathalla, S., Auer, S., & Lange, C. (2020, March). Towards the semantic formalization of science. In *Proceedings of the 35th Annual ACM Symposium on Applied Computing* (pp. 2057-2059). ACM. 10.1145/3341105.3374132

Freedman, D. A. (2009). *Statistical models: theory and practice*. Cambridge University Press. doi:10.1017/CBO9780511815867

Huser, R., & Wadsworth, J. L. (2022). Advances in statistical modeling of spatial extremes. *Wiley Interdisciplinary Reviews: Computational Statistics*, *14*(1), e1537. doi:10.1002/wics.1537

Jena, D., & Ramana, V. V. (2015). Modeling of photovoltaic system for uniform and non-uniform irradiance: A critical review. *Renewable & Sustainable Energy Reviews*, *52*, 400–417. doi:10.1016/j.rser.2015.07.079

Law, A. M. (2019, December). How to build valid and credible simulation models. In 2019 Winter Simulation Conference (WSC), USA (pp. 1402-1414). IEEE. doi:10.1109/WSC40007.2019.9004789

Lemiale, V., Huston, C., Mead, S., Alexander, D. L., Cleary, P. W., Adhikary, D., & Delaney, G. W. (2022). Combining statistical design with deterministic modelling to assess the effect of site-specific factors on the extent of landslides. *Rock Mechanics and Rock Engineering*, *55*(1), 259–273. doi:10.100700603-021-02674-x

McIntire, M. G., Keshavarzi, E., Tumer, I. Y., & Hoyle, C. (2016). Functional models with inherent behavior: Towards a framework for safety analysis early in the design of complex systems. In *ASME International Mechanical Engineering Congress and Exposition,* Volume 11*: Systems, Design, and Complexity*. IEEE. 10.1115/IMECE2016-67040

Ramos, T. M. F., Almeida, A. A., & Ayala-Rincón, M. (2022). Formalization of the computational theory of a Turing complete functional language model. *Journal of Automated Reasoning*, *66*(4), 1031–1063. doi:10.100710817-021-09615-x

Romansky, R. (2020a). An approach for mathematical modelling and investigation of computer processes at a macro level. *Mathematics*, *8*(10), 1838. doi:10.3390/math8101838

Romansky, R. (2020c). Formalization and discrete modelling of communication in the digital age by using graph theory. In M. Pal, S. Samanta, & A. Pal (eds) Handbook of Research on Advanced Applications of Graph Theory in Modern Society. IGI Global. doi:10.4018/978-1-5225-9380-5.ch013

Romansky, R. (2021b). Mathematical modelling and study of stochastic parameters of computer data processing. *Mathematics*, *9*(18), 2240. Advance online publication. doi:10.3390/math9182240

Romansky, R. (2022a). Stochastic approach to investigate protected access to information resources in combined e-learning environment. *Mathematics*, *10*(16), 2909. doi:10.3390/math10162909

Romansky, R., & Noninska, I. (2022). Deterministic model investigation of processes in a heterogeneous e-learning environment. *International Journal of Human Capital and Information Technology Professionals*, *13*(1), 1–16. doi:10.4018/IJHCITP.293228

Saatkamp, K., Breitenbücher, U., Kopp, O., & Leymann, F. (2020). Method, formalization, and algorithms to split topology models for distributed cloud application deployments. *Computing*, *102*(2), 343–363. doi:10.100700607-019-00721-8

Shukla, S. (2023c). Enhancing healthcare insights, exploring diverse use-cases with K-means clustering. *International Journal of Management IT and Engineering*, *13*(8), 60–68.

Taifa, I. W., Hayes, S. G., & Stalker, I. D. (2020, June). Computer modelling and simulation of an equitable order distribution in manufacturing through the Industry 4.0 framework. *2020 IEEE International Conference on Electrical, Communication, and Computer Engineering,* (pp. 1-6). IEEE. 10.1109/ICECCE49384.2020.9179275

Varenne, F., & Turnbull, K. (2018). *From models to simulations* (1st ed.). Routledge. doi:10.4324/9781315159904

Williams, A. E. (2020, April 16). A human-centric functional modeling framework for defining and comparing models of consciousness and cognition, PsyArXiv. doi:10.31234/osf.io/94gw3osf.io/94gw3

Yoshikawa, H., Lind, M., Matsuoka, T., Hashim, M., Yang, M., & Zhang, Z. (2013). A new functional modeling framework of risk monitor system. *International Electronic Journal of Nuclear Safety and Simulation*, *4*(3), 192–202.

Chapter 4
Technological Organization of Model Investigation

ABSTRACT

The chapter is dedicated to the technological organization of model research, discussing two important characteristics of the computer model – adequacy (how correctly it represents the investigated object) and efficiency (requirements to the machine time and ensuring convergence of the model function to the real one). It is accepted that a model is adequate if it represents the investigated object with an acceptable approximation. Several successive steps for evaluation of the adequacy are determined during its development. The concept of the model time, which should represent the real time in the studied processes, is also discussed with determining two main variants (synchronous time and asynchronous time).The last section discusses the technical and software tools for developing computer models. The technical means are mainly hardware modeling systems (stochastic machines, multiprocessor and hybrid complexes, distributed systems). Programming tools are divided into universal and specialized programming languages are presented.

4.1. ADEQUACY AND EFFICIENCY OF COMPUTER MODELS

Model Adequacy

In most cases, the model is a mathematical description of the processes in a partially or fully completed object. One of the main requirements for modeling is to give correct and understandable data obtained as experimental results. Thus, a model becomes a means of supporting decision-making based on scientific assumptions. The usefulness of a given model is assessed by its sustainability for a specific purpose, and one of the main tasks of computer modeling is the creation of an adequate model that reflects the structure of the modeled system and the processes taking place in it with sufficient accuracy (Parker, 2020). The article seeks answers to three main questions: "*What does it mean for a model to be adequate-for-purpose? What makes a model adequate-for-purpose? How does assessing a model's adequacy-for-purpose differ from assessing its representational accuracy?*"

DOI: 10.4018/978-1-6684-8947-5.ch004

It is generally understood that a model is adequate when it faithfully fulfills its purpose and describes the modeled object with a permissible approximation in a structural, logical and mathematical aspect. In this reason, the adequacy of a given model in the general sense means the correspondence between the model and the modeled object and determines how accurately and completely it reflects the structure and functioning of the original. A reliable assessment of the adequacy of a given model is possible with a correctly defined main objective and formulated criteria for performing the assessment, and it is usually recommended to use a combination of several appropriately selected different techniques. This is especially important when processing large data sets and complex models, where an incorrectly developed model "*could mislead researchers and compromise their inference*" (Carstens et al., 2022). Three main challenges for researchers are defined: (i) correct choice of modeling method and tools; (ii) correct assessment of analytical results or their interpretation; (iii) evaluation of the model and it's fit with the studied object. The paper states that a variety of approaches can be used to assess model adequacy, from simple visual inspections to statistical fit assessment tests, with the increased power and interpretability of statistical approaches justifying their increased complexity. In addition, there is a review of existing software packages for testing model adequacy. The question of the suitability of the statistical evaluation of a model is also advocated in (Fonseca et al., 2022), where it is confirmed that "*naive analysis of data ... may lead to spurious results*". To overcome such a shortcoming, the paper proposes a package to complement the log data exploration capabilities of Bayesian Skyline analyses. The goal is to simulate predictive data sets to compare with statistics calculated from the empirical data to check for model violations.

It should be noted that the model cannot completely replace the physical object. In the process of designing the model, its adequacy is artificially distorted due to the idealization of several components and influencing factors. This necessitates continuous assessment of adequacy, which can be achieved by comparing with real data (estimates) or results from another type of modeling (or monitoring), recommending the following sequential steps.

The initial assessment of model adequacy should be made already when formulating the conceptual model. It is related to the analysis of the real environment and the correct reflection of its structure and dynamics in the developed model. Such an assessment is too subjective, as it depends on the skill of the developer to reflect the real world in abstract forms and correctly form the questions whose answers will determine the level of adequacy.

The second stage in assessing adequacy is related to the specific implementation of the mathematical model and its programming in a selected language environment. Since the design of a computer model involves several successive steps, each of them affects the overall adequacy through its specific results. In this reason, the assessment of adequacy is based on a comparison with the previous state of the model for each step. For example, mathematical dependencies are compared to the concept of the model, the logical structure to the mathematical description, the descriptive functional algorithm to the logical structure, etc.

The final assessment of the adequacy of the program model can be obtained through test variants prepared independently of the implementation of the model. The coincidence of the results of the control modeling with those of the test variants testifies to the adequacy of the model for a given type of tasks. This is also the task of the model validation stage, which aims to achieve high reliability and correctness of the output results when working with real data. Test cases can be developed based on other types of models or by data on the behavior of a real system. The second approach would produce better results but is more difficult to implement.

To achieve high reliability of the computer model, it is necessary to assess the current adequacy to accompany the sequence of model development.

Model Efficiency

Modeling efficiency is determined by the machine time requirements of the program model. This is due to two reasons:

- The significant amount of processing in the model increases the machine time, and hence its cost, reducing its efficiency.
- Performing dynamic checks on the effect of minor model configuration changes is limited by response time.

To evaluate the model efficiency, the relative average time for model operation is used according to the formula below, where: T_M – average time for model implementation of a given system function; T_P – average real time to execute this function.

$$, = \frac{T_M}{T_P}$$

To increase the efficiency of modeling, it is important to quickly specify the model and reach convergence – a repetition by the model of the original regularity with sufficient accuracy. When building a model, preliminary data are collected and subjected to appropriate approximation. This allows for a real system S_O to create multiple model descriptions $\{S_{M1}, S_{M2}, \ldots, S_{Mk}\}$ that fulfill the initial hypothesis within certain limits. In this case, when choosing an optimal model, in addition to analyzing the accuracy, it is also necessary to analyze its convergence. An example of such a situation is given in Figure 1. For the real object S_P, a tolerance $\pm\Delta S$ is defined for the permissible accuracy of the created model. As can be seen, the S_M model falls within this range for most of the registrations (exception is only the final values). However, the S_M model does not repeat the behavior of the real object and does not fulfill the convergence requirements.

Figure 1. A model convergence study

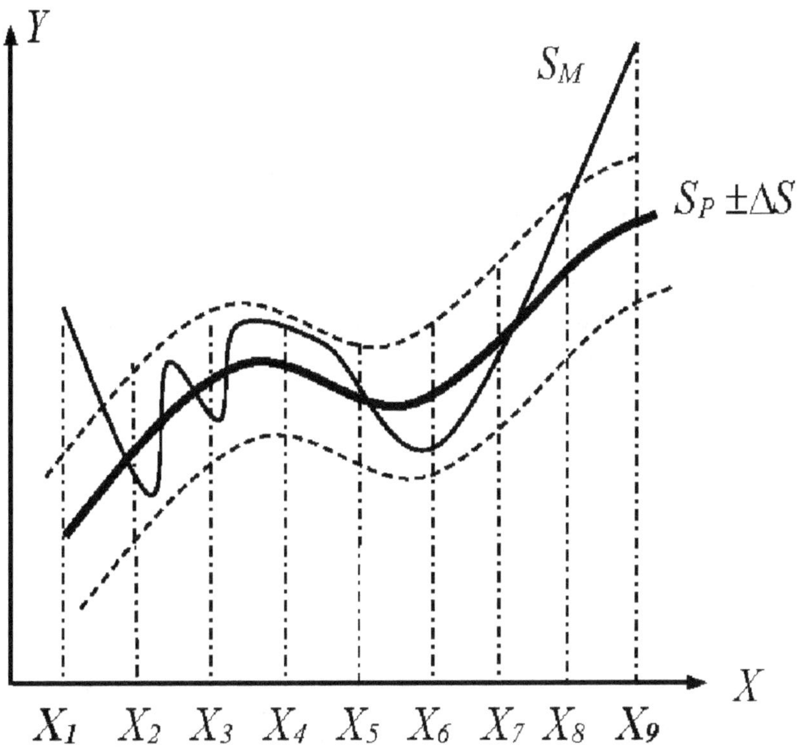

Investigation of the model convergence is a frequently discussed problem in conducting model experiments. For example, in (Kürkçü et al., 2017) this was done when developing models based on linear, non-linear differential and integral equations, using Dixon polynomials. To examine the convergence of the model solution, a residual function is used to obtain accurate Dixon polynomial solutions, which are compared with other well-known methods in tables. Another example is an analysis of the convergence of probability density functions in the study of parametric certainty of a given system using approximate models presented in (Butler et al., 2018). The aim is to update an initial probability density assumed for the input parameters of the model so that the subsequent evolution corresponds to a given probability density. Numerical results are presented to demonstrate degrees of convergence of the results. The subject of discussion in (Jin, 2020) is a model for representing the invasion and metastatic spread of cancer cells, initially establishing the level of existence of uniformly bounded global strong solutions and comparing further improvement of the resulting solutions. The result of the conducted study of the behavior of the solutions has shown a strong convergence with the steady state in a large time range.

4.2. CONCEPT OF MODEL TIME

Processes in real systems are probabilistic in nature and can be discrete or continuous. Each process can be represented as a sequence of events that occur at certain time intervals. Each event can be considered as a discrete unit, but in general the process of their realization has a stochastic character. The fulfillment of events is caused by a certain functional impact determining the conditions for the fulfillment.

Figure 2 presents an example situation for the realization of a real process $F_i(t)$ from a given system component K and the possibilities for its model description. It is assumed that the investigated object consists of different components, each of which supports different functional processes $F_i(t)$, realized as a sequence of functional impacts $F_{ij}(t)$ within time intervals τ_{ij} (j=1,2,3,…). The realization of each functional impact $F_{ij}(t)$ leads to the occurrence of event e_{ij} at time t_{ij}. Thus, the real behavior of the object, in particular represented by a process $F_i(t)$, can be represented as a sequence of functional effects $<F_{i1}, F_{i2}, F_{i3}, F_{i4}, ..., F_{ij}, ... >$, implementing a chain of events $<e_{i1}, e_{i2}, e_{i3}, e_{i4}, ..., e_{ij}, ... >$ in time.

In the model description of a process $F_i(t)$, it is important to model with sufficient accuracy the main events and their realization at certain moments in time. This is related to two tasks – functional impact modeling and system time modeling.

Figure 2. Model description of a real process

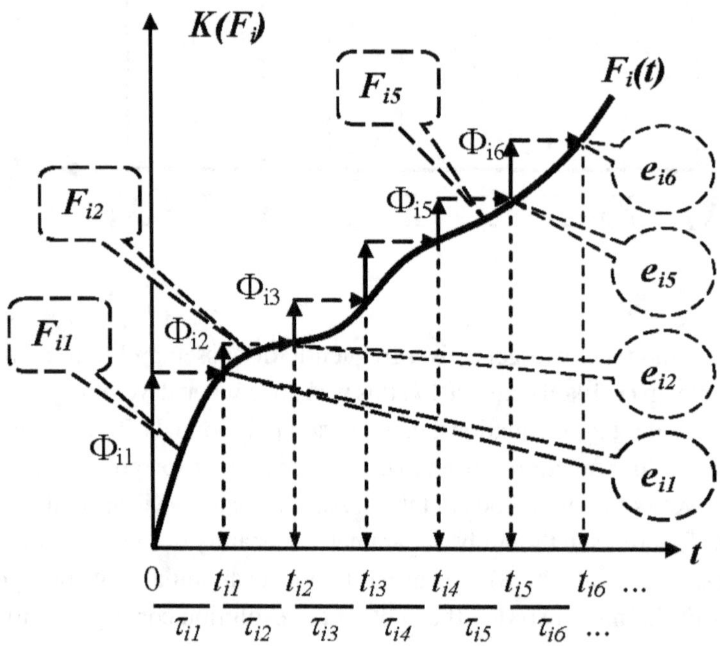

Functional impact modeling is a change in value levels for the modeling process. This change is generally instantaneous and is represented by the model impacts F_{ij} establishing a process level corresponding to the event e_{ij} (j=1,2,3,…).

The modeling of the system time is performed by a model timer that records the time instants t_{ij} for the events to occur. Changing the timer with the magnitude $\tau_{ij} = t_{ij} - t_{i(j-1)}$ can be done before or after (as is shown in Figure 3) establishing the level of the corresponding modeling impact, since the important thing is to respect the sequence of events in the time. As a result of the above, the real process $F_i(t)$ is modeled by the step function $<0, \Phi_{i1}, \tau_{i1}, \Phi_{i2}, \tau_{i2}, \Phi_{i3}, \tau_{i3}, ... \Phi_{ij}, \tau_{ij}, ... >$.

It is known that computer time is discrete, and the management of model time is an essential point in the computer implementation of the model. This is related to the determination of the quantization step and the implementation of the model timer. There are two main approaches to model time management presented in the Figure 3:

(a) Asynchronous (different quantization for the time-intervals) – moments t_{ij} are determined by the occurrence of events in the model and intervals τ_{ij} are of different lengths (it is called event-oriented model time or event management);
(b) Synchronous (uniform quantization) – a constant quantum τ_{ij} is maintained between registrations, which consider the current levels of the developing model process (control during the model timer).

Figure 3. Types of model time

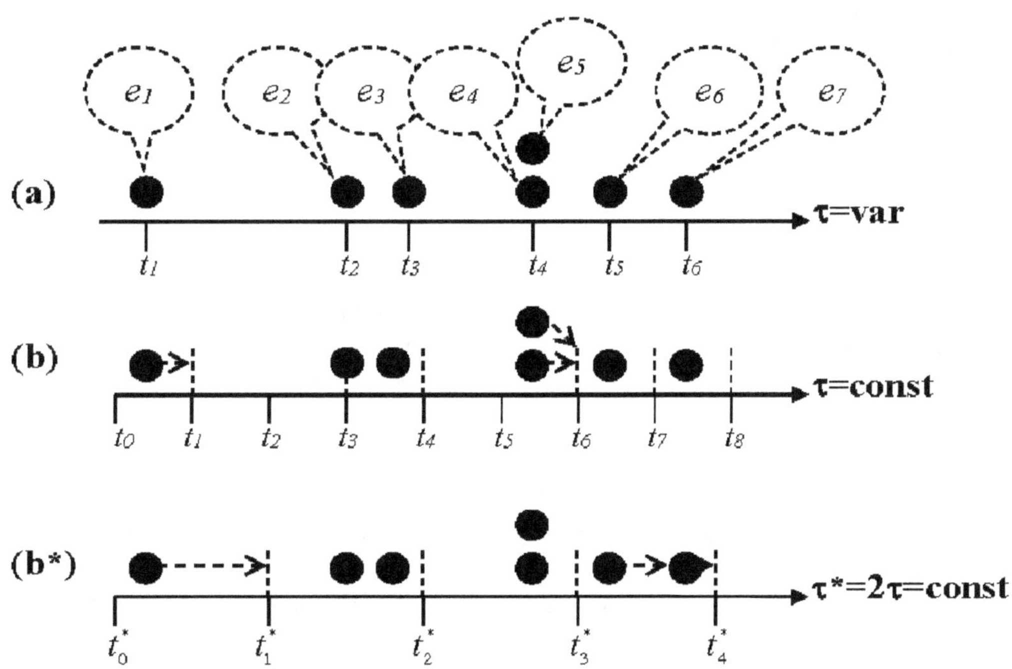

Synchronous model time is easier to manage than asynchronous, but it has several disadvantages, such as (see Figure 3):

- The accuracy of the model registrations for the events depends on the selected quantum time: comparison between options (b) and (b*).
- Empty intervals are possible, which reduced the efficiency of the model test (e.g. $[t_1,t_2]$ and $[t_4,t_5]$).
- Possible loss of the real sequence of events (e.g., events e_4 and e_5 will be processed in an order depending on the programming environment).
- Too large an event registration error is possible depending on the discretization (e.g., depending on the quantum event e_1 can be registered at time t_1 or t_1*).

- It is possible for events occurring at different times to be registered at the same time (e.g., events e_6 and e_7 will be registered at the same time t_4*).

As can be seen from Figure 3, the selected quantum of the model timer has a significant influence on the synchronous time. With twice the length (option b*), different events appear to occur simultaneously, which affects the adequacy and convergence of the model. On the other hand, with a very small quantum, many “empty” intervals are possible, which in turn reduces the effectiveness of model research. In this regard, when planning model experiments, it is recommended to conduct a so-called control stage for choosing an optimal quantum time consistent with real processes.

4.3. TECHNOLOGICAL MEANS FOR COMPUTING MODELING

Various technical and software tools are used to implement the computer models, which can be universal or specialized. In principle, universal tools are more accessible and easier to apply, but specialized ones provide higher adequacy and accuracy of model research. The choice of a specific tool should be consistent with the purpose of the model study and the requirement for the accuracy of the final estimates.

Technical Means and Tools

In principle, all modeling methods can use universal computer systems, both for creating the model and for its implementation. Analytical modeling, which is based on the application of numerical methods for solving the models, does not place any special requirements on the technique. Simulation modeling is associated with constantly referring to the model parameters in memory and storing large amounts of data, but modern computer systems satisfy this requirement. The same applies to statistical modeling, where the need to accumulate a representative sample requires higher performance and sufficient memory.

Specialized machines provide an environment for defining and developing the model process adequate to the chosen method. This also increases modeling efficiency. Examples of specialized technical means are the following.

Stochastic machines – computer environments oriented to probabilistic processes and providing the opportunity for hardware generation of random numbers and stochastic sequences, implementation of stochastic processes, etc. In principle, these are machines with a different architecture than the classical Von Neumann, encoding information with stochastic bit streams and rely only on stochastic arithmetic (Faix et al., 2016). This type of machine is designed to handle samples from the joint probability distribution of interest.

Machines for computer simulation - computer environments for the implementation of simulation modeling, the hardware configuration being tailored to the need for specific actions – load generation, monitoring of observable parameters, maintenance of model time, accumulation of statistics, analysis of accumulated data, etc. Essentially, a CS for simulation is a multiprocessor system (Figure 4) with the distribution of functions among separate specialized subsystems: SIM – implementation of simulation experiments; STAT – accumulation and processing of statistical data on observable variables and monitored processes; PLAN – planning of model experiments; OPT – analysis of results and optimization; CONTROL – general control of the computer simulation and the dialogue with the user.

Figure 4. Specialized multiprocessor system for simulation research

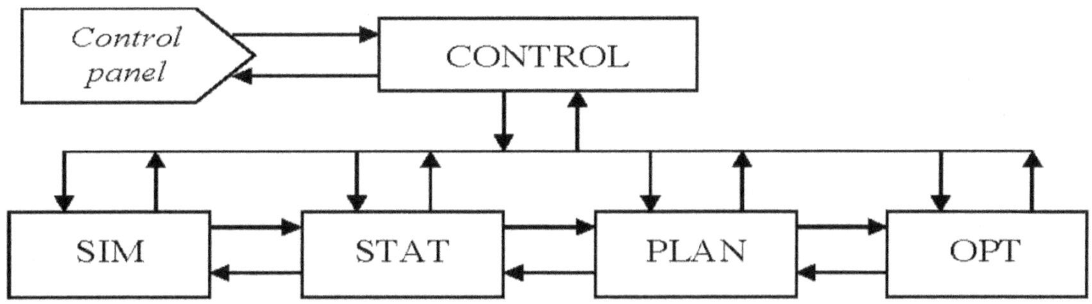

The central CONTROL subsystem prepares a model program in the selected language for simulation, performs its interpretation in an internal machine language, manages parallel processes and their synchronization, manages the dialogue with the user, etc.

The main SIM subsystem is a multiprocessor for the parallel implementation of several functional processes: ✓ generation of stochastic sequences; ✓ definition of random distributions for the input stream; ✓ model time management; ✓ arrangement of events in chains, etc.

A design of a stochastic simulation engine based on neural networks is presented in (Gabrielli & Wüthrich, 2018). It is designed to generate individual histories of claims for general insurance benefits and is a fully calibrated stochastic scenario generator that is based on real data. This stochastic simulation engine allows you to simulate your own synthetic insurance portfolio, providing the most realistic results possible.

Hybrid model complexes – combine the advantages of discrete and analog machines and are characterized by speed, accuracy, flexibility and the ability to process mixed (analog and discrete) information. Increasing requirements for modeling accuracy and providing reliable experimental results lead to a growing demand for computing power to significantly accelerate software. An effective way to achieve this is to use modern, hybrid computing architectures with new data structures and code refactoring (Schneider, 2016). In the contemporary digital society, more attention is being paid to network technologies and in particular to cloud computing and IoT. Despite the widespread use of modeling in (Batool & Niazi, 2017) it is stated that there is no standard methodology available for modeling such real-world complex IoT-based scenarios. The paper proposes a novel hybrid approach using simulation through Cognitive Agent-Based Computing applied to the modeling of several standard complex network topologies such as lattice, random, small-world, and scale-free networks. In addition, a new algorithm for autonomous monitoring in networked IoT devices is presented and simulation experiments using several network configurations are performed.

Distributed modeling systems – unite remote nodes for model investigation of processes and systems at different levels in a hierarchical plan. They are characterized by distribution of resources and virtuality of computer modeling tools in distributed collective access. One example is the framework for parallel simulations of climate impact models presented in (Elliott et al., 2013). The framework includes tools for ingesting large amounts of data and transforming them into a standard, modeling-friendly form, as well as a scalable parallel framework for performing large-ensemble simulations on diverse computing systems, from small local clusters to supercomputers, including distributed networks and clouds.

Software Means and Tools

The software tools used in the implementation of the computer models are also universal and specialized. Requirements are placed on them, such as: ✓ maintain data organization providing simple and efficient modeling; ✓ to have convenient tools for formally describing the dynamics of processes; ✓ to have opportunities for modeling stochastic processes, etc.

Universal programming languages are affordable and easy to use, but require more time to develop and set up the models. They are more suitable for creating analytical and statistical models of a general nature, while simulation modeling will require more effort in creating, tuning and refining the model.

An example of universal program environment suitable for developing analytical models within different directions is TryAPL2 which support language APL2. One main feature is that it can support graph theory-oriented processes and summary of opportunities of this spice will be presented in the next sub-section.

Specialized languages provide macro-level solutions for describing the structure and dynamics of modeled structures and processes as individual statement blocks. The main disadvantage is the significant machine time to execute the model and the limitations to the data format, but they have significant advantages:

- Little time to program the model.
- Accelerated settings and refinement of the model.
- Automatic definition of certain data types.
- Convenience in accumulating the information and forming the results.
- Possibility to unify individual elements of the model.

The simulation model represents a system of objects with their own attributes that determine the state of the system at successive moments in time. State changes are managed by events occurring based on defined rules. Usually, several events are realized in a system, which must be monitored, and the correct order of their fulfillment must be guaranteed. This also determines the importance of time synchronization, as well as input and output of data (workflow and workload). These requirements must be supported by the relevant simulation languages, an example classification of specialized languages is presented in Figure 5. Typically, simulation languages are used in the computer simulation of discrete systems and processes (due to the discrete nature of computer processing). Their features are as follows:

- *Activity-oriented languages* – the functional algorithm is represented as a sequence of activities managed by a special scheduling mechanism that maintains a list of activity start and end.

- *Event-oriented languages* – the model is a set of event handling procedures, the execution of the procedures being synchronized by a dispatcher registering the time of the event and the name of the procedure serving it.
- *Process-oriented languages* – the model is a set of process descriptions, each corresponding to a certain process class; each description is a procedure executed by the class representative existing at the current time.
- *Transaction-oriented languages* – the dynamics of the investigated object is described by active elements (transactions) with assigned attributes (parameters), which change when passing through the blocks of the static structure of the model.
- *Aggregate-oriented languages* – the model is presented as a set of aggregates (dynamic objects supporting multiple events), and the interaction between them is synchronized by a control program that recognizes which of the events is implemented in the aggregate.

Figure 5. Example specialized modeling languages

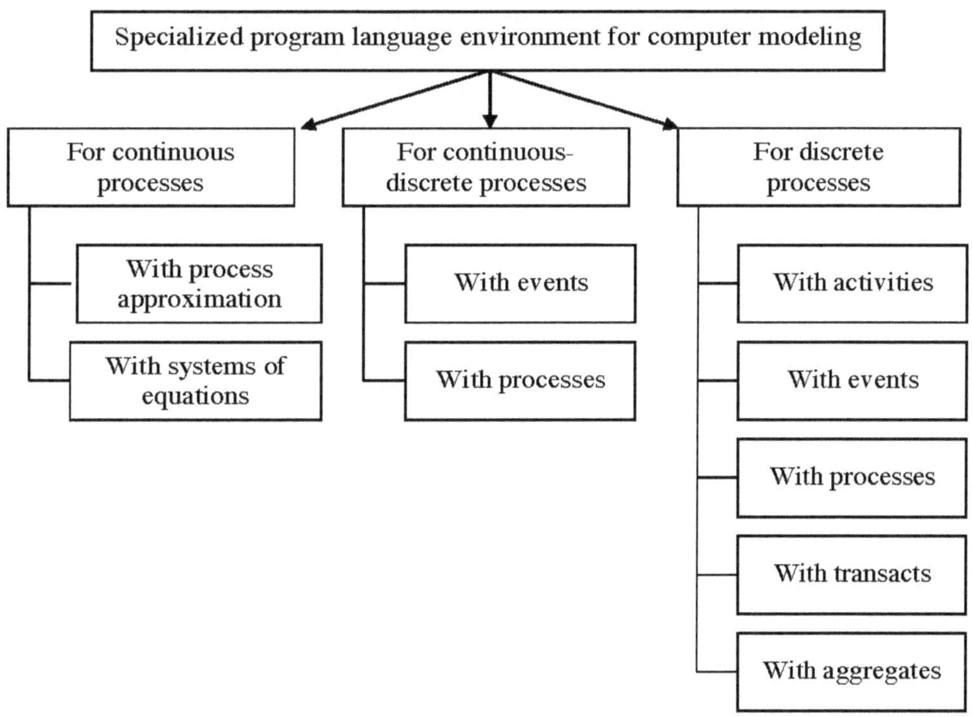

An example of a transaction-oriented language environment is *GPSS (General Purpose Simulation System)*. It is applied in simulating spatial movement of physical and logical objects with a fixed functional structure of the modeled object and using FCFS-type queues (Karian & Dudewicz, 2020). Some simulation languages (e.g. SIMSCRIPT II.5, SIMAN, GASP) provide functions that allow users to develop continuous or discrete models. In such languages it is possible to use both discrete and continuous functions within a simple model. The implementation of simulation models based on the GPSS language renews the theory of the machine service, through which the complex objects are represented as Queue Networks (QN). The GPSS language has different versions and modifications, keeping its basic idea and basic set of operators but also including specific functions (GPSS H, GPSS VX, GPSS/PC, WebGPSS for Windows). For example, GPSS/PC is an interactive implementation of GPSS for a family of microcomputers compatible with the IBM-PC and the DOS operating system. The universal GPSS environment is designed to perform discrete simulations of reliability of mass service systems, which is achieved by using the functionalities included in the software package (Popov, 2017). The article states that the simulation environment offers better flexibility than competing load and service parameter certification programs. Furthermore, it allows simulating the behavior of Markov and semi-Markov processes and the associated Poisson flows.

A simple example of applying the GPSS language to create a simulation model is shown in Figure 6 in two forms: (a) block diagram; (b) program model; (c) simulation result. This example simulates a single-server queue (SSQ) which corresponding to the formal description of CS presented in Figure 6 (Chapter 3). Memory is modeled by the FCFS buffer for tasks waiting MEMO and processor – by service unit PROC with time for processing 15±7 time units (UMT). The input workflow is formed by task each generated after time period 10±3 (from 7 to 13 UMT). Additional statements for simulation beginning/finishing control (SIMULATE, END) and for number of model executions (START) are presented. Selected part of experimental results for the components MEMO and PROC are presented in Figure 6(c).

Figure 6. Example of GPSS block diagram for a single-server queue

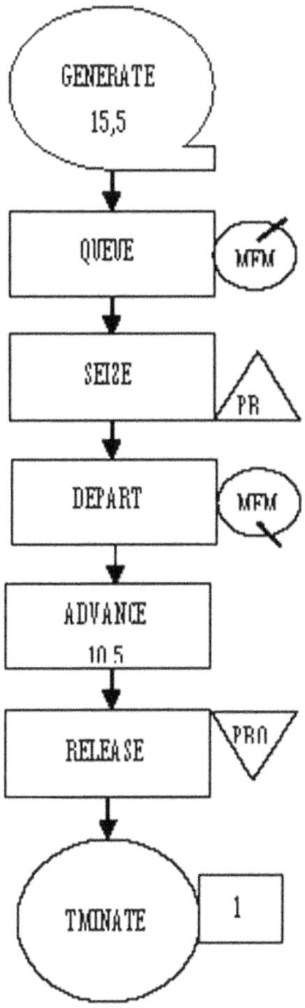

(a)

```
SIMULATE
GENERATE      10,3     Task arriving for time 10-20 UMT
QUEUE         MEM      Input to wait buffer W
SEIZE         PROC     Access workstation
DEPART        MEM      Leave the wait buffer W
ADVANCE       15,7     Service time 5-15 UMT
RELEASE       PROC     Return work station
TERMINATE     1        Leave the system
START         100      Simulate for 100 tasks
END
```

(b)

```
SIMULATION
      START TIME=0              END TIME=1547.533

PROC
      ENTRIES                   101
      UTILIZATION               0.992
      AVERAGE TIME              15.203
      DELAY                     54

MEMO
      MAX.CONTENT               55
      ENTRY                     155
      ZERO ENTRY                1
      AVER. CONTENT             26.866
      AVERAGE TIME              268.23
```

(c)

Simulation by using GPSS language is presented in (Romansky & Noninska, 2021a) where communication parameters of distributed memory MIMD architecture with ring topology are investigated. A comparative analysis of experimental data with varying the number of nodes and the degree of connection is made. The study was conducted by simulation experiments with initial formalization of primary factors and by varying the controllable factors.

Summary of the TryAPL2 Functionality for Analytical Modeling

The program environment TryAPL2 is for universal use, but it can be applied in analytical modeling, which is based on the capabilities of the parallel language APL2 oriented to working with matrices and the graph theory (Romansky, 2021c). The organization is based on the realization of functional modules (separate functions) executed independently or as a series of realization of f calls. The system workspace OR can be used to perform the experiments which support the functions presented in Table 1.

Table 1. Main supported functions by OR

Function	***Purpose***
SETUP *matrix*	Defines a specific STN (State Transition Network), described by a numerical matrix of connections matrix. Defines: *NODES* - list of node numbers; *SIZE* – number of nodes; *NETWORK* – a copy of the matrix; *CM* – connection matrix. To illustrate the functionality, the OR supports a matrix called SPM, describing the weight connections in a seven-node network.
PATHSFROM *node*	Calculates all paths in the STN from the specified *node* to the last node. In an STN with one input node and one final node, all paths are calculated by ρPATH←PATHSFROM 1
ARCS *path*	Determines the weight of each connection (arc) of the attribute *path*. Examples: ARCS ↑PATH – displays information about the first of the paths stored in PATH; +/ARCS ↑PATH – determines the length of this path; +/¨ARCS¨PATH – determines the length of all paths in STN.
VALUE *path*	Similar to +/ARCS *path*, where *path* may contain one (↑PATH), more or all (¨PATH) possible paths.

Investigation of graph structure (as a formal model of system or program/microprogram execution) is based on a presentation as a set of states (nodes in a graph) and possible transitions between them. It can use presentation by STN (State Transition Network) if the arcs present transition probabilities (weight graph with probabilities for transitions between nodes). In this case, an initial quantitative analysis of a created formal description can be performed through the soy using standard OR-functions from Table 2. Determining the reachability between the initial and final node and storing all defined paths in an internal structured variable for further study with the possibility of visualization of a selected path can be made by using command from Table 3.

Table 2. Standard functions in OR for SETUP MATRIX

SETUP MATRIX	***Comment***
NODES	Determines the numbers of the nodes in the STN
SIZE	Displays dimensions of the connection matrix
NETWORK	Displays defined STN
CM	Displays the connection matrix defined by MATRIX

Table 3. Commands for investigation of reachability

Command	*Comment*
PATHSFROM 1	Determining possible paths from input (node 1) to the output
ρPATH←PATHSFROM 1	Stores all defined paths in the structural variable
↑PATH	Displays the first path in the structure PATH

If the model describes a stochastic process, the spice OR proposes operators for quantitative investigation of the weights of the possible paths based on the arcs between nodes and calculation of probabilistic estimates for one or more paths in the structure. In addition, an analysis of probabilistic parameters and determining minimal and maximal values can be made. A summary of these opportunities is presented in the Table. 4.

Table 4. Determining the parameters of the paths and probability analysis

Command	*Comment*
ARCS ↑PATH	Determines the weights of the arcs for the first path in PATH
x/ARCS ↑PATH	Calculates the multiplication of the probabilities between nodes in the first path
A←ARCS¨PATH	Stores in A the weights (probabilities) for all determined paths
x/¨ARCS¨PATH	Determines the multiplication of all probabilities in each path in STN
VALUE¨PATH	Determines the sum of the probabilities for the each path in the STN
+/¨ARCS¨PATH	Similar to the previous one
V←x/¨ARCS¨PATH	Stores the result of the operation in a work variable V
MIN←⌈/V	Calculates the minimal probability multiplication for a path
MAX←⌊/V	Calculates the maximal probability multiplication for a path
(V=⌈/V)/PATH	Determines the path in the STN with maximum result of the probabilities multiplication

The language APL2 permits to create collections of program models of computer structures in different levels and such example for analytical models of sample micro-operations (machine algorithms or digital schemes) is presented below. These are models of elementary transformations supported in a hardware environment as a device of a traditional processor, some of which are presented below and their analytical program models are presented in Figure 7.

a) ***INVERS*** – model of a digital code invertor of the binary code entering in the inputs.

b) ***DCOD*** – model of a code converter.

c) ***COMPR*** – model of a comparator for comparing input binary codes X and Y (with equal length) to forming output result Z on the principles Z = 0 (*if* X≥Y) or Z = 1 (*if* X <Y).

d) ***ADDER*** – model of a parallel adder for unsigned multi-digit binary numbers.

When organizing model experiments for the study of basic operations for computer processing at a low (micro) level, the following is performed:

1. Specifying the register structure for the operation.
2. Development of the machine algorithm.
3. Formalization of the machine algorithm for representation of the elementary operations in canonical mathematical form.
4. Creation of a program model of the formalized algorithm in APL2 environment.
5. Conducting test performances with the model.

Figure 7. Program models of micro-operations

```
∇INVERS  P; I
[1]    I←ρP
[2]   ET1:→(P[I]=1)/ET2
[3]    P[I]←1
[4]    →ET3
[5]   ET2:P[I]←0
[6]   ET3:I←I-1
[7]    →(I≥1)/ET1
[8]    'RESULT IS: '
[9]    P∇

X←1 0 1 1 0 1 1 0
      INVERS  X
RESULT IS: 0 1 0 0 1 0 0 1
```

(a)

```
∇Y←DCOD X
[1]    N←ρX
[2]    Y←X
[3]    →(X[1]=0)/OUT
[4]    Y←~X
[5]    Y[1]←1
[6]   ET:Y[N]←Y[N]+1
[7]    →(Y[N]=1)/OUT
[8]    Y[N]←0
[9]    N←N-1
[10]   →ET
[11]   OUT: 'OPERAND IS:' , X
[12]   'RESULT IS: ',  Y∇

X←1 0 1 1 0 1 1 0
      Y←DCOD  X
OPERAND IS: 1 0 1 1 0 1 1 0
RESULT IS:  1 1 0 0 1 0 1 0
```

(b)

```
∇Z←X COMPR Y
[1]  Z←I←0
[2] REP:I←I+1
[3]  →(I>ρX)/OUT
[4]  →(X[I]=Y[I])/REP
[5]  →(X[I]>Y[I])/CUT
[6]  Z←1
[7] CUT: 'IF X≥Y => Z=0; IF X<Y => Z=1'
[8]  ' FLAG  Z  IS: ', Z∇

Z←10110110 COMPR 11001010
IF X≥Y => Z=0;  IF X<Y => Z=1
FLAG  Z  IS:  1
```

(c)

```
      ∇ S←X ADDER Y; C
[1]   →((ρX)≠(ρY))/0
[2]   X←0,X
[3]   Y←0,Y
[4] LOOP:C←X^Y
[5]   X←(((~X)^Y)∨(X^(~Y)))
[6]   Y←(1↓C),0
[7]   →(C≠0)/LOOP
[8]   S←X∇
```

(d)

To illustrate the described approach two APL software models for study of machine operations are presented in the Figure 8: (a) normalization of floating-point numbers (NORM function); (b) multiplication of fixed-point binary numbers by the method of lower digits (MULT function). They are developed based on traditional register structures for the implementation of the respective operations at a low level, and in the formalization of the algorithms the correspondence of an operator order with an elementary operation (micro-operation) is sought.

The presented procedures are fully compliant with the supported functions of the experimental environment APL2 and with the requirements for mutual communication between them when constructing firmware.

Figure 8. Program APL2-models of basic machine algorithms (microprograms)

```
     ∇ M  NORM  P
[1]   D1: ⎕←'ENTER START = 1'
[2]   START←⎕
[3]   →(START≠1)/D1
[4]   P1: START←0
[5]   P2: FINI←0
[6]   P3: ZERO← 0
[7]   P4: R←M
[8]   P5: S←P
[9]   D2: →((+/R)=0)/P8
[10]  D3: →(R[2]=1)/P9
[11]  P6: R←R[1],(2↓R),0
[12]  P7: S←((ρS)ρ2) ⊤ ((2⊥S)-1)
[13]  →D2
[14]  P8: ZERO←1
[15]  P9: FINI←1
[16]  'RESULT  IS:'
[17]  'FINI = ', FINI
[18]  'ZERO = ', ZERO
[19]  'M = ', R
[20]  'P = ', S    ∇
```

```
Program execution:
M←1 0 0 0 1 1 0 0 1 0 1 1
P←0 1 1 0
M  NORM  P
```

(a) Normalization

```
     ∇ X  MULT  Y
[1]   D1: ⎕←'ENTER START = 1'
[2]   START←⎕
[3]   →(START≠1)/D1
[4]   ⎕IO←0
[5]   P1: START←0
[6]   P2: FINI←0
[7]   P3: R1←X
[8]   P4: R2←Y
[9]   P5: S←8ρ0
[10]  P6: B←7
[11]  D2: →(R2[7]=0)/P8
[12]  D7: S←(8ρ2) ⊤ ((2⊥S)+(2⊥(1↓R)))
[13]  P8: R2←R2[0],0,R2[1 2 3 4 5 6]
[14]  P9: R2[1]←S[7]
[15]  P10: S←0,S[0 1 2 3 4 5 6]
[16]  B←B-1
[17]  D3: →(B≠0)/D2
[18]  P11: S[0]←(R1[0]≠R2[0])
[19]  P12: FINI←1
[20]  'RESULT  IS:'
[21]  'FINI = ', FINI
[22]  'X . Y - ',  S,(1↓R2)
[23]  ⎕IO←1    ∇
```

```
Program execution:
X←1 0 0 0 1 0 1 0
Y←0 0 0 0 0 0 1 0
X  MULT  Y
```

(b) Multiplication

A final comment. The continuous development and improvement of software technologies also affects modeling language environments. The trend is to introduce a hybrid approach by bringing together different tools for programming automation, as discussed in (Narang & Mittal, 2022). The paper discusses a random model of a chain of automation tools and proposes a hybrid model and evaluating software quality metrics. Another aspect of hybrid modeling is discussed in (Pushpalatha & Math, 2022), where the human activity recognition task is analyzed with a focus on machine learning. According to the authors, the reason for the conducted research is that the existing systems "*could not identify the activity when there is a sudden change in the activity*". The result is a proposal of a model using a hybrid approach, the aim of which is to achieve a better level in predicting activity and sudden transitions from one activity to another.

REFERENCES

Batool, K., & Niazi, M. A. (2017). Modeling the internet of things: A hybrid modeling approach using complex networks and agent-based models. *Complex Adaptive Systems Modeling*, *5*(4), 4. Advance online publication. doi:10.118640294-017-0043-1

Butler, T., Jakeman, J., & Wildey, T. (2018). Convergence of probability densities using approximate models for forward and inverse problems in uncertainty quantification. *SIAM Journal on Scientific Computing*, *40*(5), A3523–A3548. doi:10.1137/18M1181675

Carstens, B. C., Smith, M. L., Duckett, D. J., Fonseca, E. M., & Thomé, M. T. C. (2022). Assessing model adequacy leads to more robust phylogeographic inference. *Trends in Ecology & Evolution*, *37*(5), 402–410. doi:10.1016/j.tree.2021.12.007 PMID:35027224

Elliott, J., Kelly, D., Best, N., Wilde, M., Glotter, M., & Foster, I. (2013, July). The parallel system for integrating impact models and sectors (pSIMS). In *Proceedings of the Conference on Extreme Science and Engineering Discovery Environment: Gateway to Discovery* (pp. 1-8). 10.1145/2484762.2484814

Faix, M., Laurent, R., Bessière, P., Mazer, E., & Droulez, J. (2016). Design of stochastic machines dedicated to approximate Bayesian inferences. *IEEE Transactions on Emerging Topics in Computing*, *7*(1), 60–66. doi:10.1109/TETC.2016.2609926

Fonseca, E. M., Duckett, D. J., Almeida, F. G., Smith, M. L., Thomé, M. T. C., & Carstens, B. C. (2022). Assessing model adequacy for Bayesian Skyline plots using posterior predictive simulation. *PLoS One*, *17*(7), e0269438. doi:10.1371/journal.pone.0269438 PMID:35877611

Gabrielli, A., & Wüthrich, V., M. (. (2018). An individual claims history simulation machine. *Risks*, *6*(2), 29. doi:10.3390/risks6020029

Jin, C. (2020). Global classical solutions and convergence to a mathematical model for cancer cells invasion and metastatic spread. *Journal of Differential Equations*, *269*(4), 3987–4021. doi:10.1016/j.jde.2020.03.018

Karian, Z. A., & Dudewicz, E. J. (2020). Modern statistical, systems, and GPSS simulation (2nd ed.), CRC press, Taylor & Francise group (94 p.). doi:10.1201/9781003067993

Kürkçü, Ö. K., Aslan, E., & Sezer, M. (2017). A numerical method for solving some model problems arising in science and convergence analysis based on residual function. *Applied Numerical Mathematics*, *121*, 134–148. doi:10.1016/j.apnum.2017.06.015

Narang, P., & Mittal, P. (2022). Performance analysis of DevOps based hybrid models integrated with different automation tool chains for quality software development. *International Journal on Information Technologies and Security*, *14*(4), 25–32.

Parker, W. S. (2020). Model evaluation: An adequacy-for-purpose view. *Philosophy of Science*, *87*(3), 457–477. doi:10.1086/708691

Popov, G. (2017). GPSS language as tool for reliability simulations. *2017 15th International Conference on Electrical Machines, Drives and Power Systems (ELMA)*, Sofia, Bulgaria, 2017, pp. 461-463, 10.1109/ELMA.2017.7955486

Pushpalatha, S. S., & Math, S. (2022). Human activity recognition using hybrid model. *International Journal on Information Technologies and Security*, *14*(4), 55–66.

Romansky, R. (2021c). Program environment for investigation of micro-level computer processing. *International Journal on Information Technologies and Security*, *13*(1), 83–92.

Romansky, R., & Noninska, I. (2021a). Investigation of communication parameters in multicomputer architecture with ring topology. *Proceedings of the 2021 IEEE International Conference on Information Technologies (InfoTech-2021)*, 16-17 Sept 2021, Bulgaria, pp. 119-123. 10.1109/InfoTech52438.2021.9548514

Schneider, A. (2016). *Modelling of data uncertainties on hybrid computers* (No. GRS-392). Gesellschaft fuer Anlagen-und Reaktorsicherheit (GRS) gGmbH.

Chapter 5
Mathematical Formalization and Abstract Description

ABSTRACT

The chapter is dedicated to the possibilities of formalization of an investigated object and its abstract description with appropriate mathematical means. At the beginning, the need to decompose complex systems is discussed to facilitate the following model study. The appropriate elements of a standard formal theory, applied in mathematical formalization and the abstract description of the investigated object, are formulated. A major place is devoted to the application of discrete and stochastic structures for the formalization of computer systems and processes for the purposes of further modeling. In the part of discrete means, the possibilities of formalization based on set theory, binary and weighted graphs, theory of finite automata are discussed. Some examples of specific applications are also given. The discussion of stochastic means is mainly focused on the application of elements of probability theory (random processes, streams of random events), the theory of Markov processes (in particular the application of Markov chains), and the theory of queues.

1. DECOMPOSITION OF COMPLEX SYSTEMS AND FORMALIZTION

Modeling is an abstraction that seeks to provide insight into processes and structures in the real environment. Technological development has allowed many types of modeling to be developed – from completely intuitive models to highly controlled ones. The basis of each approach is the use of specific terms and semantic tools that define the model language with rules for its application. Defining such an alphabet can be done based on natural language symbols, but assigning meaning to linguistic elements can lead to ambiguity and changing syntactic rules. The introduction of strict formal languages that mathematics offers, with the definition of strict semantic rules, allows us to overcome this shortcoming (Mayr & Thalheim, 2021). The article states that in the natural and engineering sciences, the first step is to conceptualize the object (the problem to be solved) and then move on to modeling and studying behavior. The very fact that abstract elements such as class, attribute, relation, domain, state, etc. are defined

DOI: 10.4018/978-1-6684-8947-5.ch005

confirm the importance of this initial stage in model design. In this reason, the conceptual model of a real system of process represents some image of it in an environment of terms, assumptions and hypotheses.

Conceptual modeling is applied in different fields and has different meanings related to the problem being solved. Nevertheless, in most cases it refers to the development of a model for the organization of a research process or the development of a prototype or system. An example in this direction that can be used as an illustration of the applicability of conceptual modeling is presented in (Fernandes et al., 2019). There, the concept of developing a system of information systems, representing dynamic associations of independent but interoperable information systems, is discussed. The idea is to set a common goal to achieve by sharing resources. To clarify the characteristics of such a complex system of systems, a conceptual model is proposed in the article, which, in addition to assisting the research and development of a description of the unified system, allows to make a correct classification of the various types.

As it was specified, for the creation of a conceptual model, mathematical means are applied, through which a formal description is created in a selected mathematical system. In the case of complex systems, a preliminary ***decomposition of the system*** into a finite number of subsystems is necessary, preserving the general functional algorithm. If necessary, subsystems can also be decomposed into elements convenient for formal description. Thus, relatively independent components of the general system with minimal information connections and interactions are identified. These components can easily be formalized independently.

Decomposition is a frequently used approach in system design because it allows the functions of individual sub-systems to be more easily realized. There are enough examples of active application of decomposition, some of which are presented below.

- Application in tree decomposition of a constraint system with adaptation of tree decomposition to work with parametric variables (Thibault, 2022).
- Development of a distributed framework based on the graph algorithm for calculating a control invariant set for non-linear cascade systems using the structure of interconnections in the process network (Decardi-Nelson & Liu, 2022).
- Use in the development of an algorithm for the assessment of the static state of energy systems with the expansion of the limits of resistance to deviations and cyber attacks (Ahmadi et al., 2021).

From a functional point of view, decomposition is the opposite of composition. In discrete mathematics, composition is an operation on representations and sets. Let us consider the set of functions over a set of elements M: $F_M = \{f \,/\, f\!: M \to M\}$.

The operation φ_o: $F_M \times F_M \to F_M$ is defined by the expression $f \circ g = h(x) = f(g(x))$, $h \in F_M$ and is called the composition of the functions f and g. The composition of functions is an associative operation, i.e., if $f, g, h \in F_M$, then:

$$((f \circ g) \circ h)(\mathrm{x}) = (f \circ g)(h(\mathrm{x})) = f(g(h(\mathrm{x}))) = f((g \circ h)(\mathrm{x})) = (f \circ (g \circ h))(\mathrm{x})$$

or for short:

$$f \circ g \circ h = (f \circ g) \circ h = f \circ (g \circ h) = f(g(h(\mathrm{x})))$$

Applying the inverse theory, the decomposition consists in determining the composite functions f, g and h involved in the composition of a given output function $\varphi = f \circ g \circ h$ (Figure 1).

Figure 1. Structural scheme for functional decomposition

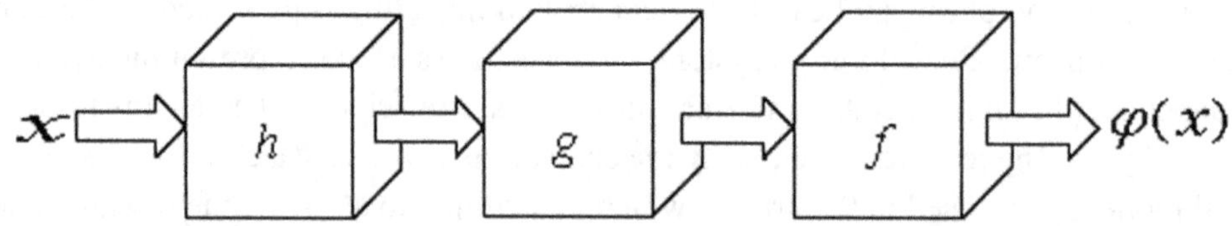

In decomposition, the choice of level of detail is important. Depending on the purpose of the model description, the real object can be represented as a single system or decomposed into separate components (blocks, nodes, details, functional units). For example, when formalizing the system resource, it is appropriate for the elements to be relatively independent and complete blocks that can be functionally described. Such a formalization was done in (Lecci et al., 2020) in modeling 5G networks using millimeter waves (mmWave) to evaluate performance and reliability in mobile communications. In particular, the paper introduces a detailed mathematical formulation of quasi-deterministic channel models characterizing propagation in terms of clusters of multipath components. When formalizing a workload, it is appropriate to select a separate task or user service request for a base unit. In all cases, however, the eventual decomposition of the object must preserve its functional behavior.

Decomposition should be considered as a stage of the formalization of the object applied in the transition from the conceptual model to the mathematical model. ***Formalization*** is based on defining a formal theory through which the descriptions of the object or its components are created (Gazzari, 2022). Discrete mathematics considers a formal theory T = <A, F, B, R> as a set of: ✓ alphabet A – set of symbols; ✓ formulas F – set of words above the alphabet A; ✓ axioms B – subset B ⊆ F; ✓ a set of relations R called inference rules.

The alphabet A can be a finite or an infinite set. A finite set of characters (letters and digits) is usually used. The set of formulas F is usually given inductively (via formal grammar). As a rule, this set is infinite. The sets A and F define the language of the formal theory. The set of axioms B can also be finite or infinite. If it is infinite, a finite set of axiom schemes and rules for generating specific axioms from these schemes are defined for its assignment. The set of inference rules R is in principle infinite.

Infinitely many languages can be defined over a finite alphabet (the set is uncountable). In formalization, a countable set is considered – the formal languages.

The application of the formal theory is associated with the following concepts.

1. *Deducibility* – the ability to draw conclusions (conclusions) of one formula from several others. Let $F_1, \ldots, F_n, G$ be formulas of the formal theory T. If there exists an inference rule $R \in$ R such that $(F_1, \ldots, F_n, G) \in R$, it is said that G is directly deducible from the formulas $F_1, \ldots, F_n$, which are called the hypotheses of the inference.
2. *Interpretation* – representation of the formulas from the set F in another set M, called an algebraic system. It is written $I: F \rightarrow M$ and means that each formula from the theory T is uniquely matched

to a specific expression (proposition) over the set *M*. If the corresponding proposition is true, the formula is said to be fulfilled in the given interpretation.

3. *Universality and non-contradiction* – a given formula is universally significant if it is true in every interpretation and contradictory – if it is always false. A formal theory T in which there is no contradictory theorem (formula) is called semantically consistent.
4. *Completeness* – a formal theory T is called complete if every expression over *M* corresponds to a theorem of it.

Any formal theory is based on a grammar of rules that distinguish between regular and irregular forms in building well-formed structures. According to (Larsen-Freeman & DeCarrico, 2019) "*models of grammar differ greatly, depending on whether they are formal grammars or functional grammars*" stating that formal gramma "*is concerned with the forms themselves and with how they operate within the overall system of grammar*".

In the formalization of computer objects, a finite set of elements S is usually defined, over which a system of base operations O for conversion and logical conditions L is defined. Each operation $o_i \in O$ assigns a mathematical representation $o_i : S \rightarrow S$ to the set of states in itself, and every elementary logical condition $l_j \in L$ defines the image $l_i : S \rightarrow \{\text{True, False}\}$. It should be noted that not every representation $S \rightarrow S$ assigns an operation to the formal system, which also applies to logical conditions. In this reason, an operating device can be thought of as a set of abstract registers containing an ordered sequence of words $<w_1, w_2, \ldots, w_k,>$ over a finite alphabet (in particular, the alphabet may be $\{0, 1\}$). These words define the current state of the operating device, and each change causes a transition to a new state. Any such functional representation $f{:}S \rightarrow S$ defines for each word w_i $(i=1 \div k)$ the functions $y_{ij} = f_{ij}(x_{i1}, \ldots, x_{ij}, \ldots, x_{im})$, where x_{ij} is the content of the jth bit before the conversion, and y_{ij} – after him.

The formal approach is implemented in (Bargiela & Pedrycz, 2022) in human-centered information processing research. The apparatus of fuzzy logic was used and an alternative framework for the formalization of granular computing is proposed. The publication is aimed at researching the basics of granular computing and its representation as a structured formalized combination of algorithmic and non-algorithmic information processing, intelligent synthesis of knowledge from imprecise and/or fuzzy information.

The abstract description allows a clear representation of the formal model by means of an appropriate topological system. The main task of such a topology is to allow a clear visualization of the main components of the structure, as well as what the inputs and outputs of the task are. The topological scheme proposed in (Brehmer & Munzner, 2013) allows complex tasks to be expressed as a series of interdependent simpler tasks using represented abstract descriptions. An approach model paradigm is proposed in (Atanasov, 2021) which allows the design to be made based on the abstract hierarchy. It can be used in the system development process, including development of networking systems and the main characteristics of the proposed approach as part of the development process are discussed.

Most often, the formalization is based on the theory of finite automata applied to describe individual subsystems or operating devices. This allows for their abstract representation through graphs, structural or block diagrams, algorithmic logic diagrams, flowcharts, Petri nets, etc.

The ability of a system to recover its functions is an important property, especially in control systems. The main task is to be able to determine the current state of the system based on set input and output sequences of impacts. In deterministic systems, recoverability defines the possibility of determining successive next states given a known current state. This corresponds to the behavior of a state machine operating with a Boolean network, as discussed in (Gao et al., 2021) where the investigation of reconstructibility of switched Boolean control networks is presented. The paper states that if the input and output sequences can accurately determine the current state, a deterministic finite automata can be applied to conduct the study. A weighted dual graph describing all pairs of states is defined and developed to determine the recoverability of the system.

When formalizing the behavior of a discrete device, three basic forms of information conversion can be defined, and their abstract description is presented in Figure 2:

(a) Transmission (**T**) – transferring the input information X(t) to the output with a delay:

$Y(t+\Delta t) = \mathbf{T}[X(t)] = X(t);$

(b) Processing (**P**) – conversion of input information flows $X_j(t)$ into a new information flow:

$Y(t+\Delta t) = \mathbf{P}[X_1(t), \ldots, X_1(t)];$

(c) Control (**C**) – formation of control (*) output flow based on input control (*) and information flows: $Y^*(t+\Delta t) = \mathbf{C}[X(t), X^*(t)]$.

Figure 2. Basic forms in the formalization of information conversion

(a) transmission *(b) processing* *(c) control*

2. APPLICATION OF DISCRETE STRUCTURES FOR FORMALIZATION AND ABSTRACT DESCRIBING

2.1. Set-Theoretic Formalization

In discrete mathematics, the concept of "set" is fundamental and there is no strict definition for it. A set *M* is assumed to be defined if its constituent objects (they are called elements of the set) are defined. Any defined set can be included as an element of a new set, not allowing the set to be an element of itself (Russell's paradox). A convenient way to visualize sets is *Venn diagrams*, by which each set is represented as part of a plane represented by a closed curve (Figure 3).

Figure 3. Set determining and basic operations

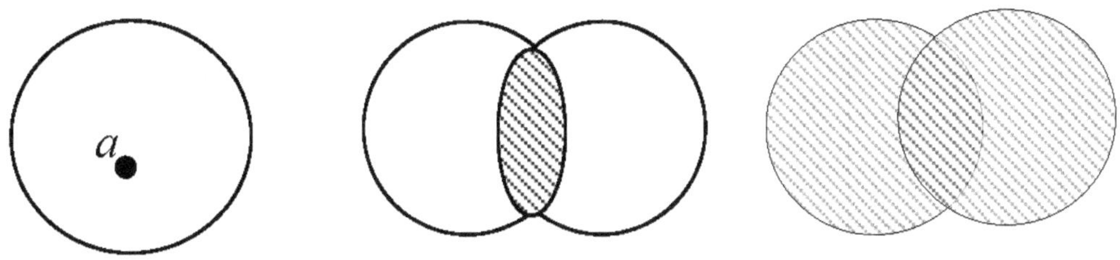

Set and element	Intersection of sets	Union of sets
$a \in M$	$A \cap B$	$A \cup B$

Possible ways to set a set are as follows: ✓ By listing the elements: M:= $\{a_1, a_2, \ldots, a_n\}$; ✓ By characteristic predicate: M:= $\{x\,/\,P(x)\}$; ✓ By generating procedure: M:= $\{\,x\,/\,x{:=}f\}$.

If the set A is contained in the set B, we say that B includes A which is written $A \subset B := x \in A \Rightarrow x \in B$ (A is a subset of B). If complete matching of the elements is also allowed, the entry $A \subseteq B$. is used. The number of elements in each set M determines its cardinality and is denoted by ú*M*ú. For a finite set ú*M*ú=n; for empty set ú∅ú=0, but ú{∅}ú=1. If ú*A*ú=ú*B*ú, then the two sets are equal. When operating with sets, the following basic operations are usually applied:

Union $A \cup B$:= {*x* ú $x \in A \vee x \in B$};
Section $A \cap B$:= {*x* ú $x \in A \wedge x \in B$};
Difference $A \backslash B$:= {*x* ú $x \in A \wedge x \notin B$};
Symmetric difference $A \Delta B := (A \cup B) \backslash (A \cap B)$
$A \Delta B$:= {*x* ú $x \in A \wedge x \notin B$} ∨ {*x* ú $x \notin A \wedge x \in B$};
Complement $\bar{A}$:= {*x* ú $x \notin A$ }.

By definition, if a, b are arbitrary elements, (a, b) denotes the set $\{a, \{a, b\}\}$ and is called an ordered pair of the elements a, b. Also, by definition, for two sets A and B, the set $A \times B$ = {(a, b) ú $a \in A$, $b \in B$} is called the Cartesian multiplication of the sets A and B. According to the adopted constructive principle, since the sets A and B are already constructed, an ordered pair (a, b), $\forall a \in A$, $b \in B$ can be formed.

A classic example of a Cartesian product is the set of points on the Euclidean plane, in which a rectangular coordinate system is introduced by two perpendicular real lines R_x and R_y (Figure 4). The two orthogonal projections x and y along the axes of the point P form an ordered pair (x, y) that uniquely defines the point P. Thus, for the entire Euclidean plane E, it turns out that $E = R_x \times R_y$. The Cartesian product $M \times M = M^2$ is called the Cartesian square of M. For example, for $B_2 = \{0, 1\}$, the Cartesian square will be: $B_2 \times B_2 = \{(0, 0), (0, 1), (1, 0), (1, 1)\}$.

Figure 4. Cartesian square

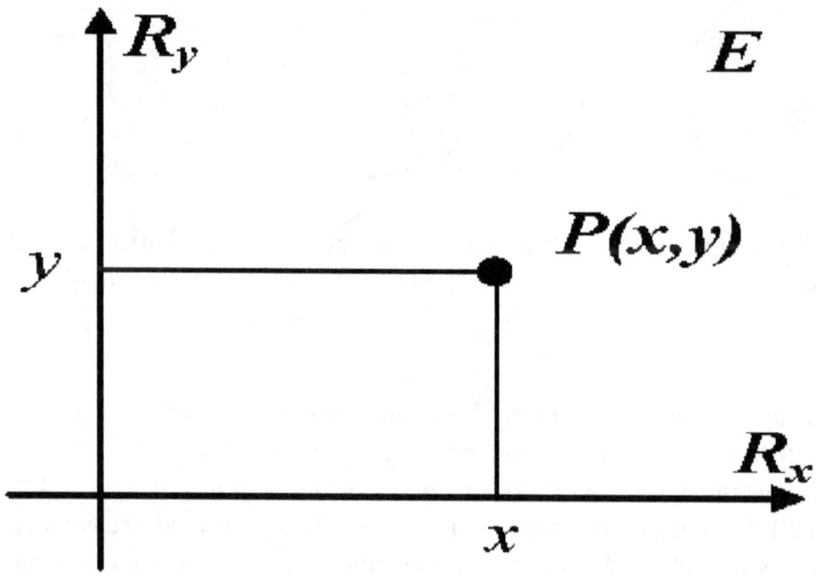

The Cartesian product operation is not associative in principle. For example, the ordered pair ((*a*, *b*), *c*) is significantly different from the ordered pair (*a*, (*b*, *c*)). For several applications, however, it is interesting to consider an associative version of the operation $(A\times B)\times C = A\times(B\times C) = A\times B\times C$. In this case, the elements of the Cartesian product (*a*, *b*, *c*) are called ordered triples. Extending the latter one can reach the notion of an *n*-tuple corresponding to an *n*-multiple Cartesian products. For a set *M*, write $M^n := M \times M \times .... \times M$.

Another basic concept in discrete mathematics is "relation", both in theoretical and applied aspects. From a practical point of view, a relation is a mathematical model of various real-world problems. For two sets *A* and *B*, a subset *R* of their Cartesian product is called a binary relation *R* from set *A* to set *B* and is written $R\subset A\times B$ or by infix notation: $aRb := (a, b) \in R\subset A\times B$. Similarly, when working with *n*-dimensional Cartesian products, an *n*-place relation *R* is defined – a set of ordered tuples (*n*-tuples):

$$R \subset A_1 \times A_2 \times .... \times A_n = \{(a_1, a_2, ..., a_n) \text{ ú } a_1\in A_1 \ \&\ a_2\in A_2 \ \&\ ... \ \&\ a_n\in A_n\}.$$

The fact that binary relations are most often used in practice is not a significant limitation, since the two domains *A* and *B* in the relation $R\subset A\times B$ can be represented as n/2-place Cartesian products and thus realize *n*-place relation. An example of a typical two-place relation over the sets of real numbers is the straight-line L: $y = x$. Other such examples are the „=", „≠", „<", „≤", „≥", etc.

The set-theoretic formalization is a mathematical representation of the relationship between the components of a system based on multiple descriptions and definition of relations. For a set *M*, a relation of rank "*n*" is a law that forms ordered groups (*n*-tuples) of *n* elements each of *M*. We consider tuples $(a_1, a_2, ..., a_n)$ whose elements a_j in the general case belong to different sets A_j (j=1,2, ..., *n*) and satisfy the condition $a_j\in R$ – the elements are in relation *R* and it is written $(a_1, a_2, ..., a_n, R)$ or $R(a_1, a_2, ..., a_n)$.

The theoretical-multiple apparatus is used in the formalization of information processes related to the construction of databases and information systems (Maggiora & Vogt, 2021), as well as in tasks requiring the definition of relations in which individual elements enter (Gutman, 2021). It is also applied in the choice of topology in distributed systems (Romansky & Noninska, 2021a), as well as in the study of the transfer of information between nodes in a communication network and the formation of connection channels in the topological structure.

It is known, for example, that in multiprocessor systems and computer networks, the presence of a reduced number of connections leads to a delay in the transfer of data. This also applies to the connection topology. The analysis of data transfer delays can be done by examining the binary relations $R(a_i, a_j)$ between the elements of a finite set of devices in the multiprocessor or nodes in the computer network. To reduce this delay, it is expedient for the communication environment to be multi-connected, i.e. to allow a larger number of binary relations $R(a_i, a_j)$. In the maximum version, a fully connected network is reached, in which a binary relationship with all other elements is defined for each element (node). For a communication environment with n nodes (a set with n elements), the number of relations (connection lines) will be $k=n(n-1)/2$, which makes it quite expensive. This necessitates the selection of an optimal connection topology meeting a selected criterion of economic efficiency.

Figure 5 shows two connection options between processor nodes – ring topology and incompletely connected topology. The separate nodes enter into certain binary relationships, some of which participate in both topologies. For bidirectional lines, $R(a, b) = R(b, a)$ is valid, otherwise the ordered pair determines the direction of transmission. Through operations on the set $\{a, b, c, d\}$ and examination of the possible ordered pairs, an optimal topology can be determined for a given criterion.

Figure 5. Binary relationships in a connection topology

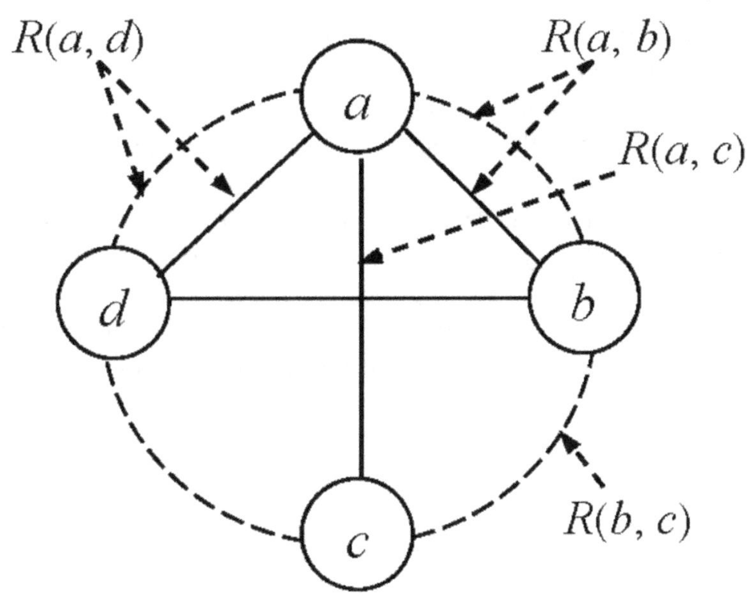

The investigation presented in (Romansky & Noninska, 2021a) discuses an approach for evaluation of the communication parameters of Distributed memory (DM) MIMD architecture with ring topology in the presence of chord connections between certain nodes (Chord ring – HRING), making a comparative analysis of experimental data with varying the number of nodes and the degree of connection. The used approach is simulation by using GPSS spice, but an initial set formalization is made. In a ring topology, the connection is serial, with each processor node (PN_J) having a direct connection to its two neighbors. In "pure" ring topology with one-way transmission, the maximum communication distance is $\chi = N\text{-}1$ (in two-way connection $\chi = N/2$). To reduce communication losses, chord connections at distances *K* positions are added, usually of a regular nature (Figure 6).

Figure 6. An example of connections for a node

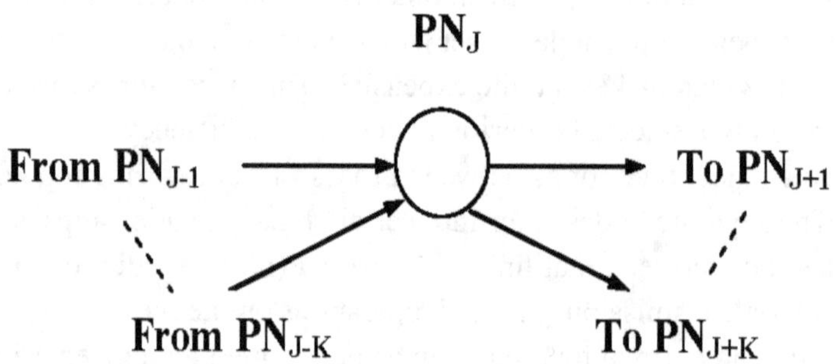

The information packets are routed by a routing algorithm from the source-transmitter to the respective receiver on a circular route with possibilities for using chord communication connections. As a problem can be pointed out is the increasing complexity of the interconnection network, which increases the risk of damage, as well as complicating the management of communications. In this case three multicomputer architectures with topology connection "chord ring" type is investigated, and their set formalization is shown in Figure 7. For each of them the degree of connection is fixed ($\alpha = 4$).

An application of the set theory is so called "rough set" which is toll in mathematics and uses the set theory for classification of objects in decision support systems (Savyanavar & Ghorpade, 2021). It uses approximate reasoning to classify objects into three regions: positive, negative and boundary. The research presents a developed classification approach in a mobile grid by using rough set theory. The authors point out that "*the novelty of our approach is that we applied non-computationally intensive rough set theory for efficient classification of nodes*" and the approach can be used in devices with Android.

Figure 7. Topological structures for investigation with α=4

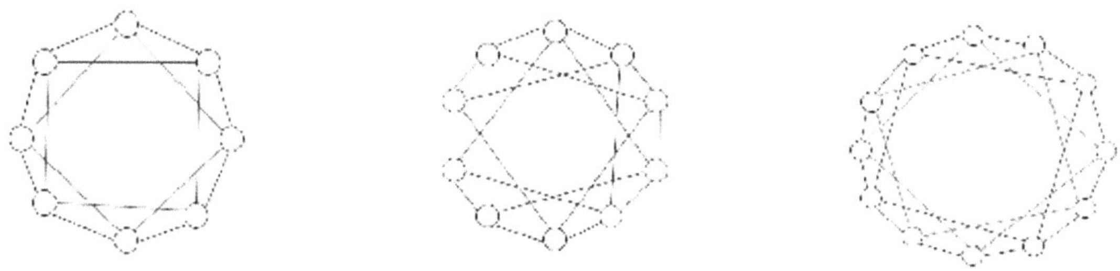

HRING-8 (N=8, K=2) HRING-10 (N=10, K=3) HRING-12 (N=12, K=3)

2.2. Graph Formalization

Graph is discrete structure $G(V, A)$, where $V=\{V_1,..., V_n\}$ is a finite nonempty set ($V\neq\varnothing$) of the nodes, and $A=\{a_{ij}\}\subset V\times V$ is finite set of arcs connected couples nodes (V_i, V_j). Graph theory determines the next opportunities:

- Depending on a defined direction of the arcs: non-oriented (undirected) if $a_{ij}=a_{ji}$ or oriented (directed) if $a_{ij}=<V_i\rightarrow V_j> \neq a_{ji}$.
- Depending on the associated values of the arcs:

Binary graph: $$AM[i,j]=\begin{Bmatrix} 0; if \nexists a_{ij}, \forall i,j \\ 1; if \exists a_{ij}, \forall i,j \end{Bmatrix}$$

Weigh graph: $$AM[i,j]=\begin{Bmatrix} 0; & \forall i=j \\ k; & \exists a_{ij},\ \forall i\neq j;\ k\in N^{+} \\ \infty; & \nexists a_{ij},\ \forall i\neq j \end{Bmatrix}$$

Graph is determined by Adjacency Matrix (AM) which describes the existing arcs a_{ij} in a binary graph with AM[i,j]=1, and the missing arcs with 0. For a weighted graph, the corresponding values (weights) are noted. In a general case, function f_G: $A\rightarrow V\times V$ is discussed, which maps to each arc an ordered pair of nodes and it is denoted by $G(V, A, f_G)$, where $f_G(a_{ij}) = (V_i, V_j)$. In a broad sense, the definition of a graph allows the existence of more than 1 arc per pair of nodes, which defines the structure "multigraph" (it will be used in Petri net modelling). The function f_G is unique, and each ordered pair $(V_i, V_j) \in V\times V$ can participate at most once in the definition of a graph. In the multigraph, the function f_G : $A\rightarrow V\times V$ can generate several realizations (arcs) for a given ordered pair (V_i, V_j).

Some basic concepts and definitions for graphs are as follows (Figure 8):

- For each node V_i in the graph, incoming $\{a_{ki}\}$ and outgoing $\{a_{iq}\}$ arcs can be defined, and two nodes are adjacent if they are connected by an arc.
- A path in a graph is a sequence of arcs allowing non-adjacent nodes to be connected. A simple path is one that passes through consecutive nodes once, and a cycle is a path with the same starting and ending node.
- A fully connected graph is an undirected graph with arcs between each pair of nodes. If this holds for a directed graph, it is defined as a strictly connected graph (Figure 8a).
- A reflexive graph is a graph composed only of reflexive nodes, where an outgoing arc is also an incoming one, i.e., for $\forall V_i$, $i=1,2,\ldots,n \Rightarrow \exists a_{ii}$ (Figure 8b).
- Complete graph – a graph where for $\forall <V_i, V_j>$ there is at least one directed path in one of the two directions (Figure 8c).

Figure 8. Types of graph strictures

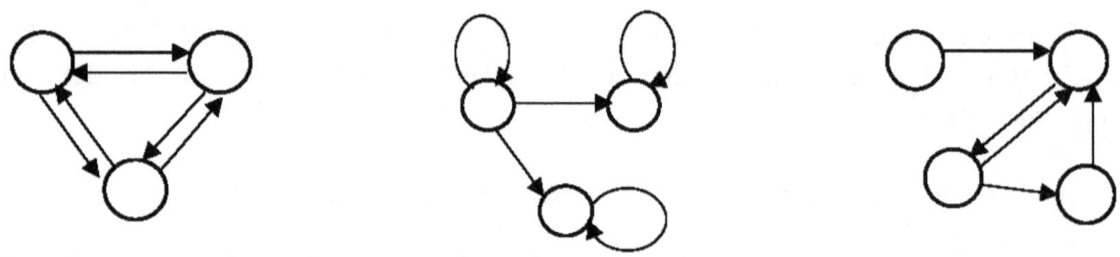

(a) Strictly connected graph *(b) Reflexive graph* *(c) Complete graph*

A key point in the study of computer processing is the problem of investigating reachability. By definition, a node V_j is reachable from a node V_i in a given graph if there exists a directed path $<V_i \rightarrow V_j>$. If there is also a reverse path, then node V_j is counter-reachable by node V_i. This information is contained in the elements RM[i,j] of the two-dimensional reachability matrix (RM), marking the presence of a path from V_i to V_j. For a binary graph, 0 and 1 are used, and for a weighted graph, the elements of RM are the values for the paths of minimum length.

Basic operations on graphs $G(V_i, A_i)$ are:

- Union of graphs G_1 and G_2 – graph $G_1 \cup G_2$ with set of nodes $V_1 \cup V_2$ and set of arcs $A_1 \cup A_2$ (under conditions $V_1 \cap V_2 = \varnothing$ and $A_1 \cap A_2 = \varnothing$). If, under the same conditions, a sum $G_1 + G_2$ is set, the operation is called a compound.
- Adding and removing nodes in a graph.
- Adding and removing arcs in a graph, etc.

The application of graph formalization is based on the fact that the different structure and functioning of computer objects can be represented by a finite set of states $S=\{s_i, s_2,\ldots, s_n\}$ (nodes of a given graph) and transitions between them (arcs in the graph). Most often, it is a directed graph, which is defined by the adjacency matrix (AM) between the basic states of the described object. The elements of this matrix (for a binary or weighted graph) can describe possible transitions between structural components, exchange channels with transmission delays, connection lines with information transfer times, etc.

Graf formalization is applied for presenting technological frame to support the quick communications between workers and staff in industrial environment discussed in (Munir et al., 2020). The relationship between persons in real time is realized by using opportunities of IoT, Big data and semantic web. Article determines the important role of semantic and profiling information as a key to a successful implementation of Industry 4.0. In this reason, the article proposes "*a semantic framework which integrates semantic information of Employee/worker through Knowledge Graph, temporal profiling information and facial recognition*". A graph formalization of relations is proposed with the goal to present the transformation of four tuples into two integrated triples.

Intelligent information processing and pattern recognition offer a good opportunity for graph formalization (Fuchs & Riesen, 2022). The reason is that complex data can hardly be represented correctly by linear vector data structures. To overcome this problem, the paper proposes a new option for encoding specific information using a graph formalization. The aim is to formally describe the stable cores of individual classes of graphs and to evaluate the utility of matches based on the study of two classification approaches: ✓a distance-based classifier for analysis of matches in graphs; ✓ creating sets of matching graphs to embed in a vector space.

Below are discussed some applications of graph formalization in computing.

Formalization of network environment. Simple abstract model of a network environment with 4 workstations is shown in Figure 9: a) directed weight graph to describe the network topology with arcs to represent the direct connection lines; b) matrix of adjacency with elements the average time-delay when transferring the information along the communication lines.

Figure 9. Simple formalization of network environment

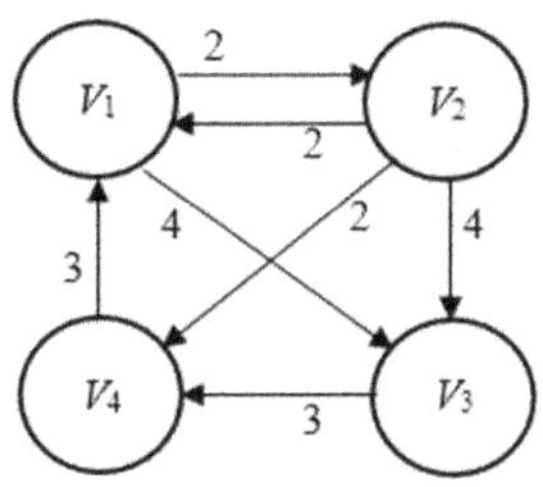

a) Graph

A	1	2	3	4
1	0	2	4	∞
2	2	0	4	2
3	∞	∞	0	3
4	3	∞	∞	0

b) Adjacency matrix

Graph formalization is used in model investigation of processes in a Distributed Learning Environment (DLE) discussed in (Romansky, 2017). There are three groups of objects – "Users", "Information Resources" and "Communication Medium", which can be considered as independent geographically distributed units with their own internal structure and functionality (Figure 10).

Figure 10. General structure of the investigated DLE (abstract model)

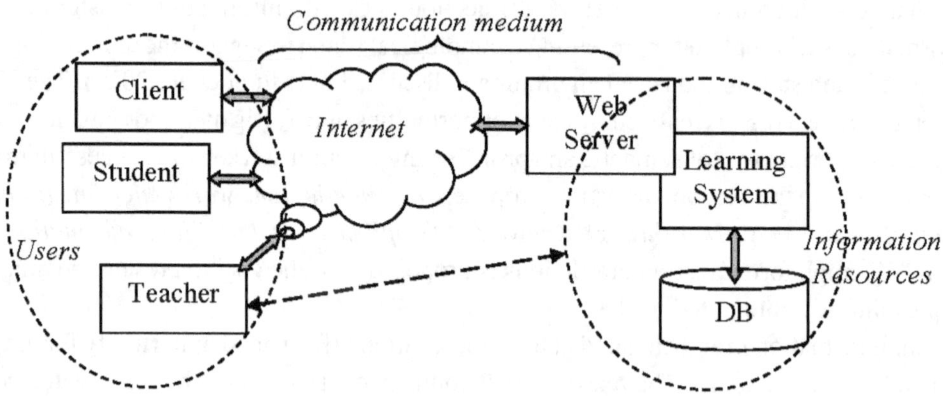

This allows DLE to be described as a discrete set of nodes representing the individual geographical points entering into relations with each other. The initial formalization is by directed weight graph with discrete and finite sets: V – set of nodes to represent the distributed components of the environment; A – a set of connecting lines between pairs of independent nodes, represented by weight coefficients d_{ij}, corresponding to the communication parameters between individual locations of the objects (Figure 11).

The real localization of the components in the individual physical nodes of $V = \{V_1, \ldots, V_n\}$ is important for the concretization of the processes in the information service. It is possible the set V to be divided to several subsets for presentation different relatively independent parts of DLE. Relations in the network environment are based on the connections between nodes which are presented by the set $A = \{a_{ij} / i,j = 1,2,\ldots, n\}$. In addition, the two sets presented in Figure 11 are the set of users (U) and the finite set of resources accessed by users (R). Connections between them are realized by using transmitters $T = \{T_q/q=1 \div K\}$ and distributors (D) for which rout the requests and information resources in the network medium. The last two interactions can be presented by relations ***req***: $U_i \rightarrow R_j$ and ***Inf***: $R_j \rightarrow U_i$, for $\forall U_i \in U$ and $\forall R_j \in R$.

Figure 11. Formal description of DLE abstract model

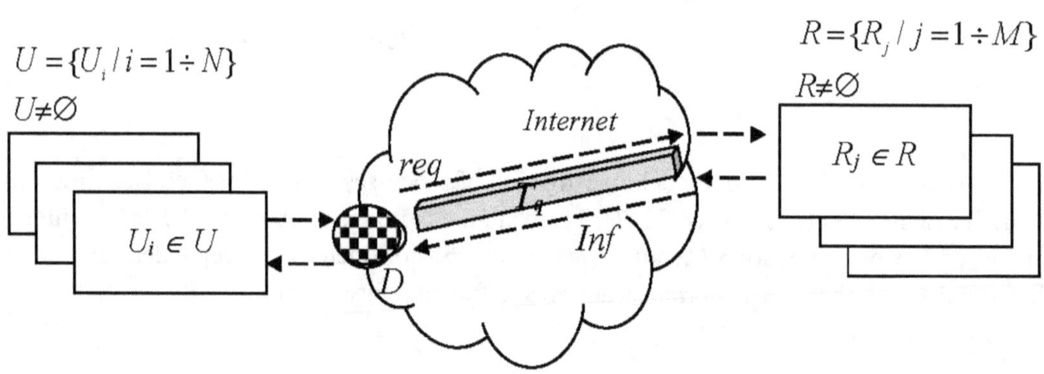

For the general situation, it is assumed that $U \cap R \neq \varnothing$ and one or more nodes of V may contain objects from both groups (the sets U and R). This allows the formation of two types of relations in DLE (Figure 12).

$$(1)\ \exists V_k \in V \Rightarrow (V_k \in U)\ and\ (V_k \in R)$$

$$(2)\ \exists V_k \in V \Rightarrow [(V_k \in U)\ and\ (V_k \notin R)]\ or\ [(V_k \notin U)\ and\ (V_k \in R)]$$

Figure 12. Graphical presentation of relations

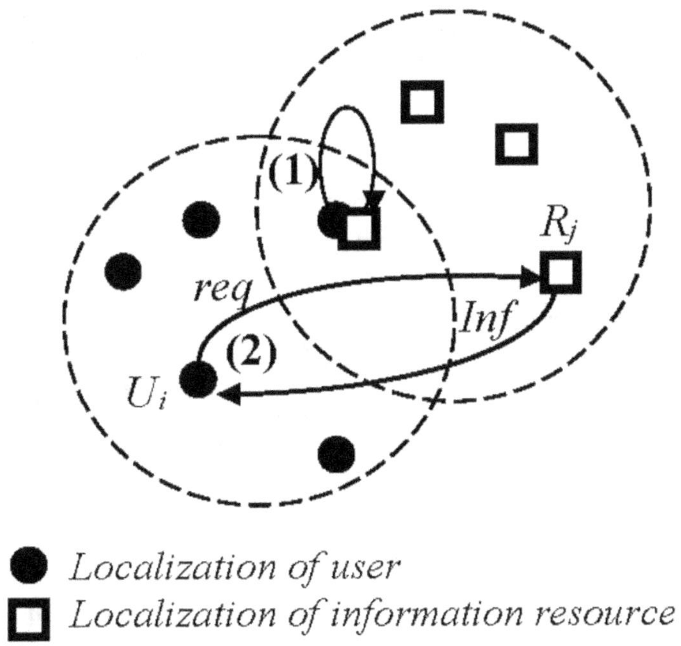

An example of determining the relational scheme is presented in Figure 13, where is assumed that relations are type (2) only and notations are: U – users, R – resources, VL – relation type vector.

Figure 13. Scheme for relations determining

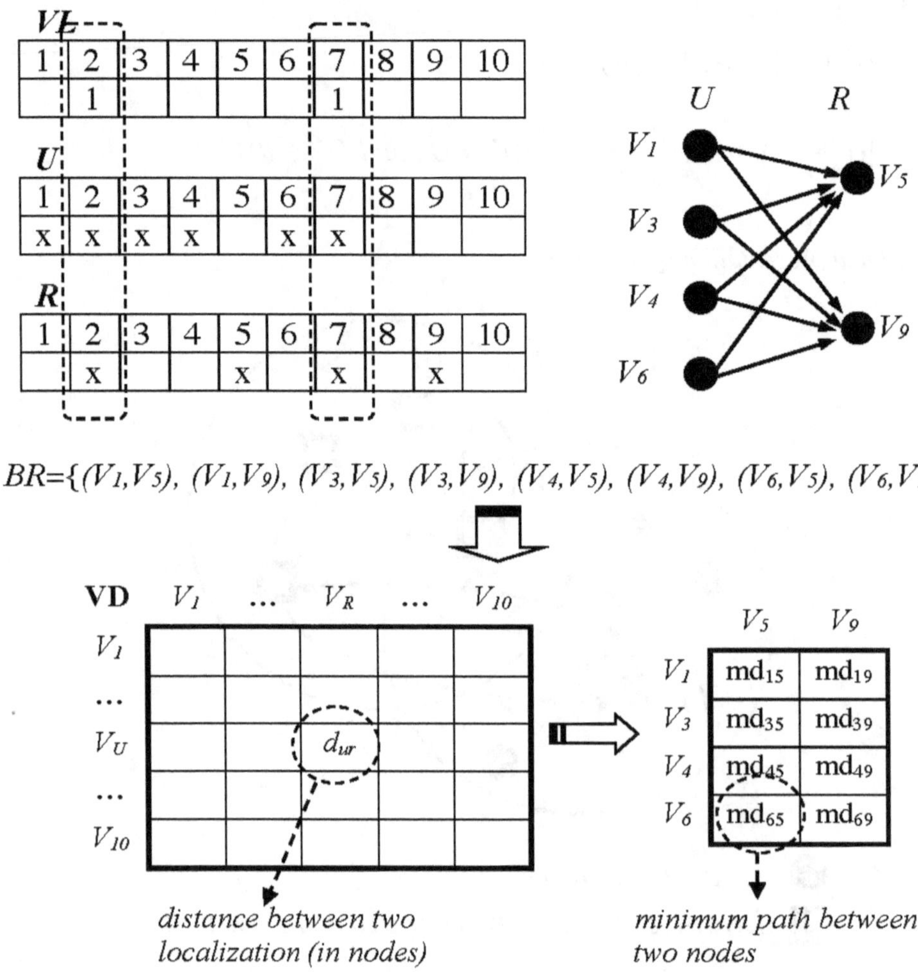

Determine the maximum network flow. It is related to finding the maximum flow in a network specified by a throughput matrix on the link channels. There is a developed algorithm related to Ford and Falkerson's theorem, by which, given an initial flux approximation (usually zero initial flux is assumed), a set of nodes to which there is a path from the respective sources is determined. This task is applicable in the formalization: ✓ of network traffic in a distributed environment; ✓ of message exchange for multiprocessor systems; ✓ for a multi-machine manufacturing system; ✓ for the throughput of a transport system, etc.

To this field, but in a more extended aspect, can be referred the formalized representation of geographic knowledge called GeoKG, which is discussed in (Wang et al., 2019). The formalization of knowledge is applied in Big Data Analytics, mining, and visualization, where information is regarded as elements related to objects or concepts presented by using tree or graph structures. The article justifies the new look at the representation of knowledge because "*discrete knowledge items are difficult to represent geographic states, evolutions, and mechanisms*". Research has been conducted on the possibilities of the proposed formalized model based on a comparative analysis with another environment, and the results are in favor of the proposal.

Determining a minimum path. This is a basic task with various practical applications. There are different algorithms for the solution, such as: Warshall's (detects the existence of chains of arcs); Floyd's (finding a minimal path for each pair of nodes in a graph); of Dijkstra (minimum path between two given nodes). A practical determination of a minimum path with specified start ("1") and end ("4") nodes is shown in Figure 14. The algorithm consists of determining all possible paths by describing the successors of a given node and minimization by rank.

Figure 14. Determination of minimal path in an undirected graph

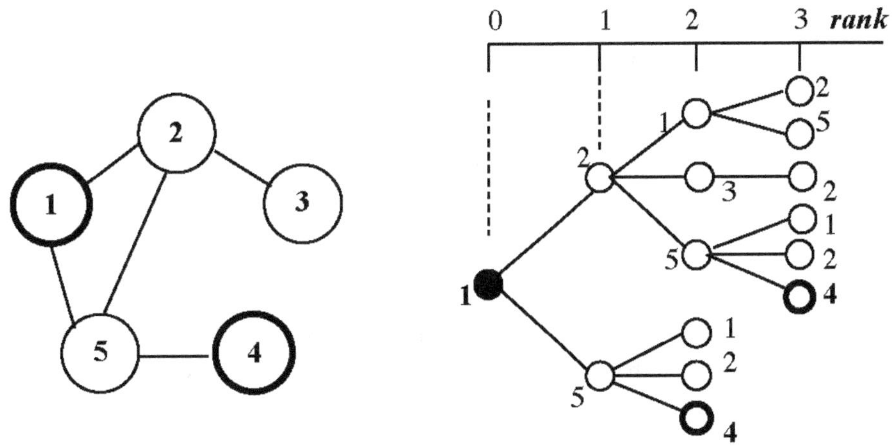

For the starting node "1" a rank (0) is defined and its successors "2" and "5" are defined, which are placed in the next rank (1). The successors of each node included at the current rank are placed at the next rank. After enumerating all the nodes, the minimum rank of the given terminal node is determined, which also defines the length of the minimum path. For example, two ranks (2 and 3) are defined for the final node "4", which defines a minimum path length of 2.

Digital scheme formalization. Each node of a directed binary graph is associated with a logical variable x_i. A directed arc between two variables with a direction from x_i to x_j will exist if the variable x_i (from the source node) is an argument of the function f_j implementing the variable x_j (Figure 15).

Figure 15. Graph formalization of a digital element

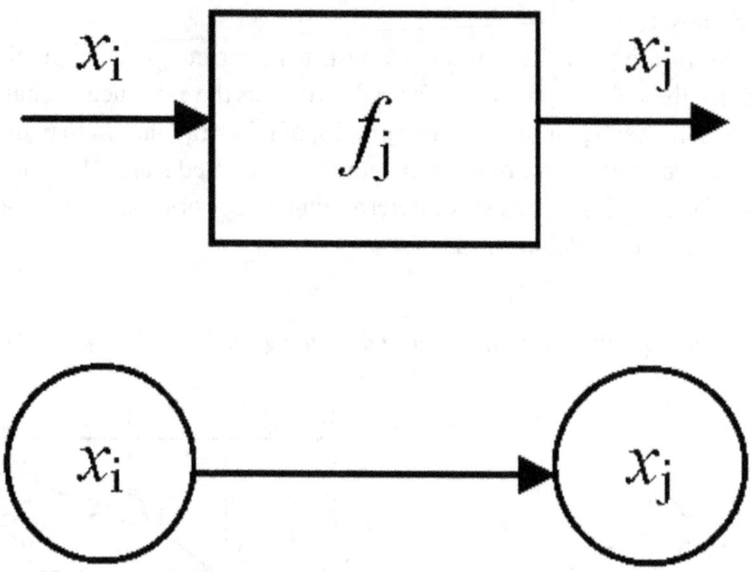

In the formalization of a digital circuit set with a system of logic equations, start (no input arcs), internal and end (no output arcs) nodes will be defined. As a result, the study of the digital scheme can be reduced to a transformation of the directed graph in order to obtain a cyclic graph – it is claimed that then the system has a solution.

Formalization of microprogram processes. It is concerned with finding all possible paths in a microprogram from start to end micro-instruction. It allows estimation of microprogram execution time on different inputs. The abstract representation (Figure 16) uses a directed graph, the nodes of which represent the individual microinstructions, and the arcs are carriers of the governing conditions during a transition. Each particular implementation of the microprogram is an implementation of a path from the input to the output, and the set of possible paths defines all possible implementations of the microprogram. The length of each path can be uniquely related to the time for the particular implementation of the microprogram, and the multiplicity of paths allows a minimum, average, and maximum execution time to be specified.

Figure 16. An example for graph formalization of microprogram

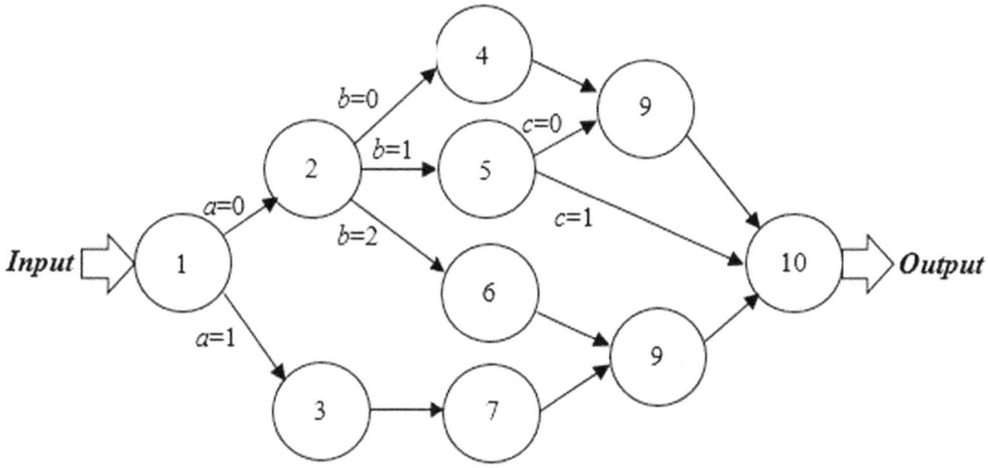

2.3. Finite Automata Formalization

Finite automata formalization of a discrete system describes the structure and functionality as a sequence of states in the separate time moments (Kohavi & Jha, 2009). The time can be regarded as discrete (automata) or continuous (aggregate). The Finite automata approach is applied in (Zozulya et al., 2022) for investigation and testing algorithms and software during the design of microcontroller. A block diagram of the testing system by using the theory of finite automata is presented, which analyses the two levels – software and hardware. For the purpose of the testing itself, a detailed finite state machine is defined that describes the successive transitions between the discrete states of the process.

In discrete mathematics, the so-called *abstract mathematical machine*, presented as a "black box", which reads from the input words over a given alphabet and can output words over another alphabet (not necessarily different from the input). It is characteristic that at each moment of its operation the machine is in a certain state z belonging to a finite set of states Z. The change of one state to another occurs in a discrete (countable) set of moments of time, called clocks. Between two strokes the machine remains in the same state. The new state is determined uniquely by the current state and the input letter that the machine reads at clock time. The output letter (if any) is also a function of the current state and the input letter (Figure 17).

Figure 17. Abstract mathematical machine

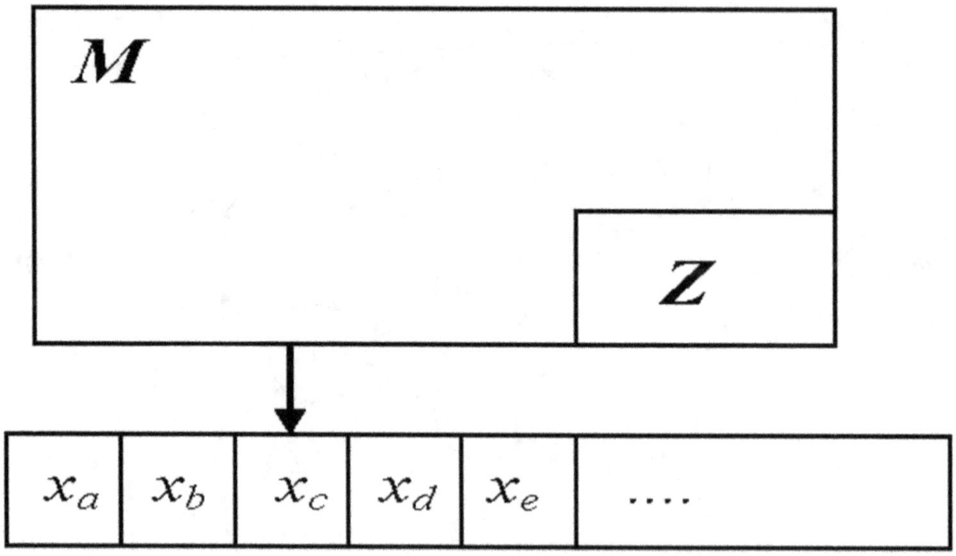

At the start of operation, the abstract machine is always in the same initial state and can terminate when it reaches one of the predefined final states. There are machines that are capable under certain circumstances of never completing their work. Two simple abstract machines are defined – finite deterministic automaton and finite nondeterministic automaton.

Finite deterministic automata (FDA) is called the ordered structure $A = <X, Z, z_0, F_Z, Z^*>$, where: X – finite input alphabet; Z – finite set of states; $z_0 \in Z$ – initial state; F_Z: $Z \times X \to Z$ – partial function of the transitions for the internal states; $Z^* \subseteq Z$ – sets of the final (final) states of the automaton. Visually, FDA is represented by a finite directed multigraph where the elements of Z are the individual nodes and the arcs $x \in X$ connect elements $z_i, z_j \in Z$ for which $F_z,(z_i, x) = z_j$. Figure 18 shows the graph abstract representation of FDA for which $Z = \{z_0, z_1, z_2\}$, $X = [a, b]$, $Z^* = \{z_2\} \subset Z$. The presented formalization of the operation of the automata does not produce an output. This is a special case of the more general concept of "*finite automata*" used in the computer field, which includes both an output alphabet and a function for determining (calculating) the outputs (Vayadande et al., 2022).

Figure 18. An example of finite deterministic automata

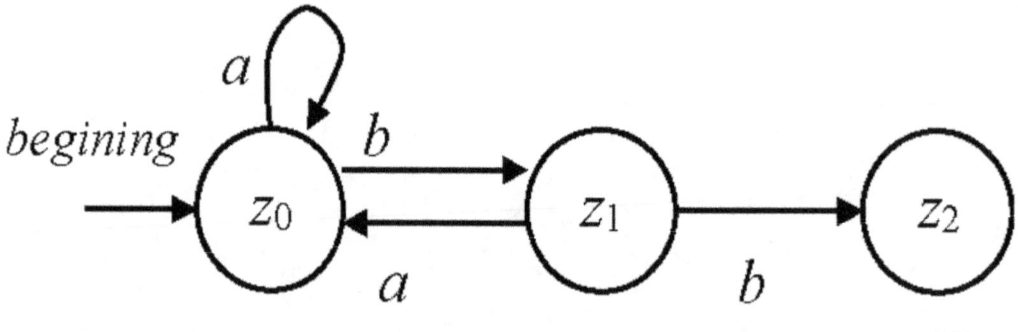

Finite nondeterministic automata (FNA) is defined by an analogous ordered structure, the difference being in setting the function of the transitions: F_Z: $Z \times X \to 2^Z$. His work is associated with the following. When, as a result of calculation of the function F_Z, a set Z' of states to which the FNA must pass is obtained, it is multiplied into úZ'ú copies and each of them passes into one of the states of Z'. Mentally, FNA is found in all states of Z'. With each new read of an input letter, a new FNA propagation follows.

In automatic formalization, a finite state machine with an output function and a time count is usually used. A general definition is $A = <X, Y, Z, z_0, Z^* F_Z, F_Y, T>$, where the new elements are: Y – finite set of output reactions; F_Y – partial function of output reactions; T – a finite set of time instants. If the initial state $z_0 = Z(t_0)$, $t_0 \in T$ is known, it is possible for $\forall t_i \in T$ to determine:

- The current state of the system $z(t) = z_i \in Z = F_Z[X, z_0, Z, T]$.
- The output reactions of the system $Y = F_Y[X, Z, T]$.

In the automatic approach, a set of elements is defined for a given class of systems – *functional automata* Φ_i $(i=1 \div k)$, forming an abstract formal scheme of basic descriptions (elements). At any moment in time, the basic automata is in one of the allowed states Z. For the computer domain, basic computer devices (processor, memory, terminal, peripheral device, channel, etc.) as well as various mathematical objects can be defined as basic automata to generate and convert processes. The action of an automatic function Φ_i (Figure 19) can be described by the sets: $x(t) \in X$ – input signals (external influences); $y(t) \in Y$ – output signals (characteristics) determined by the current state of the automaton through the function of outputs F_Y depending on the mode of operation; $g(t)$ – control signals determining the mode of operation.

In the general case, the functioning of each individual automaton F_i depends on the two functions F_Z and F_Y, which are determined at the current moment by the control effects $g(t)$. The input signals form the so-called **x**-messages over the input alphabet, and the time-ordered set of output signals (**y**-messages).

Figure 19. Formal description of an element in automata formalization

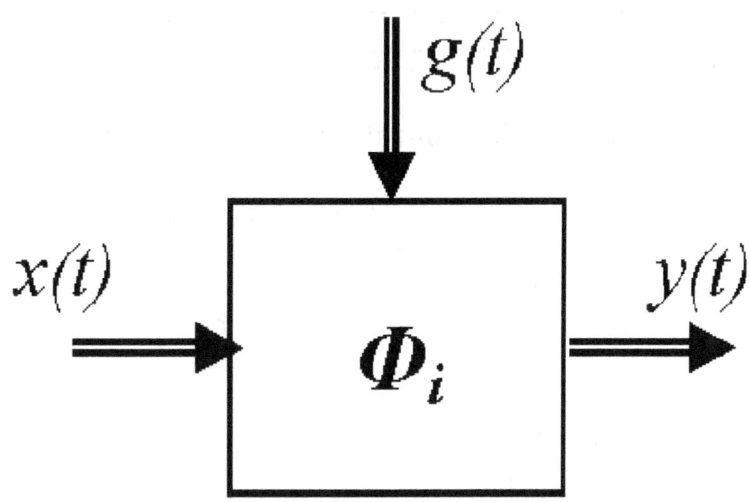

The automata formalization of a system is related to the formation of a set of non-repeating basic automata $\{\Phi_i / i=1,2,\ldots, n\}$, allowing description of the functional behavior. The signals $x(t)$, $y(t)$ and $g(t)$ are considered as multicomponent (composed of elementary signals) transmitted over elementary channels.

An example of abstract description of finite automata formalization of a system is presented in Figure 20. Two streams of output messages Y_1 and Y_2 are formed, and according to their purpose, the individual elements can be defined as follows: Φ_1, Φ_2 – generators of the input stream of tasks forming the general workload (they differ according to the generating algorithm); Φ_3 – dispatcher implementing the scheduling algorithm and resource allocation; Φ_4, Φ_5, Φ_6 – various service computer devices.

Figure 20. Finite automata formalization of a system

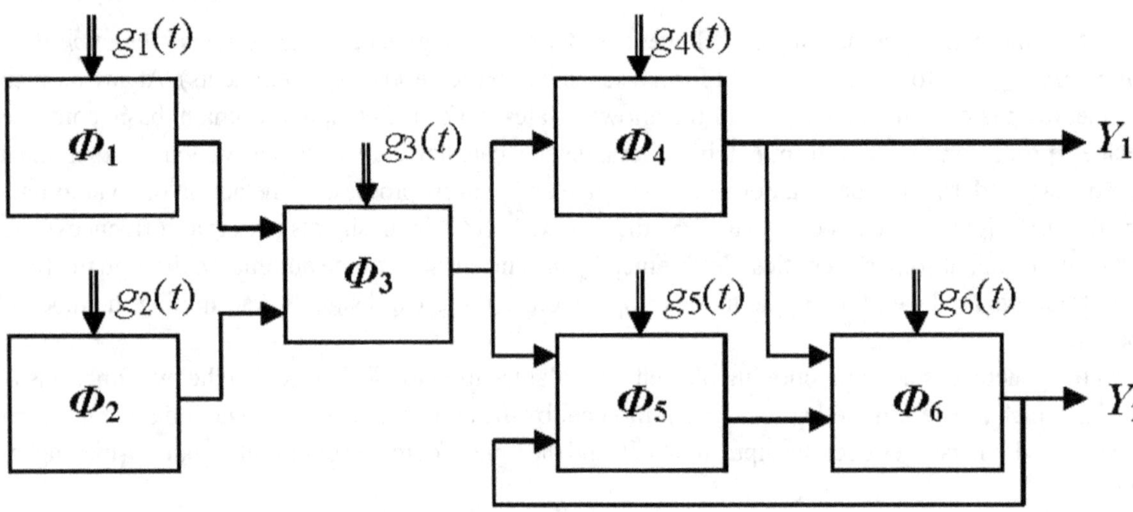

3. APPLICATION OF STOCHASTIC STRUCTURES FOR FORMALIZATION AND ABSTRACT DESCRIBING

3.1. Probabilistic Approach in Formalization of Systems and Processes

Real systems usually have a discrete structure, but the processes occurring in them are probabilistic in nature. That is why the probabilistic approach is applicable in their formalization and allows a more accurate abstract representation of structural connections, transitions between states and the movement of processes. The essence consists in defining probabilistic parameters to reflect the stochastic nature of the processes.

The theory of random processes is a branch of the probability theory for studying regularities in random phenomena in the dynamics of their development (Koralov & Sinai, 2007). Basically, in nature, all processes are more or less random. For processes that are defined as deterministic, the random factor is so small that it can be neglected. In other processes, the probabilistic nature is clearly expressed, and it is necessary to apply a probabilistic approach to the description. In theoretical terms, the random method is a formalized group of sets with a random sequence (Singh et al., 2021), and the paper explores the properties and possibilities for constructing exact conditional tests.

A *random process (RP)* occurring in a system S is a sequence of random transitions from one state of the system to another (Rosenblatt, 2012). Thus, the behavior of the RP is manifested in time and can be written as $X(t)$. Each RP $X(t)$ at a given time $t = t_i$ assumes a specific value $X(t_i)$, representing a random variable (RV), called the section of the RP. Since an RV X is associated with an elementary event $\omega \in \Omega$ (Ω is the space of elementary events), then an RP can be represented as the function $X(t)=\varphi(t,\omega)$, for $\omega \in \Omega$, $t \in T$, where: ω – elementary event; Ω – space of elementary events; T – area (set) of the values of the time argument; Σ – set of possible values of the process $X(t)$.

A particular behavior of $X(t)$ in time is considered as one realization of RP, representing already a non-random function $x(t)$ obtained from successive experiments in time (Figure 21). In the case $x(t)$ is the set of successive realizations $x(k)$ of RV X in time. In several different implementations of RP, the so-called a family of realizations represents the basic experimental material for random process analysis. A study of the evolution of the cumulative distribution function of RPs in dynamical systems is presented in (Lucchesi et al., 2022). A method for determining the evolution is proposed and some uncertainties in the characterizing parameters are determined, and it is proved that the distribution functions are a solution of a partial differential equation.

Figure 21. Random Process Realization

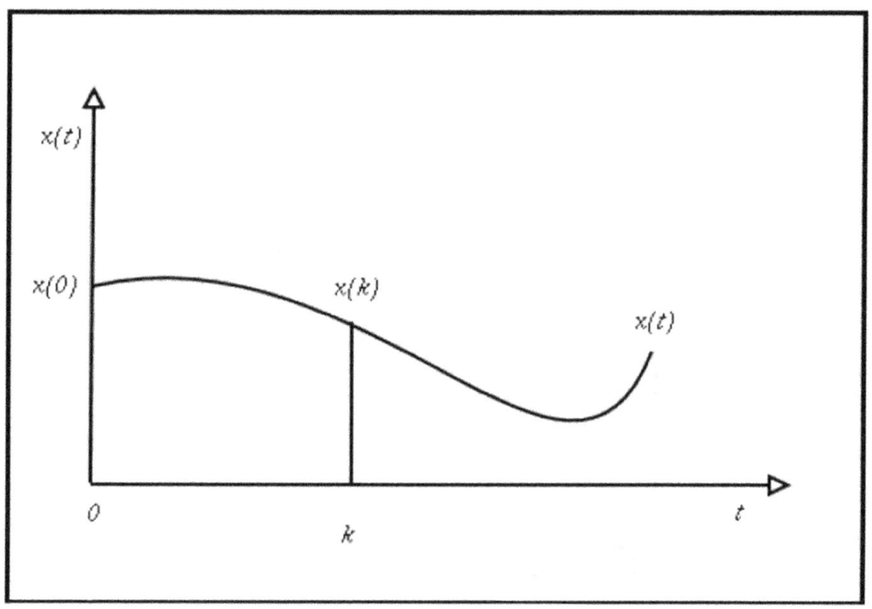

Classification of RPs can be made based on different features, the one presented below is based on the two parameters “time” and “states” (in discrete and continuous forms). Combining the presented 4 categories gives rise to 4 types of random processes in time and states as follows:

- RP $X(t)$ with discrete time – the system in which the process takes place can change its state only at discrete moments of time $T=\{t_1, t_2, \ldots, t_n\}$.
- RP $X(t)$ with continuous time – a transition between states of the system can take place at any moment t of the observed period τ (the set T is uncountable and infinite).
- RP $X(t)$ with discrete states – at each moment of time t the set of its admissible states is finite and countable (at each moment its section is a discrete RP).
- RP $X(t)$ with continuous states – at each moment of time t the set of its admissible states is uncountable (at each moment its section is a continuous RP).

Combining these 4 categories gives rise to 4 types of RP depending on time and states.

In a contemporary aspect, new challenges to stochastic processes are emerging that reflect in other classifications, such as the recently introduced "underspread/overspread" on process spectra with quadratic integrable covariance functions (Kozek & Riedel, 2018). The article reviews the most known definitions of time-varying power spectrum and analyzes the applicability to non-stationary processes. An example is given with the non-stationary Wiener filter, where a time-varying power spectrum can be used analogously to the time-invariant power spectrum of stationary processes.

In practical problems, it is convenient to represent a RP by an elementary random function with argument t, where the dependence on t is usually a non-random function and one or several RVs independent of t are involved as a parameter. A simple example is shown in Figure 22. The elementary random function is $Y(t)=at+X$, where X is RV independent of t; a is non-random variable. Each realization is a straight line parallel to the line $y=at$, at $X=0$.

Figure 22. Realization of an elementary random function

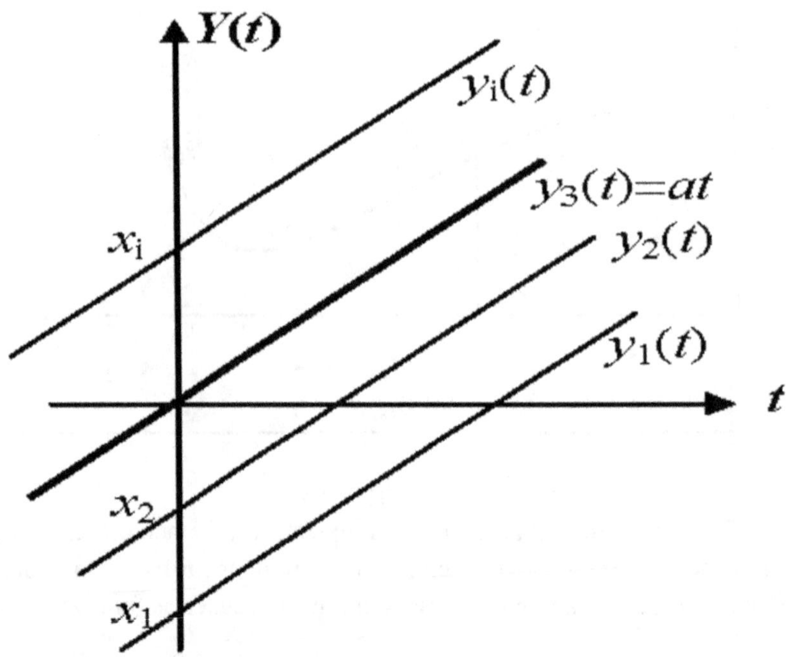

Random streams with discrete states and continuous time are related to *event streams*. In a homogeneous flow of events, the latter differ only in the time of realization (completion). In the flow of events, the concept of "event" is fundamentally different from the concept of "random event" in probability theory. The latter may or may not happen, while with the flow of events they always happen, but depending on the distribution function $F(t)=P(T<t)$. It determines the occurrence of the event within an interval $T<t$ (Figure 23).

Figure 23. Fulfillment of an event by time T<t

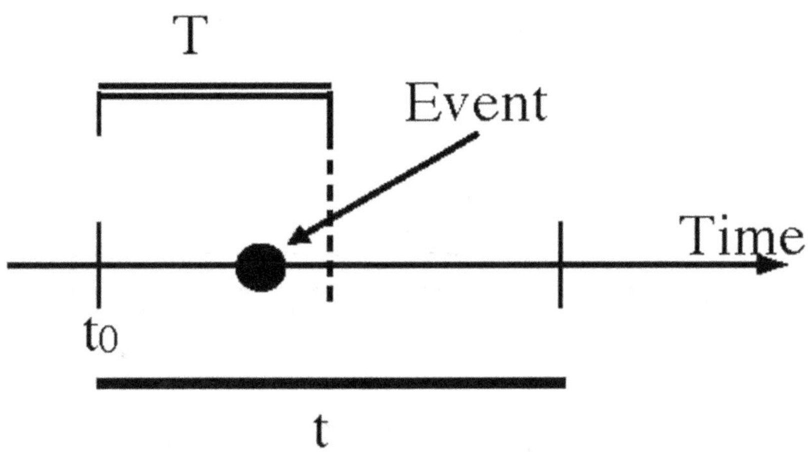

The simplest structure is the regular flow, in which the fulfillment of events takes place in constant time intervals ($\tau=const$). More often, however, one considers a random flow described by the density function $f(t)$ and the distribution function $F(t)$. In the *stochastic formalization* of computer objects, the following type flows apply:

- Stationary flow – the probability of a given number of events occurring in an interval τ depends only on its length without being affected by its location on the time axis (i.e., it does not depend on the start t_0). For a steady flow, the intensity is $\lambda(t)=\lambda=const$ and is independent of time.
- Flow without consequences – events occur at successive moments in time, independent of the others.
- Ordinary flow – the probability that more than one event will occur at the same time is negligibly small.

The behavior of a CS can be formalized by RP $X(t)$, assuming a finite and countable set of states $S = \{S_1, S_2, \ldots, S_i, \ldots, S_n\}$ at discrete moments of time $t_i \in T$. In the successive steps of its implementation, the process will fall into one of the admissible states $S_i \in S$. The abstract description of such a formalization can be done by a *state graph* with vertices covering the set S and edges corresponding to probabilistic transitions.

Figure 24 shows an example of a probabilistic formalization of a multiprogramming CS, including several subsystems – central (processor block) and peripheral. The following stochastic parameters are defined:

μ_0 – service status of task z_i in the processor.
μ_j – service status of task z_i in peripheral subsystem "*j*" (external device).
τ_j – random variables reflecting the service time in the corresponding device j.
p_{ij} – probability of returning task *i* to device *j* (*j*=1÷*k*) after being serviced by the processor; is valid:

$\sum_{j=1}^{k} p_{ij} = 1$ and $p_{i0}=1$

Figure 24. Stochastic formalization of a system

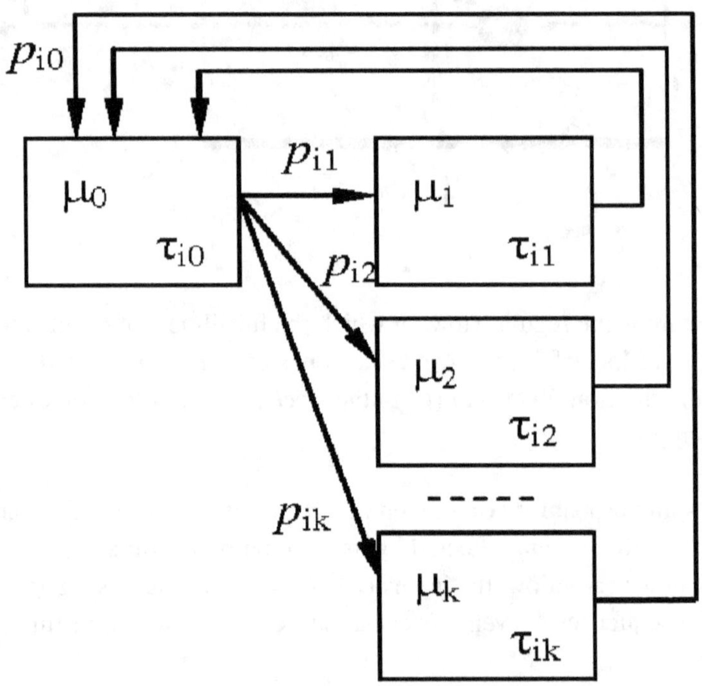

The index "*i*" is used for the sequence number of a task (in particular for dispatching priority). According to the scheme, each current request is first directed to the processor, where it is served for time τ_{0i}, and then (if necessary) it can go to some peripheral subsystem with a certain probability. If the input stochastic flow of tasks $Z=\{z_1, \ldots, z_n\}$ is also considered, it is necessary to define a probability law for its distribution.

An algorithm for developing a mathematical formal model of the functioning of an economic system is proposed in (Schislyaeva et al, 2022). A reason for directing the research to the field of probability is that the activities, or at least one of them, in a given economic environment is related to a random flow of applications. This gives reason to model the operation of such a system as a sequence of operations using the mathematical apparatus of the theory of random processes. The effect of the built mathematical model is an opportunity to forecast the costs and profits in the process of functioning of the system.

3.2. Formalization and Abstract Describing by using Stochastic Pocesses

3.2.1 Application of Markov Processes Theory

The theory of Markov processes is a branch of the theory of probabilities and discusses discrete or continuous stochastic processes and apparatus (Dynkin, 2012; Stroock, 2013). Its application in the description and investigation of computer objects is determined by the fact that their functional behavior is characterized by Markov properties. This is concept that at any moment in time, the probabilistic characteristics of the transition depend on the current state of the process and do not depend on when and how the system got into it. The Markov RP $X(t)$ has discrete states $S=\{s_1, ..., s_n\}$ and is represented by a graph of states (Figure 25), which is like a discrete graph (nodes can be circles or squares), but the arcs have probabilistic values. A state that has only outgoing arcs is called a source (s_1), and a state with only incoming arcs is called an end state (passive/absorbing – s_5). States between which a transition is possible are called adjacent (for example, s3 and s4 are adjacent to s2. States that a process can enter, and exit are called transitive (s_2, s_3, and s_4).

Figure 25. Graph of the states

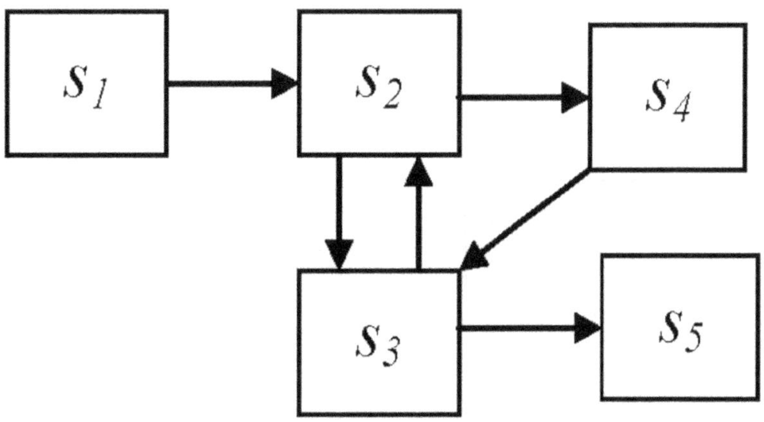

The Markov random process $X(t)$ is a process running in a system with states $s_1, s_2, ..., s_i, ...$, where the probability of transaction $s_i(t_k) \rightarrow s_j(t_{k+1})$ depends only on the probability characteristics of current state. Due to the nature of computer systems and processes, the probabilistic formalization uses RPs with a finite number of states $S = \{s_1, s_2, ..., s_n\}$, for which time is a discrete or continuous quantity.

Markov process with discrete states and discrete time is called a Markov chain (MC), which at each discrete time (step) k the system can be in only one of the allowed discrete states $S(k)=s_j$. Getting into it is determined by the state at the previous moment $S(k\text{-}1)=s_i$ and the transition probabilities $p_{ij}(k)=P[S(k)=s_j$

/ $S(k\text{-}1)=s_i$], which are conditional probabilities for each step k=1,2,... . This permits to formalize the behavior of a process as a discrete sequence <$S(0)$, $S(1)$, $S(2)$,..., $S(k)$,...> for the individual moments (steps) t=0, 1,...,k,... . The definition of a MC includes the next components:

1) Set of the states $S = \{s_1, s_2, ..., s_n\}$.

2) Matrix of transition probabilities $P = \begin{Vmatrix} p_{11} & p_{12} & \cdots & p_{1n} \\ p_{21} & p_{22} & \cdots & p_{2n} \\ \vdots & & & \\ p_{n1} & p_{n2} & \cdots & p_{nn} \end{Vmatrix}$.

3) Vector of initial probabilities $P_0 = \{p_1(0), p_2(0), ..., p_n(0)\}$.

State s_j is reachable from the other state s_i if $\exists k \geq 0$, so that $p_{ij}(k)>0$. In the study of discrete systems, the following are used:

- Homogeneous MC for which is valid $p_{ij}(k)=p_{ij}=const$.
- Ergodic MC – the matrix of transition probabilities allows, after a finite number of steps, to make a transition between each pair of states.
- Regular MC – for step k, the matrix of transition probabilities does not contain a zero element, i.e. $\forall p_{ij}(k)>0$, for i,j=1чn .

An example for MC with set of states $S = \{s_1, s_2, s_3, s_4, s_5, s_6\}$ and vector $P_0 = \{1, 0, 0, 0, 0, 0\}$ is presented in Figure 26 with its matrix of transition probabilities and the graph of the states.

Figure 26. An example for Markov chain

states	1	2	3	4	5	6
1	0	1	0	0	0	0
2	0	0	1	0	0	0
3	0	0	0	0,2	0,5	0,3
4	0	0,6	0,4	0	0	0
5	0	1	0	0	0	0
6	0,7	0	0	0	0	0,3

Matrix of transition probabilities

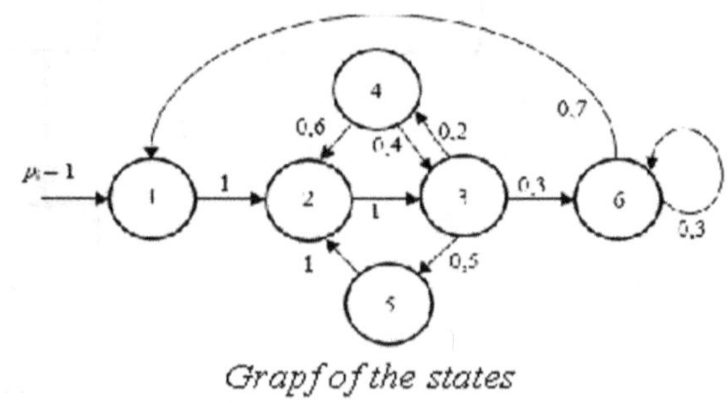

Grapf of the states

An example for stochastic formalization of the processes in DLE from Figures 10 and 11 by using the MC apparatus is presented below.

Formal model is defined as a MC with 7 states – $S = \{s_1, s_2, s_3, s_4, s_5, s_6, s_7\}$ and vector of initial probabilities P_0=(1,0,0,0,0,0,0) which determines the start from the first state. The states have the following meaning for the formalized process: s_1 – generation of a user request for access to an information resource; s_2 – shaping and packaging of the request to send; s_3 – routing and transport through the communication medium; s_4 – processing of received request; s_5 – provision of an information resource for use; s_6 – transfer of information block; s_7 – receiving a block of information and working with the content. The matrix $P[i,j]$ and graph of states are presented in Figure 27.

Figure 27. Formal description of processes in DLE

P	S1	S2	S3	S4	S5	S6	S7
S1	0,3	0,7	0	0	0	0	0
S2	0,1	0	0,9	0	0	0	0
S3	0	0,2	0	0,8	0	0	0
S4	0,3	0	0	0	0,7	0	0
S5	0	0	0	0	0	1	0
S6	0	0	0	0	0,3	0	0,7
S7	0,8	0	0	0	0,2	0	0

Matrix of transition probabilities

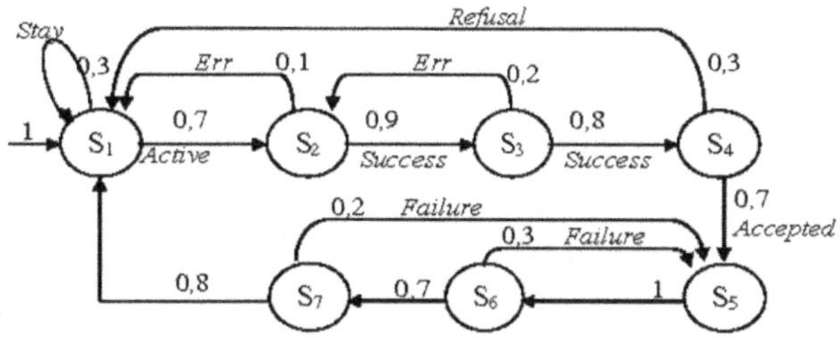

Graph of the states

Markovian RP with discrete states and continuous time is characterized by transactions between states at any time $t > t_0$. They depend only on the probabilistic characteristics of the current state in which the process is located. The transition $S_i \rightarrow S_j$ takes place under the influence of a flow of events with intensity λ_{ij} (most often a Poisson flow). It is assumed that the transition is realized when the first event of the stream occurs.

If the events in the random stream are distributed according to the Poisson law, then the intervals between their successive occurrences are exponentially distributed $F(t)=1-e^{-\lambda\tau}$ (the reverse is also true). A random flow with an exponential distribution is without consequences and fulfills the requirements of stationarity and ordinariness.

Palm flow is also a steady flow of limited consequence. For him, the intervals τ_j between consecutive events are independent uniformly distributed random variables.

3.2.2 Application of Queuing Theory

Queuing theory (QT) deals with the behavior of systems that receive a certain flow of service requests, most often with random time characteristics (Afolalu et al., 2021). The service of a request is realized by providing a system resource for a certain amount of time, after which it is redirected to the next resource or leaves the system if it has been served. An approach based on the QT is suitable for analysis of functionality of multi-stage production line as is made in (Marsudi & Shafeek, 2014). The main goal of this article is to present an analytical approach based on real life data from a company. The research carried out queuing analysis of arriving and leaving distributed data to make an evaluation of the production.

The QT regards the researched object as a service system based on one or more queuing nodes, each of which can be described by the formalization from Figure 28. Such an environment (Queuing System – QS) for service is a time-logical scheme uniting the three categories:

(a) Input flow of requests for service (λ).

(b) Parameters of all service units (servers SUs).
(c) Strategies (disciplines) for request servicing in the separate QS.

The first two categories are temporal and are described by regular or probabilistic dependencies defined by the time between successive requests entering the system and the service time. A service discipline is a logical category that defines the criteria for selecting a current service request, need for buffering, routing of requests, and other features of the QS.

Figure 28. Descritpion of a standard queuing node (QS)

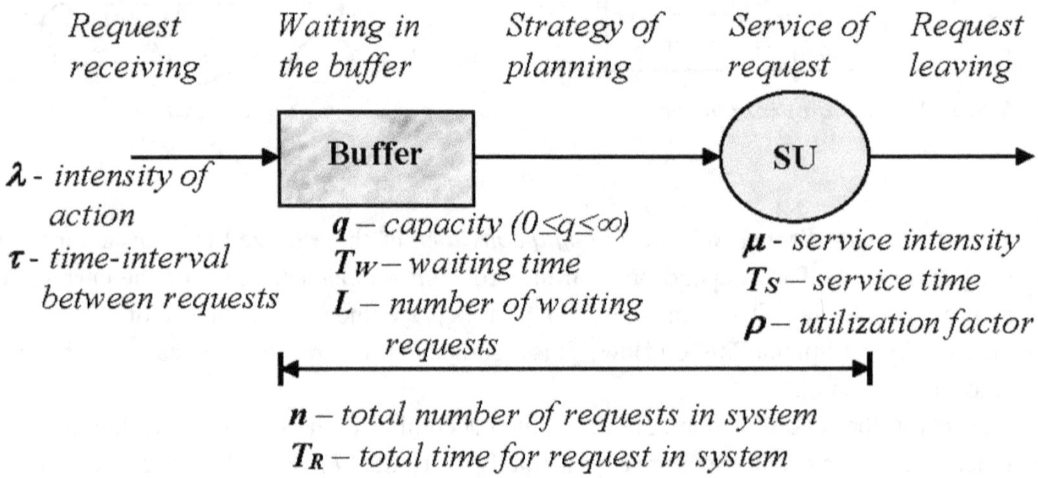

The formalization based on the QT in the investigation of real systems is based on the stochastic nature of their behavior over time. The input stream is considered as a random stream of events $X(t)$, and a single request z_j is represented by $X(j, t_j)$, where j is the number of the request and t_j is the starting time of its entry into the QS. A Poisson random flow with exponentially distributed inter-request intervals is most used. To simplify the investigation, as well as under some specific operating conditions of real systems, service can be considered as a deterministic (regular) flow.

The formalization and abstract representation through the means of QT requires building the static structure as a network of QSs (so called queuing network QN) and defining the actual parameters for all participating components. The second stage is defining the dynamic structure formed basically by the intensity of workflows λ_j (if they are more than one) and the intensity of service flow μ_i for each included QS(i). To specify each queuing node in the structured network, the ordered group of symbols A/B/k/N/z is used (the last two being for extended description). The meaning of the symbols is as follows: A – input flow distribution law; B – service distribution law; k – number of service channels in QS; N – total number of requests in the QS; z – number of request sources. The distribution laws for fields A and B are denoted: M – exponential distribution; Er – Erlangen distribution of the r-th order; D – deterministic (regular) distribution; G – distribution of general type.

The structure and behavior of each QS included in the network is mainly determined by two basic parameters. The first is the capacity q of the buffer, and the following options are possible (Figure 29):

QS with failures (q=0); QS with limited buffer at fixed value $q=n=const>0$; QS with infinite buffer (queue type FCFS/FIFO) at $q=\infty$.

Figure 29. Types of QS determined by the buffer capacity q

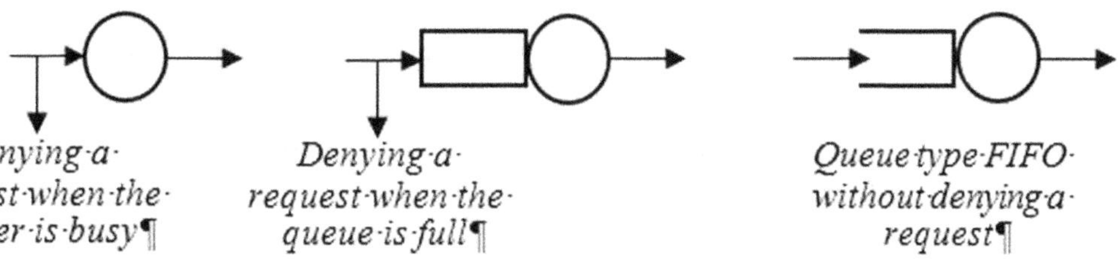

The second parameter is the number k of service channels – single-channel QS (k=1) or multi-channel QS (k>1). In a multi-channel QS (Figure 30), all servers have the same probabilistic parameters and behavior, and their workload is determined by the chosen strategy of serving requests from the common buffer.

Figure 30. Multi-channel (multi-server) QS with infinite buffer

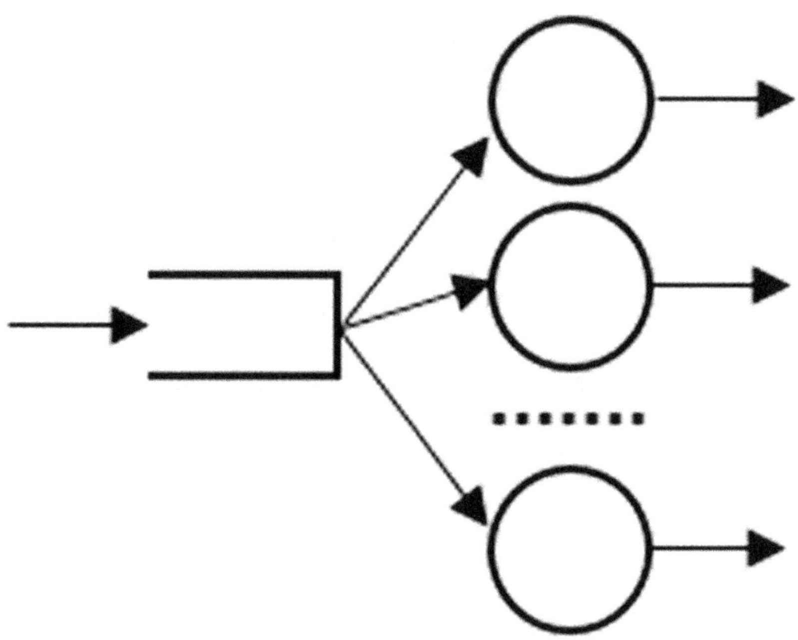

The formalization of a system is implemented as a stochastic network with set H of QSs where $|H|=m\geq 2$ and determined relations R_i between elements. It is written in the form $<H, R_1, R_2, ..., R_m>$ and represents the topology of the network and the relationships between individual QSs. Two main types of stochastic networks are defined (Figure 31):

- Open network, where the input stream (workflow) is generated from an external source without considering the current time of the stochastic network. Discussion of this type networks is made in part 20.10 of (Winston, 2022).
- Closed networks, in which the intensity of the input flow is determined by the internal state of the network, maintaining a constant number of units (requests) – a new request is generated only after completing the service of another request in the network. Discussion of this type network is made in part 20.13 of (Winston, 2022).

Figure 31. Open and closed stochastic networks

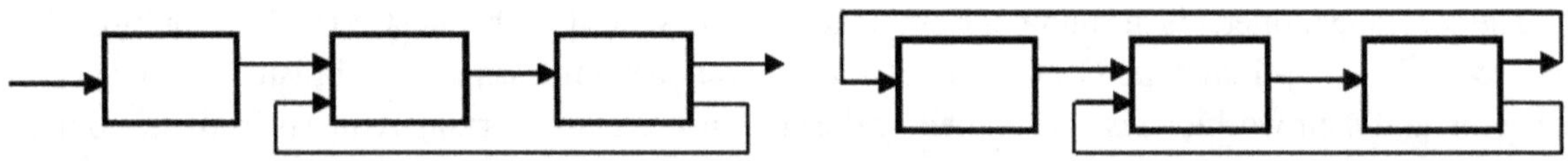

For each CS or individual device, an abstract formal description can be drawn up as a stochastic network with a finite and countable set H of single-channel and/or multi-channel QSs and certain relations R_1, assumptions, and limitations.

There is a lot of research that applies the QS apparatus and queuing networks to initial formalization of the object or process under study. The rationale is that they are probabilistic in nature and are based on the arrival of requests that are served in a server device according to appropriate rules. Typically, MC theory is applied to model and conduct experiments on such systems, which offers a good basis for this. One such case is discussed in (Callegati et al., 2022), where a QS model based on Markov chain with a “Death and Reproduction” (“Birth–Death”) process is presented. The paper confirms that the use of queuing networks is suitable for studying various processes and scenarios where the request flows are of Poisson type and the service is an exponential process. This apparatus will be discussed in Chapter 8 of this book.

The same approach has been applied in the formalization of autonomous air traffic management for urban air mobility presented in (Paul et al., 2021). It is pointed out that in these processes, time is a critical component, and to ensure the correctness of the study, a two-phase distributed knowledge dissemination protocol is used. The formalization and subsequent modeling of delay research is through single-server (single-channel) QS with exponential flows of incoming requests and service (QS type M/M/1). This allowed the probabilistic bound on the development time to be examined as a function of the probabilities of the total delays for which the tolerances were set.

REFERENCES

Afolalu, S. A., Ikumapayi, O. M., Abdulkareem, A., Emetere, M. E., & Adejumo, O. (2021). A short review on queuing theory as a deterministic tool in sustainable telecommunication system. *Materials Today: Proceedings*, *44*, 2884–2888. doi:10.1016/j.matpr.2021.01.092

Ahmadi, N., Chakhchoukh, Y., & Ishii, H. (2021). Power systems decomposition for robustifying state estimation under cyber attacks. *IEEE Transactions on Power Systems*, *36*(3), 1922–1933. doi:10.1109/TPWRS.2020.3026951

Atanasov, V. (2021). RFC in approach model paradigm. *International Journal on Information Technologies and Security*, *13*(4), 15–24.

Bargiela, A., & Pedrycz, W. (2022). Granular computing. In *Handbook on computer Learning and Intelligence* (Vol. 2, pp. 97–132). Deep Learning, Intelligent Control and Evolutionary Computation.

Brehmer, M., & Munzner, T. (2013). A multi-level typology of abstract visualization tasks. *IEEE Transactions on Visualization and Computer Graphics*, *19*(12), 2376–2385. doi:10.1109/TVCG.2013.124 PMID:24051804

Callegati, F., Cerroni, W., & Raffaelli, C. (2022). Formalizing the queuing system: State diagrams and birth–death processes. In *Traffic Engineering: A Practical Approach* (pp. 45–63). Springer International Publishing.

Decardi-Nelson, B., & Liu, J. (2022). Computing control invariant sets of nonlinear systems: decomposition and distributed computing. *arXiv preprint arXiv:2205.05622*.

Dynkin, E. B. (2012). *Theory of Markov processes*. Courier Corporation.

Fernandes, J., Ferreira, F., Cordeiro, F., Neto, V. V. G., & dos Santos, R. P. (2019, July). A conceptual model for systems-of-information systems. In *2019 IEEE 20th International Conference on Information Reuse and Integration for Data Science (IRI)* (pp. 364-371). IEEE. 10.1109/IRI.2019.00063

Fuchs, M., & Riesen, K. (2022). A novel way to formalize stable graph cores by using matching-graphs. *Pattern Recognition*, *131*, 108846. doi:10.1016/j.patcog.2022.108846

Gao, Z., Wang, B., Feng, J. E., & Li, T. (2021). Finite automata approach to reconstructibility of switched Boolean control networks. *Neurocomputing*, *454*, 34–44. doi:10.1016/j.neucom.2021.05.019

Gazzari, R. (2022). Formal theories of occurrences and substitutions. *The Bulletin of Symbolic Logic*, *28*(2), 261–263. doi:10.1017/bsl.2021.53

Gutman, A. (2021). Boolean-valued set-theoretic systems: General formalism and basic technique. *Mathematics*, *9*(9), 1056. doi:10.3390/math9091056

Kohavi, Z., & Jha, N. K. (2009). *Switching and finite automata theory*. Cambridge University Press. doi:10.1017/CBO9780511816239

Koralov, L., & Sinai, Y. G. (2007). *Theory of probability and random processes*. Springer Science & Business Media. doi:10.1007/978-3-540-68829-7

Kozek, W., & Riedel, K. (2018). On the underspread/overspread classification of random processes. *arXiv preprint arXiv:1803.05582.*

Larsen-Freeman, D., & DeCarrico, J. (2019). Grammar. In *An introduction to applied linguistics* (pp. 19–34). Routledge. doi:10.4324/9780429424465-2

Lecci, M., Polese, M., Lai, C., Wang, J., Gentile, C., Golmie, N., & Zorzi, M. (2020, December). Quasi-deterministic channel model for mmWaves: Mathematical formalization and validation. In *GLOBECOM 2020-2020 IEEE Global Communications Conference* (pp. 1-6). IEEE.

Lucchesi, M., Pintucchi, B., & Zani, N. (2022). The evolution of the law of random processes in the analysis of dynamic systems. *Meccanica, 57*(10), 2553–2565. doi:10.100711012-022-01589-3

Maggiora, G., & Vogt, M. (2021). Set-theoretic formalism for treating ligand-target datasets. *Molecules (Basel, Switzerland), 26*(24), 7419. doi:10.3390/molecules26247419 PMID:34946500

Marsudi, M., & Shafeek, H. (2014, January). The application of queuing theory in multi-stage production line. In *Proceedings of the 2014 International Conference on Industrial Engineering and Operations Management Bali*, Indonesia.

Mayr, H. C., & Thalheim, B. (2021). The triptych of conceptual modeling: A framework for a better understanding of conceptual modeling. *Software & Systems Modeling, 20*(1), 7–24. doi:10.100710270-020-00836-z

Munir, S., Jami, S. I., & Wasi, S. (2020). Knowledge graph based semantic modeling for profiling in Industry 4.0. *International Journal on Information Technologies and Security, 12*(1), 37–50.

Paul, S., Patterson, S., & Varela, C. (2021). Formal guarantees of timely progress for distributed knowledge propagation. *arXiv preprint arXiv:2110.12587.*

Romansky, R. (2017). *Information servicing in distributed learning environments. Formalization and model investigation.* LAP LAMBERT Academic Publishing.

Romansky, R., & Noninska, I. (2021a). Investigation of communication parameters in multicomputer architecture with ring topology. *Proceedings of the 2021 IEEE International Conference on Information Technologies (InfoTech-2021)*, (pp. 119-123). IEEE. 10.1109/InfoTech52438.2021.9548514

Rosenblatt, M. (2012). *Random processes* (Vol. 17). Springer Science & Business Media.

Savyanavar, A. S., & Ghorpade, V. R. (2021). Node classification model for on-the-fly computing based mobile GRIDS using rough set theory. *International Journal on Information Technologies and Security, 13*(3), 15–26.

Schislyaeva, E., Vasileva, N., Grigoriev-Golubev, V., Evgrafova, I., & Belopolsky, V. (2022). *Modeling the management of an economic system based on random processes. Networked Control Systems for Connected and Automated Vehicles* (Vol. 2). Springer International Publishing.

Singh, B., Jalil, N. A., Sharma, D. K., Steffi, R., & Kumar, K. (2021, March). Computational systems overview and Random Process with Theoretical analysis. In *2021 7th International Conference on Advanced Computing and Communication Systems (ICACCS)* (*Vol. 1*, pp. 1999-2005). IEEE. 10.1109/ICACCS51430.2021.9441739

Stroock, D. W. (2013). *An introduction to Markov processes* (Vol. 230). Springer Science & Business Media.

Thibault, J. (2022). *Constraint System Decomposition* [Doctoral dissertation, Inria Rennes]. doi:10.13140/RG.2.2.13004.49285

Vayadande, K. B., Sheth, P., Shelke, A., Patil, V., Shevate, S., & Sawakare, C. (2022). Simulation and testing of deterministic finite automata machine. *International Journal on Computer Science and Engineering*, *10*(1), 13–17.

Wang, S., Zhang, X., Ye, P., Du, M., Lu, Y., & Xue, H. (2019). Geographic knowledge graph (GeoKG): A formalized geographic knowledge representation. *ISPRS International Journal of Geo-Information*, *8*(4), 184. doi:10.3390/ijgi8040184

Winston, W. L. (2022). *Operations research: applications and algorithms* (4th ed.). Cengage Learning.

Zozulya, M.M., & Kravets, O.Ja., Atlasov, I.V., Aksenov, I.A., Bozhko, L.M. & Rahman, P.A. (2022). Algorithmization of the software testing system based on finite automata. *International Journal on Information Technologies and Security*, *14*(1), 77–86.

Chapter 6
Features and Aspects of Functional Modeling

ABSTRACT

The chapter discusses the possibilities for functional modeling of systems and processes, presenting the features of the used tools. It is indicated that this is an important element of the conceptual analysis of behavior performed in system design. The objects of discussion are functional modeling of system resources, study of algorithmic structures and information processes. The first part is based on the understanding that the structure of a computer device or system can be represented as a functional diagram (synchronous or asynchronous network of functional blocks). Functional modeling of algorithmic structures is based on the study of the logical scheme of an algorithm and its transformation into an ordered form. In addition, the application of D-cards proposed by Dijkstra, and discrete and stochastic graph description are presented. The third part deals with functional modeling of information processes and structures and discusses the opportunities of IDEF technology, data flow diagrams, dialog transition networks, generalized net, and UML apparatus.

1. FUNCTIONAL MODELING IN SYSTEM RESOURCE INVESTIGATION

Functional modeling is an important element of the conceptual analysis of the behavior of systems and processes. It is related to the representation of the general architecture (the static structure of the object) and the dynamics of the developing processes, the flows of information objects and the events that occur. In this sense, functional modeling is related to the study of the functional behavior of the object, considering it as a set of actions. It is accepted that this is necessary to ensure that the design process is carried out correctly (McIntire et al., 2016). The formalism in the description of the action implies the definition of basic structural elements (functional blocks) through which the descriptive functional diagram is built. The latter graphically represents the functional behavior of the object through functional blocks connected by pointed arrows. This is especially useful when designing a new object for which many of the features are a process of refinement and development.

DOI: 10.4018/978-1-6684-8947-5.ch006

The main tasks facing functional modeling are in two directions. The first is related to the study of the system computer resource as an interconnected set of hardware components. The second is the functional study of computer processing, i.e., of algorithmic and program components for organization and management of data processing. In essence, these are tasks of system analysis, considering the mutual influence of individual subsystems and devices. As a stage of the design of a system or a technological process, in the functional modeling it is necessary to reflect mainly the significant characteristics and regularities of the behavior, while idealizing others considered secondary. The rationale for this approach lies in the purpose of the research itself – to confirm the correct functionality of the object being developed.

The consideration of two main directions in functional modeling is also confirmed in (McIntire et al., 2016), where a behavioral model and a functional model are defined and an explanation of the relationship between them is given. The behavioral model describes the behavior of specific system components relative to functions in the functional model. On each structural element of the functional flow, a behavior is defined according to the investigated operating modes. A functional model is viewed as a graph in which each functional element is a state machine. The ability to construct a custom functional model with consistent behavior as graphical Python NetworkX description objects is indicated.

To provide a common mechanism and its easy application, a functional modeling framework has been defined (Williams, 2020). Complex systems research and software engineering techniques were used in the development, with the goal being to reflect the full range of important site functions. The goal is to be able to represent even functions for which the implementation mechanisms are not known. The paper discusses the application of the framework for modeling biological functionality that can be assumed to be dynamically stable. This assumption makes it possible to formulate a set of characteristics by which to determine the commonality between functional components in a wide range with other, including non-biological, systems. The discovery of such similarity allows physical and mathematical means to be applied to represent a research model.

The functional study of computer processing is discussed in (Atoum, 2019), where it is specified that successful software development depends on the adequacy and capabilities of the defined requirements. It is emphasized that the latter is a voluminous process, especially for complex software applications, which takes considerable time and can lead to error-prone solutions. In order to overcome this problem, the article proposes "*a scalable operational framework to learn, predict, and recognize requirements defects using semantic similarity models and the Integration Functional Definition methods*". The framework enables automation of the model validation process and ensures better performance of the software engineering solution. The development of IT in the direction of artificial intelligence requires an extension of the notion of functional modeling, which was done in (Rao & Reimherr, 2023) with the introduction of a new class of nonlinear models for functional data. Neural networks have been used, stating that deep learning is also very successful in non-linear modeling. Two variants of a functional neural network framework are proposed – with continuous hidden layers and with base extensions of hidden layers. Both frameworks for oriented cam structures using functional data, with functional gradient based optimization algorithm used for verification.

Structurally, a computer device or system can be represented as a functional diagram built as a synchronous or asynchronous network of basic functional blocks. Each block operates with input and output variables of two types - information and control, which can be represented as Boolean vectors. Blocks can perform: ✓ transfer – input information vectors and output information vectors; ✓ processing / storage – input information and control vectors and output information vectors; ✓ control – input information and control vectors and output control vectors.

An example of a functional memory description is given in Figure 1. Addresses are assumed to have a fixed length, measurable by the capacity of the address register, and data is *k* bytes in size and can be located in the data I/O buffer (register). In this case, the two ADDR and BUF blocks process input and output information vectors, and the MEM block, in addition to information vectors, also processes control vectors R and W (the vectors have one element – Boolean variables).

Figure 1. Functional model of memory

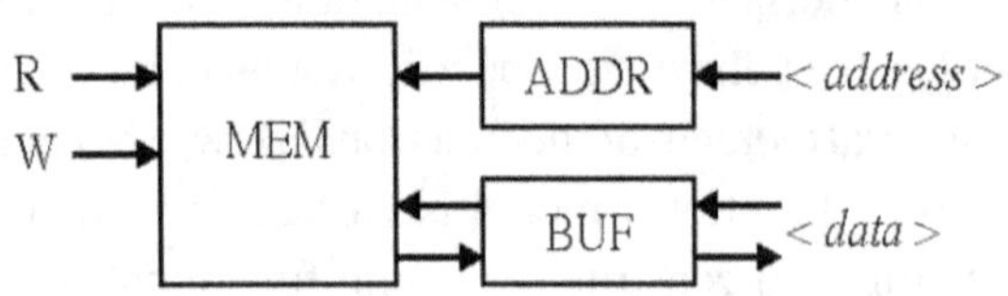


```
MEMORY
  LOOP:  if W∨R=1 then START else LOOP
  START: ADDR:=<address>
         BUF:= <data>
         if R=1 then BUF := MEM(ADDR÷ADDR+k-1)
         if W=1 then MEM(ADDR÷ADDR+k-1) := (BUF)
end_MEMORY
```

In a functional aspect, a CS is viewed as a sequence of transitions between a finite number of states. Transitions between states are assumed to be instantaneous, with the specific values for system parameters $\{s_i\}$ and characteristics $\{y_i\}$ being determined by the functional algorithm A_F and the set of input effects $\{x_i\}$. A suitable apparatus for this is the State Transition Network (STN) discussed in Chapter 2 as a toll for preliminary formalization. There was presented an example of the formalization of the process for executing a program object (task) under the control of the system software (Figure 8, Chapter 2). This can be taken as the initial phase of further functional modeling to analyze the correctness of processing.

The graph approach in functional modeling is used in the description and study of topological problems in parallel and distributed systems, as well as in the description of routing in computer networks. One of the commonly implemented routing protocols is OSPF (Open Shortest Path First), where the shortest routes are discovered. The main part of it is the reporting of the state of the connection line. During the initialization the router determines the values (weights) of the lines for all its own network interfaces and sends them to the other routers in the network. Thus, each router has information about the load on all lines and can make a current assessment of the network topology, allowing the determination of the shortest route (path). Topology information is maintained in routers in the form of a database that is continuously updated by sending and receiving packets. Upon receiving a packet, the router must send an acknowledgment.

The functional behavior of the network under the OSPF protocol can be described by a directed weight graph with two types of nodes and arcs:

- Nodes:
  a) type "router"(R - ■) and type "host" (H - □);
  b) type "network" (N - ●).
- Arcs:
  a) for connection between two nodes type "router";

b) for connection between node type “router” and node type “network”.

An example network topology is given in Figure 2a, and its graph functional description with representation of only part of the weights (σ, τ, ρ) is given in Figure 2b.

Figure 2. Application of graph functional modelling in routing investigation

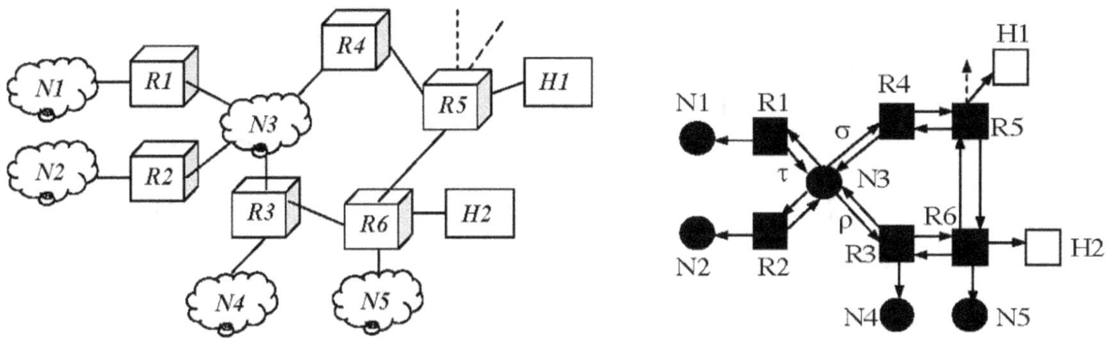

(a) network topology *(b) graph model based on OSPF*

In the transition from the topological model to the graph functional description, the following transformations are used:

- two point-to-point routers are represented as nodes connected by a pair of oppositely directed arcs (one for each direction) – for example, R3 and R6;
- if several routers are connected to a given network, the links are represented by pairs of arcs – for example, N3 and (R1, R2, R3, R4);
- if one router is connected to a given network, the network is represented as an end subscriber with a single directed arc – for example, N4;
- an end subscriber of the “host” type can connect directly to a router - for example, H1 and H2;

A database corresponding to the formed directed weight graph is stored in each router and allows to determine the minimum cost path (via Dijkstra’s algorithm) to each network.

An application of Dijkstra’s algorithm for selecting the shortest path in a network is illustrated by the following example. The network topology is represented by an undirected weight graph (Figure 3), with the values of the corresponding connection lines given on the arcs.

Figure 3. An example for investigated network topology

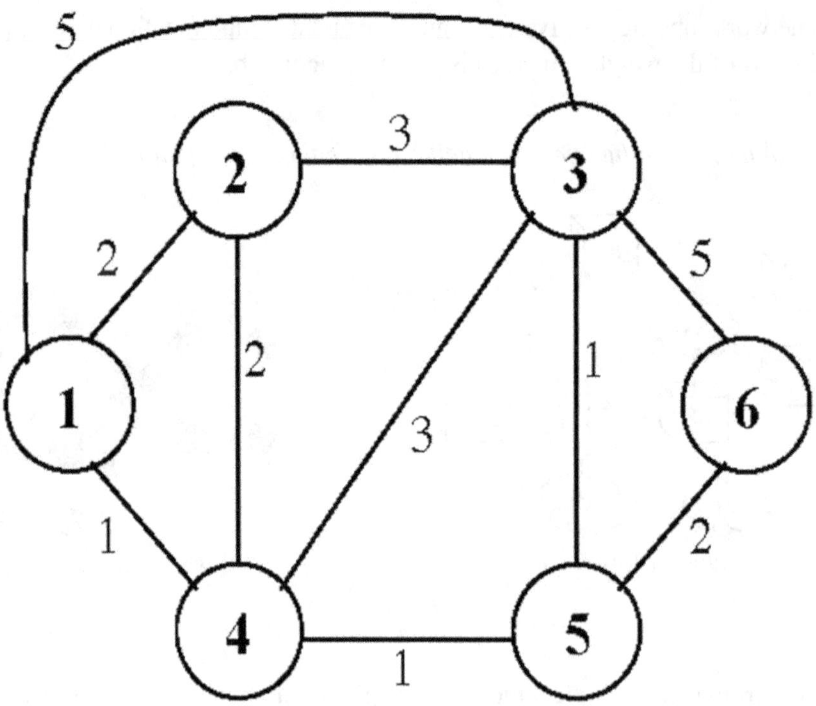

The task is to determine the shortest paths from node 1 (source) to the remaining nodes in the structure. Let the distance from node (1) to node (v) be denoted by $D(v)$ and $L(i,j)$ be the value of the path between a pair of nodes <*i,j*>. The algorithm is carried out in two parts - initial and iterative.

- Initial establishment – $N=\{1\}$ and for $\forall v \notin N$ we define $D(v)=L(1,v)$;
- Sequence of steps – define a node $w \notin N$ with $D(w)$=min and include it in N, then the value $D(w)$ is updated for all other nodes not belonging to N by the operator

$$D(v) \leftarrow \min[D(v), D(w) + L(w, V)].$$

Termination condition – all nodes are included in N.

The result of the algorithm is summarized in Table 1, with the absence of a direct relationship marked by ∞. For each step, a node with $D(v)$=min is marked, which is connected to N. When updating the values, it starts from the current composition of N.

Table 1. Results from functional modeling

Step	D(2)	D(3)	D(4)	D(5)	D(6)	N
Beginning:(1)→N	2	5	1	∞	∞	{1}
1	2	4	**1**	2	∞	{1, 4}
2	2	3	1	**2**	4	{1, 4, 5}
3	**2**	3	1	2	4	{1, 2, 4, 5}
4	2	**3**	1	2	4	{1, 2, 3, 4, 5}
5	2	3	1	2	**4**	{1, 2, 3, 4, 5,6}

Based on the results of the algorithm operation, the minimal path tree shown in Figure 4 is built. This happens, as when a given node is included in *N*, it is joined with the already included nodes. Through the formed tree, a table of routes is defined, showing in which direction the packet should be directed when it is transmitted to a certain recipient. For example, in transmission (1)→(3), packets must be routed to node (4) first.

Figure 4. Tree of minimum paths

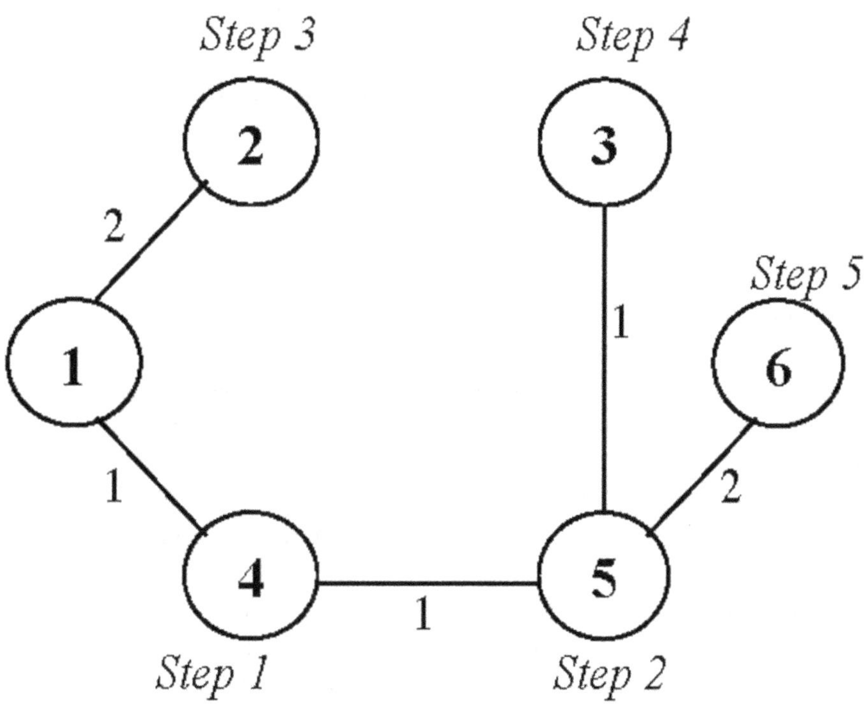

Another point of view is the application of functional modeling in non-traditional structures for e-learning, for example heterogeneous environment based on cloud computing, social media, etc. (Figure 5). The main groups of participants in such an environment are teachers and learners. To them should be added the group of employees supporting learning processes and profiles with learning and personal data, as well as processes in the network space. Representatives from this third group are directed to the main tasks of data protection in personal profiles, which is the object of a system for managing access to information resources (Romansky & Noniska, 2019).

Figure 5. Conceptual model for functionality of combined e-learning environment

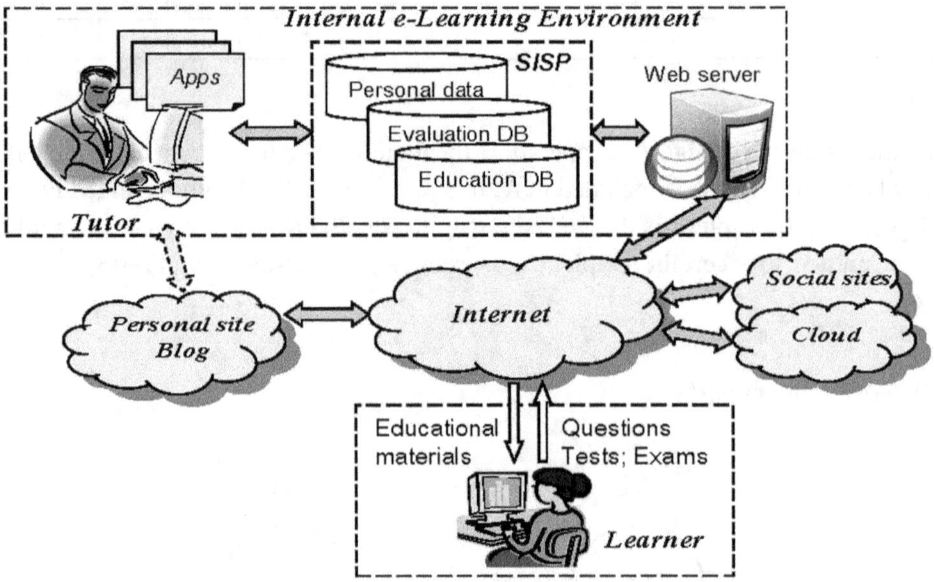

The main participants in the educational processes have access to various learning resources (media, educational files, digital library, etc.), and the supported functional processes are the following categories: ✓ communication between participants; ✓ remote access of learners to learning resources; ✓ activities of the moderator to prepare learning resources; ✓ assessment procedures. All these processes use databases that must be reliably protected against unauthorized attacks and destruction. For this purpose, the access control functions are distributed between the two main sub-systems. Front office sub-system is a web portal that supports user's access to e-learning services. The main functions are user's identification, legitimate access checking, registration procedure for a new user and authentication. Additional supported procedures are: ✓ registration of each login (time, IP address and relative attributes); ✓ supporting audit file for any access; ✓ collecting statistical data and access parameters. Back-office deals with basic administrative procedures directing requests to a chosen component of the system for processing. The input gate for this sub-system is the module for analysis of requests and their distribution. Other functions are checking right for access and resources use, and processing of all correct requests. Functional model of the combined e-learning system is presented in Figure 6.

Figure 6. Functional model of the access to the system resources

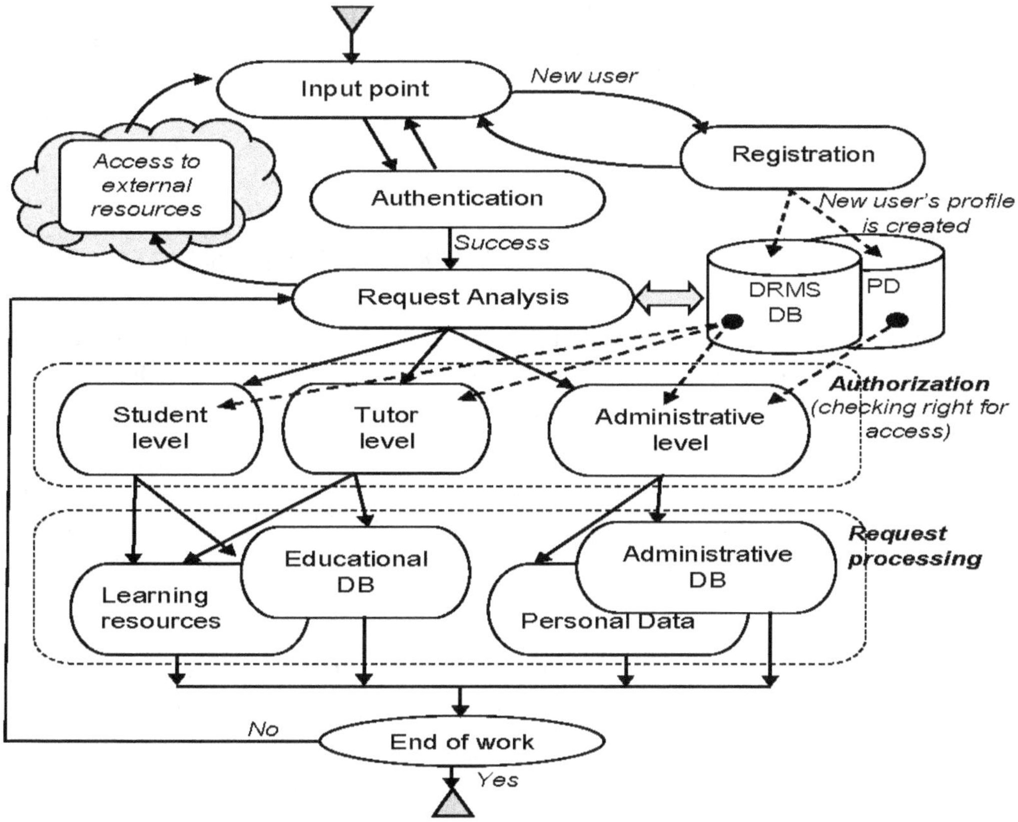

The following groups determine the general functionality of the studied system.

-Functionalities of the Front Office sub-system: *input point* (access to official portal of e-learning environment); *registration* (every new user should make a registration before accessing and using different resources and services); *authentication* (procedure able to guarantee secure and reliable access to environment for registered users);

-Functionalities of the Back Office sub-system: *request analysis* (preliminary analyses of user's request based on checking user's profile with Personal Data – PD and level of rights for accessing internal or external learning resources based on Digital Right Management System – DRMS principles); *authorization* (determining user's status and defined rights to use requested information – learning resource, evaluation information, administrative information, personal data processing, etc.); *request processing* (realization of a permitted request after successful authorization);

-Functionalities of collaborated external resources: they unite spaces in the field of cloud computing (services IaaS, SaaS, PaaS, data centers), social computing (social media, social networks, social aggregators, blog/microblogs, forums, etc.) and personal specialized web sites. All these spaces and technological opportunities could be used by rent, so it is accepted that they are reliable protected, applying recommendable information security and privacy protection standards as an obligation of service owner and/or service provider.

The functionality of formalized combined e-learning environment can be generalized as a family of discrete-time stochastic processes each of them described by a random process *X(t, Φ, Θ)*, were *t* is an ordered increased set of discrete time points $t_0 < t_1 < \ldots < t_n \in T$ (*T* is a parametric space), *Φ* represents all random factors of the system and *Θ* is a set with all possible design parameters. The realization of discrete-time stochastic process is a sequence of states $X(t_i)$ for discrete moments $t_i \in T$, for i=0,1,2,…,n. This will determine the next phase of investigation – discrete or stochastic modeling.

2. FUNCTIONAL MODELING IN THE STUDY OF ALGORITHMIC STRUCTURES

The organization of computer processing is related to providing computing resources to active applications (tasks). Functions for this are taken over by system programs. In execution, each task goes through different states depending on the functional behavior of the computer. This is determined by the so-called complete functional algorithm A_F – set of separate individual algorithms $A_j \in A_F$. For the functional modeling of computer processing (modeling of the global algorithm) it is necessary to give an unambiguous description of the functional relationship between the individual algorithms, their sequence and the conditions of execution. An abstract description of computer processing at the algorithm level can be done through the diagrams below.

Logical scheme of algorithm (LSA) – the execution of each individual algorithm in the composition of the global A_F is associated with a certain condition a_{ij}: $A_i \to A_j$. The condition a_{ij} evaluates the possibility that an algorithm A_j will be executed after an algorithm Ai has already been executed. The formal record of the LSA is as follows ($A_i \in A_F$, for $\forall i = 1 \div N$; and N=úA_Fú):

$$A_i \to \bigvee_{j=1}^{N} a_{ij} A_j = a_{i1} A_1 \vee a_{i2} A_2 \vee \ldots \vee a_{iN} A_N .$$

Matrix scheme of algorithm (MSA) – square matrix with dimension N=úA_Fú, containing the conditions a_{ij} for each pair $A_i \to A_j$ (Figure 7a).

Graph scheme of algorithm (GSA) – is set by an adjacency matrix which is essentially MSA (Figure 7b).

Figure 7. Schemes for general functional algorithm describing

	A_1	A_2	A_3	A_4	A_5	A_6	A_7	A_8
A_1		a_{12}					a_{17}	
A_2			a_{23}	a_{24}	a_{25}	a_{26}		a_{28}
A_3						a_{36}		
A_4								
A_5								
A_6								
A_7								
A_8					a_{85}			

(a) MSA (matrix of connections)

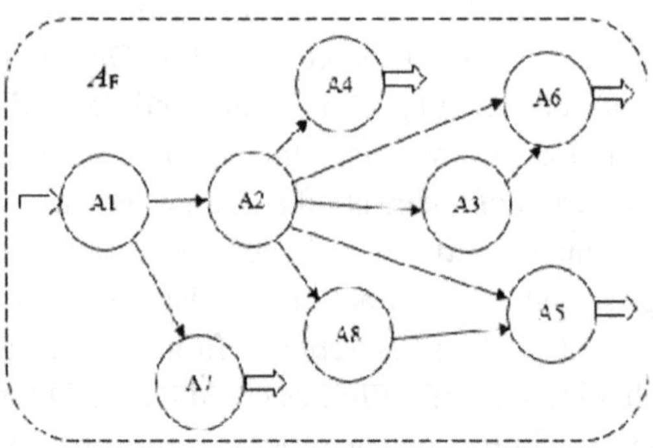

(b) GSA

The matrix of existing connections (MSA) $\{a_{ij}\}$ can be treated in stochastic and deterministic aspects. In the stochastic approach the elements of MSA are probabilities of making the corresponding transition $a_{ij}=p_{ij}$; $0\leq p_{ij}\leq 1$.

In the deterministic approach, the elements $a_{ij}\equiv b_{ij}\in\{0, 1\}$ indicate the absence or presence of a transition (connection) between a given pair $A_i \rightarrow A_j$. This initial GSA can be transformed in so called Ordered GSA (OGSA) – Figure 8. It is determined by nodes ranking in separate layers (ranks) based on the heredity and their informational independence. The result is that algorithms (tasks) included in a common layer will be able to be executed in parallel on independent devices.

Figure 8. Ordered GSA

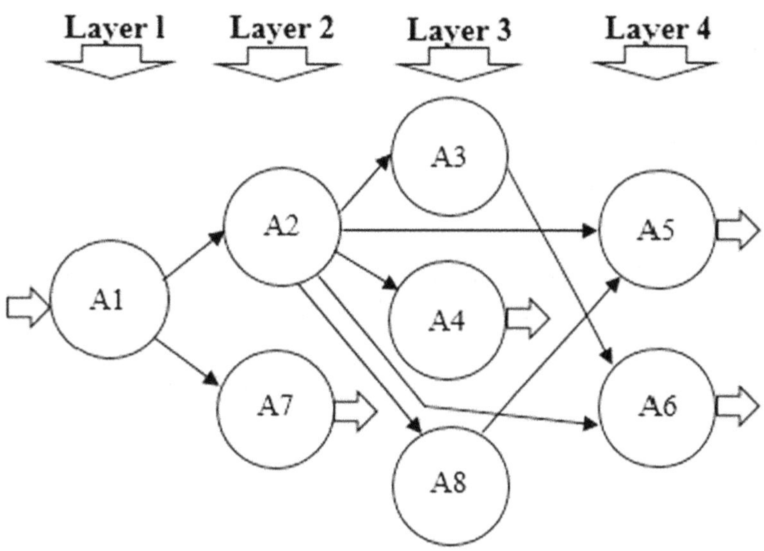

The arrangement of the nodes from the graph is done according to the rules:

- Nodes from the first layer have no predecessors, and nodes from the last layer have no successors;
- Nodes included in a common layer have no connecting arcs.

The layers in the ordered GSA are formed according to the procedure:

1. All initial nodes (no predecessors) are numbered first and included in layer (1).
2. A consecutive vertex is numbered (subject to inclusion to current layer) if all its predecessors are already numbered (included in previous layers).
3. Nodes to be numbered are immediate successors of already numbered nodes.

The functional modeling of the output GSA ranking is based on the following steps, and the program implementation in the APL2 language environment is shown in Figure 9.

Step (1): Forming a transposed matrix *LT* from the input Boolean matrix $L=\{a_{ij}\}$. The pillars $i=1,2,...,n$ of *LT* are treated as vectors describing the successors $LT[i,j]\neq 0$ ($j=1\div n$) for a given vertex A_i (the pillars of *LT* are named V_{Aj}).

Step (2): Calculation of the elements of vector $V_1 = \Sigma\, V_{Aj}$ (for each row [Ai] of *LT*, summation of the elements by j=1, 2,…, 8 is done):

$$V_1[j] = LT[j,1] + LT[j,2] + ... + LT[j,n] \text{ ; for } j=1\div n.$$

If $V_1[j] = 0 \Rightarrow A_j \in Level(1)$, i.e., vertex A_j falls into layer (1), which is denoted in the layer matrix by $AL[j,1]=1$.

Step (3): Calculating the elements of a vector V_2:

$$V_2[j] = V_1[j] - \sum_{A_k \in V_1} LT[j,k]\text{ ; for } j=1\div n,$$

where $A_k \in V_1$ reflects participation of the elements of row *j* of *LT* corresponding to the nodes already found in vector V_1. Elements corresponding to $V_2[j]=0$, fall into a layer *Level*(2) and $AL[j,2]=1$.

Step (4) and other: Analogous calculation of successive vectors V_q (q=3,4,...) for the elements of *Level(q)* at $j=1,2,...,n$:

$$V_q[j] = V_{q-1}[j] - \sum_{A_k \in V_{q-1}} LT[j,k]$$

until obtaining a zero vector, i.e.,$\forall V_q[j]=0$ ($j=1\div n$). For each layer "*q*" the presence of the elements is reflected by $AL[j,q]=1$, if fulfilled $V_q[j]=0$.

Figure 9. Program realization of GSA ranking

```
[0] S←GSA  L
[1] 'MATRIX  L:'
[2] ⎕←LT←L
[3] N←ρL[1;]
[4] LT←Ø⊃L
[5] 'MATRIX  LT:'
[6] ⎕←LT
[7] V←AL←(N,N)ρ0
[8] K←I←1
[9] QLIST←Q←ρ0
[10]ET1:V[I;K]←(+/LT[I;J← ⍳N])
[11]→(V[I;K]≠0)/ET2
[12]Q←Q,I
[13]AL[I;K]←1
[14]ET2: →(N≥I←I+1)/ET1
[15]ET3:'VECTOR V[',K,'] : ',V[;K]
[16]'  TASKS IN LEVEL ',K,' : ',Q
[17]QLIST←QLIST,Q
[18]→(0=(+/V[;K]))/OUT
[19]K←K+I←1
[20]QNEW←ρ0
[21]ET4:VA←0
[22]J←1
[23]LOOP:→((+/(J=Q))=0)/JUMP
[24]VA←VA+LT[I;J]
[25]JUMP:→(N≥J←J+1)/LOOP
[26]V[I;K]←V[I;K-1]-VA
[27]→(V[I;K]≠0)/ET5
[28]→((+/I=QLIST) ≠0)/ET5
[29]QNEW←QNEW,I
[30]AL[I;K]←1
[31]ET5: →(N≥I←I+1)/ET4
[32]Q←QNEW
[33]→ET3
[34]OUT:'MATRIX AL OF TASK I IN LEVEL J:'
[35]AL
```

(a) Program module

```
MATRIX  L:
0  1  0  0  0  0  1  0
0  0  1  1  1  1  0  1
0  0  0  0  0  1  0  0
0  0  0  0  0  0  0  0
0  0  0  0  0  0  0  0
0  0  0  0  0  0  0  0
0  0  0  0  0  0  0  0
0  0  0  0  1  0  0  0

MATRIX  LT:
0  0  0  0  0  0  0  0
1  0  0  0  0  0  0  0
0  1  0  0  0  0  0  0
0  1  0  0  0  0  0  0
0  1  0  0  0  0  0  1
0  1  1  0  0  0  0  0
1  0  0  0  0  0  0  0
0  1  0  0  0  0  0  0

VECTOR V[1]. 0 1 1 1 2 2 1 1
  TASKS IN LEVEL 1 : 1

VECTOR V[2]: 0 0 1 1 2 2 0 1
  TASKS IN LEVEL 2 : 2 7

VECTOR V[3]: 0 0 0 0 1 1 0 0
  TASKS IN LEVEL 3 : 3 4 8

VECTOR V[4]: 0 0 0 0 0 0 0 0
  TASKS IN LEVEL 4 : 5 6

MATRIX AL OF TASK I IN LEVEL J:
1  0  0  0  0  0  0  0
0  1  0  0  0  0  0  0
0  0  1  0  0  0  0  0
0  0  1  0  0  0  0  0
0  0  0  1  0  0  0  0
0  0  0  1  0  0  0  0
0  1  0  0  0  0  0  0
0  0  1  0  0  0  0  0
```

(b) Experimental result

In OGSA, 4 layers are defined, comprising tasks that can be performed independently of each other. For each vertex of layer *Layer(i>1)*, a heredity equation is defined, indicating the presence of information dependency (Table 2).

Table 2. Defined levels in the ordered GSA

Layer(1)	*Layer(2)*	*Layer(3)*	*Layer(4)*
$A_1 = f(0)$	$A_2 = a_{12}.A_1$ $A_7 = a_{17}.A_1$	$A_3 = a_{23}.A_2$ $A_4 = a_{24}.A_2$ $A_8 = a_{28}.A_2$	$A_5 = a_{25}.A_2 + a_{85}.A_8$ $A_6 = a_{26}.A_2 + a_{36}.A_3$

The set of final tasks $\{A_4, A_5, A_6, A_7\}$ allows determination of all paths in the graph for realization of computational processes. Through successive transformations (replacing a successor vertex with a corresponding algebraic expression), the aim is to obtain an equation describing the dependence of a final vertex on one or several initial ones. For example, for vertex A_5 the transformation is: $A_5 = a_{25}.A_2 + a_{85}.A_8 = a_{25}.[a_{12}.A_1] + a_{85}.[a_{28}.A_2] = a_{25}.a_{12}.A_1 + a_{85}.a_{28}.a_{12}.A_1$. The two components of the final expression define the presence of two paths to the final task A_5, which are: $<A_1 \rightarrow A_2 \rightarrow A_5>$ and $<A_1 \rightarrow A_2 \rightarrow A_8 \rightarrow A_5>$.

Functional description of algorithmic structures can be done by various means, most of which are associated with graphs. Some of the possibilities are presented below.

D-card is a method proposed by Edsger W. Dijkstra in 1972 (Dijkstra, 1972) for representation and transformation of algorithms. This is a directed graph with nodes of the types shown in Table 3.

Table 3. Blocks used for D-card preparation

□	◇	○	▽	△
Operator, consecutive operator, or other D-cards	Selection block	Junction	Start of algorithm	End of algorithm

The functional description of an algorithmic structure such as a D-map is built through the elementary D-maps shown in Figure 10 – they show the possibility of forming a new D-map Z by combining already defined D-maps X and Y.

Figure 10. Basic simple D-cards

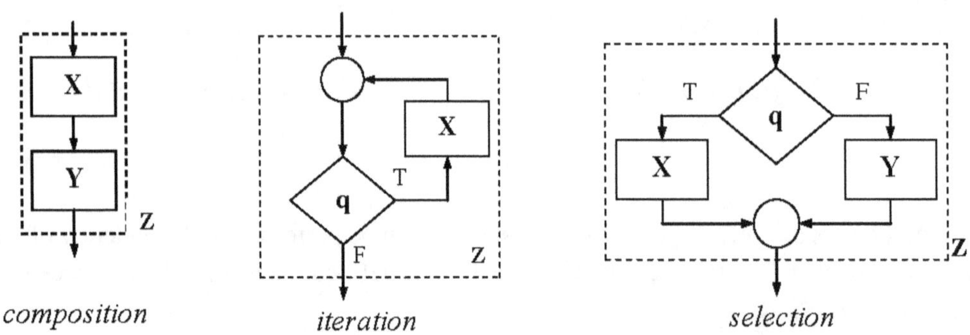

Discrete graph description of an algorithmic structure can be obtained directly from the block diagram by directly replacing an operator or logic block with a graph node (Figure 11)

The study of such a functional model is associated with a transformation of the directed graph according to the following principle. Initially, for each operator of the algorithm, a vertex is formed in the graph. The nodes are connected by directed arcs, corresponding to the transfer of control between operators. All consecutively connected nodes (with single connecting arcs between them) are then merged into a common vertex. The point of the last conversion is that sequentially linked statements are always executed the same number of times (unless an error interrupt occurs).

A graph functional model can answer the following questions:

- What is the implementation time of the algorithm?
- What fraction of the total time is used to execute a subroutine?
- How many reversals to a given data structure are implemented?
- What is the calculation profile?
- What are the possible paths (of the computation) from the beginning to the end of the algorithm and what are their time parameters?

Figure 11. Algorithmic structure: Block diagram and graph representation

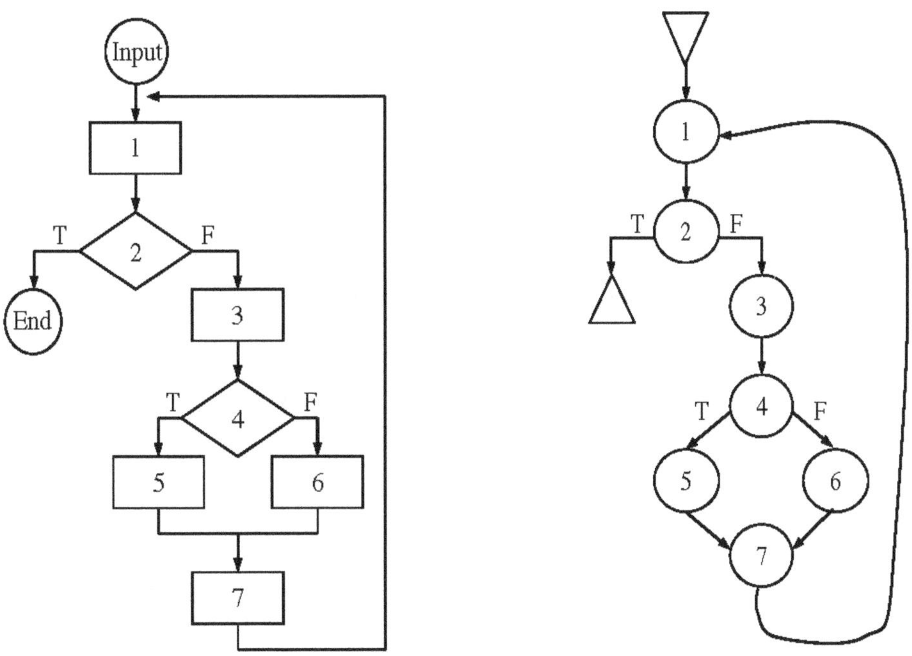

Stochastic graph description of an algorithmic structure is realized when the object of the functional representation is the transfer of control between operators, described as a process with a stochastic nature. A directed graph is also used, but each arc is represented by an ordered triple <*p, t, v*>, where: *p* – conditional probability of control transmission along the corresponding arc; *t* – average transition execution time (mathematical expectation); *v* – dispersion of that time.

Figure 12 presents a fragment of stochastic graph functional model, which includes first operators of the algorithm shown in Figure 11. The nodes, which are of the “merge” and “branch” type, are represented by a circle, and the processing is represented in the arcs by the values of the stochastic parameters <*p, t, v*>. To study the characteristics of the algorithm, the model is subjected to successive transformations of the type A→B→C ⇒ A→C by defining new values: $p_{AC}=p_{AB}\times p_{BC}$; $t_{AC}=t_{AB}+t_{BC}$; $\vartheta A_{C=}\vartheta AB_{+\vartheta}BC$.

Successive vertex exclusions continue until the stochastic graph model is reduced to only input and output nodes.

Figure 12. Stochastic graph description of algorithmic segment

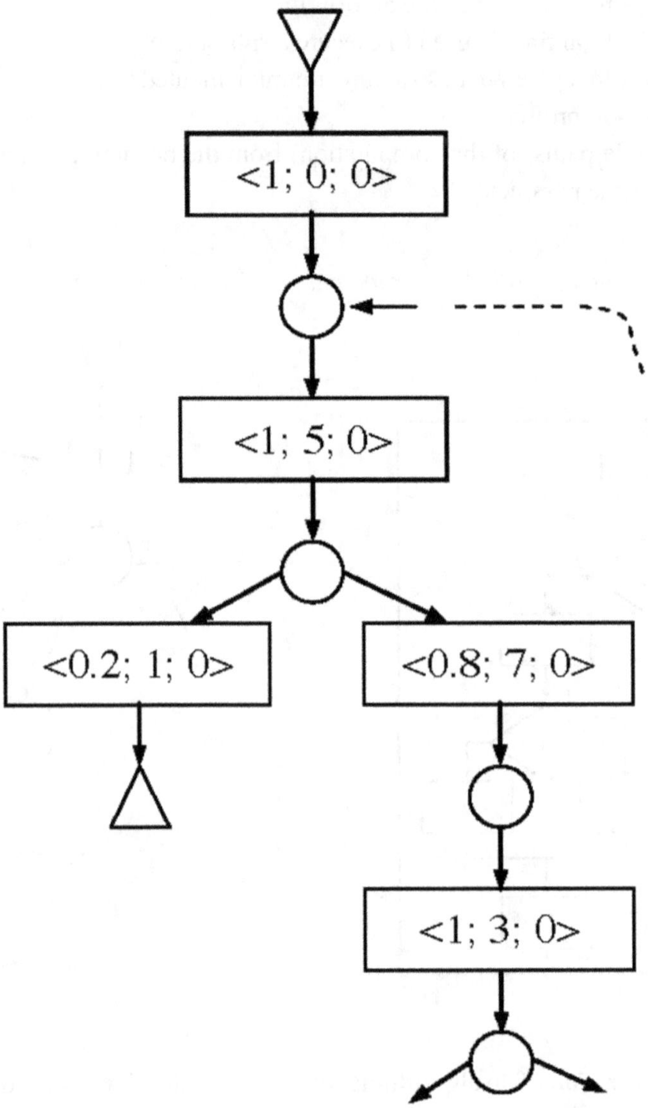

3. TECHNOLOGIES FOR FUNCTIONAL MODELING IN DATA PROCESSING

IDEF Technology

IDEF (Integrated DEFinition) modeling and conceptual design technology was developed under the American program Integrated Computer-Aided Manufacturing (IDEF, 2023). Certain methods of technology find application in different fields due to the possibility of very good representation of key activities in system design and a high level of detail, as stated in (Kermanshachi et al., 2019). Examples of specific applications are in defining semantic relationships to determine relevancies (Muminov & Bekmurodov, 2020), in modeling the relationship between the logical and temporal sequence between operational tasks (Fu et al., 2018), in developing a project to identify unidentified goals (Baek et al., 2019), etc. The technology unites fifteen methods, including at the beginning (in 1995) the traditional IDEF0, IDEF1/IDEF1X and IDEF2, and in subsequent stages the others were successively developed, starting with the IDEF3, IDEF4, IDEF5 and IDEF9 methods. The rest are in various stages of assembly, with some methods at the initial definition level. A brief presentation of the most commonly used methods (standards) is given in Table 4.

Individual IDEF technology standards are widely used in modeling business processes in systems engineering, as well as in communications research, management of authorization procedures, regulation of information flows, etc. (Johnson et al., 2022). The article combines the two main standards, with IDEF0 applied to risk identification in an organization's activities, and IDEF1 describing the attributes of objects as a set of rules. The goal is to achieve greater comprehensiveness in tracking the sources of risk. It is achieved by developing an extended IDEF1-model that identifies and tracks sources of risk as features of the model. As a future task, the study is set to integrate other business process models, including IDEF 1X, 2, 3, 4, 5, etc.

Table 4. Methods of IDEF technology standard

Method	Purpose
IDEF0	Functional modeling – implements a methodology for functional modeling of sequential systems for management, production, and business processes, with the aim of providing an answer to the question "what to do".
IDEF1 IDEF1X	Information/Data Modeling – for structuring the basic information about objects in a given production and information modeling of business processes. One of its main components is the graphic language ERD (Entity-Relations Diagrams), allowing description of the objects (entities) and the relationships between them when developing applications. IDEF1 is not a database development method, it only allows to understand business processes and the information related to them more clearly.
IDEF2	Simulation Model Design – for behavioral (simulation) modeling of business processes by considering each of them as a finite sequence of states with application of queuing theory, Petri Nets theory, theory of finite automata, etc.
IDEF3	Process Description Capture – for documenting technological processes and modeling activities.
IDEF4	Object-Oriented Design – a graphically oriented method for designing object-oriented programming systems. It was developed as a design tool and supports language environments such as Common LISP Object System, Flavors, C++, SmallTalk, Objective C and others.
IDEF5	Ontology Description Capture – to systematize the applied objects and present the information in a user-friendly form. Symbolic designations (descriptors) of objects and their associations are used for this purpose.
IDEF6	Design Rational Capture – using rational design experience.
IDEF7	Information System Auditing – verification of information systems.
IDEF8	User Interface Modeling – development of dialogue procedures for human communication with technical equipment.
IDEF9	Business Constrain Discovery – analysis of existing conditions and arrangements (including physical, legal, and political) and their influence on decision-making in re-investment in production.
IDEF10	Implementation Architecture Modeling – modeling of the executive architecture for the implementation of business processes.
IDEF11	Information Artefact Modeling – information modeling.
IDEF12	Organizational Modeling – organizational modeling.
IDEF13	Three Schema Mapping Design – designing a planning three-level scheme.
IDEF14	Network Design - designing computer networks by graphically describing configuration, network components, queues, and reliability requirements.

The **IDEF0** standard offers a technology to describe a system as a set of interdependent actions or functions, and a functional model is a diagram of arrow-connected type functional blocks (Mora et al., 2022; Tseng et al., 2021). It is closely related to the so-called structural approach to the analysis and design of systems. The application of the methodology is in various fields, such as in strategic planning (Spanidis et al., 2021; Spanidis et al., 2022), in manufacturing risk analysis (Collier et al., 2022), in the development of software-defined manufacturing equipment (Barwasser et al., 2022), etc. Before building an IDEF0-model itself, the following preliminary steps must be performed:

(a) determination of its purpose (questions to which the model must answer);
(b) defining the boundaries of the modeling (scope of the subject area and level of detail);
(c) specifying the target audience for whose needs the model is created;
(d) specifying the available information about the modeled object, what additional materials or documents will be needed for the target audience, what language and style will be most appropriate;
(e) choosing an appropriate point of view when modeling the object, which remains constant for all elements of the model

In Figure 13 the main structural element of the IDEF0-diagram with the possible four types of connecting arrows is presented. The function block represents a specific action (process) for converting input parameters (Input) into output (Output). Through control, instructions or restrictions affecting the execution of the process are set, and the executing mechanism (Mechanism) presents elements necessary for the implementation of the process, which are not subject to change. The direction of each arrow is fixed, not all of them are required (there must be at least one C and O type arrow each).

Figure 13. Basic structural element of IDEF0 functional diagram

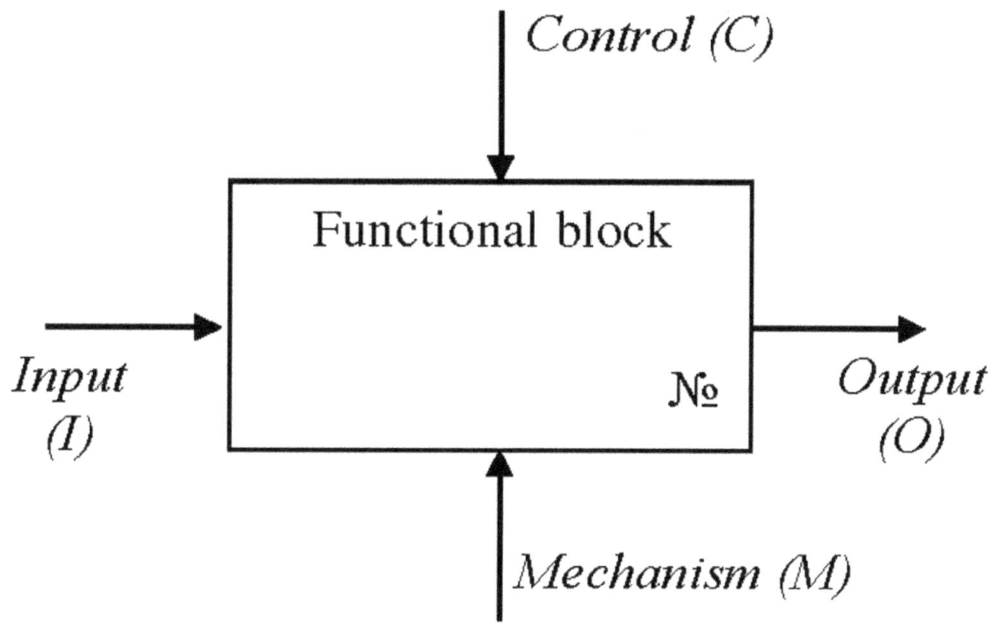

The IDEF0 functional model is a diagram of functional blocks connected by the so-called compound arrows, which can be of five main types:

- output-input (O-I) – when the completion of the action from a given block is required before starting the next one.
- output-control (O-C) – reflects the control impact of a given block to the next one.
- output-mechanism (O-M) – the results of the action of a given block is a toolkit for the next block.
- output-feedback to control (O-C*) – represents an O-C type impact, but to a previous block.
- output-feedback to input (O-I*) – like the previous impact, but to the input of a block.

In the diagram, splitting and joining arrows are allowed when the corresponding parameters or part of them point to several different blocks.

Figure 14 shows an example functional model in the IDEF0 standard of a financial accounting process in a company. The individual function blocks are numbered consecutively depending on the sequential actions. It is also allowed to use a letter (for example A) that is repeated for all blocks in a pattern. The numbering should also reflect the successive levels of decomposition of the system. The block of the highest level (context block) always has the number A0, and the ones generated during its decomposition - with the numbers A1, A2, A3, etc. For the next level, A11, A12, A21, A22, etc. are used.

Figure 14. An example for IDEF0 functional diagram

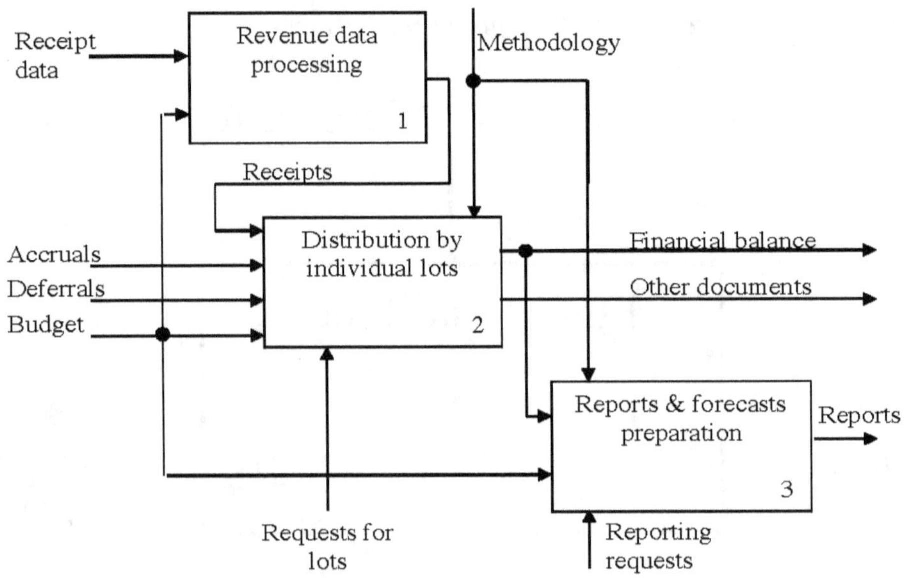

The **IDEF3** standard is a way of describing processes using a structural approach, allowing objects to be represented as an ordered sequence of events. The technology is suitable for modeling and designing business processes, as there are no rigid syntactic and semantic restrictions. IDEF3-modeling complements traditional functional modeling through IDEF0. For example, the IDEF3 standard was used in the development of the methodology proposed in (Evgenev, 2021) for creating intelligent systems for automated process planning. When presenting the knowledge in a business language (accessible to non-programmers), the standard was used to describe the knowledge pyramid and form the process flows. Another application of IDEF3 for air defense and anti-missile combat mission modeling in complex and variable environments is presented in (Yu et al., 2022). A method for modeling a mixed combat mission by combining the IDEF0 and IDEF3 modeling methods is proposed based on IDEF-standards. Some results of a research project for building information modeling are presented in (Zanni et al., 2021). The initial data is collected based on expert knowledge to specify the information requirements. As a next step, an IDEF3 structured diagram for process mapping of information flows was developed. The functional model provides the necessary information to develop a decision support tool and enables the automation of prescribed tasks.

The basic IDEF3-model represents a process scenario defining the sequence of actions (sub-processes) of the analyzed system. It is represented by a diagram composed of separate "units of work" (Unit Of Work - UOW), distinguished by a proper name (verb) and a unique number (identifier)– Figure 15.

Individual units (UOW) represented by a rectangle are connected by directional arrows – reflecting the presence of a certain type of relationship between actions (Table 5).

Figure 15. An example for IDEF3 functional diagram

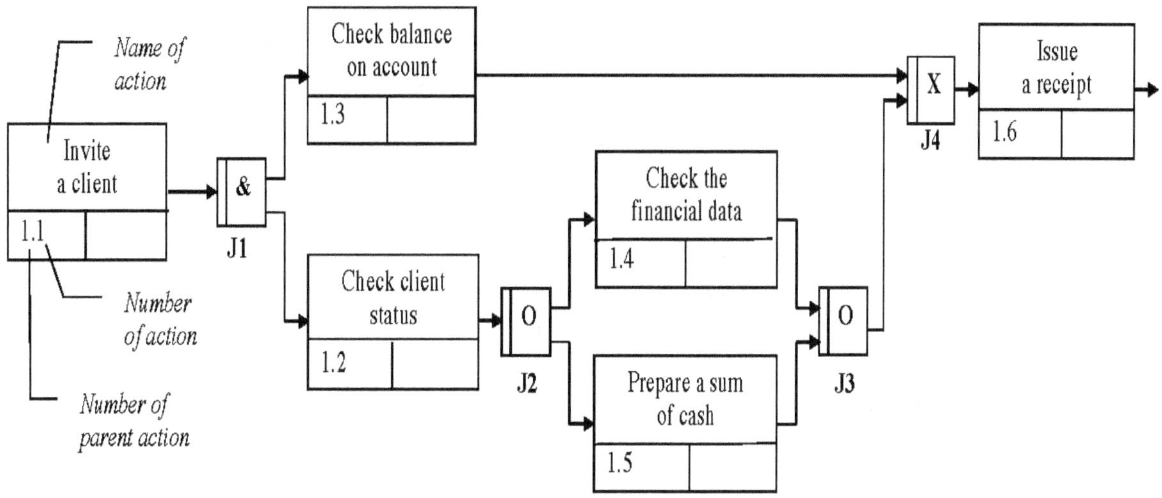

Table 5. Types of connections between actions

Type	*Name*	*Comment*
⟶	Temporal Precedence	The next action can only be started after the previous one is completed
⟶→	Object Flow	The output of an action is the input for the next action
···→	Undefined Relationship	The type of interaction between two actions is set for each individual use case

For branching and joining of the process in the functional diagram, the so-called compounds are used (Table 6). Through them, the internal stream branches into several new streams or several streams are combined into one.

Table 6.Types of elements

Graphical element	*Name*	*Type*	*Rules for initialization*
&	AND"	Branch	Several actions are initiated that start at the same time
		Compound	Several actions are initiated that start at the same time
O	OR"	Branch	One or more actions can start at the same time
		Compound	One or more actions can complete at the same time
X	EXOR"	Branch	Only one of the intended actions is initiated
		Compound	Only one of the initial actions must complete

IDEF3 is a way to functionally model business processes as an ordered sequence of concurrently described objects directly related to the process. This approach allows easy collection of data needed to perform a structural analysis and value assessment of the studied system.

Data Flow Diagram (DFD)

The flowchart has a similar organization to the previous diagrams - a set of actions connected by directed arrows, the latter showing the movement of data from one action to another. DFD is a modeling notation aimed at the important functional aspects of software development in the initial stages (Seifermann et al., 2022). The article indicates two main possibilities of the technology - control of information flow or control of access to information resources. To combine these advantages, an extended DFD syntax through clauses and first-order logic is proposed. Although DFDs are mainly used to model functional properties of a given system, (Alshareef et al., 2022) discusses the possibility of their application to model non-functional properties as well. Acknowledging this possibility, a formal framework for applying DFD to security and privacy is proposed. It includes three main elements: *"(1) annotating DFDs with purpose labels and privacy signatures, (2) checking the consistency of labels and signatures, and (3) inferring labels from signatures*". The DFD application in privacy is an object of extended discussion which is confirmed by a previous publication of these authors (Alshareef et al., 2021). In this respect, an example in this field can be found at the end of the DFD section.

Flowcharts were originally developed as a means of modeling and researching software development processes and have since developed their application in other areas. The method is known by the names of its authors (Chris Gane and Trish Sarson), who also proposed graphical representations for the main objects (Table 7a). Another Yourdon-DeMarco notation exists (Table 7b).

Table 7. DFD notations

	Process (Functional block)	*Source / Receiver of data*	*Storage (Data file)*
(a)	№ *<action>*	№ *<name>*	№ *<name>*
(b)	*Process*	*External entity*	*File*

In the functional modeling of information processes, DFDs represent the flow of data between three types of objects - a process (a functional block with a specific action on the data), a source or receiver (user) of data (external entities providing the necessary inputs to the system or appearing users of source data) and repository (data storage object).

The function block in the DFD is the main “transformer” of the data that moves through the system (it models some function affecting the flow of data). In this sense, action names are essential. When entering a process, the data undergoes changes.

External entities provide the input data flow (sources) or model the users of the output data (sinks). An external entity can be both a source and receiver of data.

Repositories serve to model objects for temporary “storage”, as in the field of information technology it is any mechanism for storing data in the form of files, directories, databases, archives, etc.

Arrows in a DFD model the movement of data between system components. Unidirectional (directional flow) and bidirectional (two-way exchange) arrows are allowed.

An example DFD-diagram modeling a warehouse-accounting system is presented in Figure 16.

There are two approaches when creating a DFD. The first is based on the structural approach to design and provides for the initial construction of a model of the physical structure of an existing system and subsequent creation of a logical model for assessing its requirements. As the next stage, a new logical model is defined to represent the requirements of the designed real system, and the physical model for its implementation is built on its basis.

The second approach, known as “event splitting”, is oriented towards the development of information systems. It creates several DFD models in sequence:

- Logical model representing the system as a set of actions.
- A model of the surrounding environment, describing the system as an object reacting to external influences.
- A behavior model showing how the system processes different events.

Figure 16. An example for DFD functional model

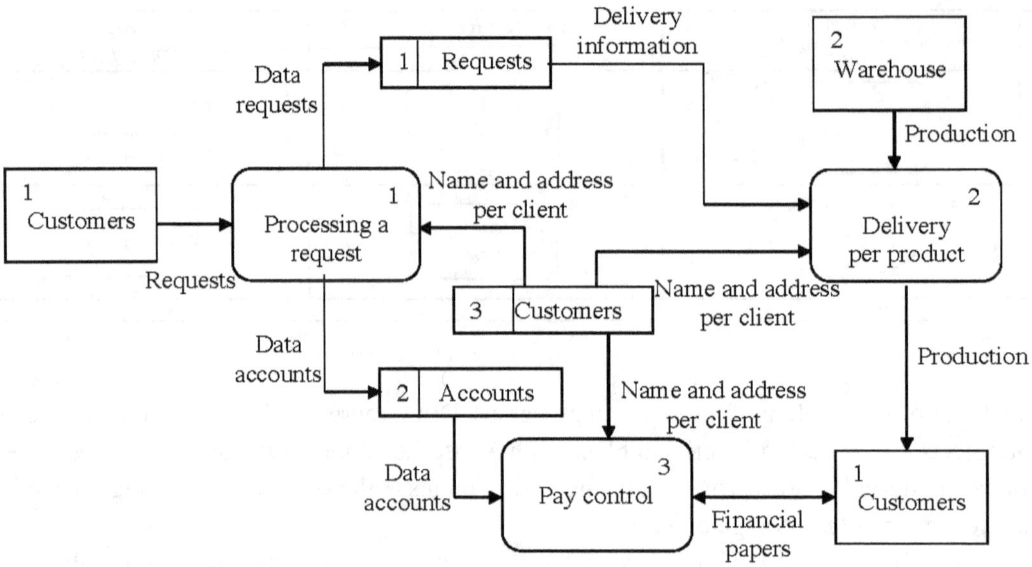

DFD defines operational constraints based on the "paths" formed by the data flow, notes the moments for storing current content, handling errors, communications between objects, etc. It is suitable for structuring data and individual objects, as well as for describing communications between them. It allows studying the static characteristics of program modules and the organization of information elements - dictionaries, files, and data. They mark the moments of saving current content, handling errors, etc.

Various activities related to access to information resources and their processing pose the important question of ensuring adequate and functional policies for data security and confidentiality. Any such organization must use an effective information security and access rights management system. This is especially important when working with personal data in order to ensure the privacy of employees and participants in the activities. At the core of any such system are authentication and authorization processes to ensure regulated access to data by those who have the corresponding granted rights.

The legal framework for the protection of personal data defines as observed any action related to them, starting with the requirement for their informed provision by the Data subject, passing through any processing activities and ending with the necessary destruction in fulfillment of the set purpose. For the correct maintenance of the procedures for processing available data, it is necessary to introduce strict organizational and technological measures and rules. An example functional DFD-model of an environment for personal data processing is presented in Figure 17, where 5 external entities (source/receiver), 9 basic information procedures and 3 storage units are defined (Romansky, 2023). This functional model is used for designing stochastic model for investigation of the procedures by using Markov chain theory, including extension of the processes based on cloud computing.

Figure 17. An example for DFD functional model

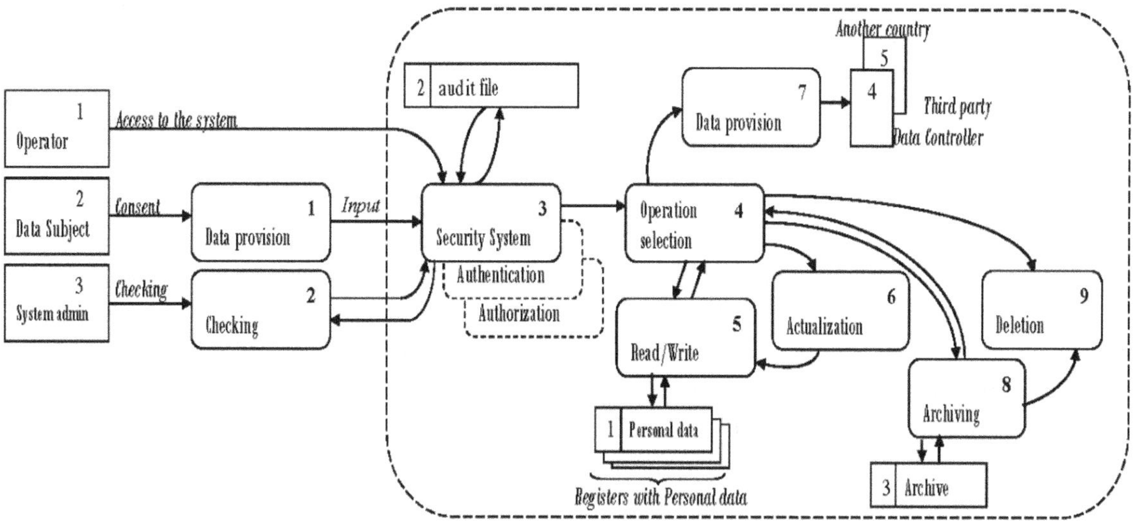

Dialog Transition Network (DTN)

The correct organization of dialogue in the development of computing systems and software applications is an important requirement for modern technologies (Allen et al., 2020). The article confirms the effectiveness of techniques for developing interactive programming environments with a developed user interface, but also points to the need to create more complex dialogue techniques corresponding to the human ability to solve tasks and artificial intelligence. In this vein, a model of dialogue to solve this challenge is described. Development of the topic in the direction of improving the abstractions and procedures for generating the user dialogue in the applications was done in (Yu et al., 2015). A method is proposed to solve the stated problem by giving a definition and presentation of dialog user interface models. The result is the provision of an opportunity to automate the conversion from an abstract dialog level to a concrete dialog logic.

Dialog Transition Network (DTN) is an approach applied in the development of dialog software to functionally describe the process of human-program communication, and one solution is presented in (Kang et al., 2013). It is a new practical system to support the requirements of stable dialog management. It uses the joint work and cooperation of two components – the hull technology framework and the DTN dynamic model. In such cases, the capabilities of DTN are used to study the dynamics of synchronous processes from information processing when designing dialog program systems. The choice of DTN in the last presented study is based on these capabilities, which provide robust dialogue management.

DTN allows description of the functionality of a developed product through transitions between individual states in the user's dialogue with the environment. In essence, it is a directed graph of transitions between states (nodes in the graph) that represent the individual phases of the ongoing dialogue (a sequence of screen forms for human-program communication). The main descriptive elements of the DTN are given in Figure 18.

Figure 18. Basic DTN elements

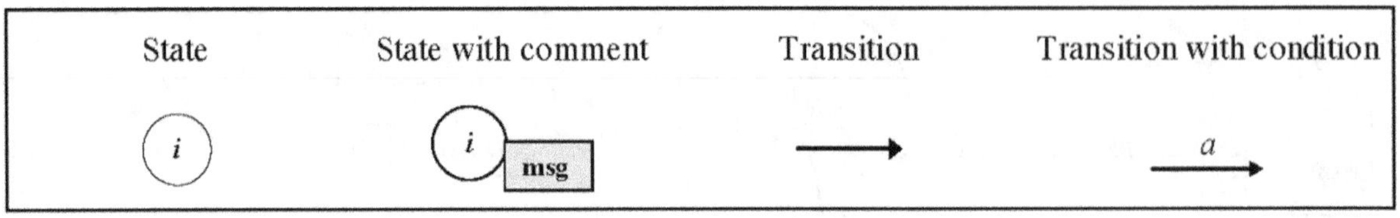

Nodes in the DTN-model are of two types - simple and with an accompanying message. The role of the latter is to mark the need to output a specific text when entering this state. In this sense, the name of the accompanying messages must be different for individual nodes. The arcs follow the dialog process and if there are several transitions from a given state, they need to be accompanied by the corresponding logical condition to activate the transition.

An example segment of a DTN-model of a dialog program is given in Figure 19. The DTN-model must describe all the possibilities of the possible dialogue, and a particular implementation will correspond to a path in the network from the starting node (starting the program) to one of the intended outputs. In the example, several dialog branches are formed, which are governed by the conditions {*a, b, c, d, e, f, g, h*}. These conditions are tailored to the purpose of the individual dialog states. An exemplary interpretation of the messages to the individual nodes and the transition conditions are presented in Table 8.

Table 8. Interpretation of messages and conditions

Interpretation of messages to nodes	
in	Initial entry in the designed program system with a requirement to enter access rights information
hlp	Output of help information
err	Output of message for error
menu	Main menu of the developed program system with 3 opportunities
op1, op2, op3	Information for entry in the relative sub-system
Interpretation of conditions for transfer	
a	Availability of a request for help information
b	Incorrectly entered username and/or password
c	Correct entry in the system
d, e, f, …	Selected entry code to the corresponding menu option, etc.

Figure 19. An example for segment of DTN model

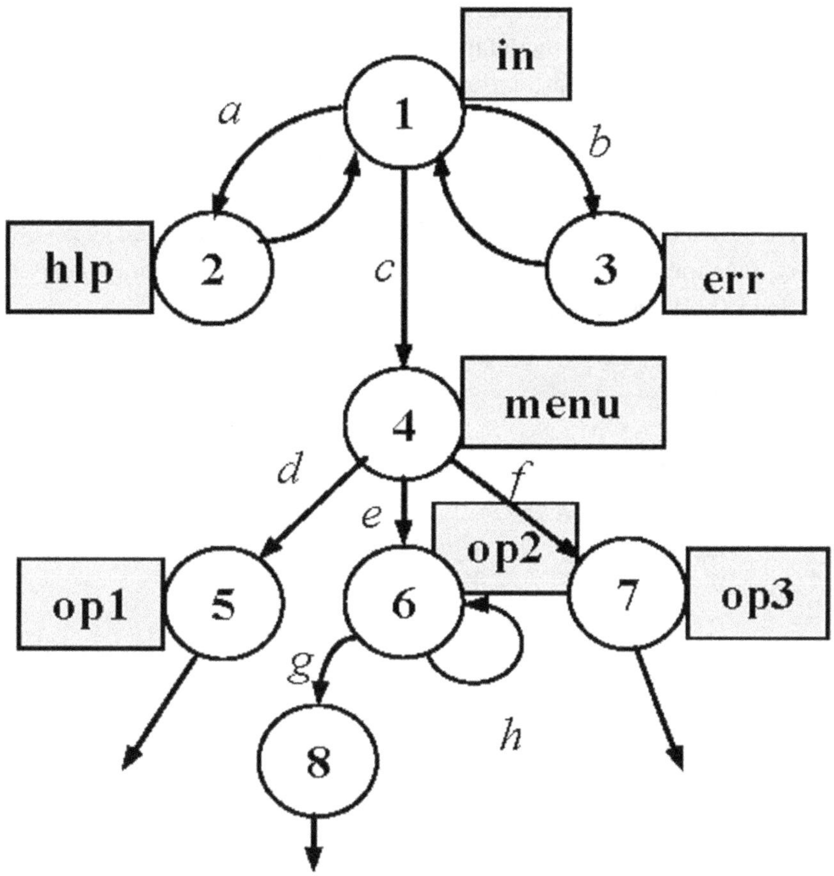

Unified Modeling Language (UML)

The language UML (Unified Modeling Language) appeared at the end of the 20th century as a result of various methodologies for designing software products, including notation systems. The goal is to create a standard for visualizing and documenting the functional nature of the object-oriented system being developed, which will bring together the main elements of analysis and design – a semantic model, syntax notation, and a development diagram (Rumpe, 2016). An analysis of the application of UML models as a software technology based on identification and classification is done in (Ciccozzi et al., 2019). A systematic review of the topic was conducted, and a classification framework was defined to be able to determine the main characteristics of solutions using UML. As a result, the potential opportunities, and challenges of using UML as a functional modeling tool are discussed.

The significant level of applicability of UML in developing software architectures is discussed in (Ozkaya & Erata, 2020), stating "*Unified Modeling Language (UML) is so popular among practitioners for modeling software architectures from different viewpoints*". The survey made in (Mkhinini et al., 2020) on the relationship of UML models with ontologies and the possibilities for their combination can be taken as a confirmation of this statement. The study explores the relationship between the two techniques in both a theoretical and practical sense and presents a detailed classification of existing solutions in the field. The relevance of UML in the functional modeling of systems and processes is also confirmed by the discussed conversion from IDEF3-model into UML-diagram, presented in (Khubaev et al., 2015).

UML allows description and documentation of the main behavior characteristics of the developed system. Through a created model, the following can be presented: ✓ the functionality of the product (by use cases); ✓ environmental impacts (by actors); ✓ to determine the relationships between use cases and actors (use case diagrams). There are two basic elements in a UML functional diagram (Figure 20).

Actor (external active subject) in UML is related to a thing which communicates with the system and can perform various functions such as: ✓ only input data in the system; ✓ only receiving data from the system; ✓ interaction in both directions.

Figure 20. Basic UML elements

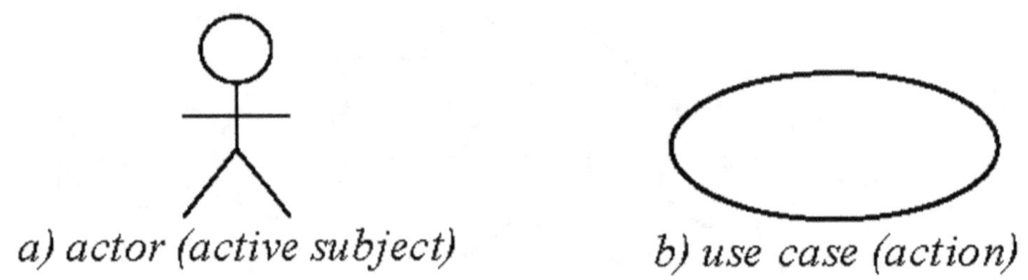

Use cases allow to model the dialogue between the active subjects and the system and reflect the functions of the latter. In UML, they represent in a formalized form the sequence of transactions performed by the system. Each use case is associated with a specific flow of events corresponding to a system function. In its description, it is determined what is to be realized, ignoring the question of how to do it.

Association relationships describe the communications between actors and use cases. They model the communications of the subject (actor) with the system within a certain use case. They are represented by directional arrows, with the direction indicating who initiated the connection.

Dependency relationships can exist between individual use cases, which are of two types:

- inclusive connection – connect variants of use performing the same functions;
- extensive connections – are used to describe additional capabilities, such as optional functions, atypical system situations, etc.

An example of UML diagram is presented in Figure 21.

Figure 21. Simple UML diagram

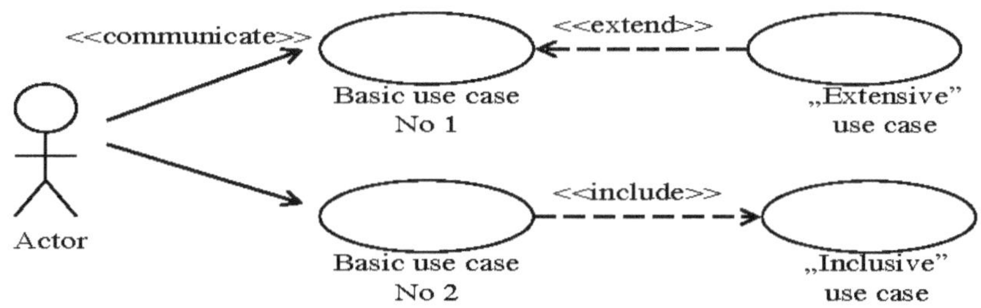

The main essence of the model is determined in the initial phase of development when the main actors and use cases of the system are identified. This is reflected in the so called "*use case diagram*" (Figure 22), which is a graphical description of subset of actors interacting with the system through various use cases.

Figure 22. UML use case diagram for functional model of university system

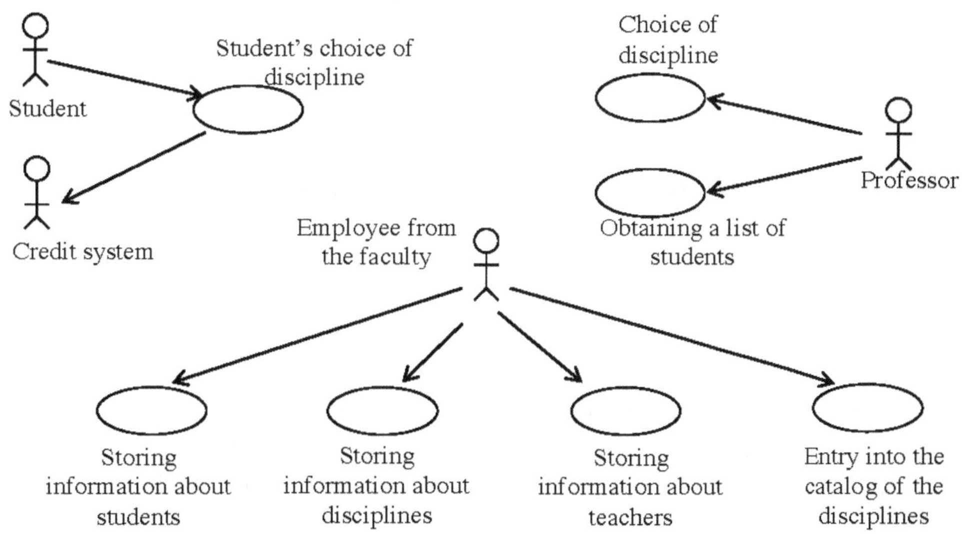

Activities diagram depicts the dynamic characteristics of the system by depicting the flow of control, possible parallel branches in the process, alternative possibilities to achieve the goal, etc. It is developed in the initial phases of the life cycle of the system and shows the flows of events (actions) during implementation of one or more usage options. The main elements of the action diagram are shown in Figure 23, and an example segment is shown in Figure 24.

Figure 23. Basic element in an UML activities diagram

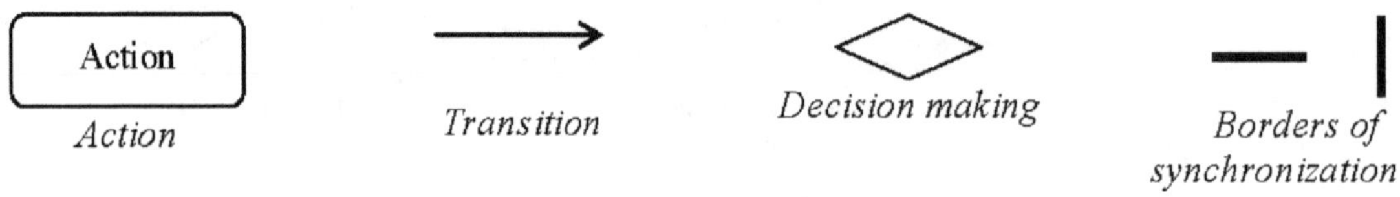

Figure 24. An example for UML activities diagram

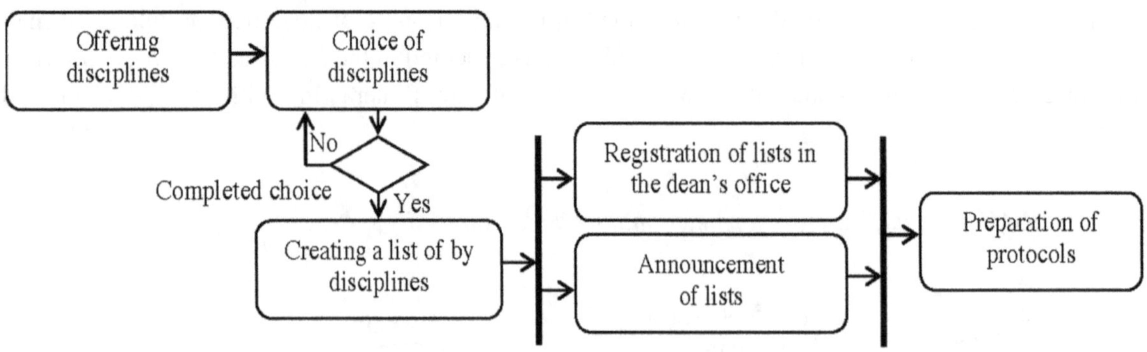

4. THEORY OF GENERALIZED NETS AND FUNCTIONAL MODELING

Basic Theoretical Aspects

Generalized nets (GNs) is another approach for preliminary fictional modeling in the field of programming and it is discussed in (Gochev & Hinov, 2022). They are an extension of graph theory and of the Petri net theory in particular. Due to the significant possibilities for describing the functionalities of a given object, they are used in various areas, including software development. Their descriptive capabilities allow a preliminary study of elements of software systems and their behavior by building functional models. Generalized nets were created by Prof. Krasimir Atanasov and over time have become an established formalism that represents the functioning of all modifications of Petri nets through the development and application of similar apparatus. On the other hand, they also have fundamentally new components that provide them with significantly greater opportunities for modeling real systems and processes. In this way, the field related to generalized networks has gradually developed and distinguished itself in its own theory and practice. An application to investigate processes in electronic components is presented in (Goheva et al., 2018b).

Generalized networks, as well as object-oriented programming, have their beginnings at the same time - from the 1980s. This coincidence is not accidental and is part of the correspondence (including in terminology) between Petri nets (see Chapter 7 for details) and generalized networks on the one hand, and programming paradigms on the other.

On the other hand, the quality of the software designed and developed for it is of key importance for the application of a given modeling, simulation, and control tool in as many areas as possible. Often, models with relatively little potential are preferred over others, theoretically more powerful, but insufficiently provided with software packages. In the case of generalized networks, it can be argued that the software for them has always used and continues to use up-to-date technologies. Figure 25 presents the functional model of a program module of a generalized network.

Figure 25. An example for GN functional model of program module

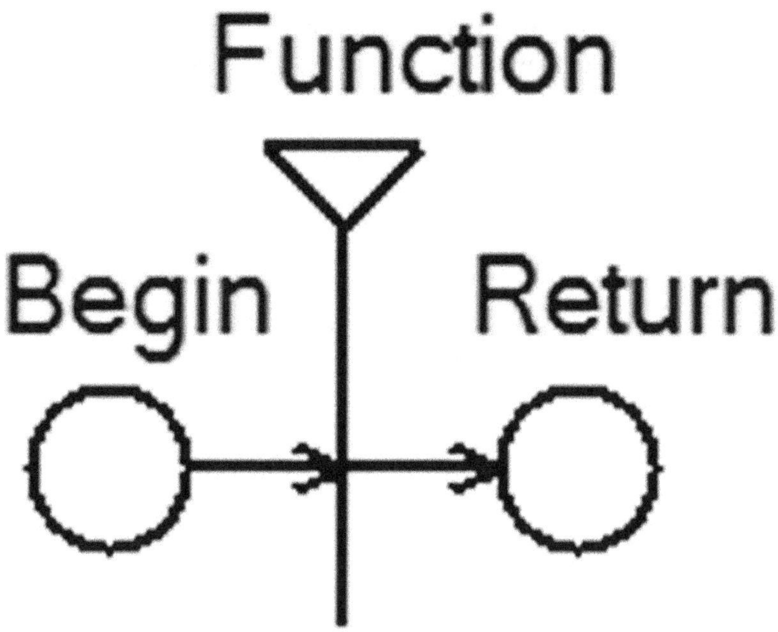

Generalized nets, like all modifications of Petri nets (PNs), are objects that have positions and transitions through which nuclei pass. Each transition in a GN is ordered set of parameters:

$$Z=\left\langle L', \quad L'', \quad t_1, \quad t_2, \quad R, \quad Q, \quad ' \right\rangle,$$

where:

L' and L'' are finite nonempty sets of its input and output positions, respectively.

t_1 is the next point in time at which the transition can be activated, and the value of this parameter is subject to update.

t_2 is the duration of the transition's active state, which is also updatable.

R is a transition condition, which represents the indexed matrix containing predicates associated with the transition arcs, the fidelity values of the latter determining the possibility or impossibility of passing nuclei.
M is an indexed matrix containing the capacitances of the transition arcs.
δ has a form similar to a Boolean expression and indicates in which of the input positions there must be kernels to enable the transition to be possible.

If any of the components of a transition or the entire GN is not specified, it is assumed by default and denoted by the symbol "*"; in this case the GN is called reduced. The transition condition is a fundamentally new component compared to other modifications of GNs. Its importance also stems from the fact that any GN can be replaced by another GN fully representing the functioning of the first one, but containing only transitions of the type:

$$Z=\langle L', \ L'', \ *, \ *, \ R, \ *, \ *\rangle.$$

On the other hand, each GN represents an ordered quad:

$$E=\langle\langle A, \grave{A}_{A}, \grave{A}_{L}, c, f, {}_{1}, {}_{2}\rangle, \langle K, \grave{A}_{K}, {}_{K}\rangle, \langle T, t_0, t^{*}\rangle, \langle X, ¦, b\rangle\rangle,$$

where:

A is a set of transitions.
π_A is a function setting the priority of transitions.
π_L is a function setting the priority of position.
c is a function specifying the capacity of the positions.
f is a function that calculates the fidelity values of the already described predicates at any moment of the operation of the GN.
θ_1 is a function specifying for each transition the next point in time at which it can be activated.
θ_2 is a function setting the active state duration of transitions.
K is a set of cores.
π_K is a function setting the priority of the cores.
θ_K is a function specifying the moments of arrival of the cores in the network.
T is the point in time at which the GN begins its operation.
t_0 is an elementary time step.
t* is duration of functioning of GN.
X is a function specifying the initial characteristics of the cores entering the OM.
Φ is a characteristic function specifying a subsequent characteristic of each core passing from the input to the output position of a transition.
b is the maximum number of characteristics that a core can acquire, moving in GN.

Most of the components of the transition definition in GN are fundamentally new compared to PN. The time components t_1 and t_2 are bound to a *global time scale* (described in the definition of the particular GN).

Indexed matrices are created for GN purposes, with the condition of each transition having the following form:

$$R = \begin{array}{c|cccc} & l''_1 & l''_2 & \cdots & l''_n \\ \hline l'_1 & r_{11} & r_{21} & \cdots & r_{1n} \\ l'_2 & r_{12} & r_{21} & \cdots & r_{2n} \\ \vdots & \vdots & \vdots & \ddots & \vdots \\ l'_m & r_{m1} & r_{m2} & \cdots & r_{mn} \end{array}$$

where r_{ij}, $i \in \{1,2,\ldots,m\}$, $j \in \{1,2,\ldots,n\}$, in the general case is a predicate (some of the elements of the indexed matrix may be constant fidelity values). At any moment of the functioning of the GN, the fidelity value of r_{ij} determines the possibility or impossibility of moving a given core from position l'_i to position l''_j. Using predicates is an approach that greatly facilitates the modeling of real processes.

An indexed matrix is similarly set:

$$Q = \begin{array}{c|cccc} & l''_1 & l''_2 & \cdots & l''_n \\ \hline l'_1 & q_{11} & q_{21} & \cdots & q_{1n} \\ l'_2 & q_{12} & q_{21} & \cdots & q_{2n} \\ \vdots & \vdots & \vdots & \ddots & \vdots \\ l'_m & q_{m1} & q_{m2} & \cdots & q_{mn} \end{array}$$

where q_{ij}, $i \in \{1,2,\ldots,m\}, j \in \{1,2,\ldots,n\}$, is the maximum number of cores that can transition from position l'_i to position l''_j within a single transition activation.

The expression δ contains identifiers of the transition's input positions and the Boolean symbols for logical addition and logical multiplication.

As noted in GN theory, the above components are divided into static, dynamic, temporal, and memory. They are presented in this order in the ordered four.

The first group of components actually determines a static (unchanging) structure, and the dynamics of the processes taking place in the generalized network is determined by moving cores. The latter occur at times specified by θ_K and have a different priority according to π_K. Each generalized network has time components and in particular a global time scale with an elementary time step, denoted above by t_0. Important from the point of view of the temporal positioning of the processes are the moment in time when the generalized network starts functioning (T) and the duration of its functioning (t^*)

The characteristics of cores contribute perhaps to the greatest extent to the possibility of modeling a variety of processes. In the general case, each nucleus receives an initial characteristic when it enters the GN (through the function X) and new characteristics when it moves in it (through the function Φ), forming a kind of memory. Theoretically, cores can acquire an unlimited number of features, but in a program implementation there is a danger that such an assumption will lead to a memory shortage. To prevent such a scenario, a constraint is introduced through *b*.

The definition is not fully formalized but allows each specialist to define components in an appropriate way depending on the problem area in which he works. PNs preceding generalized networks usually operate with relatively homogeneous information, and high-level ones can relatively successfully represent more heterogeneous processes. Generalized networks, on the contrary, make it possible to drop many restrictions related to considering essential features of modeled objects. In the context of this manuscript, the type of system in the .NET Framework facilitates the definition of complex kernel characteristics and associated predicates.

As a rule, generalized networks used in practice are reduced, which means that some of their components are taken by default. An important property of Reduced GNs is that they can be classified. In (Atanassov, 2001) it was shown that generalized networks obeying the definition:

$$E = \langle\langle \mathrm{A},\ *,\ *,\ *,\ \mathrm{f},\ *,\ * \rangle, \langle \mathrm{K},\ *,\ * \rangle, \langle *,\ *,\ * \rangle, \langle \mathrm{X},\ \vdots\ ,\ * \rangle\rangle.$$

and containing the transitions described above fully represent the operation and performance of each GN. The latter means that any GN can be functionally replaced by a generalized network of this class.

Of the components presented in this way, the function calculating the fidelity values of the predicates and the characteristic function are fundamentally new. The presence of characteristics of kernels, as well as predicates associated with transition arcs, determine the significantly greater opportunities that GN provides for working with real systems compared to all other modifications of PN. In particular, there are many publications proving the representability of the latter by GN and in this sense all results obtained on the basis of PN can also be obtained by GN (Zoteva & Angelova, 2021). In this manuscript, not all components of GN will be used, but the above provides additional arguments regarding the applicability of GN in service systems based on queuing theory (Queuing Sistem – QS).

It is important to emphasize that the above definition of generalized networks does not give a complete formalization but provides an opportunity for each specialist to define the components in an appropriate way depending on the field of application. In this regard, numerous GN simulations have been developed, most of which use specialized software (Atanassov, 2019).

Queuing Systems Modeling by Using Generalized Nets

Queuing systems (QS) are so ubiquitous that it is impossible to find an area in which they are not applied. The first significant results in QS-theory, obtained at the beginning of the last century in connection with the study of telephone networks, marked the beginning of a very large number of studies. While the early advances in this field were entirely due to the use of purely analytical methods, the modern trend, also advocated in the present manuscript, is the use of simulation modeling theory (Gocheva & Gochev, 2018a; Tomov et al., 2019).

In this sense, results obtained by modeling with GN are compared to limit values calculated based on probability theory. QS with waiting is used as an example (see Chapter 5, part 5.3.2 and Chapter 9 for details). The following restrictions are imposed: ✓ all requests are of the same type; ✓ the incoming request and service flows are Poisson.

The reason for the two limitations is that some of the most popular results in QS research with analytical methods have been obtained under these conditions. The case of single-channel QS with equal intensity of incoming request flows λ and service flows μ (λ=μ) is chosen. The relative throughput in this case is calculated by the formula:

$$q = \frac{1}{d+2},$$

where d is the number of waiting places.

Poisson event flows are related to the Poisson distribution as follows:

$$P_m = \frac{(\lambda t)^m .e^{-\lambda t}}{m!} = \frac{a^m .e^{-a}}{m!},$$

where P_m denotes the probability of occurrence of exactly m events in a time interval of duration t, λ – the intensity of occurrence of events, and a – the average number of occurrences within an interval of duration t.

Two GN-models have been developed. They have the same graphical structure, presented in Figure 26, and cores that do not have characteristics (due to the uniformity of requests). The cores in position L_1 represent incoming requests, the cores in L_2 represent requests being serviced, the cores in L_3 represent pending requests, the cores in L_4 represent rejected requests, and finally, the cores in L_5 represent serviced requests. The capacities of positions L_2 and L_3 are equal to the number of channels (in this case one) and the number of waiting places, respectively. The Z_1 transition models the arrival of requests for service or at a waiting location, as well as the receipt of a denial. In it, the L_3 position is both input and output, because pending requests, like those entering the system, can begin to be served or be denied. The priority of L_2 is higher than that of L_3, which in turn is higher than that of L_4. Transition Z_2 represents completion of request servicing. The difference between the models is in the different functions calculating the fidelity values of the predicates.

Figure 26. GN-model of a QS

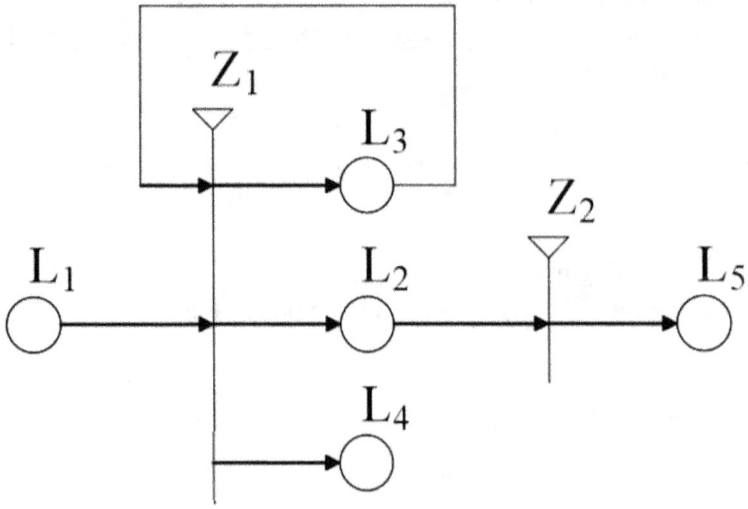

GN-model No. 1. This model uses the ordinariness property of Poisson flows. It reflects the fact that the smaller the interval, the less the probability of more than one event occurring compared to the probability of exactly one event occurring, and is written as follows:

$$lim_{t\to 0} = \frac{\sum_{i=2}^{+\infty} P_i}{P_1}.$$

Only the probability P_1 is considered in the model, and it is assumed that at most one request is received and/or served per time slot. Another assumption is that if within a time step the service of a request (if there is one pending) precedes the arrival of a new request. These two assumptions lead to an accumulation of systematic error in one direction. For the purposes of the calculation process, based on the probabilities P_0 and P_1 for the two flows (of arrival and service), two pseudo-random variables are constructed, representing respectively the number of received and serviced requests for each time step. The values of the two quantities determine the number of cores entering L_l and the number of cores passing from L_2 to L_5, respectively. Simulations were carried out with the simulator for generalized networks SIM2000 (Nikolov, 2004) for CMOs with different number of waiting places (from 0 to 3), and for each of the latter 1,109 incoming requests were considered. Relative throughput is calculated as the relative proportion of successfully served requests. The results are given in Table 9. As can be seen, the error depends on the average number of incoming requests per time step.

Table 9. Errors made in calculating relative throughput of single-channel mass service systems using GN-model Nº 1

Average number of requests arriving in one time step	Number of waiting places			
	0	1	2	3
1,00	11,27%	9,31%	7,60%	6,365
0,10	4,57%	2,94%	2,21%	1,78%
0,01	0,31%	0,17%	0,16%	0,14%

GN-model No. 2. Probabilities P_2 through P_{10} are also considered in this model. In this way, the two already described pseudo-random values take values from 0 to 10. Using pseudo-random values, the moments of request arrival and channel release are also determined. All this leads to a substantial reduction in systematic error (which in this case is due to neglecting probabilities with indices above 10).

This is clearly seen from the results presented in Table 10. The error is mostly due to the pseudo-random elements in the operation of the generalized network. The simulations with both models give information mainly about the precision of the modeling of stochastic processes with generalized networks. The second model is accurate enough, while the first has the advantage of its simplicity. Below are guidelines for the appropriate use of GN components when working with real systems.

Table 10. Errors made in calculating relative throughput of single-channel mass service systems using GN-model Nº 2

Average number of requests arriving in one time step	Number of waiting places			
	0	1	2	3
1,00	0,0001%	-0,0025%	0,0013%	-0,0020%

If the requests differ from each other, then it is appropriate that each core entering the L1 takes as an initial characteristic all the characteristics of the corresponding request (among them the moment of arrival, the type, and the priority of the request).

-The conditions of the transitions must be defined in agreement with the rules under which service or denial occurs. For example, the service time of each request, which is a pseudorandom variable, can be modeled depending on the type and priority of the request.

-In order to provide the most reliable representation possible, especially in complex systems, it is necessary in some initial position that the kernels receive appropriate characteristics. A simple example is the use of the moment of entry of a nucleus into the starting position as a characteristic. Complex QSs can be represented as a set of subsystems and for each of them a GN-model can be constructed. In the latter case, it is appropriate to construct a GN-model representing the functioning of the entire system.

The extremely good applicability of generalized networks as a tool for modeling of the QS functional features stems from their definition and the presented new mechanisms for working with parallel and competitive processes. The obtained results show good accuracy.

Generalized Nets Representing C-based Programming Constructs

In C-based programming languages, the functions (and in object-oriented programming – the methods (member functions)) are basic structural units. (Gocheva et al., 2018b; Gochev & Hinov, 2022). The most common type of GN, representing execution of the program code of any function, has one transition only (Figure 25). In it the entry of a single token in input position *Begin* represents a call of a function, and this token contains in its initial characteristic values associated to the function's arguments. Passing through transition *Function* represents either a returning value via a return operator, or (in the absence of a return value, usually indicated by keyword void) – an execution of either the return operator without operand, or an execution of the last program line in the body of the function. The token passing through the transition receives the corresponding return value as a new characteristic or does not receive any characteristic (for a function that does not return a value). This GN reflects only two main actions – calling and ending the execution (possibly with returning a value). Indeed, software developers very often are interested not so much in the way the functions are implemented, but rather in their arguments and return values. In this regard, the described transition should be used as part of GN, which represents execution of a program code in which the function is called. With properly working software, as a rule, each called function gives a result within a user-friendly time interval. An example of the opposite is the endless cycle – the repeated and non-alternative execution of a single program fragment; in this case, in the proposed GN it will not be possible for the token to move from the input to the output position.

The return value, although most naturally constructive, is far from the only result of performing a function (method). Other examples are updating values of variables which are external to the called function and changing values of arguments. The results can also be reflected in files, databases, etc. In general case, the GNs, representing execution of program code in functions, has external memory. The latter should be considered when calculation of truth values of predicates is performed, as well as when new characteristics of tokens are given. The GN representation of execution of program code of a function can be detailed and this is shown below. Oppositely, results of work of the following GNs can be used in transitions with the graphic structure from Figure 25. Some constructions used in a program code realization and their GN-models are discussed below.

Representation of conditional operators. In all programming languages branching computational processes are realized through the conditional operators. The popular if/else program construct, for example, provides two alternatives and has the following syntax:

```
if(boolExpression)
   ifOperator;
else
   elseOperator
```

boolExpression is a Boolean expression, and ifOperator and elseOperator are correct operators. Significant (from the point of view of the present paper too) is the fact that the latter ones can be complex, i.e., to include a large amount of program code. For this reason, in the literature there are terms such as "program construction", "program structure", "program fragment", "composite operator" and others (Van der Linden, 1994).

The most general graphical view of GN, representing the implementation of 'if/else' operator, is given in Figure 27. A single token enters the input position *Begin*. As in the theory of programming, branching of a computational process takes place, in this case – by transition condition R_{If} ', which has form:

R_{If}' =		*IfOp*	*ElseOp*
	Begin	r_{If}	$\neg r_{If}$

with predicate r_{If} = "boolExpression *has value "true"* ", where "¬" is a logical negation operator. Passing the token through *If* ' to one of its output positions represents a choice between two alternatives based on the truth value of boolExpression. The calculation of values of predicates in R_{If} ' can be a complex process involving calling multiple functions. Upon entering *IfOp* the token receives as a new characteristic the result of execution of ifOperator, and in *ElseOp* – the result of execution of elseOperator. Each of these two results may include a return statement or an unconditional goto statement execution, both of which are not possible at same time. Position *Next* simulates reaching program line after the if/else operator, position *Return* – reaching operator return or the last operator from the body of the function containing the presented construction if/else, and N output positions *Goto 1*, *Goto 2*, ..., *Goto N*, where N is the number of different labels after operator goto within *ifOperator* and *elseOperator* – reaching operator goto with the corresponding label.

Figure 27. GN model of an if/else operator

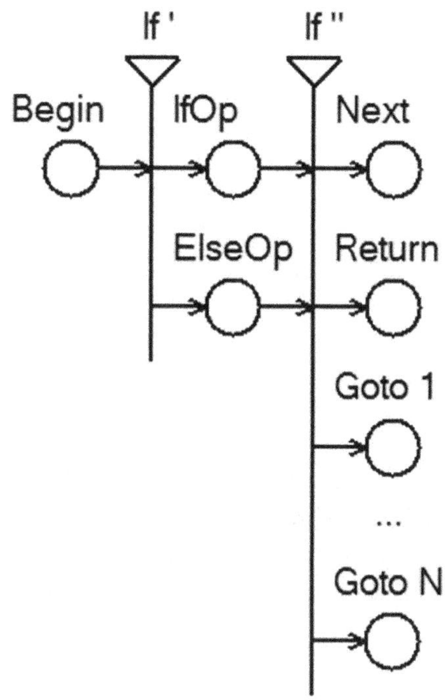

The condition R_{If}'' of transition *If* '' has form:

	Next	*Return*	*Goto 1*	...	*Goto N*
IfOp	r_{Next}	r_{Return}	$r_{Goto\ 1}$	...	$r_{Goto\ N}$
ElseOp	r_{Next}	r_{Return}	$r_{Goto\ 1}$	...	$r_{Goto\ N}$

with predicates:

r_{Return} = *"operator* return *or the last operator of the current function is executed"* ;
$r_{Goto\ i}$ = *"operator* goto *with* i-*th label is executed"* ;
$r_{Next} = \neg r_{Return} \,\&\, \neg r_{Goto\ 1} \,\&\, \ldots \,\&\, \neg r_{Goto\ N}$.

Upon entering the output positions of transition *If* '', the token does not receive a new characteristic. Operator **goto** (whose use in C and in its derived languages is not recommended) is very likely not to be presented in **ifOperator** and **elseOperator**; in this case the output positions of transition *If* '' can be reduced to two positions: *Next* and *Return.*

Similar considerations can be made to present operator switch, which takes part in the following source code:

```
switch(x)
{
   case a_1:
      operator_1;
      break;
   ...
   case a_M:
      operator_M;
      break;
   default:
      operator_D;
      break;
}
```

The advantage of switch over the if/else construct is the lack of a limit on the number of alternatives. The most general graphical view of GN, representing operator switch, is given in Figure 28. It checks variable *x* for M values (a_1, a_2, …, a_M). In the absence of operator goto, the nimber of output positions can also be reduced to two (this remark is valid in next section too).

Figure 28. GN model of a switch operator

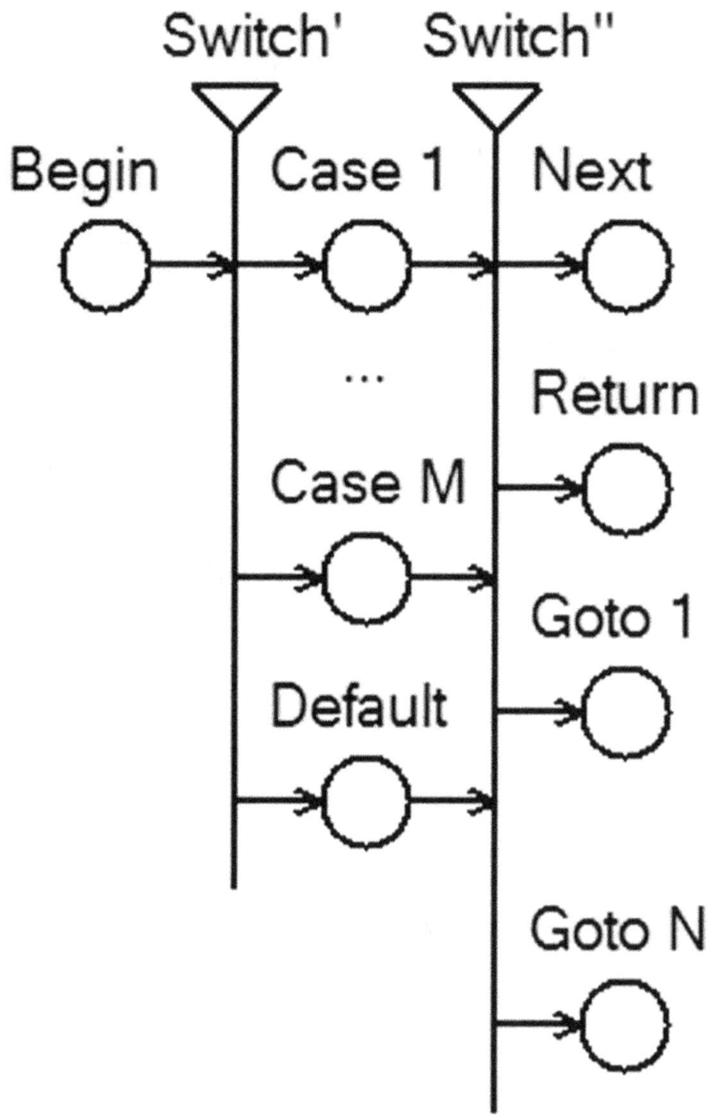

Representation of loops. The main constructs for loops in C programming language are while, do/ while and for. In case of while the respective syntax is

```
while(boolExpression)
{
   doOperator;
}
```

The most general graphical view of GN, representing the operator while is given on Figure 29.

Figure 29. GN model of a while operator

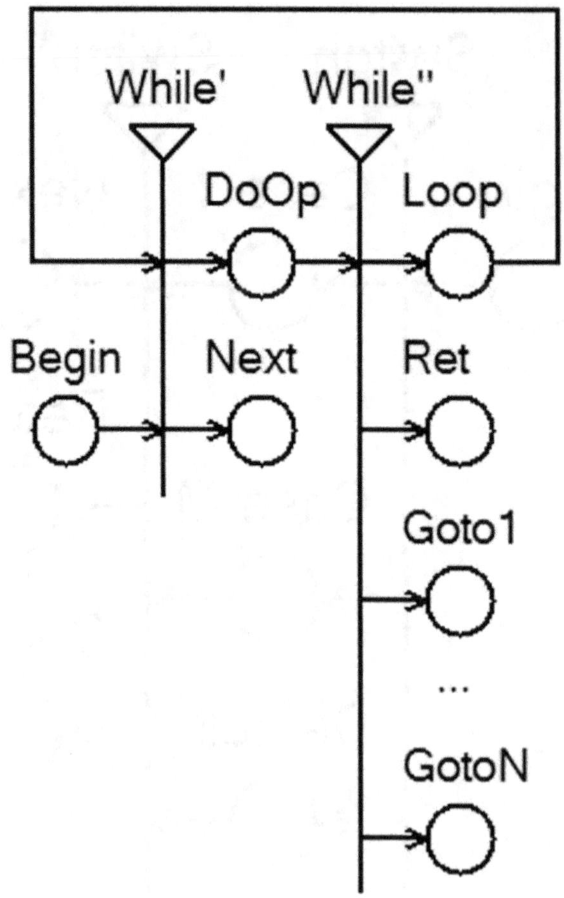

One token enters input position *Begin.* The branching of the computational process is done by a condition of transition *While* ‘, which has form:

R_{While} ‘ =		*DoOp*	*Next*
	Begin	r_{If}	$\neg r_{If}$
	Loop	r_{If}	$\neg r_{If}$

Upon entering *DoOp*, the token receives as a characteristic the result of the execution of doOperator, and in *Next* – the token does not receive any characteristic.

The condition R_{While} ‘’ of transition *While* ‘’ has form:

	Loop	*Return*	*Goto 1*	…	*Goto N*
DoOp	r_{Next}	r_{Return}	$r_{Goto\ 1}$	…	$r_{Goto\ N}$

and upon entering the output positions of this transition the token does not receive a new characteristic.

The components of the above GN correspond to the semantics of operator while. Similarly, operator do/while can be represented by a GN with the same components, but with one difference – a value "true" in the first transition condition; it reflects the fact that doOperator is executed before the boolExpression check. In this case transition *While* ' has condition:

$R_{While}' =$		*DoOp*	*Next*
	Begin	*true*	$\neg r_{If}$
	Loop	r_{If}	$\neg r_{If}$

which reflects syntax:

```
do
{
   doOperator;
}
while(boolExpression);
```

The upper considerations can easily be used in a GN which represents a loop based on operator for.

Representation of a program code. While Figure 30 shows the graphical structure of the most general GN with one transition, representing the execution of an arbitrary function, an approach for modifying this GN to include details is described below by way of example. It is possible due to the fact that each of the control programming constructs available in the C language can be represented by a GN which has a single input position named *Begin*, and a set of output positions.

{Next; Return; Goto 1; Goto 2; …; Goto N}

with reduced version

{Next; Return}.

Each function can contain many different operators, as well as calling other functions or itself (the latter technique is called "recursion"). In particular, in order to take into account control structures, the transitions shown in Figure 27, Figure 28 and Figure 29 can be added to the transition of Figure 25.

Example 1: *GN representing execution of a program code*

Let us the following function is given:

```
static long fact(int n)
{
    if (n < 0 || n > 20)
    {
        return -1;
    }
    int i = 1;
    long p = 1;
    while(i <= n)
    {
        p *= i;
        i++;
    }
    return p;
}
```

It calculates mathematical function

$$n! = \begin{cases} 1,2,...,n; & n \in \{1,2,...,20\}; \\ 1, & n = 0; \\ -1, & otherwise. \end{cases}$$

The limit for a maximum value of 20 is imposed due to the fact that 21! is out of range for type long.

The GN from Figure 25 is taken as a basis for representation of static method long fact (int n), and transitions for the operators presented in the method are added to it (Figure 30). The fact that the operator

return -1;

corresponds to ifOperator, and both assignments

int i = 1; long p = 1;

can be combined in its alternative composite elseOperator, is taking into account. Similarly, the block

{p * = i; i ++;}

can be presented with a composite doOperator.

Thus are applied the models in previous sections. In order to make clear the accordance between these GNs and the resultant GN for example, the names of the transitions and the positions are preserved maximally in the latter with respective indexing. However, a modification of the model on Figure 27 is made, as two transitions are combined in *If1* (this is possible because a value is returned in the if block, and the next operator is executed essentially in an else block). New in context of the paper is transition *Operator3* only, representing the last operator in the body of fact(), namely – return p.

Figure 30. GN model of the source code from example one

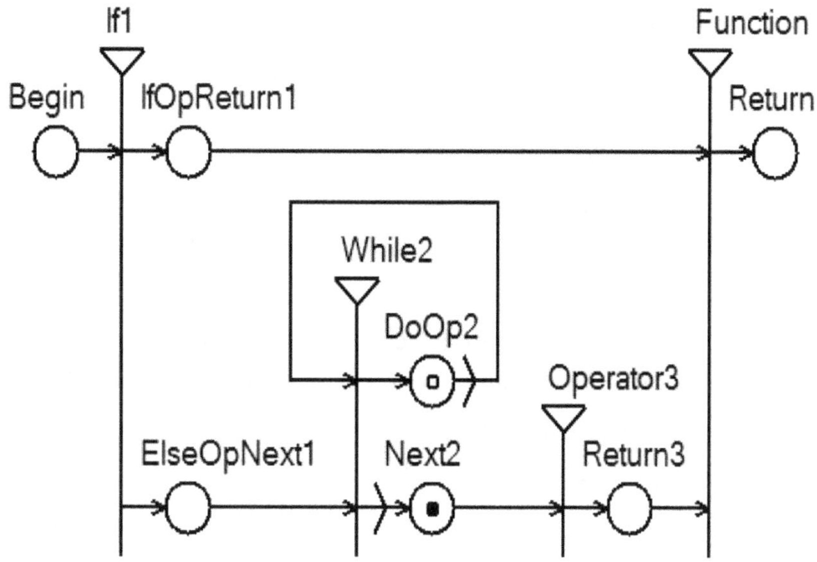

The generalized nets, which are described in the paper, model control structures in proper way. All images are generated using specialized software for generalized nets. The main difference in the proposed method is the possibility for generalization and uniform presentation of different classes of schemes and devices: those with a fixed structure and those with a variable structure within the period of work.

REFERENCES

Allen, J., Galescu, L., Teng, C. M., & Perera, I. (2020). Conversational agents for complex collaborative tasks. *AI Magazine*, *41*(4), 54–78. doi:10.1609/aimag.v41i4.7384

Alshareef, H., Stucki, S., & Schneider, G. (2021, December). Refining privacy-aware Data Flow Diagrams. In *Software Engineering and Formal Methods: 19th International Conference, SEFM 2021, Virtual Event, December 6–10, 2021, Proceedings* (pp. 121-140). Cham: Springer International Publishing. 10.1007/978-3-030-92124-8_8

Alshareef, H., Tuma, K., Stucki, S., Schneider, G., & Scandariato, R. (2022, August). Precise analysis of purpose limitation in Data Flow Diagrams. In *Proceedings of the 17th International Conference on Availability, Reliability and Security* (pp. 1-11). 10.1145/3538969.3539010

Atanassov K., (2001). Generalized nets as tools for modelling in the area of the artificial intelligence. *Advanced Studies in Contemporary Mathematics*, Vol. 3, 2001, No. 1, 21-42.

Atanassov, K. (2019). On two-way generalized nets. In *Advanced Computing in Industrial Mathematics: 12th Annual Meeting of the Bulgarian Section of SIAM,* (pp. 51-62). Springer International Publishing.

Atoum, I. (2019, January). A scalable operational framework for requirements validation using semantic and functional models. In *Proceedings of the 2nd International Conference on Software Engineering and Information Management* (pp. 1-6). 10.1145/3305160.3305166

Baek, H., Jun, B. H., Yoon, S. M., & Noh, M. (2019). Study on identification procedure for unidentified underwater targets uing small ROV based on IDEF method. *Journal of Ocean Engineering and Technology, 33*(3), 289–299. doi:10.26748/KSOE.2019.022

Barwasser, A., Lentes, J., Riedel, O., Zimmermann, N., Dangelmaier, M., & Zhang, J. (2022). Method for the development of Software-Defined Manufacturing equipment. *International Journal of Production Research*, 1–18.

Ciccozzi, F., Malavolta, I., & Selic, B. (2019). Execution of UML models: A systematic review of research and practice. *Software & Systems Modeling, 18*(3), 2313–2360. doi:10.100710270-018-0675-4

Collier, Z. A., Gaskins, A., & Lambert, J. H. (2022). Business process modeling for semiconductor production risk analysis using [*IEEE Engineering Management Review.*]. *Ide (São Paulo)*, F0.

Dijkstra, E. W. (1972). Notes on structured programing. In O. J. Dahl, E. W. Dijkstra, & C. A. R. Hoare (Eds.), *Structured programming* (pp. 1–82). Academic Press Ltd. https://dl.acm.org/doi/pdf/10.5555/1243380

Evgenev, G. B. (2021). Intelligent system of computer aided processes planning. In Z. Hu, B. Wang, S. Petoukhov, & M. He (Eds.), *Advances in Artificial Systems for Power Engineering. AIPE 2020. Advances in Intelligent Systems and Computing* (Vol. 1403, pp. 26–39). Springer Int. Publ. doi:10.1007/978-3-030-80531-9_3

Fu, M., Wang, D., Wang, J., & Li, M. (2018, October). Modeling method of operational task combined with IDEF and UML. In *2018 IEEE 3rd Advanced Information Technology, Electronic and Automation Control Conference (IAEAC)* (pp. 1443-1447). IEEE. 10.1109/IAEAC.2018.8577660

Gochev, V. P., & Hinov, N. L. (2022). Generalized nets representing C based programming constructs. In *2022 IEEE International Conference on Information Technologies (InfoTech-2022)*, (pp.13-16). IEEE. 10.1109/InfoTech55606.2022.9897111

Gocheva, P. V., & Gochev, V. P. (2018a), Application of generalized networks for modeling mass service systems. In *National Forum Electronics 2018.*

Gocheva, P. V., Hinov, N. L., & Gochev, V. P. (2018b). Modeling of electronic circuits with Generalized Nets. In *2018 IX National Conference with International Participation ELECTRONICA,* (pp. 1-4). IEEE. 10.1109/ELECTRONICA.2018.8439168

Johnson, D. A., Melo, V., & Lambert, J. H. (2022, October). Risk identification with entity attributes diagrams in business process modeling. In *2022 IEEE International Symposium on Systems Engineering (ISSE)* (pp. 1-8). IEEE.

Kang, S., Ko, Y., & Seo, J. (2013). A dialogue management system using a corpus-based framework and a dynamic dialogue transition model. *AI Communications, 26*(2), 145–159. doi:10.3233/AIC-130552

Kermanshachi, S., Safapour, E., Anderson, S., Goodrum, P., Taylor, T., & Sadatsafavi, H. (2019, January). Development of multi-level scoping process framework for transportation infrastructure projects using IDEF modeling technique. In *Proceedings of Transportation Research Board 98th Annual Conference*. IEEE.

Khubaev, G. N., Scherbakov, S. M., & Shirobokova, S. N. (2015). Conversion of IDEF3 models into UML-diagrams for the simulation in the SIM system-UML. *European science review*, (11-12), 20-25.

McIntire, M. G., Keshavarzi, E., Tumer, I. Y., & Hoyle, C. (2016, November). Functional models with inherent behavior: Towards a framework for safety analysis early in the design of complex systems. In *ASME International Mechanical Engineering Congress and Exposition*. American Society of Mechanical Engineers. 10.1115/IMECE2016-67040

Mkhinini, M. M., Labbani-Narsis, O., & Nicolle, C. (2020). Combining UML and ontology: An exploratory survey. *Computer Science Review*, *35*, 100223. doi:10.1016/j.cosrev.2019.100223

Mora, M., Adelakun, O., Galvan-Cruz, S., & Wang, F. (2022). Impacts of IDEF0-based models on the usefulness, learning, and value metrics of Scrum and XP project management guides. *Engineering Management Journal*, *34*(4), 574–590. doi:10.1080/10429247.2021.1958631

Muminov, B. B., & Bekmurodov, U. B. (2020, October). IDEF models and innovative system for search data in stochastic information environment. In *2020 IEEE 14th International Conference on Application of Information and Communication Technologies (AICT)* (pp. 1-6). IEEE. 10.1109/AICT50176.2020.9368581

Nikolov, N. (2004). The SIM2000 simulation package for generalized nets: Architecture and Language. Issues in Intuitionistic Fuzzy Sets and Generalized Nets. WSISiZ, Warsaw.

Ozkaya, M., & Erata, F. (2020). A survey on the practical use of UML for different software architecture viewpoints. *Information and Software Technology*, *121*, 106275. doi:10.1016/j.infsof.2020.106275

Rao, A. R., & Reimherr, M. (2023). Non-linear functional modeling using neural networks. *Journal of Computational and Graphical Statistics*, 1–20. doi:10.1080/10618600.2023.2165498

Romansky, R. (2023). Mathematical model investigation of a technological structure for personal data protection. Axioms, 12(2), 102. doi:10.3390/axioms12020102

Romansky, R., & Noninska, I. (2019). Technological organization of the access management to information resources in a combined e-learning environment. *International Journal on Information Technologies and Security*, *11*(4), 51–62.

Rumpe, B. (2016). *Modeling with UML*. Springer. doi:10.1007/978-3-319-33933-7

Seifermann, S., Heinrich, R., Werle, D., & Reussner, R. (2022). Detecting violations of access control and information flow policies in data flow diagrams. *Journal of Systems and Software*, *184*, 111138. doi:10.1016/j.jss.2021.111138

Spanidis, P. M., Pavloudakis, F., & Roumpos, C. (2021). Introducing the IDEF0 methodology in the strategic planning of projects for reclamation and repurposing of surface mines. *Materials Proceedings*, *5*(1), 26.

Spanidis, P. M., Roumpos, C., & Pavloudakis, F. (2022). A Methodology combining IDEF0 and weighted risk factor analysis for the strategic planning of mine reclamation. *Minerals (Basel)*, *12*(6), 713. doi:10.3390/min12060713

Tomov, Z., Krawczak, M., Andonov, V., Atanassov, K., & Simeonov, S. (2019). Generalized net models of queueing disciplines in finite buffer queueing systems with intuitionistic fuzzy evaluations of the tasks. *Notes on Intuitionistic Fuzzy Sets*, *25*(2), 115–122. doi:10.7546/nifs.2019.25.2.115-122

Tserng, H. P., Cho, I. C., Chen, C. H., & Liu, Y. F. (2021). Developing a risk management process for infrastructure projects using IDEF0. *Sustainability (Basel)*, *13*(12), 6958. doi:10.3390u13126958

Van der Linden, P. (1994). *Expert C programming: deep C secrets*. Prentice Hall Professional.

WilliamsA. E. (2020). A Human-Centric Functional Modeling Framework for Defining and Comparing Models of Consciousness and Cognition. PsyArXiv, April 16. doi:10.31234/osf.io/94gw3

Yu, J., Xiao, B., & Liang, H. (2022, September). Decomposition modeling of uncertain combat missions for air defense and anti-missile based on IDEF. In *2022 6th International Conference on Automation, Control and Robots (ICACR)* (pp. 176-179). IEEE. 10.1109/ICACR55854.2022.9935553

Yu, K., Hua, Q., Wang, S., Li, N., & Zhang, Y. (2015, September). An user interface dialog control model based on UI patterns. In *2015 6th IEEE International Conference on Software Engineering and Service Science (ICSESS)* (pp. 702-705). IEEE. 10.1109/ICSESS.2015.7339154

Zanni, M., Sharpe, T., Lammers, P., Arnold, L., & Pickard, J. (2021). Towards a BIM-based decision support system for integrating whole life cost estimation into design development. In *Proceedings of the 18th International Conference on Computing in Civil and Building Engineering: ICCCBE 2020* (pp. 197-206). Springer International Publishing. 10.1007/978-3-030-51295-8_16

Zoteva, D., & Angelova, N. (2021). An overview of the main results and applications. Research in Computer Science in the Bulgarian Academy of Sciences, 177-226.

Chapter 7
Features and Application of Deterministic Analytical Modeling

ABSTRACT

The main subject of discussion is deterministic analytical modeling, including its features and possible applications. The first section discusses the general organization of analytical modeling and presents its main implementation in the direction of deterministic, stochastic, and combined approaches. The second section deals with the use of mathematical approximations as an apparatus for discrete model investigation, presenting two sub-parts: the analytical approximation with a mathematical description and the application of analytical modeling in the study of power electronic converter. The next two sections are devoted to the application of graphs and Petri nets for the realization of deterministic models and their investigation, with presenting selected examples for illustration. The last section of the chapter discusses pseudo-stochastic analytical modeling, which is a deterministic variant of the description of processes that are stochastic in nature. This part makes a transition to the stochastic approaches in the next chapter.

1. GENERAL ORGANIZATION OF ANALYTICAL MODELING AND APPROACHES

Analytical modeling is related to creating a mathematical model of the studied object to represent its physical properties in a selected mathematical system (Hazır, 2015). For this purpose, two main sets of parameters are considered (Figure 1) – of the input impacts $\{X_i\}$ and of the output reactions $\{Y_j\}$, and the mathematical description of the object is most often a system of equations.

DOI: 10.4018/978-1-6684-8947-5.ch007

Figure 1. Analytical model defining

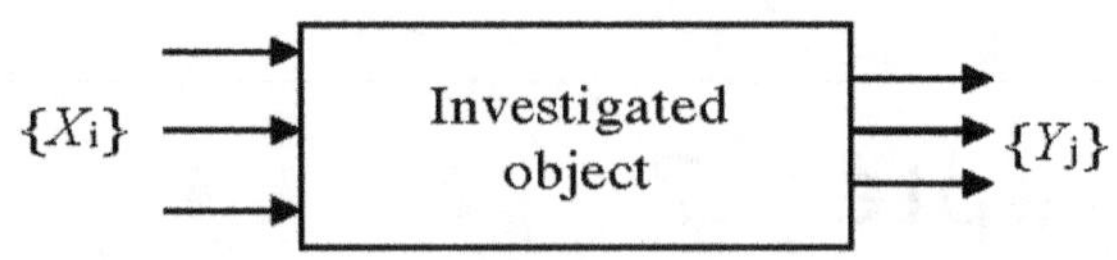

$$\begin{cases} Y_1 = F_1(X_1, X_2, \ldots, X_n) \\ Y_2 = F_2(X_1, X_2, \ldots, X_n) \\ \ldots\ldots \\ Y_m = F_m(X_1, X_2, \ldots, X_n) \end{cases}$$

The advantages of analytical modeling are determined by the requirements for precision in formulating the mathematical relations, assumptions, and approximations. In this reason, analytic dependencies are strictly provable and imply analytic credibility. Analytical models have a great cognitive value because they allow an arbitrary combination of the analyzed parameters. In addition, they are characterized by a minimum complexity of calculations, because for their solution known numerical methods and approaches are usually applied.

The disadvantages are primarily associated with the influence of the subjective factor when defining the mathematical description, which can lead to an overly simplified image of the real object due to the introduction of unjustified simplifications and approximations. As a rule, the creation of an analytical model is associated with certain assumptions – for example, independence of individual factors, linear approximation of dependencies, instantaneous transitions, etc. However, they must be well thought out and justified in order not to violate the adequacy of the model process.

Computer systems and processes have characteristic features that enable the successful application of analytical modeling. For example, in the investigation of cloud services (Antonelli et al., 2020), edge and fog computing (Pereira et al., 2021), machine learning (Nai-Zhi et al., 2022), etc. Research framework for computer systems engineering and modeling is proposed in (Kurniawan, 2019), which is based on service-oriented architecture (SOA) principles. The framework is designed to develop model-based computing systems for process analysis and optimization. A summary of the main features of computing is presented below.

- Regardless of the discrete nature of the structure of computer devices and systems, the processes taking place in them are of a stochastic nature.
- The competition between the active processes for occupying the free resource and their service is oriented to the theory of mass service (queuing theory).
- Computer processes are characterized by Markov properties.
- When studying the computer parameters, average values for the estimates are usually sought, which implies stationarity of the processes.

The development of an analytical model follows the generalized technological scheme, and a simplified adaptation of the procedure is shown in Figure 2.

Figure 2. Organization of analytical modeling

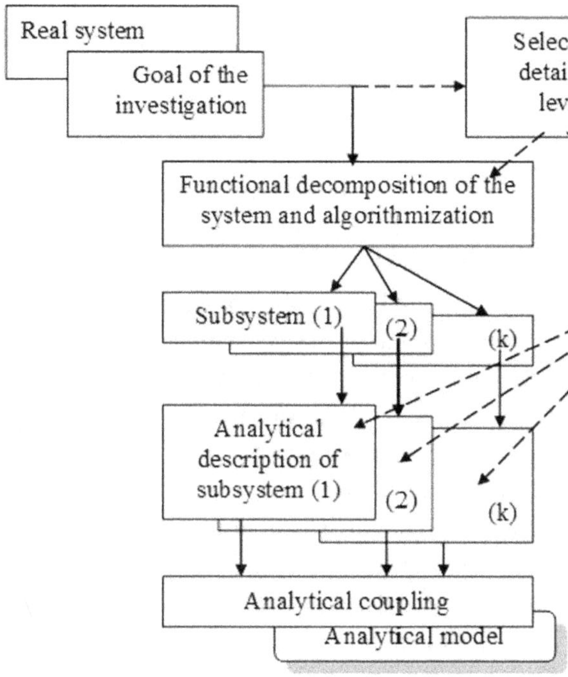

As can be seen from the organizational structural framework, the particularly important point is the decomposition of the research object and its mathematical formulation to create an analytical description. For this purpose, an appropriate mathematical system must be chosen, which is suitable for the correct formulation of the research object and is consistent with the sphere of this existence and field of application. One example in this direction is the research in the field of image processing conducted in (Kravets et al., 2022). It is stated that in contemporary monitoring systems and operational decision-making based on imaging requirements is very strict. On the other hand, the high complexity of the images places serious limitations on the rapid action of their processing. An analytical approach is proposed by determining the search motion vectors for which an algorithm based on an analytical solution is developed. This approach consists of decomposing each frame into individual components that are compared with similar ones in adjacent frames. This allows finding motion vectors to synthesize an intermediate frame that is compared to the main reference frame of the original image. The problem with 2D motion estimation boils down to finding a similar block in the previous image $f(\vec{x},n-1)$ within a specified search area. In analytical notation $\vec{x}$ is a two-dimensional position in the image and n is the image number. This permits to formulate a vector field $\vec{d}(\vec{x},n)$ which participates in the implementation of an analytical formula $f(\vec{x}-\vec{d}(\vec{x},n),n-1)\approx f(\vec{x},n)$. The carried out this analytically investigation is based on the principle that the motion vector is the same for a given block of pixels under a valid equation $\vec{d}(\vec{x},n)=\vec{d}(\vec{x}',n)\ \forall\vec{x}'\in B(\vec{x})$, where $B(\vec{x})$ is a block of pixels at position $\vec{x}$. The analytical basis thus defined allows to develop a block search algorithm in solving the two-dimensional motion estimation problem.

Another direction with the application of analytical modeling is the minimization of power losses as an important part of the design of distribution transformers (Digalovski & Rafajlovski, 2020). To solve this problem, a direct global iterative algorithm approach was applied, and a mathematical model of the distribution transformer was defined to obtain the necessary information about energy losses. The created optimized analytical model is designed to investigate the level of energy efficiency improvement. The optimization itself was performed based on input values, behavior diagrams and parameters obtained from the project catalog and laboratory tests. A mathematical equation is defined describing the relationship between all parameters of the transformer for minimization of the objective function (total energy losses) with the determination of nine variables ($x_1 \div x_9$) at specified limits of their variation: P_γ = Total loss $= f\,[x_1\, x_2\, x_3\, x_4\, x_5\, x_6\, x_7\, x_8\, x_9]$. The selected variables represent the geometrical dimensions of the transformer and the electrical and magnetic quantities, and a detailed description of the mathematical model for minimizing losses is also presented. An optimal result was achieved, which was confirmed by laboratory testing of the designed and manufactured transformer.

The development of an analytical model implies a preliminary decomposition of the studied object and the creation of an algorithm for the transition from functional behavior to its analytical description. This approach was applied in modeling a multifunctional system for managing access to a virtual environment of information resources in digital space (Romansky & Noninska, 2021b). Typically, these resources are located at various nodes of the global network and include technical equipment, software tools, data, and other information objects, including personal data. In most cases, this requires a proper approach to access them with regulation of the powers for their use, as an example functional model is presented in Figure 3. It is accepted that resources can be classified into two groups – public resources with authorized access to them and protected resources with implemented protection tools and procedures.

Figure 3. An example for general functional stricture of the system

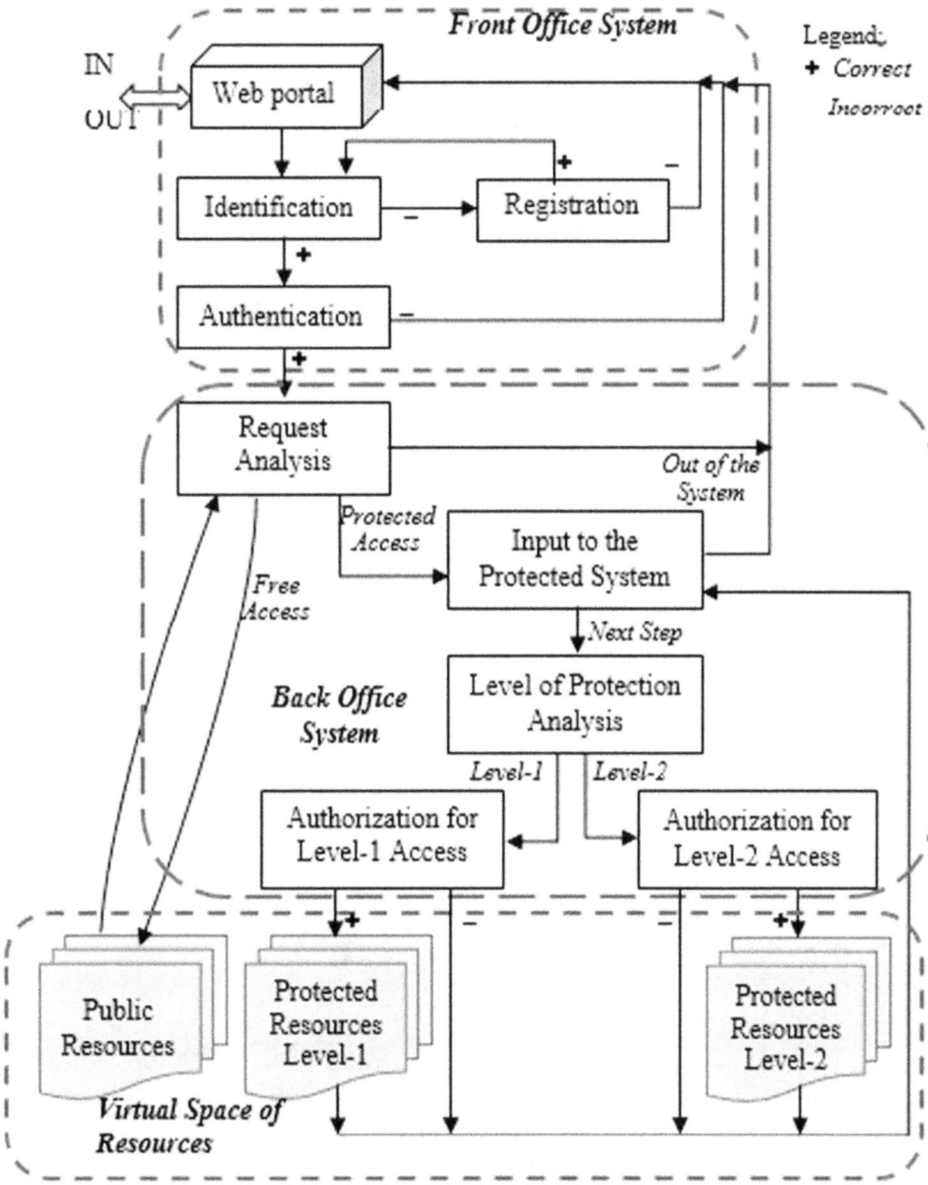

The presented access management system includes tools and measures implemented to counter all threats, attacks, and unauthorized access (internal and external). The goal is to implement a strong information security policy and ensure successful data protection during entry, processing, archiving and transfer over the Internet. The following three relatively independent but related sub-systems can be distinguished.

- *Front Office System* for receiving and preliminary processing input user's requests which must guarantee correct access to other parts of the system based on included procedures.

- *Back Office System* which is the core of information security activities with applying administrative procedures to ensure protected access and use. It must support all requirements of the CIA-Triad (Confidentiality, Integrity, Availability).
- *Virtual Space of Resources* is a collection of different information objects which could be defined as public (without restricted access) and private (with strong secure access based on the level of protection).

An algorithm supporting analytical modeling of the secure functionality is shown in Figure 4. This procedure determines the steps for user's access regulation to the information resources. Two types of private resources based on their specific requirements for protection are determined as Level-1 (High) and Level-2 (Low).

Figure 4. Algorithm for the system functionality analytical modeling

```
[1]  begin
[2]       IF identification = OK  THEN  goto [5]
[3]       IF registration = OK  THEN  goto [2]
[4]       goto [20]
[5]       IF authentication ≠ OK  THEN  goto [20]
[6]       IF request = END  THEN goto [20]
[7]       IF request = PRIVATE  THEN  goto [10]
[8]       using public resource
[9]       goto [6]
[10]      input in protected system
[11]          request_type analysis
[12]          IF request_type = FINISH  THEN  goto[6]
[13]          IF request_type = LEVEL-2  THEN  goto [17]
[14]          IF authorization ≠LEVEL-1  THEN  goto [11
[15]          using protected resource LEVEL-1
[16]          goto [12]
[17]          IF authorization ≠ LEVEL-2  THEN  got [11]
[18]          using protected resource LEVEL-2
[19]          goto [12]
[20]      out of the system
[21] end
```

The organization of analytical modeling is related to two main approaches defining both main approaches to analytical modeling.

Deterministic approach in analytical modeling of stochastic in nature computer processes most often leads to obtaining approximate estimates. The limitations are imposed by the accepted assumptions and the replacement of the probabilistic characteristics with deterministic values (averages, limits, etc.). Usually, deterministic modeling is applied in approximating real processes when probabilistic factors are missing, or when the latter can be replaced by their statistical estimates. One example of the application of the deterministic approach is the two solutions proposed in (Su & Schön, 2022) for limiting uncertainty and the comparison made with the conventional least squares method. The methods use deterministic intervals and linear propagation of uncertainty, which limits the stochasticity of the processes and allows obtaining a poly-thematic set of solutions. The following two main groups of deterministic analytical models (DAMs) are defined.

- *Formal DAMs* – they are built based on formal techniques or strict mathematical systems and can be connected to formal models. In contrast, formal DAMs are implemented (performed, solved, etc.) in the relevant mathematical environment to obtain estimates for the behavior of the object under study. Typical mathematical systems in this direction are mathematical approximations, mathematical theory of finite automata, graph theory, Petri nets theory, etc.
- *Pseudo-stochastic DAMs* – the idea of the stochastic nature of the original is preserved, but in the model the probabilistic values are replaced by their deterministic values (average or extreme).

Stochastic approach in analytical modeling corresponds to a greater extent to the real behavior of computer objects. When defining a stochastic analytical model (SAM), it is usually assumed that the behavior of the object can be described by a set of states $\{s_1, s_2, ..., s_n\}$, in which it falls at successive moments of time $t_1, t_2, ..., t_m$. Usually, the possibilities of stationarity of the random process (RP) defined in this way are investigated, looking for estimates of the possible successive transitions between the states and their probabilities, as well as the final probabilities of the process falling into each of the possible states. Depending on the probabilistic apparatus used, the presented below classes SAMs can be defined which will be discussed in the next chapters.

- *Probability analytical models* – they are implemented as stochastic approximations or based on probability equations for representing the behavior of the object under study.
- *Markovian SAMs* – they are built based on the theory of Markovian and semi-Markovian RPs, representing the behavior of the studied object as probabilistic transitions between states depending only on the probabilistic characteristics of the current state. Markov model research is concerned with solving systems of probabilistic equations for transitions between states under certain conditions, providing normalized and unbiased estimates.
- *Queuing SAMs* – they represent networks of service structures (single-channel or multi-channel QNs) through which flows of service requests pass. This determines the workload of each individual structure and the system as a whole. To describe the behavior at the different levels, probabilistic equations, most often of the Markov type, are applied.

Hybrid approach is preferred when conducting investigations of complex systems in order to preserve the interrelationship between individual processes or components. This is usually a combination of two or more techniques in model development. The creation of a common model allows applying different techniques falling into one of the above approaches, while applying techniques from the two different approaches necessitates developing a hierarchical structure of relatively independent sub-models. In both cases, the goal is to build a maximally adequate model for studying a real system, especially if it has a dynamic character. Such a task is the identification of a linear dynamic object and evaluation of its response to impacts. It is possible to determine the response $y(t)$ to an arbitrary input action $x(t)$ using the Duhamel integral if the step response is known ($\dot{x}(\tau)$ is the time derivative of the action:

$$y(t)=x_0\cdot h(t)+\int_0^t \dot{x}(\tau)\cdot h(t-\tau)d\tau .$$

To solve the given task, in (Karaseva & Semenkin, 2021) it is proposed to create an analytical model representing a differential equation in symbolic form based on data for a dynamic system. This initialization allows the identification problem to be reduced to a symbolic regression problem, the solution of which can be obtained by applying a tree-structured genetic programming algorithm. The presented hybrid approach combines the modified genetic programming functionality for encoding a differential equation in tree form with the opportunities of differential evolution method to optimize the expression's constants and initial conditions. In this case, the task of the genetic programming algorithm is to automatically generate the structure of the differential equation, and the differential evolution method is used to optimize numerical parameters (the vector of constants and initial conditions). The proposed hybrid evolutionary approach enables efficient modeling of dynamic objects with minimal error while being robust to data noise and small sample sizes.

2. DETERMINISTIC ANALYTICAL MODELING BY USING APPROXIMATION

2.1. Analytical Approximation and Mathematical Description

Using analytical approximation in DAM development requires defining a functional relationship $Y=f(X)$ between the set Y of the dependent parameters (the output responses) and the set X of the independent parameters (the external impacts). The analytical description is presented in the form of one or several (system) equations, which are implemented programmatically in a selected language environment. Such a model is used in the study of functional interrelationship in a set of processes, looking for an analytical representation of the functional dependence based on the correlation between the ongoing processes. Usually, an adequate mathematical description of the conceptual model for the researched object is sought as a mathematical dependence (equation) under certain initial or boundary conditions.

Linear approximation is one of the most used in defining DAM. An example of a market trend research model based on linear approximation is given below. The main model parameters are the offered volume of a given product on the market (S) and its price (P). They are located depending on the time T and each other. For example, as demand increases (output supplied decreases), price P will also increase, while as supply S increases, price P will decrease. This allows to assume a linear model for the market trends, defined by the linear dependencies:

$S(t)=a+b.P(t-1)$ and $P(t)=c-d.S(t)$

The model thus defined describes the market system through an iterative process over time (a sequence of discrete steps). An exemplary implementation of the model for several consecutive steps is presented in Figure 5, with the following values set for the coefficients a=40; b=0.5; c=40; d=0.5 and initial price for production $P(0)$=10

Figure 5. Realization of model based on linear approximation

t =	*1*	*2*	*3*	*4*	*5*	*6*
S(t)	45	48.75	47.81	48.05	47.99	48.05
P(t)	17.5	15.63	16,09	15.98	16.02	15.98

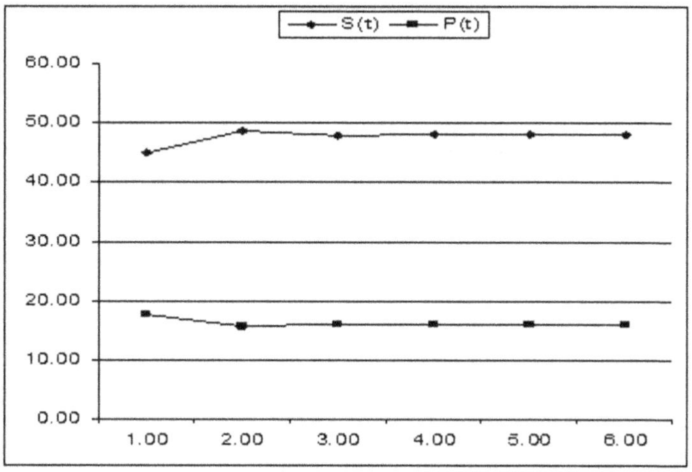

Analytical approximation is also applied in workflow (WL) modeling. It is known that, in the general case, WL in computer systems has a stochastic nature – it obeys a certain probability law (distribution function) $f(x)$. In this case, a tabular description of the function in the form $<x_i, f(x_i)>$ is used for $i=1,2,\ldots, N$, and the workflow model (WFM) is implemented according to the following procedure illustrated in Figure 6.

To determine a particular value $f(A)$ of the function $f(x)$, a check is made whether $x_{MIN} \leq A \leq x_{MAX}$ is satisfied. If the latter is provided, there are two possibilities:

a) The argument A matches a value xi set in the table and then a value B=y_i is determined for the function;
b) b) If the argument A is between two set values x_i and x_{i+1}, interpolation B=y_i+d is done, as

$$\frac{d}{''y} = \frac{a}{''x} \Rightarrow d = \frac{a.''y}{''x} = \frac{(A-x_i)(y_{i+1}-y_i)}{(x_{i+1}-x_i)} \Rightarrow B = y_i - \frac{(A-x_i)(y_{i+1}-y_i)}{(x_{i+1}-x_i)}.$$

The interpolation error is (B*-B).

Figure 6. Formation of current value f(A) when approximating

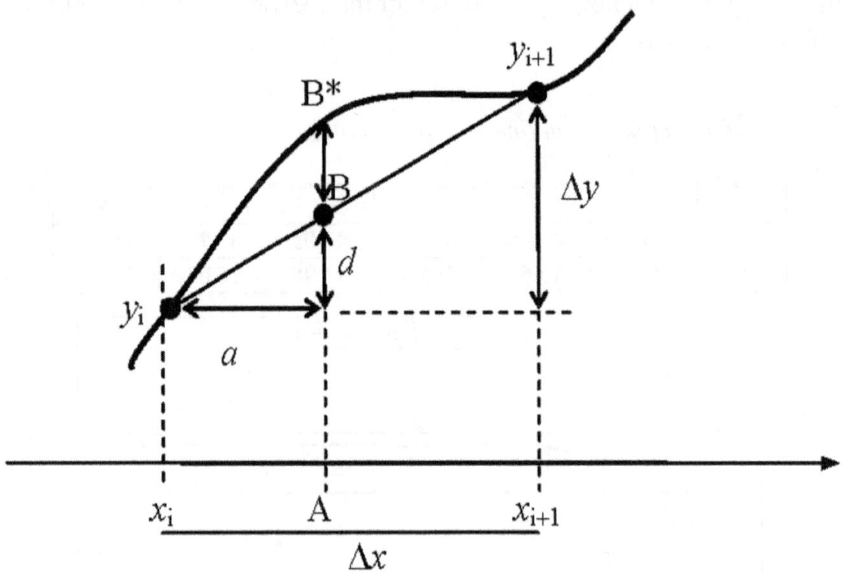

Mathematical description is another well-used technique in DAM development, and some simple examples for analytical description can be presented for information.

Mathematical description of the value of corporate management: $C=J_C+P_C+M_C+S_C$, where: J_C – amount set aside for insurance; P_C – amount for service personnel; M_C – equipment costs; S_C – service costs.

Mathematical model of linear motion and resistance, where <X_i, Y_i> are coordinates of the movement trajectory at time t_i; B is the resistance force of the moving body; V is speed; c_0, c_1, c_2, are resistance coefficients (positive values); $P=V_\tau$ refers to the time interval $\tau = t_1-t_0$; α is the azimuth measured in the north direction.

$X_1=X_0+R.\sin\alpha$ $Y1_{=}Y0_{+}R.\cos\alpha$ $B=c0+_{c}1._{V}+c2._{V}2$

Mathematical model of trajectory (g – gravity coefficient; V – speed; Θ – angle between the trajectory and the horizon; D – aerodynamic drag; M – body mass):

(a) angular rate of change: $\frac{d\theta}{dt}=\frac{g.\cos\theta}{V}$

(b) acceleration: $\frac{dV}{dt}=-\frac{D}{M}+g.\cos\theta$

Mathematical model of sensor for operating errors (W_{ij} – number of errors occurring between phases *i* & *j*; Q_j – number of operations to be executed in phase *j*; P_i – number of working nodes in phase *i*; R_{ij} – linear distance between *i* and *j*):

$$W_{ij} = \frac{Q_j / R_{ij}.P_i}{\sum_{j=1}^{m}\left[Q_j / R_{ij}\right]}; i = 1 \div n; j = 1 \div m$$

Some approaches for mathematical description in the field of computing and electronics are presented below.

Deterministic analytical model of a multiprocessor array structure. This type of architecture has the property of regularity, enabling a low-cost VLSI implementation, flexible structure and simple control of the execution and communication processes. In this reason the array architecture can be adapted to many applications to achieve high-speed computations. The array connection permits to realize in parallel structures and particular in the wave front array (WFA) structure with different strategies for calculation and scheduling (Kung, 2020). The very important problem of the WFA design is to keep the computational effectiveness and high system performance. This problem could be decided by restructuring of processor array on the base of fault detection on PE level. This can be solved by building a reconfigurable WFA (R-WFA) and the mathematical description for analytical model design is presented below.

Each WFA processor is *N*-dimensional structure built by peripheral *PE*[1,*j*] and internal *PE[i,j]* (*i*>1, *j*>1) processors. Each *PE*[*i,j*] (*i,j*=1÷*N*) communicate with the neighbors in the row *PE*[*i*+1,*j*] and in the column *PE*[*i,j*+1] and forms wave front WF_j for single calculation. The full task execution is realized by moving the wave calculation from *PE*[1,1] to *PE*[*N,N*] that defines a sequence of (2*N*-1) wave fronts $\{WF_1 \rightarrow WF_2 \ldots \rightarrow WF_{2N-1}\}$ (Figure 7a). Each processor realizes two phases – data loading (by communication time CT) and real data processing (by processing time PT). Two mathematical parameters are determined: $(t_c \pm \Delta t_c)$ for CT; $(t_p \pm \Delta t_p)$ for PT. Because each $PE \in WF_j$ send the results in different times to the neighbors it is possible to define that for each wave front the start time is defined by $min\{t_c \pm \Delta t_c\}$ and the final time – by $max\{t_c \pm \Delta t_c\}$. It is possible to define for the ideal case the next equations:

If for $\forall WF_j$, PT=const & CT=const $\Rightarrow T^*_{ideal}=(2N\text{-}1)\times$ (PT+CT).

The formal analytical time profile (Figure 7b) shows that the WF_j starts with the beginning of the first operation in any $PE \in WF_j$ and finishes after the full computation of the data in all $PE \in WF_j$. This processing realization defines the times $t_p\left(PE_{ij}\right) = \overline{t_p} \pm \Delta t_p$ and $t_c\left(PE_{ij}\right) = \overline{t_c} \pm \Delta t_c$.

Figure 7. Realization of a wave front processing in asynchronous WFA

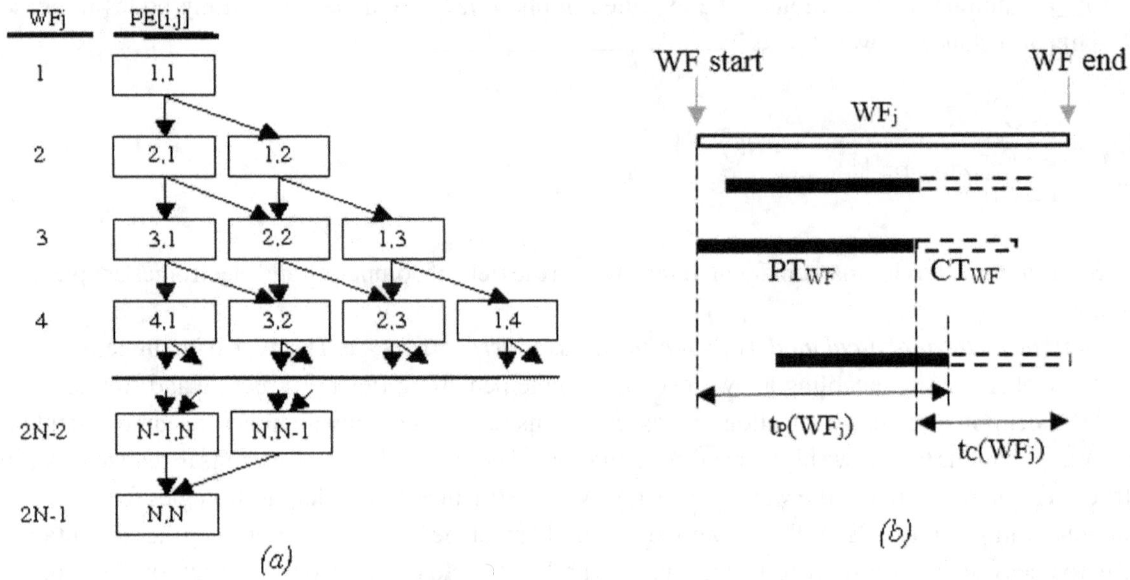

For mathematical description of the investigated processes in the WFA structure is assumed that the input task flow is regular and the ranges $t_p=[t_{p\,min}, t_{p\,max}]$ and $t_c=[t_{p\,min}, t_{c\,max}]$ could be calculated because they depend on known conditions. Some deterministic analytical parameters are defined and presented in the Table 1.

Table 1. Analytical description of investigated parameters

Processing time	$PT = \sum_{j=1}^{2N-1} t_p^{(WFj)} \equiv \sum_{j=1}^{2N-1} PT[WFj] \Rightarrow PT = (2N\text{-}1).t_p$
Communication time	$CT = \sum_{j=1}^{2N-1} t_c^{(WFj)} \equiv \sum_{j=1}^{2N-1} CT[WFj] \Rightarrow CT=t_c+(2N\text{-}1).t_c$
Total time for task execution	$T=T(Z_q)=PT+CT \Rightarrow T = (2N\text{-}1)(t_p+t_c) + t_c$
Factor of efficiency	$\chi = PT/T = (2N\text{-}1)t_p /T$
Relative workload for wave front	$\eta = \chi\,(2N\text{-}1)^{-1}$
Coefficient of granularity	$\xi = PT/CT$

2.2. Analytical Modeling of Power Electronic Converter by Means of Fuzzy Logic

Power electronics is an area of electrical engineering that deals with the creation of efficient electronic devices and systems for controlling energy flows by means of semiconductor electronics. This field includes devices for converting electrical energy from one form to another, such as changing the voltage, frequency, or form of an electrical current. Power electronics are used in many applications, such as electronics for AC and DC motors, power supplies for computers and other electronic devices, solar panels, inverters for solar systems, power regulators for electric cars, and more. Power electronics is an important field that helps to optimize the energy efficiency of both many electronic devices and realize innovative technologies. On the other hand, power electronics as an interdisciplinary field provides good opportunities for applying artificial intelligence techniques in the design and prototyping of power electronic devices and systems. The development of modern information and communication technologies, as well as computational mathematics and mathematical software, help to stimulate this process (Andries et al., 2017). In this regard, a process of development of the classic methods for designing power electronic devices and systems is observed in the direction of their automation and formalization through various innovative approaches. Fuzzy manufacturing rules are widely used in control synthesis of power electronic devices and systems. However, such rules are rarely used to guarantee their performance. The approach presented in the manuscript is based on (Dubois & Prade, 2006), the elements of fuzzy set theory (including fuzzy numbers, their cores and support).

2.2.1 Reasoning With Fussy Production Rules

The most popular fuzzy production rules have form:

IF U_1 is A_1 AND U_2 is A_2 AND … AND U_n is A_n THEN V is B,

where $U_1, U_2, \ldots, U_n, V$ are variables which take their values in base sets denoted by $X_1, X_2, \ldots, X_n, Y$ respectively, and $A_1, A_2, \ldots, A_n, B$ are fuzzy subsets of these sets (Hinov et al., 2022a). An extension of fuzzy production rules is (μ is a certainty degree of the rule):

IF U_1 is A_1 AND U_2 is A_2 AND … AND U_n is A_n THEN V is $B(\mu)$.

Let $A_1', A_2', \ldots, A_n', B'$ be fuzzy sets in respectively, and let facts U_1 is A_1', U_2 is A_2', …, U_n is A_n' be given. The following inference can be considered:

U_1 is A_1', U_2 is A_2', …, U_n is A_n'

IF U_1 is A_1 AND U_2 is A_2 AND … AND U_n is A_n THEN V is B

V is B',

where the fuzzy value B' is defined by:

$$B'(y) = \mathrm{S}_{i \in \{1, 2, \ldots, n\}} B_i'(y), \; y \in Y,$$

$$B_i'(y) = \sup_{x_i \in X_i} \left[A_i'(x_i) \text{ T } F_{A_i \to B}(x_i, y)\right],$$

$$F_{A_i \to B} : \ X_i \times Y \to [0, 1].$$

$F_{A_i \to B}$ is a fuzzy relation, *T* is a T-norm, and *S* is a S-norm; in the present paper, the relation satisfies formula:

$$B_i'(y) = \sup_{x_i \in X_i} \left[A_i'(x_i) \text{ T } F_{A_i \to B}(x_i, y)\right]$$

and the T-norm being used is the minimum operator.

Example 1: A fuzzy production rule with one variable in its antecedent

Let features are defined in Z = {1, 2, 3, 4, 5} by linguistic values in the following manner (shown visually by the data presented in Table 2).

Table 2. Fuzzy values

Value	*Membership*				
	1	**2**	**3**	**4**	**5**
very low	1.0	0.6	0.0	0.0	0.0
low	1.0	0.8	0.1	0.0	0.0
more or less low	1.0	1.0	0.3	0.0	0.0
medium	0.0	0.5	1.0	0.5	0.0
more or less high	0.0	0.0	0.3	1.0	1.0
high	0.0	0.0	0.1	0.8	1.0
very high	0.0	0.0	0.0	0.6	1.0

Let the rule "IF *K* is *high* THEN *P* is *medium*" holds, and let the following fact is given: "*K* is *more or less high*". According to:

$F_{high \to medium}(x,y) = 1 \text{ T } [1 - high(x) + medium(y)]$.

This relation can be represented by:

$$F^{high \to medium} = \begin{Vmatrix} 1 & 1 & 1 & 1 & 1 \\ 1 & 1 & 1 & 1 & 1 \\ 0.9 & 1 & 1 & 1 & 0.9 \\ 0.2 & 0.7 & 1 & 0.7 & 0.2 \\ 0 & 0.5 & 1 & 0.5 & 0 \end{Vmatrix},$$

where $f_{x,y}^{high \to medium}$, $x, y \in Z$, satisfies $f_{x,y}^{high \to medium} = F_{high \to medium}(x, y)$.

A conclusion based on using the previous equations gives:

$P(y)= \sup_{x \in Z}[more\_or\_less\_high(x) \text{ T } F_{high \to medium}(x,y)],\ y \in Z.$

The calculated membership values of P are:

$P = \langle 0.3, 0.7, 1, 0.7, 0.3 \rangle.$

Example 2: A fuzzy production rule with two variables in its antecedent

Let the following rule holds:

IF *L* is *very low* AND *M* is *low* THEN *Q* is *very high.*

The variables *L, M* and *Q* are defined in base set *Z* too. Let two facts be given:

L is *low, M* is *very low.*

Relations

$F_{very\ low \to very\ high}$ and $F_{low \to very\ high}$

can be represented by:

$$F^{very\ low \to very\ high} = \begin{Vmatrix} 0 & 0 & 0 & 0.6 & 1 \\ 0.4 & 0.4 & 0.4 & 1 & 1 \\ 1 & 1 & 1 & 1 & 1 \\ 1 & 1 & 1 & 1 & 1 \\ 1 & 1 & 1 & 1 & 1 \end{Vmatrix};$$

$$F^{low \to very\ high} = \begin{Vmatrix} 0 & 0 & 0 & 0.8 & 1 \\ 0.2 & 0.2 & 0.2 & 0.9 & 1 \\ 0.9 & 0.9 & 0.9 & 1 & 1 \\ 1 & 1 & 1 & 1 & 1 \\ 1 & 1 & 1 & 1 & 1 \end{Vmatrix}.$$

The conclusions based on the last two expressions give for each $y \in Z$

$$Q'(y) = \sup_{x \in Z}\left[low(x)\ \mathrm{T}\ F_{very\ low \to very\ high}(x, y)\right];$$

$$Q''(y) = \sup_{x \in Z}\left[very\_low(x)\ \mathrm{T}\ F_{low \to very\ high}(x, y)\right].$$

This approach is discussed in the previous example, and the respective results are:

$$Q' = \langle 0.4,\ 0.4,\ 0.4,\ 0.8,\ 1 \rangle;$$

$$Q'' = \langle 0.2,\ 0.2,\ 0.2,\ 0.6,\ 1 \rangle.$$

The common value of Q is:

$$Q = Q'\ \mathrm{S}\ Q'' = \langle 0.4,\ 0.4,\ 0.4,\ 0.8,\ 1 \rangle.$$

2.2.2 Results on a Buck DC-DC Power Converter Parameters

A Buck DC-DC power converter with PID controller is described in (Hinov et al., 2022b). The following its parameters are used in the present paper:

E = 24 V ± 35% ; Utar = 10 V ;

R = 1 Ω ± 35% ; L = 20 µF ± 35% ; C = 20 µF ± 15% ;

Kp = 2 ± 15%, K_d=0 .

where *E* denotes an input voltage, *Utar* – a target output voltage, *R* – an output resistance, *L* – an inductance, *C* – a capacitance, *Kp, Ki* and K_d – coefficients of a PID controller.

Estimations of two output parameters are shown on Figure 8 *URmax* is a maximal output voltage. *Delay* is a setup time on a step response, i.e. time after which the output voltage belongs to interval [9.9, 10.1] [V].

Figure 8. Results of simulations with authors' software based on three kinds of modeling on the inductance, the input voltage and the resistance

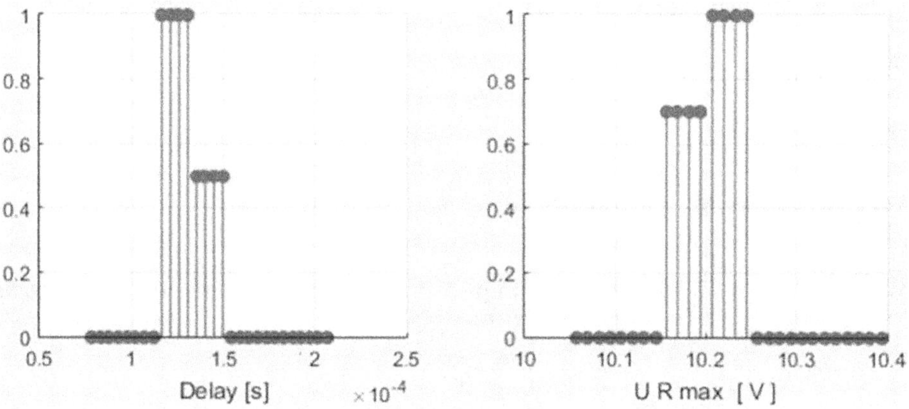

(a) with single intervals

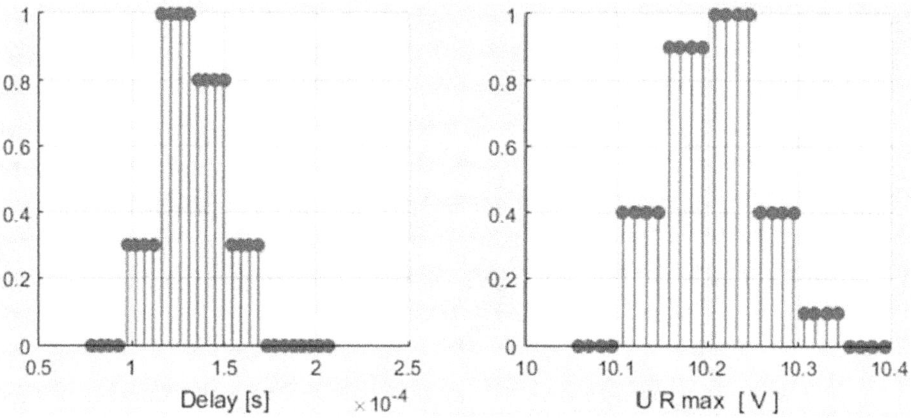

(b) with triangular fuzzy numbers

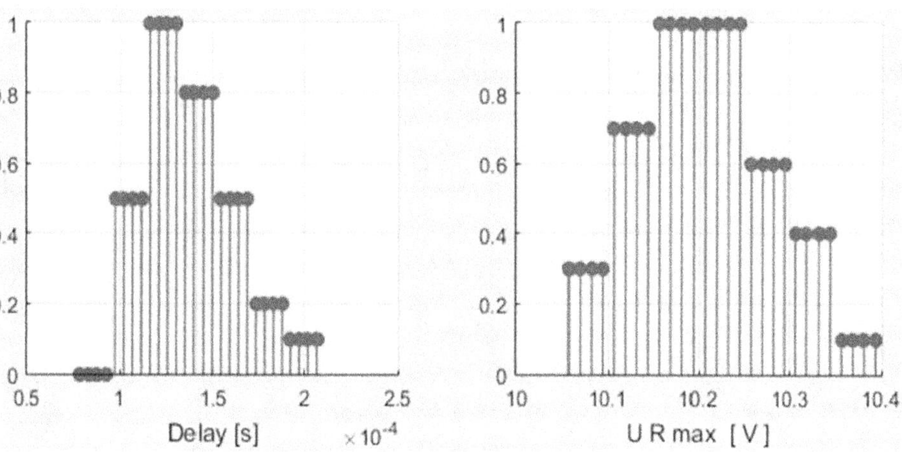

(c) with trapezoidal fuzzy numbers

Possibility distributions use the following default degrees of membership to intervals:

$I_{Par,3} \equiv [\ 0.85\ Par, 0.95\ Par\ ]$: 0.5

$I_{Par,4} \equiv [\ 0.95\ Par, 1.05\ Par\ ]$: 1,0

$I_{Par,5} \equiv [\ 1.05\ Par, 1.15\ Par\ ]$: 0.5

Par $\in$ {E, R, L, C, Kp, Ki}.

The chosen tolerance is 15% (intervals with juxtaposed zero values are not taken into account).

$I_{Par,1} \equiv [\ 0.65\ Par, 0.75\ Par\ ]$: 0

$I_{Par,2} \equiv [\ 0.75\ Par, 0.85\ Par\ ]$: 0

$I_{Par,6} \equiv [\ 1.15\ Par, 1.25\ Par\ ]$: 0

$I_{Par,7} \equiv [\ 1.25\ Par, 1.35\ Par\ ]$: 0

A greater tolerance (35%) and possibility distributions over the upper intervals is used on some parameters in order proper dependencies of output characteristics to be explored.

Example 3: Differences from these default degrees are presented in three cases:
Case A1: $I_{Par,4} \equiv [\ 0.95\ Par, 1.05\ Par\ ]$: 1, Par$\in${E, R, L}.

(intervals on *Par* with juxtaposed zero values are not noted). Respective degrees of membership on the inductance are shown on Figure 9). Proportional intervals and same degrees on the input voltage and the resistance are used. Single intervals are applied (the above defined values for $I_{par,4}$ and Par).

Figure 9. Representations in an authors' software based on three kinds of modeling on the inductance

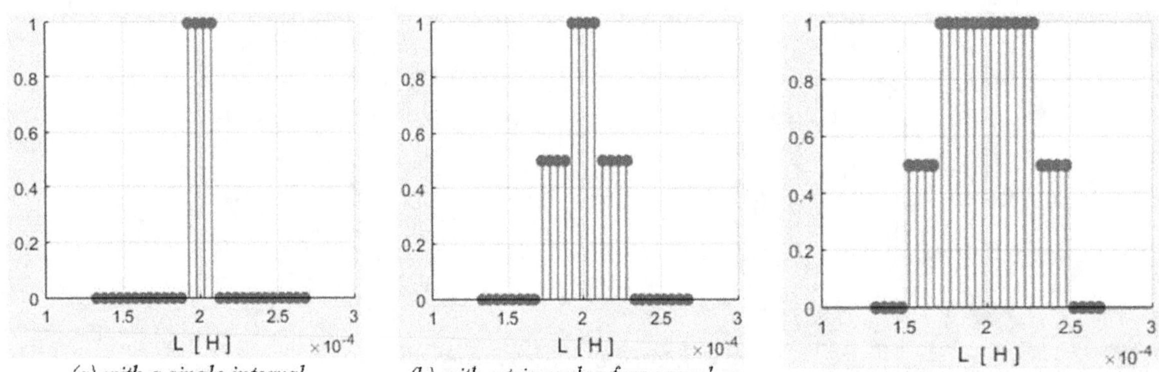

(a) with a single interval *(b) with a triangular fuzzy number* *(c) with a trapezoidal fuzzy number*

Case B1: There is no difference (see Figure 9b). Triangular fuzzy numbers are applied according to (the above defined values for $I_{par,3}$; $I_{par,4}$; $I_{par,5}$ and Par) on all parameters.

Case C1:

$I_{Par,1} \equiv [\ 0.65\ Par,\ 0.75\ Par\ ]: 0.0$

$I_{Par,2} \equiv [\ 0.75\ Par,\ 0.85\ Par\ ]: 0.5$

$I_{Par,3} \equiv [\ 0.85\ Par,\ 0.95\ Par\ ]: 1.0$

$I_{Par,4} \equiv [\ 0.95\ Par,\ 1.05\ Par\ ]: 1.0$

$I_{Par,5} \equiv [\ 1.05\ Par,\ 1.15\ Par\ ]: 1.0$

$I_{Par,6} \equiv [\ 1.15\ Par,\ 1.25\ Par\ ]: 0.5$

$I_{Par,7} \equiv [\ 1.25\ Par,\ 1.15\ Par\ ]: 0.0$

$Par \in \{E, R, L\}$.

Respective degrees of membership on the inductance are shown on Figure 9c). Trapezoidal fuzzy numbers are applied according to (the above defined values for $I_{par,1}$; $I_{par,2}$; $I_{par,3}$; $I_{par,4}$; $I_{par,5}$; $I_{par,6;}$ $I_{par,7}$ and Par) on the inductance, the input voltage, and the resistance.

These three cases determine the output voltage parameters, which depend on increasing range on *E*, *R* and *L* (Figure 8).

Example 4: Differences from the default degrees are presented in four cases.

Case A2:

$I_{L,1} \equiv [\ 0.65\ L,\ 0.75\ L\ ]: 1.0$

$I_{L,2} \equiv [\ 0.75\ L,\ 0.85\ L\ ]: 1.0$

$I_{L,3} \equiv [\ 0.85\ L,\ 0.95\ L\ ]: 1.0$

$I_{L,4} \equiv [\ 0.95\ L,\ 1.05\ L\ ]: 0.5$

$I_{R,5} \equiv [\ 1.05\ R,\ 1.15\ R\ ]: 0.5$

$I_{R,6} \equiv [\ 1.15\ R,\ 1.25\ R\ ]: 1.0$

$I_{R,7} \equiv [\ 1.25\ R,\ 1.15\ R\ ]: 1.0$

(Here and below intervals on L and R with juxtaposed zero values are not noted). Respective degrees of membership on the resistance and the inductance are shown on Figure 10. They determine the following degrees of membership on the maximal output voltage intervals via authors' software (Kumar et al., 2021). This graphical dependence is shown on Figure 11. The following values are set for changing parameters:

$I_{URmax,1} \equiv [\ 10.18, 10.24\ ]$ [V]: 0.4

$I_{URmax,2} \equiv [\ 10.24, 10.30\ ]$ [V]: 0.9

$I_{URmax,3} \equiv [\ 10.30, 10.36\ ]$ [V]: 1.0

$I_{URmax,4} \equiv [\ 10.36, 10.42\ ]$ [V]: 0.7

$I_{URmax,5} \equiv [\ 10.42, 10.48\ ]$ [V]: 0.4

$I_{URmax,6} \equiv [\ 10.48, 10.54\ ]$ [V]: 0.2

$I_{URmax,7} \equiv [\ 10.54, 10.6\ ]$ [V]: 0.1

Figure 10. Representations in the authors' software based on modeling

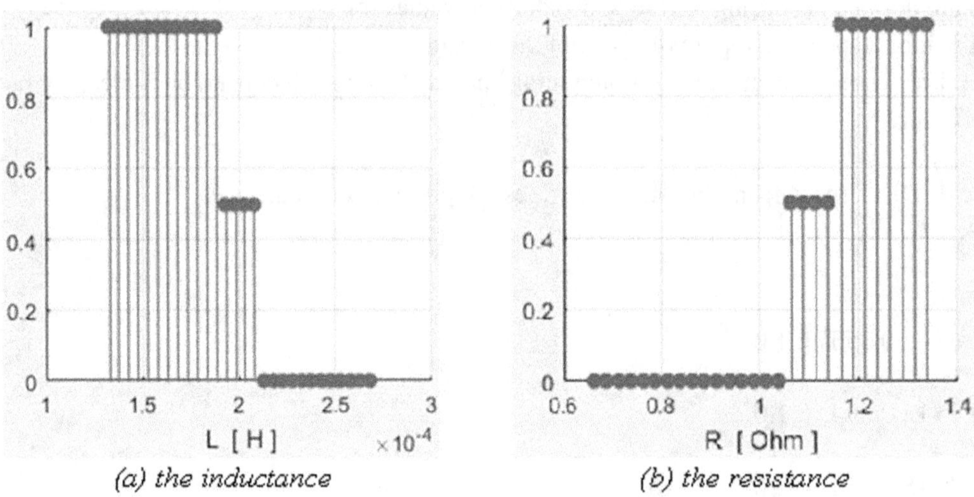

Figure 11. Results of simulations with authors' software based on membership degrees on the inductance and the resistance

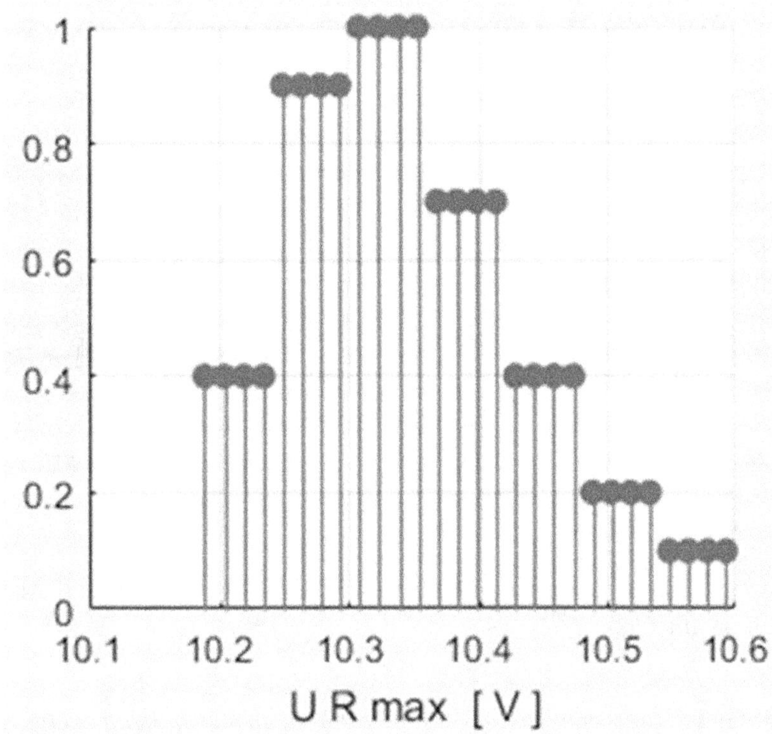

These degrees of membership form value:

$$V_{\mathrm{UR}}^{1} = \langle 0.4, \quad 0.9, \quad 1, \quad 0.7, \quad 0.4, \quad 0.2, \quad 0.1 \rangle$$

and fuzzy production rule: "IF L is *more or less low* AND R is *high* THEN *URmax* is V_{UR}^{1} " (the values on resistance and inductance based on (*case A2*) and shown on Figure 12 are interpreted as "more or less low" and "high" respectively).

Figure 12. Representations in the authors' software based on modeling

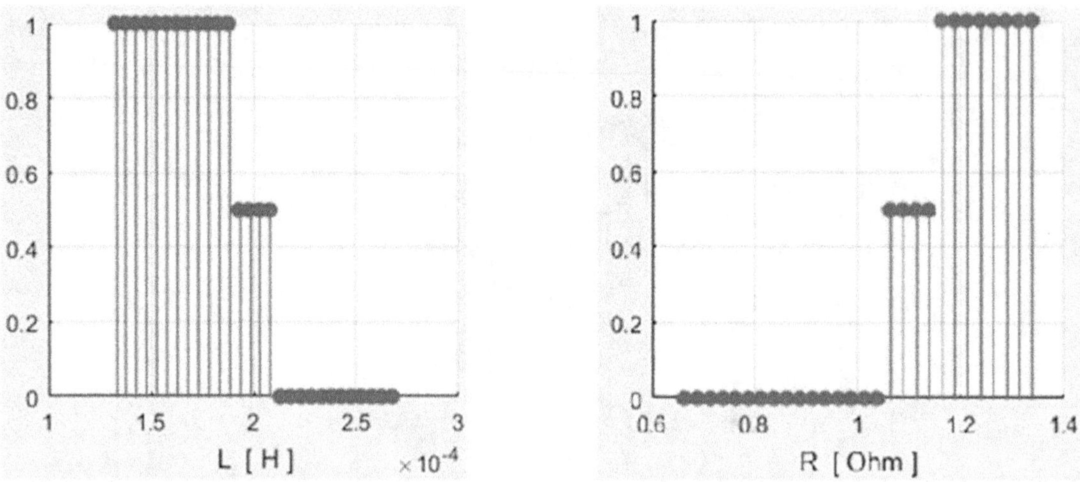

(a) the inductance *(b) the resistance*

Figure 13 shows the results regarding the output voltage of the converter at the last variation of the values of the circuit parameters.

Figure 13. Results of simulations with authors' software based on membership degrees on the inductance and the resistance

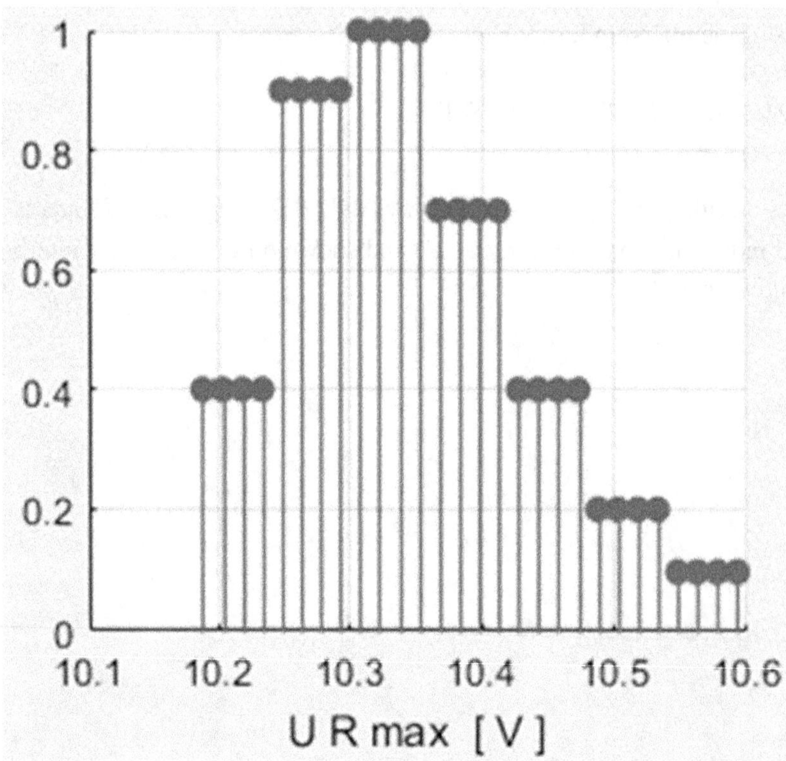

Let the following three degrees on the inductance be presented in the next three cases:

$I_{L,1} \equiv [\ 0.65\ L, 0.75\ L\ ]: 1$;

$I_{L,2} \equiv [\ 0.75\ L, 0.85\ L\ ]: 1$;

$I_{L,3} \equiv [\ 0.85\ L, 0.95\ L\ ]: 1$.

Case B2:

$I_{R,6} \equiv [\ 1.\ 15\ R, 1.25\ R\ ]: 1$;

$I_{R,7} \equiv [\ 1.25\ R, 1.35\ R\ ]: 1$.

The following degrees of membership on the maximal output voltage intervals are determined by the authors' software according to:

$$V_{\mathrm{UR}}^{2,1} = \langle 0.3,\ \ 0.9,\ \ 1,\ \ 0.6,\ \ 0.2,\ \ 0.1,\ \ 0 \rangle$$

They defer from the respective ones in:

$$V_{\mathrm{UR}}^{2,2} = \langle 0.4,\ \ 0.9,\ \ 1,\ \ 0.7,\ \ 0.4,\ \ 0.2,\ \ 0.1 \rangle$$

which is calculated in inference according to possibility distributions. The resultant mean square error on membership degrees is 11%.

Case C2:

$I_{R,5} \equiv [\ 1.05\ R, 1.15\ R\ ]: 0.2$

$I_{R,6} \equiv [\ 1.\ 15\ R, 1.25\ R\ ]: 1.0$

$I_{R,7} \equiv [\ 1.25\ R, 1.35\ R\ ]: 1.0$

The respective values on the maximal output voltage intervals determined by the authors' software and an inference are:

$$V_{\mathrm{UR}}^{3,1} = \langle 0.4,\ \ 1,\ \ 1,\ \ 0.6,\ \ 0.3,\ \ 0.1,\ \ 0 \rangle$$

$$V_{\mathrm{UR}}^{3,2} = \langle 0.4,\ \ 0.9,\ \ 1,\ \ 0.7,\ \ 0.4,\ \ 0.2,\ \ 0.2 \rangle$$

with mean square error 11% too.

Case D2:

$I_{R,5} \equiv$ [1.05 R, 1.15 R]: 0.5

$I_{R,6} \equiv$ [1. 15 R, 1.25 R]: 1.0

$I_{R,7} \equiv$ [1.25 R, 1.35 R]: 1.0

The respective values on the maximal output voltage intervals determined by the authors' software and an inference are:

$$V_{\mathrm{UR}}^{4,1} = \langle 0.5,\ \ 1,\ \ 0.9,\ \ 0.5,\ \ 0.2,\ \ 0.1,\ \ 0 \rangle$$

$$V_{\mathrm{UR}}^{4,2} = \langle 0.5,\ \ 0.9,\ \ 1,\ \ 0.7,\ \ 0.5,\ \ 0.5,\ \ 0.5 \rangle$$

with mean square error 28%.

Summary. The fuzzy production rules, which are applied on power system parameters gives satisfactory results, mostly in cases in which facts are modeled as fuzzy numbers with supports, which are subsets of cores of fuzzy numbers in these fuzzy production rules (particularly, in the presented Case B2 membership degrees on the voltage match the respective ones from the fuzzy value in the consequent of the rule). More generally, in cases in which facts are modeled as fuzzy numbers, which are subsets of fuzzy numbers in these fuzzy production rules (particularly, in the presented Case C2), the results can be satisfactory too. However, in case in which facts are modeled by a fuzzy number, which is equal to a fuzzy number in the rule, the lowest membership degree matches the lowest membership degree in the support of this fuzzy number (particularly, in the presented Case D2).

The crisp sets are a special case of the fuzzy sets. Therefore, the last ones give a more detailed view on possibilities (a given parameter can have various crisp values). Cores and supports of fuzzy sets (as interval estimations) provide less information; such intervals are not considered in the present paper, but they are popular. Indeed, in this case all membership values are equal to one or zero (in the context of the fuzzy sets theory). Particularly, in case A1 supports and cores match with length 0.1 Par. Similarly, in cases B2, C2 and D2, all inductances satisfy with intervals length 0.3 L.

Generally, the results, which are obtained with the authors' software, are based on many simulations (Gochev et al., 2021). Oppositely, the results, which are obtained with the fuzzy production rules, use simple formulae in order membership degrees to be rapidly estimated. The given mean square errors show the difference between these two approaches.

The concept of linguistic variables as a tool in approximate reasoning is widely implemented. There are such variables in the present paper too; there are fuzzy values on *Delay* and *URmax*, which can be interpreted more or less as "medium", but they are specific, and they do not cover the respective linguistic variable from the table. The used rounding in all membership values is the same.

The presented approach gives good results in the training of power electronics, as it implements in itself the application of artificial intelligence techniques and modern information and communication technologies.

3. DETERMINISTIC ANALYTICAL MODELING BY USING GRAPHS

Graph models are used in the investigation of sufficiently deterministic systems and processes, where the structure and particular values of the stochastic characteristics are known. Such a system or process can be represented by a finite number of states, and the behavior by a finite sequence of transitions between them. This allows the model to be developed as a graph with a certain topology of connections, which is described by the sets of vertices and the edges between them, and to solve it according to the rules of graph theory.

A simple example of the application of graphs in the deterministic description of objects from a given community is a representation of the functional relationships between them, as well as the relationships between different communities (sets). For this purpose, it is necessary to define the relations of the functional image between the studied sets. For example, if the relation $\boldsymbol{R} \subset A \times B = \{(0,a), (0,b), (1,a), (2,b)\}$ is defined for the sets $A=\{0,1,2\}$ and $B=\{a,b\}$, its graph representation will have the form of Figure 14(a). Another example is shown in Figure 14(b). It presents the relation $\boldsymbol{R}=\{(a,b) \mid a$ divides b without remainder$\}$, where $a,b \in A=\{1,2,3,4\}$. The solution is $\boldsymbol{R} = \{(1,1), (1,2), (1,3), (1,4), (2,2), (2,4), (3,3), (4, 4)\}$.

In an analogous way, the tasks for the graph representation of relations can be solved:

$$\boldsymbol{R}_1 = \{(a, b) \mid a \leq b\}$$

$$\boldsymbol{R}_2 = \{(a, b) \mid a < b\}$$

$$\boldsymbol{R}_3 = \{(a, b) \mid a > b\}$$

$$\boldsymbol{R}_4 = \{(a, b) \mid a + b \leq 3\}$$

Figure 14. Graph presentation of binary relations

R	*a*	*b*
0	✓	✓
1	✓	
2		✓

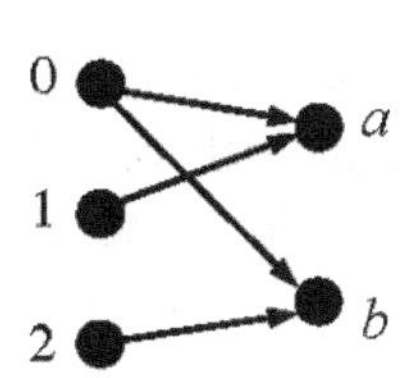

(a)

R	1	2	3	4
1	✓	✓	✓	✓
2		✓		✓
3			✓	
4				✓

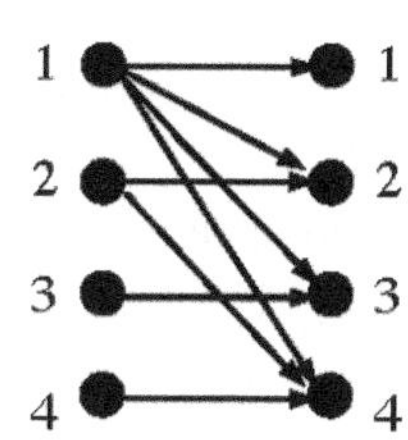

(b)

In the general case, any defined relational scheme can be represented by a directed graph with vertices - the elements of the interacting sets and edges defined by the existing relations between the elements. This is illustrated by the example of Figure 15 for the set $A=\{a,b,c,d\}$ and the relation $\boldsymbol{R}=\{(a,b), (a,d), (b,b), (b,d), (c,a), (c,b), (d,b)\}$.

Figure 15. Presentation of relations by using directed graph

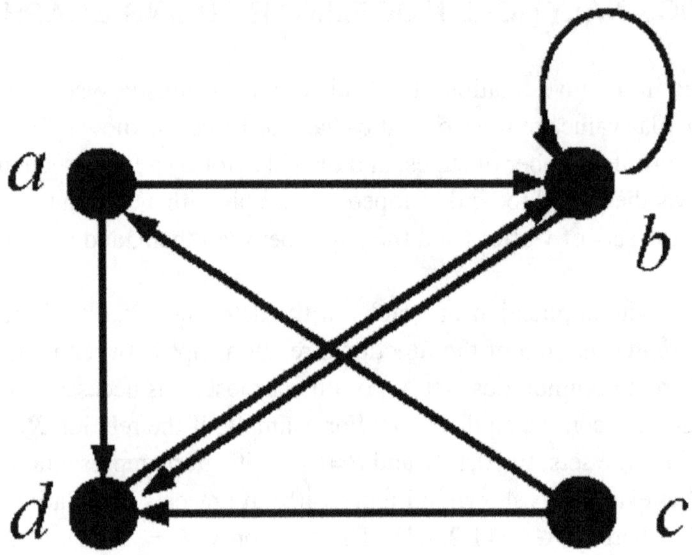

The concept of Figure 15 can be extended by associating the nodes of the directed graph with the basic states for an object under study. An illustration of this approach is presented in Figure 16 graph model for predicting the number of a given population. The following assumptions were made in its definition.

- The initial size of the population and the distribution $P(i,t=1)$ of the individuals in the different age groups $i = 1, 2,..., N$ are known, as well as the mortality and birth rates for each of them.
- There is no migration of individuals in the biological system.
- The functions of birth rate $B(i)$ and death rate $D(i)$ for each age group i are constant values for the studied period $t = 1, 2, ..., T_{max}$.
- After every interval t the individuals of the group are transferred to the next group $(i+1)$.

The graph model (Figure 16) is a directed graph with N nodes for each age group of individuals $P(1,t), ..., P(N,t)$ and a node *Dead* representing the occurrence of death for an individual from any group. This means that there exist directed arcs to the node *Dead* (absorbing node) from every other node $P(i,t)$ $(i=1,...,N)$ – the transition conditions are represented by $D(i).P(i,t)$. Each node $P(i,t)$ $(i=1,...,N)$ has two output arcs: ✓ to the next node $P(i+1, t)$ with the transition condition $[1-D(i)].P(i,t)$, and the last node $P(N,t)$ is reflexive; ✓ to the initial node $P(1,t)$ with the transition condition $B(i).P(i,t)$.

From the abstract graph description of the transitions in the biological system, a system of N equations with the possibility of an iterative solution can be compiled: ✓ newborns (i=1): $P(i,t+1)=[B(1).P(1,t)]+[B(2).P(2,t)]+...+[B(N).P(N,t)]$; ✓ groups ($1<i<N$): $P(i,t+1)=[1-D(i-1)].P(i-1,t)$; ✓ oldest group ($i=N$): $P(N,t+1)=[1-D(N-1)].P(N-1,t)+[1-D(N)].P(N,t)$.

The program model is a program description of the algorithm for solving the defined equations (the mathematical model), and for this purpose any universal language environment can be used. One possible implementation by using APL2 is presented in Figure 17. The model investigation is carried out at different values for the controllable parameters, so that positive and negative trends in the development of the population can be studied.

Figure 16. Graph DAM for the investigation of biological population

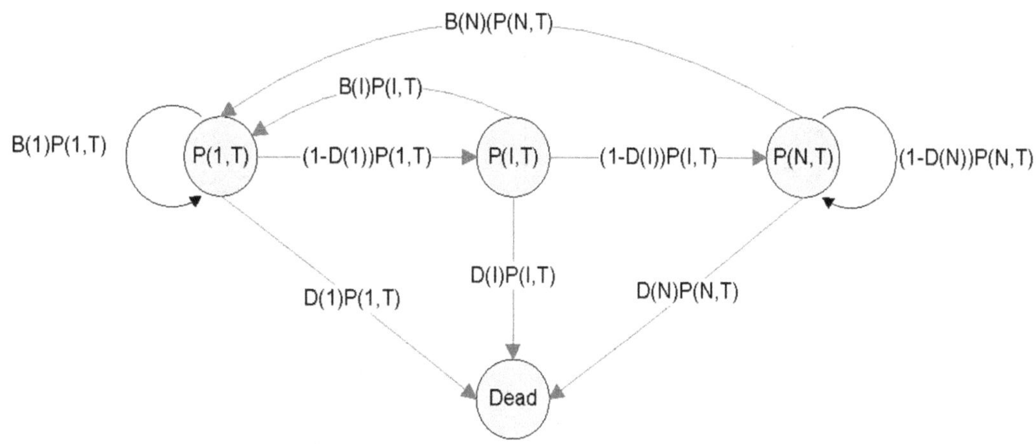

Figure 17. Program model of population

```
      ▼POPUL [□]▼
    ▼ POPUL
      'MODEL OF POPULATION
      '========================================'
      ' '
      'INPUT PARAMETERS:'
      'B '
      □←B←0  0.2  0.05  0  0
      'D:'
      □←D←0.02  0  0.05  0.15  0.25
      'P1:'
      □←P←PT1←P1←0  1000  600  300  100
      ' '
      'ENTER NUMBER OF YEARS:'
      TMAX←□←
      TABL←(TMAX,4) ρ0
      N←ρB
      NIG←N-2
      PP←TP←+/P
      T←0
    REP: →(TMAX<T←T+1)/OUT
      TP←+/P
      →(TP=0)/END
      DP←TP-PP
      PCP←DP÷TP
      □PP←5
      TABL[T; ]←T, TP, DP, PCP
      PT1[1] ← +/B×P
      PT1[N] ← ((1-D[N-1]) ×P[N-1]) + (1-D[N]) ×P[N]
      PT1[1+zNIG] ← (1-D[zNIG]) ×P[zNIG]
      P←ιPT1
      →REP
    END: 'YEAR'
      T
      'IS FINAL FOR POPULATION !'
      →0
    OUT: □PP←5
      ' '
      '     TIME      POP        ▲POP        (▲POP) ÷ POP'
      '-----------------------------------------------------'
      ' '
      TABL
    ▼
```

Model execution:

```
      POPUL
MODEL OF POPULATION
==========================================
INPUT PARAMETERS:
B :
0  0.2  0.05  0  0
D :
0.02  0  0.05  0.15  0.25
P1 :
0  1000  600  300  100

ENTER NUMBER OF YEARS:
□:
      15
```

TIME	POP	▲POP	(▲POP) ÷ POP
1	2000	0	0
2	2130	130	0.061033
3	1957	–43	–0.021972
4	1675	–325	–0.19403
5	1344	–656	–0.4881
6	1064	–936	–0.8781
7	823	–1177	–1.4301
8	631	–1369	–2.1696
9	478	–1522	–3.1841
10	359	–1641	–4.571
11	269	–1731	–6.4349
12	201	–1799	–8.9502
13	150	–1850	–12.333
14	112	–1888	–16,857
15	84	–1916	–22.81

To organize model experiments, it is necessary to define input parameters such as: N – number of age groups; B – vector of fertility rates for each of the groups i=1чN; D – vector of mortality rates for each group i=1чN; $P(i,T=0)$ – initial number of individuals by group "i"; *TMAX* – period (years) of conducting the research. The vectors B, D & $P(i,0)$ must be of equal power.

The output parameters and assessments are as follows: T (TIME) – period (years) of conducting the research; TP (POP) – total population for a given period; DP (ΔPOP) – change of the population number POP(T)-POP(T-1); PCP (ΔPOP/POP) – percentage of change.

4. DETERMINISTIC ANALYTICAL MODELING BY USING PETRI NETS

4.1 Definition of Petry Net (PN)

Petri nets (PNs) with their varieties are a simple and convenient apparatus for deterministic analytical modeling of synchronous and asynchronous processes with the possibility of parallelism. They were developed by Carl Petri (1962) as a branch of discrete mathematics (Cassandras et al., 2021).

The generalized *set-theoretic definition* of a Petri net is as an ordered triple $PN = (P, T, F)$ involving the sets: $P=\{p_1,p_2,\dots,p_m\}$ – set of elements of type "position"; $T=\{t_1,t_2,\dots,t_n\}$ – set of "transition" type elements; $F \subseteq (P \times T) \cup (T \times P)$ – set of relations (set of arcs). The sets P and T are finite and disjoint for which $m \geq 0$, $n \geq 0$ и $P \cap T = \varnothing$ hold. The set F includes ordered pairs of type (p_i, t_j) defining elements of the images $P \rightarrow T$ and $T \rightarrow P$. In this way, two functions $I, O \in \mathbf{N}^{|P| \times |T|}$ are defined, called input (function of preconditions) and output (function of post-conditions).

The input function I defines the so-called input positions for each transition, and the output function O – the output positions for the transitions. An example tabular definition of the two functions is shown in Figure 18. The corresponding input (for I) and output (for O) positions for the transitions are marked with "1".

Figure 18. Functions I and O

I	t_1	t_2	t_3	O	t_1	t_2	t_3
p_1	1			p_1			1
p_2		1		p_2	1		1
p_3		1		p_3	1		
p_4			1	p_4		1	

The extended definition of a Petri net $PN = (P, T, F, W, \mu_0)$ includes the additional elements: $W{:}F \rightarrow \{1,2,3,\ldots\}$ is function of the weight of each arc; $\mu_0{:}P \rightarrow \{0,1,2,3,\ldots\}$ – starting mark of the Petri net, through which its execution begins. The weights of W indicate how many links exist for each pair (p_i, t_j) associated with the functions I and O. According to the tables in Figure 18, the links are single. Multiple links are allowed for elements.

For the *graph definition* of PN, a directed multigraph is used with vertices - the elements of the sets P and T and arcs defined depending on the functions I and O. According to the definition of W, the graph is weighted, including two types of nodes, the so-called "conditions" (the positions) and "events" (the transitions). Links (arcs) connect only nodes of different types. Graphically, the positions are represented by a circle, and the transitions by a section (rectangle, square). Each link in this graph has a positive integer weight defined by W – it can be thought of as having several directed arcs between two nodes (a multigraph). For each transition (t_j), the incoming edges are defined by the precondition function I, and the outgoing arcs by the post-condition function O. These arcs also define the entry and exit positions for the transitions.

A set-transition and graph definitions of a PN from Figure 18 are presented in Figure 19. For example, for a transition t_2 according to the functions $I(t_2)$ and $O(t_2)$, positions p_2 and p_3 are input and position p_4 is output.

The "*marking*" μ of a PN is the image $\mu{:}P \rightarrow \mathbf{N}$ of the set of positions in the set of non-negative integers. Essentially, μ is an n-dimensional vector of integers ($n = |P|$), each of which represents the number of marks (pools) in the corresponding position. Each position can be marked with $k \geq 0$ marks, which is reflected in the graphical representation with the presence of k points in the position. A given mark μ_i

represents a specific state of the investigated system, where the presence of marks in the positions reflects the level of fulfillment of the corresponding conditions for the fulfillment of the events (realization or "ignition" of the transitions).

Figure 19. Example of PN definition

$$P = \{p_1, p_2, p_3, p_4\}$$
$$T = \{t_1, t_2, t_3\}$$

$I(t_1) = \{p_1\}$	$O(t_1) = \{p_2, p_3\}$
$I(t_2) = \{p_2, p_3\}$	$O(t_2) = \{p_4\}$
$I(t_3) = \{p_4\}$	$O(t_3) = \{p_1, p_2\}$

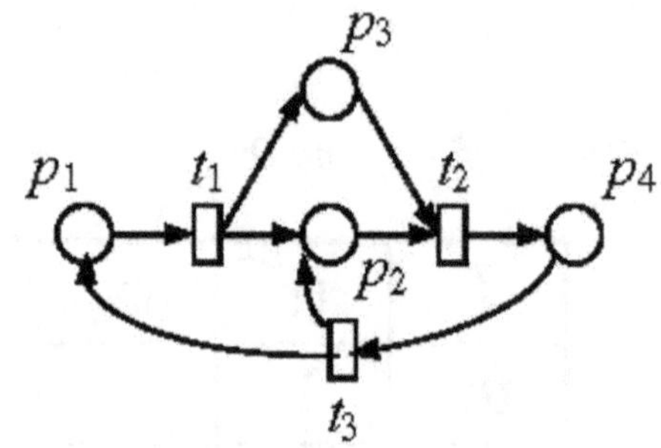

Investigation of the modeled system can be done through the so-called "execution" of the defined Petri net in which it "evolves", i.e., starting from a set initial mark μ_0, a chain of successive marks $\mu_0 \rightarrow \mu_1 \rightarrow \mu_2 \rightarrow \ldots$ is formed, reflecting each specific state. The evolution of the PN takes place upon "ignition" (activation) of a certain transition from the allowed transitions at the given time. This transition t_j is allowed for which the number of marks k_i in all its input positions pi is not less than the number of connecting input edges, i.e. . $k_i \geq W(p_i, t_j)$ – Figure 20.

Figure 20. Conditions for activation of a transaction

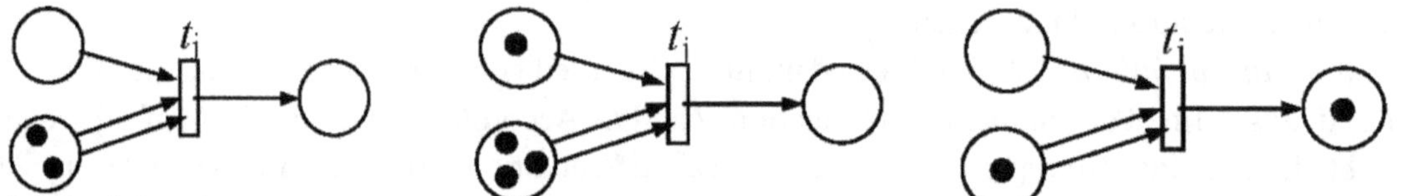

(a) Prohibited transaction *(b) Allowed transaction* *(c) Activating allowed transition*

After activating a given transition, as many marks are removed in its input positions as the weight of the incoming arcs (number of incoming arcs in a multigraph), and marks corresponding to the outgoing arcs for the transition are transferred to its output positions – Figure 20(c). The activation of a transition means that its corresponding event occurs.

An example of PN-modeling a non-distributed system of several computers (clients) connected to a common resource (printer) is shown in Figure 21. A shared resource (PRN) can only be used in monopoly mode (by one client) at a time. In the modeling network of Figure 21(b) the initial flag is placed in the PRN position, which acts as a semaphore – it allows the activation (ignition) of a transition for a given client. The internal structure of the model for a given client (Client) is given in Figure 21(c).

In order for the system to implement distribution service, it is necessary that communications be carried out through Message-Passing Channels, and the special thing is that a message can be sent only from one source and only to one recipient. Thus, a model of centralized ("client/server") service with a distributed resource RES is realized and shown in Figure 22.

It is possible during the evolution of the network to reach a state with no output ("passive" state), where no transition is allowed, and the execution is interrupted (the network is blocked).

Figure 21. Non-distributed computer systems

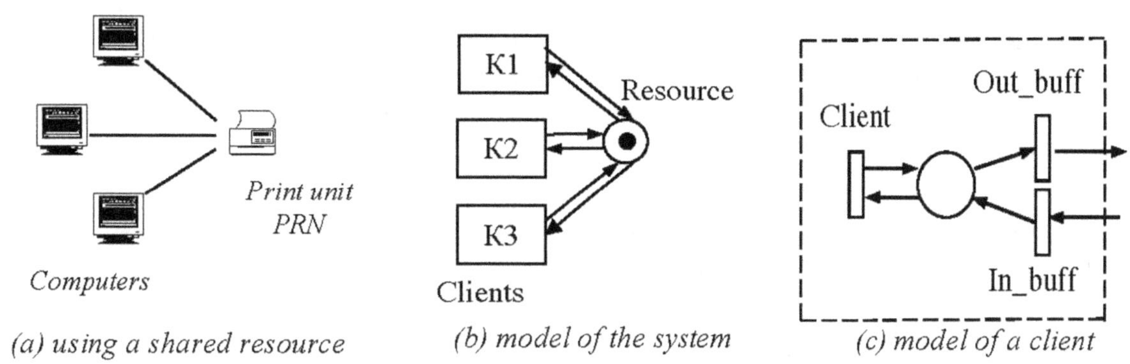

Figure 22. Model of centralized ("client/server") servicing

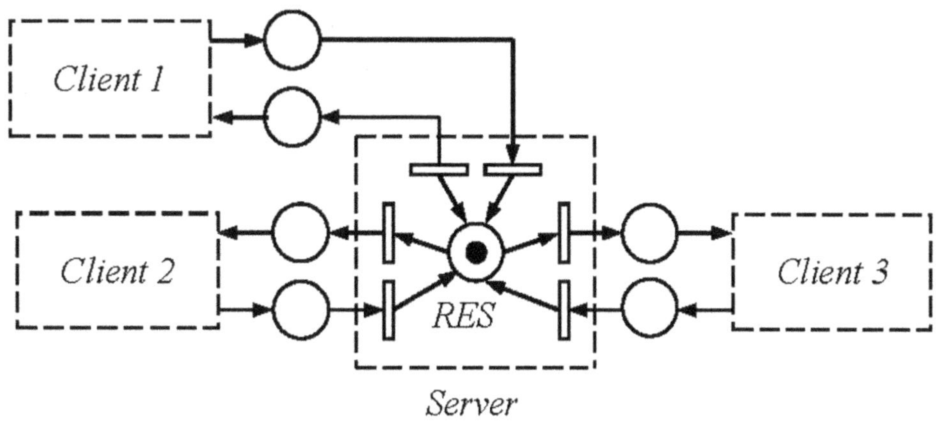

The execution of a Petri net is viewed as a sequence of discrete events (represented by transitions) and the order in which its realization (activation of transitions) is important. It is assumed that evolution is realized by instantaneous transitions (no delays are considered). A classic Petri net is asynchronous and has no means of keeping track of time.

4.2 Reachability and Properties of PN

The evolution of a given PN is realized by sequentially activating allowed transitions, where the current marking changes. Thus, the execution of the PN is represented by the chain of markings $\mu_0 \rightarrow \mu_1 \rightarrow \mu_2 \rightarrow \ldots$ because of the fulfillment of various events (activated transitions). Only one allowed transition can be activated at a time. In the presence of several allowed transitions, the analysis of the network can be carried out using the so-called *reachability tree*. It is a hierarchical tree structure with the root started marking μ_0 and branches depending on the transitions allowed at any given time. An example of such a reachability tree is presented in Figure 23. Each arc in the tree is associated with a specific allowed transition at a given current mark, and nodes with the mark after it is fired. Each path from the root (μ_0) to the leaves represents one possible evolution of the network. When a passive mark is reached, no transition is allowed and evolution along this branch is blocked.

An investigation of an analytical model implemented as a PN is performed by analyzing the main properties, which are summarized in Table 3. Through the reachability tree, various situations can be analyzed, such as the activity/inactivity of a given transition - a transition from PN is inactive if there is no arc with its name in the reachability tree. A mark is passive if no transition can be activated from it – an example is shown in Figure 23(b).

Figure 23. Investigation of a reachability

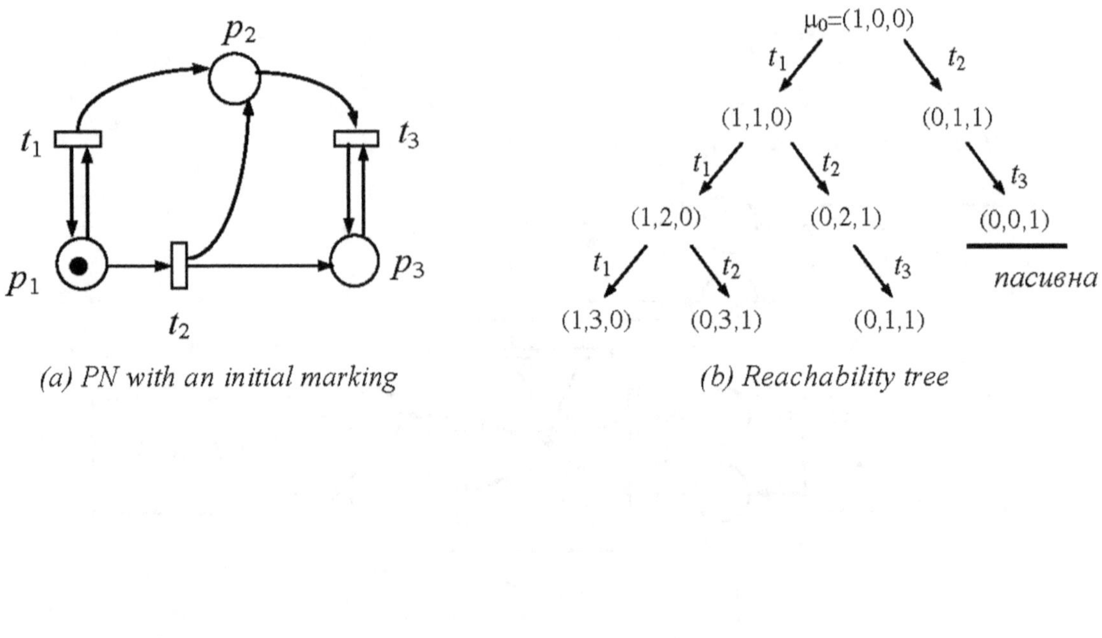

(a) PN with an initial marking *(b) Reachability tree*

Table 3. Properties of a PN

Property	*Comment*
Reliability	The total number of all marks in the positions of the grid is not greater than 1, i.e., $\Sigma\mu(p_i)\leq 1$.
Boundedness	A Petri net (PN, μ0) is *k*-bounded or simply "bounded" if the number of marks at each position does not exceed the number *k* for each mark reachable by μ_0, i.e. $\Sigma\mu(p_i)\leq k$, when $k=const$.
Safety	A PN is "safe" if all its positions are safe (a position in which there is no more than one mark – $k\mu(p_i)\leq 1$). In other words, a safe PN is 1-bounded PN.
Conservativeness	The total number of brands in the positions is constant, i.e. $\Sigma\mu(p_i)=k$.
Blocking	This is a condition where part of all PN is not work (some or all transitions are down).
Reachability	A feature for exploring the dynamic properties of PNs. During the successive "firing" of allowed transitions, markings are formed that allow from one initial state μ_i to reach another μ_j (it is said that μ_j is reachable from μ_i).
Liveness	A Petri net is said to be "live" if, at every mark reachable by μ_0, there are allowed transitions that can possibly be completed. Analysis of this important property allows detection, for example states "deadlock" in discrete systems, system software, server applications, etc.).
Persistence	The property means that the total number of marks does not change during the evolution of the PN. Strict conservation is a strong constraint on a network and can be achieved if each transition has an equal number of input and output arcs. .

An exemplary PN-model is defined in Figure 24. The goal is to determine which transitions and how many times they can be fired (activated) by analyzing the reachability tree presented in Figure 25. Results of the analysis are presented below.

- t_1 cannot be activated because the minimum condition <x,1,1,x> is never obtained;
- t_2 is activated once when there is a mark in position p_2;
- t_3 can be activated $N = 1, 2, 3, \ldots$ times consecutively also when there is a mark in position p_2, but before transition t_2 has been activated;
- t_4 is activated $N = 1, 2, 3, \ldots$ times consecutively, but after firing (activating) transition t_2 at current marking <0,1,0,*N*>.

Figure 24. PN model: Set-theoretic and graph definition

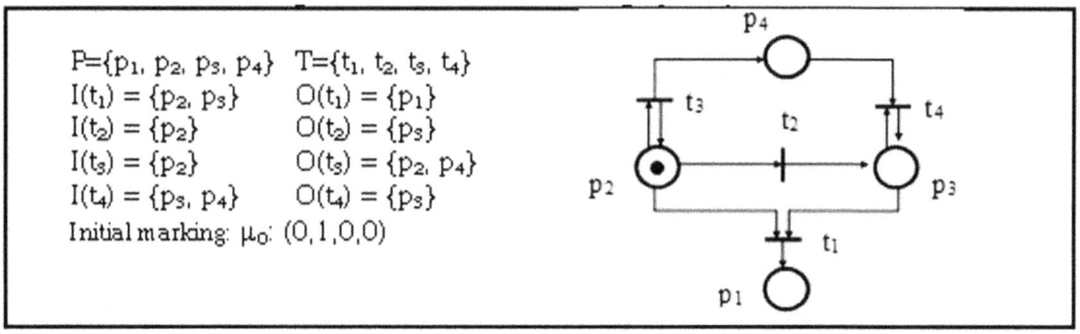

Figure 25. Reachability investigation of the PN from Figure 24

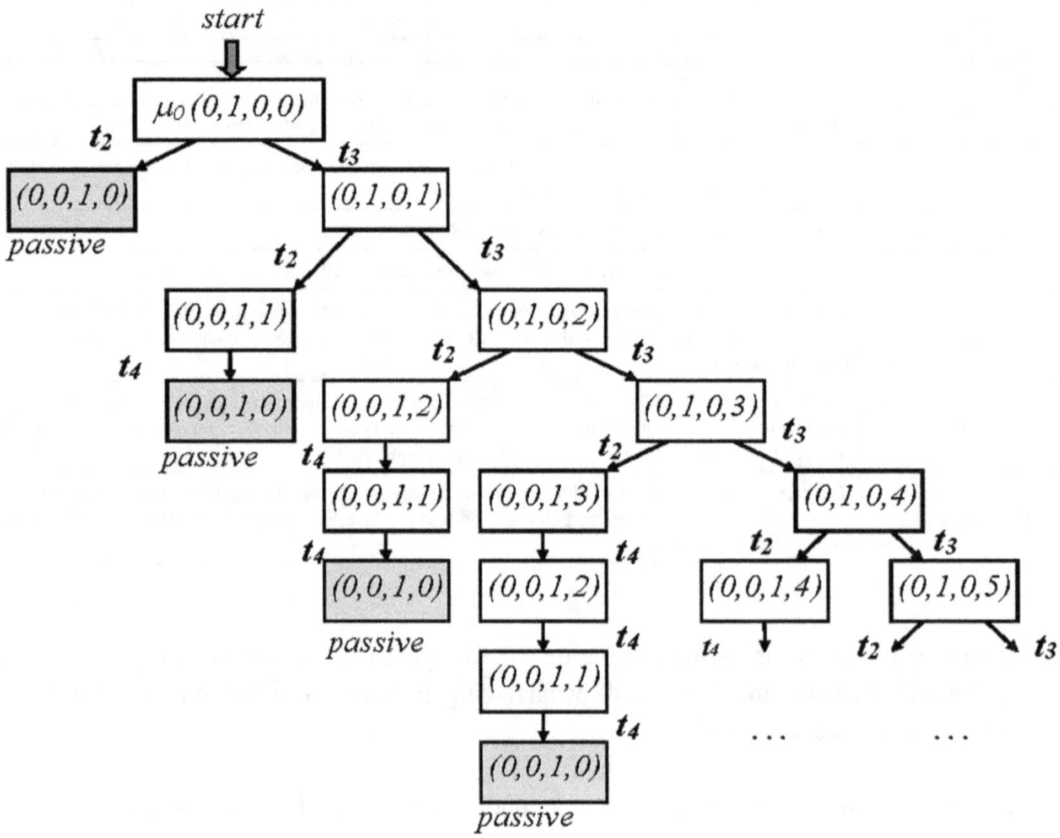

The analyzed Petri net has the following properties:

a) unreliable, because the condition $\Sigma\mu(p_i) \leq 1$ is not fulfilled during the evolution;
b) unlimited, because the condition $\Sigma\mu(p_i) \leq k$ is not fulfilled during the evolution;
c) non-conservative (similarly);
d) allows blocking at marking <0,0,1,0>.

4.3 Examples of Deterministic Models by using PN

Model of a vending machine. The PN-model definition is presented below, and the graph description is shown in Figure 26.

1. Initial description of the model components
   Events (transitions): t_1 – coin release; t_2 – door opening; t_3 – pushing the item; t_4 – door lock.
   Conditions (positions): p_1 – coin availability; p_2 – item selection and submission; p_3 – picking up the item; p_4 – expectation.
2. Definition of input and output function of the PN-model

Input and output
Prerequisites (input function):
$I(t_1) = \varnothing$
$I(t_2) = \{p_1, p_4\}$
$I(t_3) = \{p_2\}$
$I(t_4) = \{p_3\}$
Postconditions (output function):
$O(t_1) = \{p_1\}$
$O(t_2) = \{p_2\}$
$O(t_3) = \{p_3, p_4\}$
$O(t_4) = \varnothing$

Figure 26. Graph definition of PN-model with initial marking

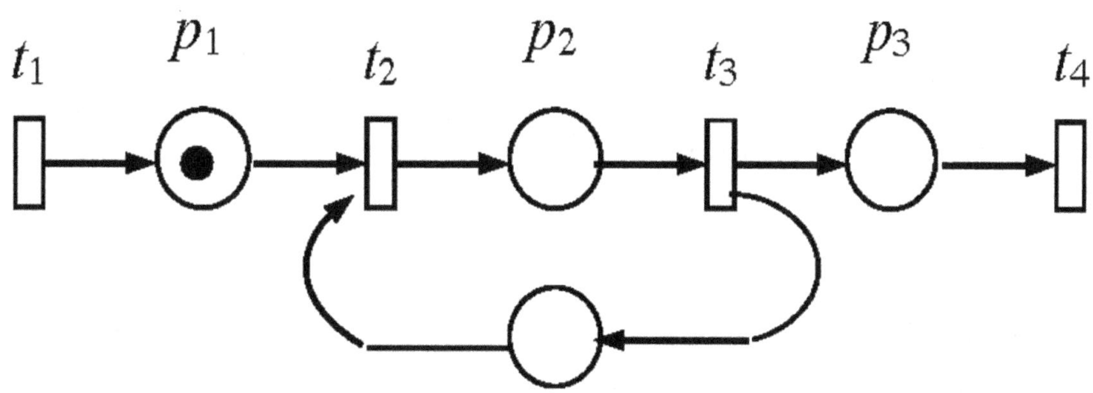

Abstract transmission protocol model. Information transmission takes place between two processes, which are determined as "source" (S) and "receiver" (D). The graph definition of the PN-model is presented in Figure 27(a), where the main events (transitions) are: t_1 – transmission of data from S to D; t_2 – preparation of process S of new data for transmission; t_3 – preparing process D to accept new data. The conditions (positions) are: p_1 – readiness of S for transmission; p_2 – readiness of D to accept; p_3 – availability of more data to transmit; p_4 – availability of data received in the buffer.

The initial conditions are represented by marks in positions p_1 and p_2. With this initial mark μ_0=(1,1,0,0) only transition t_1 is allowed and after its activation the mark changes to μ_1=(0,0,1,1). The complete PN-model evolution scheme is presented in Figure 27(b).

Figure 27. Model investigation of an abstract transmission protocol

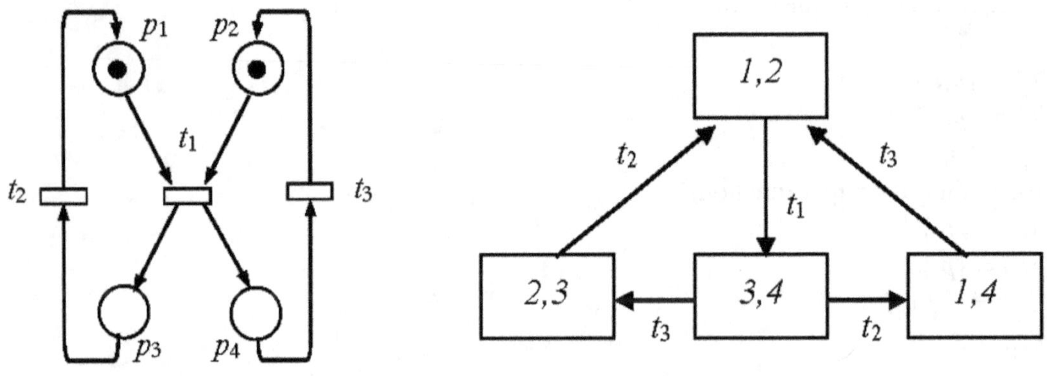

(a) PN-model *(b) Full evolution scheme*

Another option for modeling the problem is given in Figure 28. The transitions are as follows: t_1 – preparation of data for sending; t_2 – sending the data to the exchange buffer (position p_5); t_3 – extract data from the buffer; t_4 – analysis (processing) of received data. The figure shows the evolution corresponding to the transmission of a chain of blocks (data) between the two processes.

Figure 28. Another variant of the transition protocol model

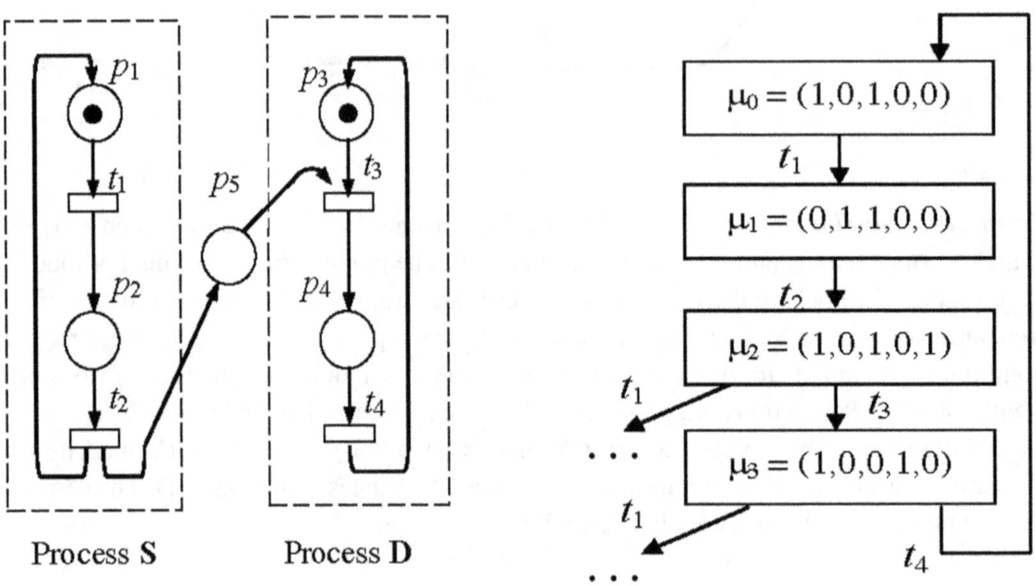

(a) Graph definition *(b) PN-model evolution*

'Token Ring' network service model. In the network with ring topology, the token moves in one direction, successively passing through the subscribers (clients) A_j. Any subscriber can receive the token and analyze it. The internal functional organization of the subscribers is the same and can be represented by the model of Figure 29.

Figure 29. Model of 'Token Ring' network

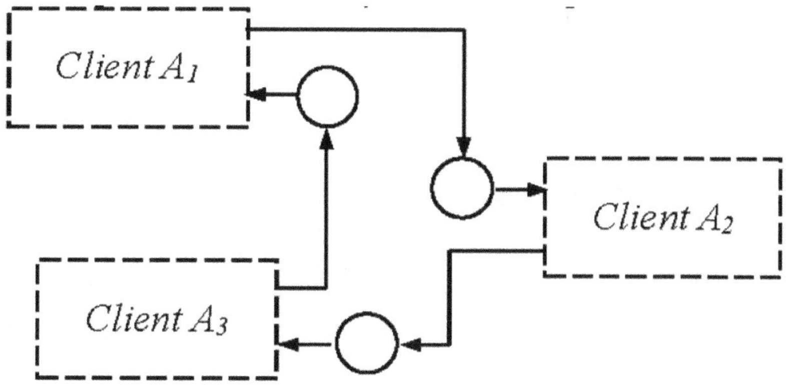

PN-modeling of processes in distributed learning environment (DLE). This example of PN-model presentation and analytical investigation is an extension of the graph formalization of DLE presented in Section 5.2.2 of Chapter 5 and Figures 10 and 11 in Chapter 5. The representation of Figure 30 is used as an abstract base segment for modeling the main components. It presents one-user access to remote information resources by using network transmission tools – distributers and transmitters. Three basic model-primitives (User, Resource, Transmitter) are defined based on this abstract conceptual model, and they are presented below. The purpose of these primitives is to develop a general PN-model of the communications in DLE.

Figure 30. Basic abstract segment for PN-model design

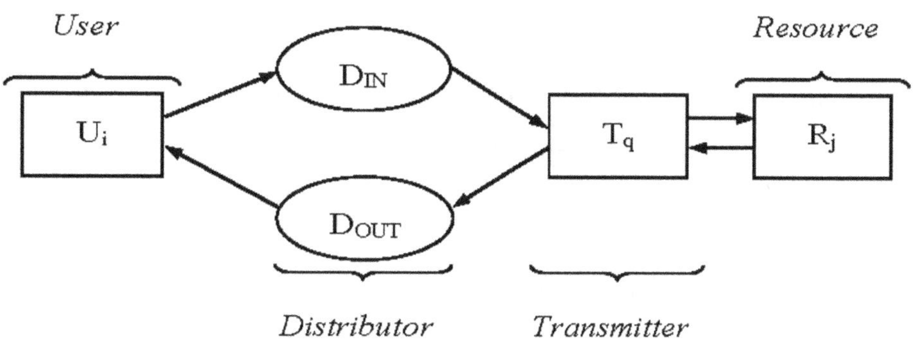

Model primitive "User". It is defined by 4 basic events and 3 conditions, presented in Table 4. Based on these predefined components, the following model primitive was created as a PN, and the graphical representation of which is given in Figure 31.

$P=\{p_1,p_2,p_3\}\Rightarrow|P|=3$ and $T=\{t_1,t_2,t_3,t_4\}\Rightarrow|T|=4$

$I(t_1)=\{p_1,p_1\}\ \ O(t_1)=\{p_1,p_2\}$

$I(t_2)=\{p_3\}\ \ O(t_2)=\{p_1\}$

$I(t_3)=\{p_2\}\ \ O(t_3)=\{Out\_U\equiv D_{IN}\}$

$I(t_4)=\{In\_U\equiv D_{OUT}\}\ \ O(t_4)=\{p_3\}$

The two matrixes – input (**D**⁻) and output (**D**⁺) are presented below.

$\mathbf{D}^-$	p_1	p_2	p_3	D_{IN}	D_{OUT}
t_1	2	0	0	0	0
t_2	0	0	1	0	0
t_3	0	1	0	0	0
t_4	0	0	0	0	1

$\mathbf{D}^+$	p_1	p_2	p_3	D_{IN}	D_{OUT}
t_1	1	1	0	0	0
t_2	1	0	0	0	0
t_3	0	0	0	1	0
t_4	0	0	1	0	0

Table 4. Events and conditions for primitive "User"

Events		Conditions	
t_1	generation of a request for access to distributed learning (information) resource	p_1	availability of the user to work with the distributed learning resource
t_2	entered information block processing	p_2	presence of a request in the input buffer
t_3	sending a request to the distributed medium for routing	p_3	presence of information block in the output buffer
t_4	an information block enters the input buffer from the distributed medium		

Figure 31. PN-primitive for "User" with initial marking μ_0 = (2,0,0)

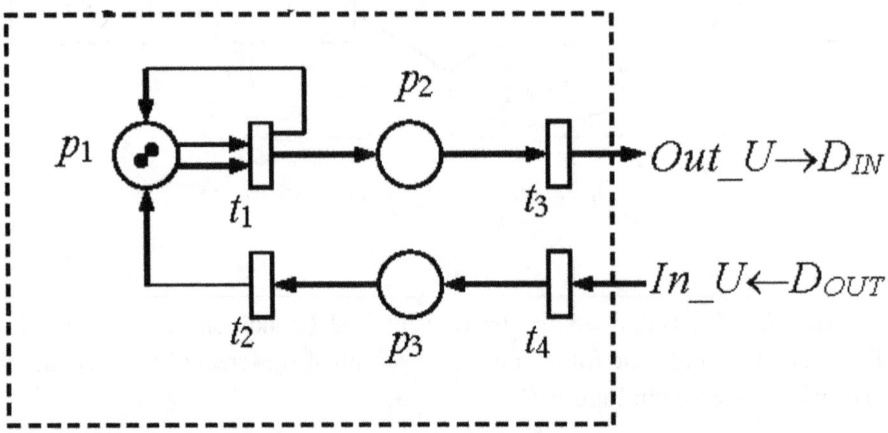

Model primitive "Resources". The defined events are: t_R – processing of an entered request for access to the information learning resource; t_S – giving the access to the learning resource (an information block sending). Conditions are: p_B – readiness to give of a distributed learning resource; p_C – presence of processed request; p_D – the access to the distributed learning resource is given and the information block is directed to the network medium. Developed PN-model for this primitive and two matrixes are presented below, and the graph definition is given in Figure 32.

$P=\{p_A,p_B,p_C,p_D\} \Rightarrow |P|=4$ and $T=\{t_R,t_S\} \Rightarrow |T|=2$

$I(t_R)=\{p_A,p_B\}$ $O(t_R)=\{p_C,p_C\}$

$I(t_S)=\{p_C,p_C\}$ $O(t_S)=\{p_B,p_D\}$

Input and output matrixes:

D⁻	p_A	p_B	p_C	p_D
t_R	1	1	0	0
t_S	0	0	2	0

D+	p_A	p_B	p_C	p_D
t_R	0	0	2	0
t_S	0	1	0	1

Figure 32. PN-primitive for "Resource" with initial marking μ_0 = (0.1,0,0)

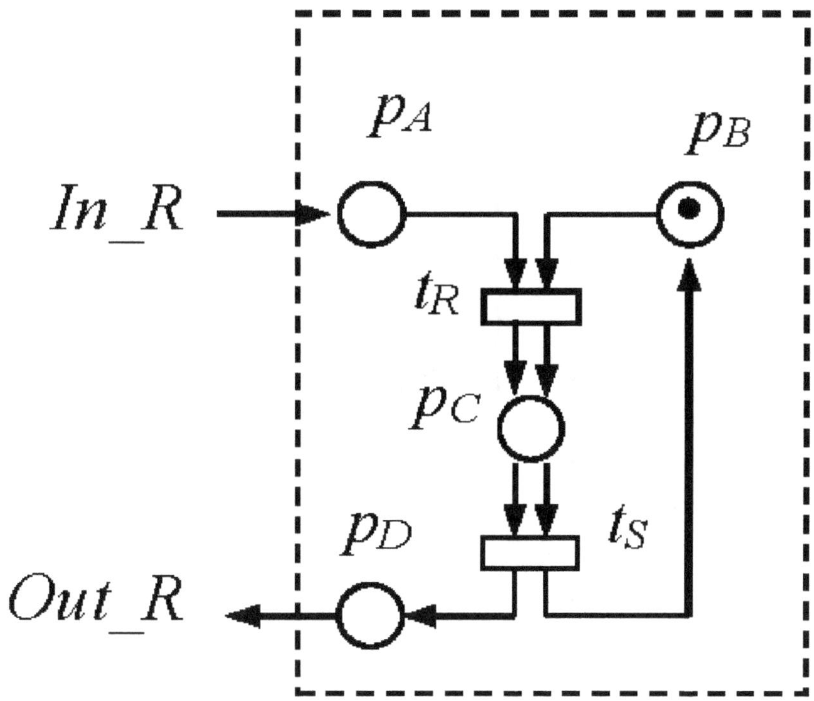

Model primitive "Transmitter". The defined events are: t_{T1} – request transmission trough the distributed (network) medium based on the routing algorithm; t_{T2} – information block (learning contents) transmission through the distributed medium to the user (respond returning). Conditions are: p_{T1} – readiness to transmit a request to node (distributed learning resource) in the DLE (availability of rout and communication resource); p_{T2} – readiness to transmit information block to the user (availability of free communication resource on the rout). Developed PN-model for this primitive and two matrixes are presented below, and the graph definition is given in Figure 33.

$P=\{p_{T1},p_{T2}\}\Rightarrow|P|=2;\ T=\{t_{T1},T_{T2}\}\Rightarrow|T|=2$

$I(t_{T1})=\{D_{IN},p_{T1}\}\ \ O(t_{T1})=\{In\_R,p_{T2}\}$

$I(t_{T2})=\{Out\_R,p_{T2}\}\ \ O(t_{T2})=\{D_{OUT},p_{T1}\}$

Input and output matrixes:

D^-	D_{IN}	D_{OUT}	p_{T1}	p_{T2}	In R	Out R
t_{T1}	1	0	1	0	0	0
t_{T2}	0	0	0	1	0	1

D^+	D_{IN}	D_{OUT}	p_{T1}	p_{T2}	In R	Out R
t_{T1}	0	0	0	1	1	0
t_{T2}	0	1	1	0	0	0

Figure 33. PN-primitive for "Transmitter" with initial marking $\mu 0 = (1,0)$

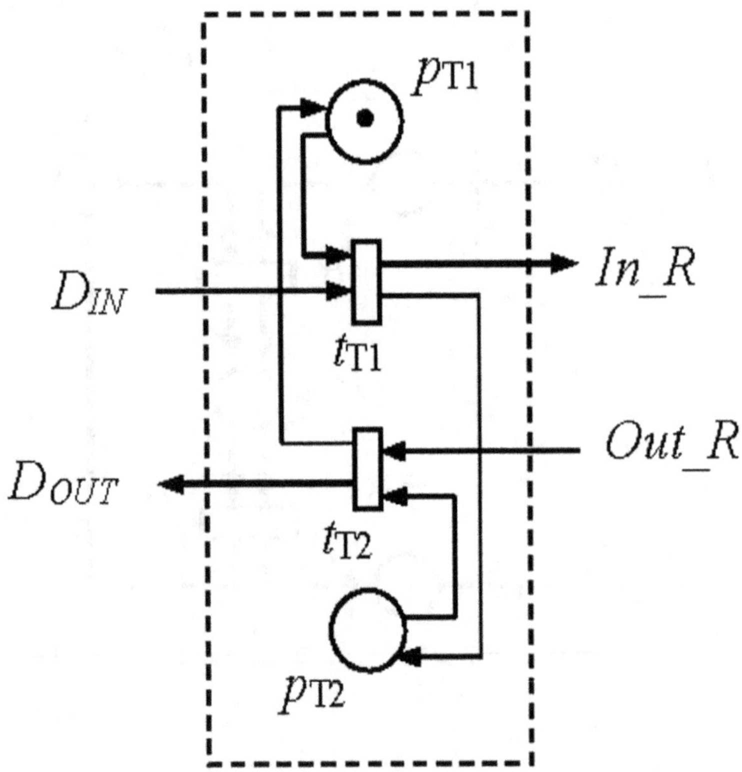

Based on the defined model primitives, each of them verified separately, can be proposed general PN model of the DLE, which uses the abstract model of the network infrastructure shown in Figure 34 and described as follows (for $j = 1, 2, \ldots, K$):

$$P=\left\{D_{IN},D_{OUT},\left\{p_{T1j},p_{T2j}\,/\,j=1\div K\right\}\right\}\quad T=\left\{t_{T1j},t_{T2j}\,/\,j=1\div K\right\},$$

$$I\left(t_{T1j}\right)=\left\{D_{IN},p_{T1j}\right\}\quad O\left(t_{T1j}\right)=\left\{In\_R_j,p_{T2j}\right\}$$

$$I\left(t_{T2j}\right)=\left\{Out\_R_j,p_{T2j}\right\}\quad O\left(t_{T2j}\right)=\left\{D_{OUT},p_{T1j}\right\}$$

The graph presentation of a segment of the general DLE-model as a PN is presented in Figure 35.

Figure 34. Abstract model of the network infrastructure

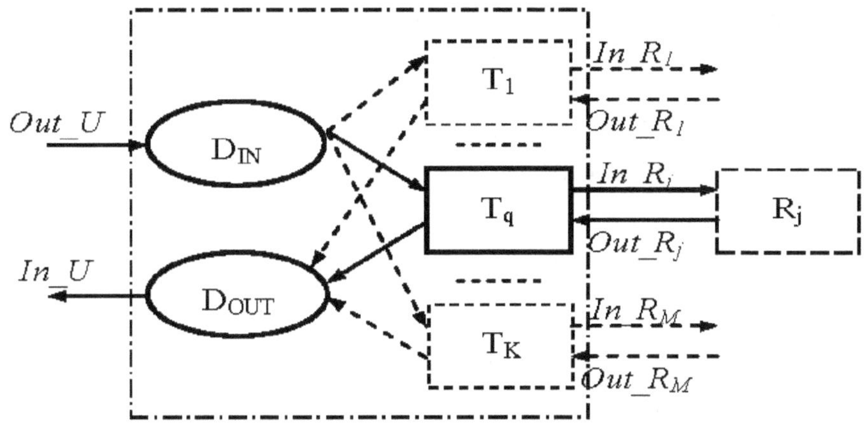

Figure 35. PN-model of the information servicing with initial marking μ_0= (2,0,0,0,0,1,0,0,1,0,0)

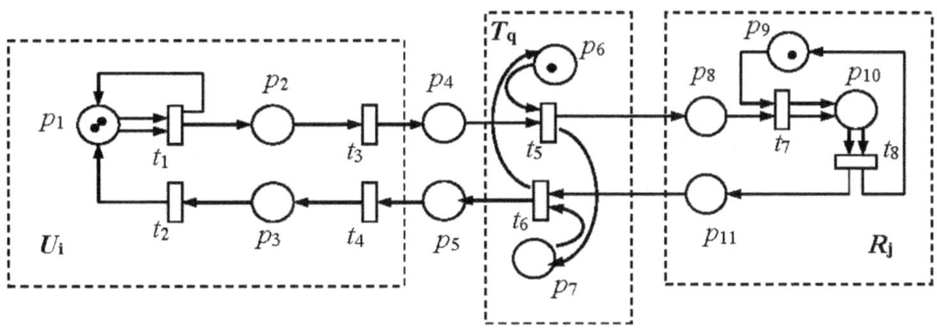

The matrix definition of the segment from Figure 35 is the following:

$$D^{-}=\begin{bmatrix}20000000000\\00100000000\\01000000000\\00001000000\\00010100000\\00000010001\\00000001100\\00000000010\end{bmatrix}\qquad D^{+}=\begin{bmatrix}11000000000\\10000000000\\00010000000\\00100000000\\00000011000\\00001100000\\00000000010\\00000000101\end{bmatrix}$$

A brief discussion of the PN-model execution and investigation is presented below. The first step is to define the permitted transactions in the model started by initial marking μ_0 – in this case only the transaction t_1 is permitted and this is defined by the checking of the condition μ0≥e[1], D-· It is realized because e[1]=(1,0,0,0,0,0,0,0) and e[1], D-=(1,0,0,0,0,0,0,0), D-=(2,0,0,0,0,0,0,0). As a second step is the calculation of incident matrix based on the follow action:

$$D=D^{+}-D^{-}=\begin{bmatrix}-1,+1,0,0,0,0,0,0,0,0,0\\+1,0,-1,0,0,0,0,0,0,0,0\\0,-1,0,+1,0,0,0,0,0,0,0\\0,0,+1,0,-1,0,0,0,0,0,0\\0,0,0,-1,0,-1,+1,+1,0,0,0\\0,0,0,0,+1,+1,-1,0,0,0,-1\\0,0,0,0,0,0,0,-1,-1,+1,0\\0,0,0,0,0,0,0,0,+1,-1,+1\end{bmatrix}$$

After these initial actions, it is possible to proceed to determine the transition $\mu_0 \rightarrow \mu_1$ to the next marking when activating an permitted transition in the model:

μ1=μ0+e[t1], D=μ0+e[1], D=(2,0,0,0,0,1,0,0,1,0,0)+(1,0,0,0,0,0,0,0,0), D=(2,0,0,0,0,1,0,0,1,0,0)+(-1,+1,0,0,0,0,0,0,0,0,0)=(1,1,0,0,0,1,0,0,1,0,0)=μ1

The model evolution for a single time execution is presented in Table 5 and the generalized scheme of this evolution (the tree of the reachability) is shown in Figure 36.

Table 5. PN-model execution (evolution)

t	μ	Places and marks										
		1	**2**	**3**	**4**	**5**	**6**	**7**	**8**	**9**	**10**	**11**
	μ_0	2	0	0	0	0	1	0	0	1	0	0
t_1	μ_1	1	1	0	0	0	1	0	0	1	0	0
t_3	μ_2	1	0	0	1	0	1	0	0	1	0	0
t_5	μ_3	1	0	0	0	0	0	1	1	1	0	0
t_7	μ_4	1	0	0	0	0	0	1	0	0	2	0
t_8	μ_5	1	0	0	0	0	0	1	0	1	0	1
t_6	μ_6	1	0	0	0	1	1	0	0	1	0	0
t_4	μ_7	1	0	1	0	0	1	0	0	1	0	0
t_2	$\mu_8 \equiv \mu_0$	2	0	0	0	0	1	0	0	1	0	0

Figure 36. Generalized form of the reachability tree

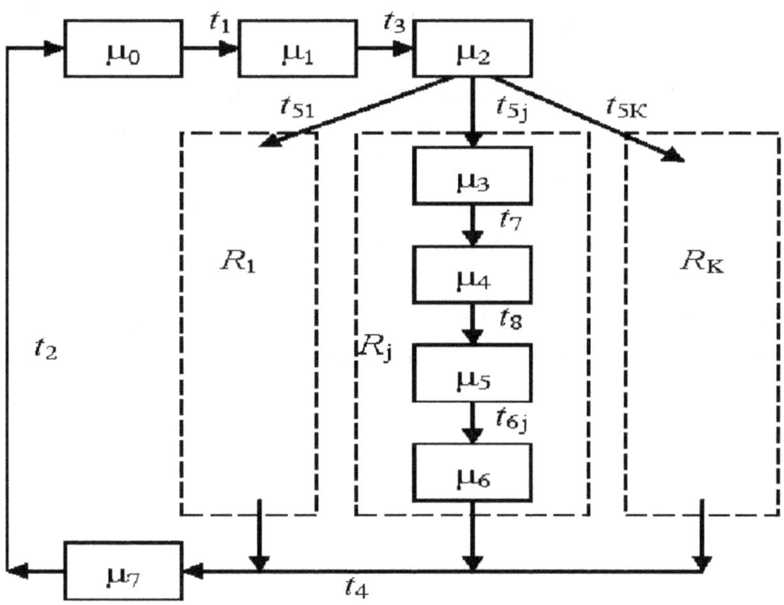

The model investigation can be realized by the analysis of the tree of the reachability from the last figure and of the basic model conditions, and the main results are summarized in Table 6.

Table 6. Basic properties of the presented model

Property	*Comment*
Reachability	The model allows cyclic execution of the basic phase of transactions fairing
Liveness	The model is alive because the evolution permits at each step to fire (activate) at least one transaction
Blocking	No situation during the evolution can block the model execution at the one-user access to the learning resource
Boundedness	The model is 2-bounded because $\Sigma\mu(p_i)\leq 2$
Safety	It may be accepted that the model is safe because for all transactions the number of input arcs is equal to the number of output arcs
Persistence	The total number of marks in the model is the same during each execution step
Conservativeness	Yes, based on the same reason

Analytical PN-modeling of virtual environment protected resources. The subject of the discussion here is a continuation of the analytical study of the functionality of a virtual environment of information resources in the digital space presented at the beginning of this chapter (see the functional structure in Figure 3). An APL2 program description of the system functionality modeling is presented in Figure 4. This algorithm supports the analytical modeling and investigation of secure access and data protection in the business system, designed by two sub-systems – Front office and Back office. The generalized structure of such a corporative system is shown in Figure 39. The software objects, administrative management tools, data structures, and specialized technical tools for data protection and secure access to resources are united in a System for Information Security (SIS). An important component of this system is Digital Right Management System (DRMS), which organizes procedures for information security and access regulation. The collection of information resources is divided into three groups: public (without protection) and two levels of protected resources (internal and external in the cloud). The Front office sub-system is an input point (portal) and support procedures for preliminary user's registration, identification and request receiving. The Back office is the important part of processes for data protection and information security based on authentication, authorization, personal profiles protection, etc.

Figure 37. Generalized structure of a corporative system for secure access to the resources

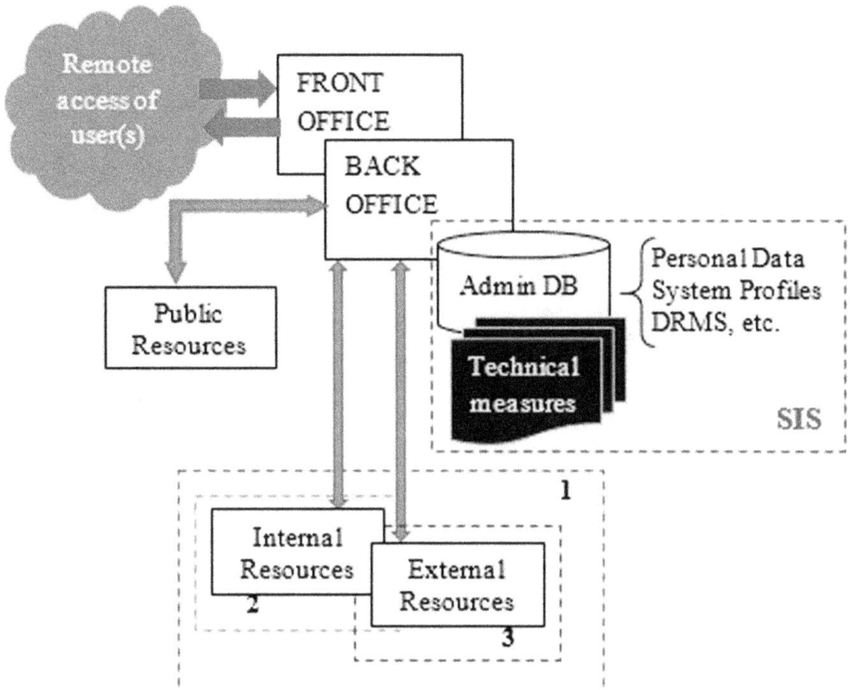

Based on the theoretical-set definition of a model $PN = (P, T, F)$, the sets of determined transitions (T) and positions (P) are presented in Table 7. The set of functions F formed based on the connections between different elements of the two sets P and T (input functions of pre-conditions I and output function of post-conditions O) are shown in Figure 38.

Table 7. Elements of sets T and P

Transitions		*Positions*	
t_0	new access generation	p_1	new request is available
t_1	identification for correct access	p_2	request (user) for registration
t_2	user's registration correct/incorrect)	p_3	access for authentication is available
t_3	procedure for authentication checking	p_4	request for processing
t_4	Type of requested resource analysis	p_5	request for accessing public resource
t_5	access & using public resource	p_6	request is available in protected system
t_6	input in the system with private resources	p_7	request for accessing private resource
t_7	determining the level of protection	p_8	determining level of protection is needed
t_8	procedure for authorization for the level 1	p_9	authorized request level 1 is available
t_9	procedure for authorization for the level 2	p_{10}	authorized request level 2 is available
t_{10}	access & using protected resource level 1		
t_{11}	access & using protected resource level 2		

Figure 38. Analytical and graph definitions of discrete analytical PN-model of the system

$T = \{t_0, t_1, t_2, \dots, t_{11}\}$ $P = \{p_1, p_2, \dots, p_{10}\}$

$I(t_0) = \{\varnothing\}$	$O(t_0) = \{p_1\}$
$I(t_1) = \{p_1\}$	$O(t_1) = \{p_3, p_2\}$
$I(t_2) = \{p_2\}$	$O(t_2) = \{p_1, \varnothing\}$
$I(t_3) = \{p_3\}$	$O(t_3) = \{p_4, \varnothing\}$
$I(t_4) = \{p_4\}$	$O(t_4) = \{p_6, p_5\}$
$I(t_5) = \{p_5\}$	$O(t_5) = \{\varnothing\}$
$I(t_6) = \{p_6\}$	$O(t_6) = \{p_7, \varnothing\}$
$I(t_7) = \{p_7\}$	$O(t_7) = \{p_8\}$
$I(t_8) = \{p_8\}$	$O(t_8) = \{p_9, p_6\}$
$I(t_9) = \{p_8\}$	$O(t_9) = \{p_{10}, p_6\}$
$I(t_{10}) = \{p_{10}\}$	$O(t_{10}) = \{p_6\}$
$I(t_{11}) = \{p_{10}\}$	$O(t_{11}) = \{p_6\}$

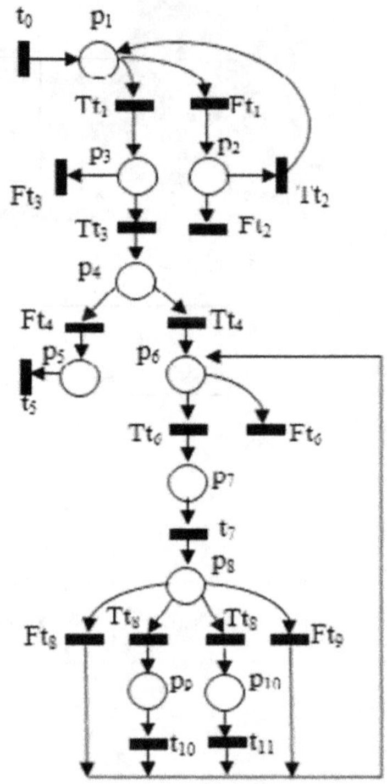

The execution (evolution) of a PN-model could be presented by constructing the tree of reachability. It can be designed by investigation of the sequence of allowed (active) transaction and constructing the chain of successive markings $\mu_0 \rightarrow \mu_1 \rightarrow \mu_2 \rightarrow \dots$ (μ_0 is an initial marking and presents the beginning of the process). Each μ_j is a numeric vector with "n" integers which present the number of marks ($k \geq 0$) for each position from 1 to n. This number of marks reflects the level of realization of the condition planned for this position. A transition can be activated if all its conditions are realized (allowed transition). PN evolution is realized by switching one or more allowed transitions in each step.

Figure 39 shows the model execution (PN evolution) by constructing the tree of reachability started by initial marking μ_0 = (0000000000) – no marks in all 10 positions. The evolution begins by switching transition t_0 without initial positions (conditions) – asynchronous activity modelled a user's access. The result after this first step is 1 mark in the position $p_1 \rightarrow \mu_1$ = **(1000000000)**. This situation activates the next step with two possible transitions Tt_1 and Ft_1 (***T***-True and ***F*** – false). The digit in each box determines the number of positions with 1 mark received in the sequence of steps (for example see block '7' – one mark in position p_7).

The reachability tree evaluation allows to determine the properties of the designed PN-model which are summarized below.

- Reliability: Yes because the total number of marks in the positions is no more than 1.
- *k*-limitation: Yes because in each position of the PN-model the number of marks is not greater than the values *k* for each marking reachable from μ_0.

- Safety: Yes, the model is safe because it is 1-limited (k=1).
- Conservativeness: Yes because the total number of marks at the model evolution is constant.
- Blocking: No state (marking) of the tree where the evolution stops (no passive marking).
- Reachability: Yes because each formed marking in the tree permits one or more allowed transitions which will form new un-passive marking(s).

Figure 39. Graph presentation of the PN-model execution as a tree of reachability

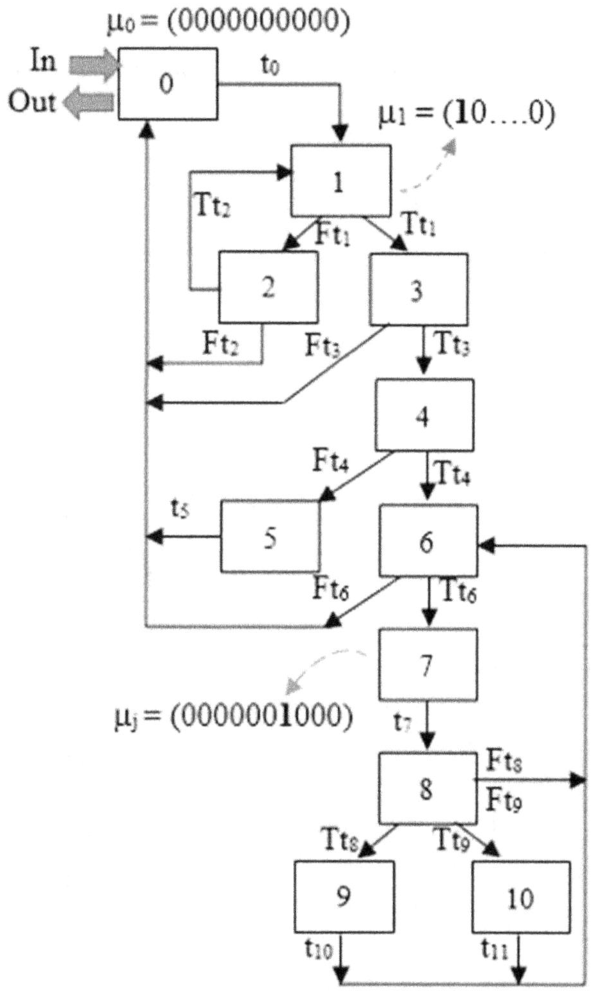

PN-modeling of algorithmic structures. When modeling algorithms and programs, the main blocks are considered as events and are represented by transitions – the operator blocks with one transition, and the logical blocks with two (for positive and negative fulfillment of the condition) – Figure 40. Usually, the start marker corresponds to the beginning (input) of the structure, and the execution of a given statement is possible only if its corresponding transition is allowed. When the model is executed, the mark is passed from start to finish, going through one of the possible paths. Through the evolution of the network, the workability of the algorithm can be investigated, and based on the reachability tree, all possible implementations can be analyzed, the presence of passive markings (situations without an exit), the time parameters for execution can be determined, etc.

An overview of the development of Petri nets, with a discussion of modeling and algorithms, is presented in (Giua & Silva, 2018). The focus of this review article is on systems theory and automatic control. This limitation is made in relation to the stated goal of answering specific conceptual questions in research in the discussed area. Some basic concepts of PN are reviewed, and the discussion focuses on the families of formalisms used as a basis for the application of PN to solving automatic control problems. The paper highlights that discrete PN modeling in the field is applied in condition assessment, monitoring and blocking control, diagnostics, etc. In the next sub-section, a more extended overview of the evolution of Petri nets is given, and the main application areas of the new extended and modified versions are also indicated.

Figure 40. Algorithmic structure and its PN-model description

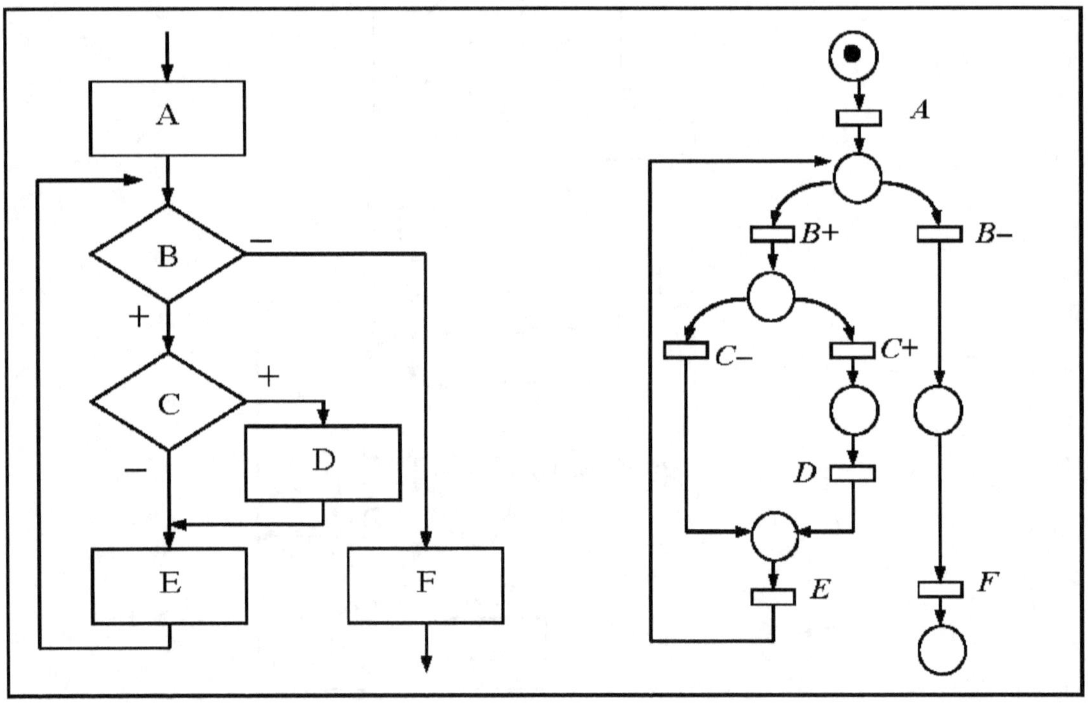

4.4 Evolution of Petri Nets

The classic version of Petri nets quickly found application in various fields, leading to modifications, extensions, and evolution of the apparatus's capabilities. One solution in this direction is generalized nets, which are presented in part 6.4 of Chapter 6. At its creation, the classic Petri net is a discrete structure with asynchronous action, without considering time parameters and probabilistic characteristics of the studied objects. This necessitated various modifications, some of which are briefly presented below.

4.4.1 Free-Choice PN (FCPN)

In classical Petri nets, it is possible to have more than one allowed transition at any given time. This requires a decision to be made as to which one to activate at the current time. In the implementation of the PN model, such a situation is defined as a conflict. To overcome this problem, the so-called "Free-Choice PN" (FCPN) are introduced (Desel & Esparza, 1995; Baccelli et al., 1996). They are characterized by the fact that each arrow represents a single output for a given position or a single input for a transition. Thus, the marker remains in the input position until the associated transition is activated. With FCPN, the reachability graph is equivalent to a given transition system and "*are more amenable for visualization and structural analysis while not being excessively simple, as in the case of state machines*" (Teren et al., 2022). Results have shown that in a certain case, for example, with a smaller number, this variant allows reducing the complexity of the model compared to classical PNs.

4.4.2 Weight PN (WPN)

Further development of Petri nets has led to the introduction of weight functions, which determine the priority of activating the currently enabled transitions. This type "Weight PN" is used for conducted research in (Kuchárik & Balogh, 2019) in case there are multiple output transitions. In this case, a random selection of one of the transitions is applied, each of which is assigned a weight, with a transition with a higher weight having a higher probability of realization. The version WPN is used in (Ventisei et al., 2022) for model investigation of transverse electromagnetic square pulse switching and routing in interconnected waveguides. The apparatus of PNs is applied, increasing their potential by introducing weights for the connecting arcs (the connections to and after the transitions) in the graphical definition of the model. To model the investigated process, a token is defined that travels from a given starting (input) location to another final (output) location, passing through transitions and arcs. In PN, the occurrence of an event is instantaneous, with the "travel" of the token being associated with a corresponding PN function. Each of the arcs involved in the defined path is assigned a weight (w), which determines the conditions for triggering the corresponding transition to which the arc belongs. These weights are involved in defining the PN model by accompanying each individual variable they are associated with, for example ($w_1.x+w_2.y+w_3.z$).

4.4.3 Timed PN (TPN)

Adding new features to the classical PN creates new variants, such as the apparatus "Timed PN" (TPN), which is an extension of PN and is defined by $TPN{:=}(P,T,I,O,\Theta)$. The standard PN-definition is extended by the set $\Theta = \{\theta_1, \theta_2, \ldots, \theta_n\}$ that presents transactions' realization delays (Silva & Del Foyo, 2012). The capabilities of TNP in the investigation of resource allocation systems in industry are discussed in (Huang et al., 2023), where it is stated "*a timed Petri net (PN) attracted much attention in the past decades to cope with RAS scheduling problems (RSPs), since PNs are very suitable to model and analyze RASs and their RGs fully reflect systems' behavior*". In particular, the ability to search the graphical representation of the reachability tree is noted as an important feature, which makes this apparatus very convenient and useful for studying the behavior of systems. To clarify the capabilities of TPNs, the article makes a review and presents a guide to their use in modeling resources and network structures, and in addition, the differences and relationships between Petri nets are given. An extension of TPN used to model synchronous electronic circuits is presented in (Parrot et al., 2023). The research was done to aid the design of pipeline systems, with an extension based on a maximal step firing rule. Reset and deferred transitions are used in modeling pipeline processing, which relaxes some timing constraints. This allows expanding the design space and resolving resynchronization tasks. In the article, a formal definition of the developed model is presented and a method for investigation of pipeline properties based on time multiplexing is proposed, which allows reducing the number of registers involved.

4.4.4 Stochastic PN (SPN)

They could be presented as determined values, but a more precise manner is to use stochastic values that extend the PN to the "Stochastic PN" (SPN). An investigation with introducing the stochastic element as an extension of the WPN is presented in (Burke et al., 2021). The formal SPN definition is by the ordered structure $SPN:=(P,T,I,O,L)$, that $L=\{\lambda_1,..., \lambda_n\}$ is a set of intensities of activation $\lambda_i=1/\theta_i$ for each transaction in the defined model. In this way each transaction t_i is associated with an intensity of activation λ_i. As a result, the analysis of TPN/SPN is reduced to an investigation of the condition for stationarity of a Markovian process and such research is presented in (Romansky & Noninska, 2016). In this case the SPN could be described as a Markovian model with discrete set of states $S=\{s_1,..., s_n\}$ and intensities as transactions between them. Each state corresponds to a reachable marking at the PN-evaluation ($s_i \equiv \mu_i$).

Even though almost immediately after their creation, classical Petri nets have established themselves as one of the most successful means of describing and analyzing systems and processes, two main disadvantages necessitate their development.

- There is no clearly formulated concept of the arrangement of the data and their presentation inside the model, which leads to an excessive growth of the network structure.
- Lacks a hierarchical concept, which prevents the construction of separate models and their unification through defined interface rules for interaction.

4.4.5 Colored Petri Net (CPN)

To overcome these shortcomings, High Level PNs are being developed, with one of the most famous representatives being the "Colored Petri Net" (CPN). They complement the qualities of classical Petri nets with the possibility of data structuring and hierarchical decomposition and allow an intuitive graphical representation of the model, easily perceived by the user (Jitmit & Vatanawood, 2021). The model consists of separate modules (pages), each representing a net of positions and transitions. Individual modules interact through interfaces known from modern programming languages. The simulation of the studied processes (the implementation of CPN) can be carried out in two modes – automatic and interactive (controlled by the user). Listed below are some of the main advantages of colored Petri nets.

- Offer an easy-to-understand and easy-to-learn system for intuitive graphic display.
- Have clearly defined semantics, allowing an unambiguous description of the network.
- Solve a wide range of analyzed objects and problems.
- Have few but powerful primitives, and the definition of the network is short and built on standard concepts from programming languages.

- Possess semantics based on real parallelism.
- Allow hierarchical description by creating and grouping constructs similar to procedures, functions and modules from programming languages.
- Integrate the description of control and synchronization with the description of data processing.
- Allow extension with a time concept implemented by a global clock and linking the markers with a flag indicating their readiness to be accepted.
- Are characterized by stability against small changes of the modeled system.
- Offer an interactive simulation, with the results presented directly in an output diagram.
- Basic and auxiliary software tools have been developed, facilitating the description, implementation, and formal analysis.

The CPN apparatus finds application in conducting research in various fields, including when combined with the quantum algorithm in the management of energy processes and their economic optimization (Liu et al, 2022). The problem of process management and use of CPN in conducting research is also discussed in (Brezovan et al., 2022), where the object of modeling is the dynamics of a controller and the structure of a control system. A preliminary discretization of the mathematical model of the controlled process was made to adapt it to the features of the CPN, and the controller itself was implemented based on a Moore automaton. Experiments were conducted on laboratory equipment using the CPN Tools toolkit.

4.4.6 Timed Colored PN (TCPN)

An interesting development of CPNs is their combination with Timed PN, defining the variant "Timed Colored PN" (TCPN) applied in various investigations. As an example, two of them will be briefly presented here. The new apparatus is applied in (He et al., 2022) in path planning of multi-type robotic systems with time windows. To conduct the research, three types of tasks have been defined - general, exclusive, and joint, and an analytical approach has been developed for planning different types of mobile robots. The multi-type of robot system and its environment is modeled with TCPN, after which logical constraints and time windows are defined. To conduct the experiments, the technique of integer linear programming was applied, and as a result, an approach was proposed to minimize the total costs of the system. The second case, which is presented in (Bożek et al., 2022), is related to the problems of planning and sizing web systems. The peculiarities of the processes and the frequent changes in the behavior of customers require timely adaptation of the applied models, most often simulation models. The article proposes TCPN-based event generators designed for web environments that can be used for performance evaluation. In a theoretical aspect, different generators with a deterministic and stochastic nature, as well as timed and non-timed generators, are considered, and then models of different classes are developed using the TCPN formalism. Again, the CPN Tools software was used to conduct the experiments.

4.4.7 The Fuzzy Petri Net (FPN)

FPN aims to aid in solving problems in the field of fuzzy sets by providing the necessary foundation for applying reasoning processes and modeling systems with uncertainty. They enable the development of knowledge representation and reasoning models of rule-based expert systems. A systematic review of FNP with emphasis on knowledge representation formalisms and reasoning algorithms is done in (Liu et al., 2017). A development of FPN for industrial application is presented in (Zhou & Zain, 2016), where a brief history of the creation and application of this extended variant of Petri nets is also presented. The formalisms used are also presented, including a reasoning algorithm. An extension of the FPN idea to the application in cloud service structures is done in (Wang et al., 2020). In particular, trust-based routing mechanisms in ad hoc network MANET are discussed. The reason is that, according to the article, there are fuzzy performance metrics when evaluating a node's trust value, which makes it impossible to quantify the performance of such metrics. To solve this shortcoming, a trust reasoning model based on cloud model and FNP is developed, and a routing algorithm based on trust entropy is proposed. The algorithm reflects the impact of the routing effect and the trust value of the nodes on the routing choices, thus improving the quality of services in the network.

5. PSEUDO-STOCHASTIC ANALYTICAL MODELS

Pseudo-stochastic DAMs represent stochastic processes through deterministic values, which allows for an unambiguous definition of the mathematical model and its unique solution under fixed operating conditions. For the implementation of these models, the probabilistic parameters are replaced by average or extreme values, most often determined by empirical observations. An example of such a model is the analytical expression presented below for the average operating time of a CPU with a cache memory.

$$t_{av} = F_{INT}.N.t_{cyc} + F_{MEM}.t_{cyc}.\left[h.t_{cache} + (1-h)(t_{cache} + t_{delay})\right]$$

A large part of the parameters has a probabilistic character, and in the expression they are replaced by deterministic (most often average) values – Table 8.

Table 8. Used parameters

Parameter	*Comment*	*Value*
F_{INT}	Fraction of processor time used for external operations	40%
N	Average number of processor cycles for external operations	2
t_{cyc}	Duration of a processor cycle	20 ns
F_{MEM}	Fraction of processor time used to access memory	60%
h	Cache performance evaluation	0,9
m=1-h	Cache failure assessment	0,1
t_{cache}	Cache access time	1 ns
t_{delay}	Additional time lost due to cache failure	3 ns

From the set numerical values in the table, a specific value can be determined for the expression:

t_{av} = 40% (2.20) + 60%[20.(0,9.1 + 0,1(1+3))] = 37,6 ns.

It is known that the queuing service is oriented towards the probabilistic investigation of systems and processes. In this sense, the ***deterministic service system model*** presented below can be considered representative of pseudo-stochastic models. In it, all stochastic parameters are replaced by deterministic ones. The service system is modeled by a single-channel QS type **D/D/1** with an infinite buffer with a regular input flow of requests $Z = \{z_i / i=1,2,\dots\}$ with a constant intensity $\lambda=1/\tau$ (τ=const) and a regular flow of service also with constant intensity $\mu=1/T$ (*T=const*).

The situation with an empty queue at the start of the service is trivial – at $T>\tau$ a stationary mode cannot be reached, and at $T\leq\tau$ one immediately enters a stationary mode.

It is interesting to study the situation when $T<\tau$ and start the service when $N>0$ requests waiting in the buffer. After the start of the service system, in parallel with servicing the requests with the sequent numbers from 1 to N accumulated in the buffer, new requests with the numbers $N+1, N+2, \dots, N+\lfloor N.T/\tau \rfloor$ continue to arrive in the queue. Since $T<\tau$ after a certain transition mode, the stationarity of the process will be reached, in which a request entering the queue is immediately directed to service (Figure 41).

Figure 41. Service at k = N requests in the buffer

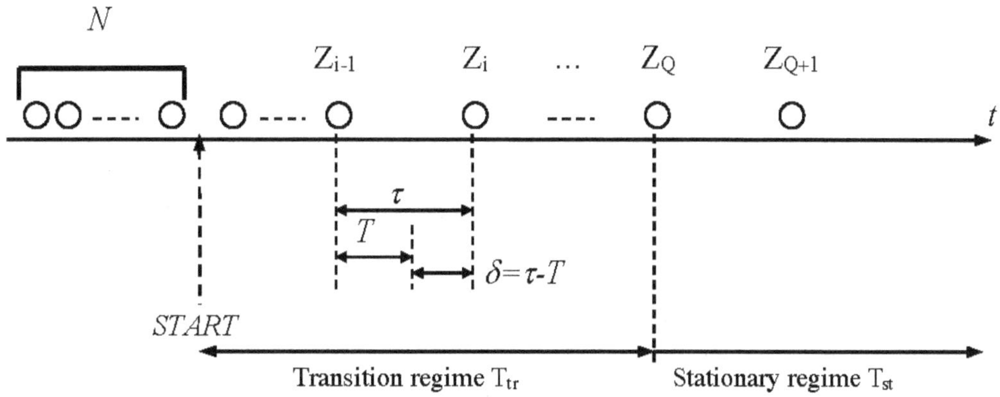

The case $N=1$ is trivial. With parameters $T<\tau$ and $\delta=\tau-T$, within one interval τ can be served $(1+1/\delta)$ requests.

At $N\geq 2$, at the start of the study, N requests have accumulated in the buffer, for the service of which it will take time $N.T$, within which new k requests will arrive, for which $k.\tau\leq N.T$ and therefore $k=\lfloor (N.T)/\tau \rfloor$. To service them, a new time $k.T=\lfloor (N.T)/\tau \rfloor .T$ will be required, during which new $l<k$ number of requests will arrive. If it is assumed that, in the general case, the generation of a new request does not coincide with the start of the service, then from the inequalities $l\leq(kT)/\tau$ and $k\leq(N.T)/\tau$ it follows that $k=(N.T)/\tau+1$ and then follows equation is valid:

$$l \leq \frac{T}{\tau}\left(\frac{N.T}{\tau}+1\right) \Rightarrow l = \frac{k.T}{\tau} = \frac{T}{\tau}\left(\frac{N.T}{\tau}+1\right) = \frac{T}{\tau}\frac{N.T}{\tau}+1$$

To service these l requests, a new time $l.T = \lfloor (k.T)/\tau \rfloor .T$ will be required, during which next $m<l$ number of requests will be generated for which $m.\tau \leq l.T$ and therefore $m \leq l.T/\tau$, i.e., $m = \lfloor (l.T)/\tau \rfloor$. The process will continue in the described manner until a request Z_Q is reached, upon generation of which all pending requests will be serviced and stationary mode will be entered. The time for the transient mode will be $T_{пр}=(Q-1)T$. The situation is illustrated by the example presented in Figure 42 with parameters $\tau=2T$. Through the transition mode, $Q=N+N^*$ requests will be served, where $N^*=\lfloor Q.T/\tau \rfloor$, and the time of this mode will be $Q.T$.

Figure 42. Example for the case N≥2

<table>
<tr><td>Arrival or service request</td><td>Before start</td><td colspan="8">Transient regime (request number)</td><td colspan="4">Stationary regime (request number)</td></tr>
<tr><td>Input in buffer</td><td>1,2,…,N</td><td colspan="2">N+1</td><td colspan="2">N+2</td><td colspan="2">…</td><td colspan="2">(N+N*)=Q</td><td colspan="2">(N+N*)+1</td><td colspan="2">(N+N*)+2</td></tr>
<tr><td>Request service</td><td></td><td>1</td><td>2</td><td>3</td><td>4</td><td>…</td><td>…</td><td>Q-2</td><td>Q-1</td><td>Q</td><td></td><td>Q+1</td><td></td></tr>
</table>

To determine the influence of the initial moment of request generation on the quantitative evaluations in the deterministic study, two situations were analyzed.

Situation (1). The start of the service coincides with the generation of a new request (for concreteness $\tau=2T$ and $N=4$).

During the service of the accumulated 4 requests, another $\lfloor N.T/\tau \rfloor = \lfloor 4/2 \rfloor = 2$ new requests (numbered "5" and "6") will be received, and during their service, another $\lfloor 2.T/\tau \rfloor + 1 = \lfloor 2/2 \rfloor + 1 = 2$ requests with numbers "7" and "8" will arrive. The request "8" entering the queue at the time request "6" is finished serving. Transition mode ends with the service of request "8" and has a duration of $8T$. An illustration is given in Figure 43.

Figure 43. Illustration of the service in Situation (1)

<table>
<tr><td>Arrival or service request</td><td>Before start</td><td colspan="8">Transient regime (request number)</td><td colspan="4">Stationary regime (request number)</td></tr>
<tr><td>Input in buffer</td><td>1,2,3,4</td><td colspan="2">5</td><td colspan="2">6</td><td colspan="2">7</td><td colspan="2">8</td><td colspan="2">9</td><td colspan="2">10</td></tr>
<tr><td>Request service</td><td></td><td>1</td><td>2</td><td>3</td><td>4</td><td>5</td><td>6</td><td>7</td><td>8</td><td>9</td><td></td><td>10</td><td></td></tr>
</table>

Situation (2). The start of the service does not coincide with of request generation and the ration of the time parameters is also $\tau=2T$ and number of requests waiting in the queue before service start is

N=10. An illustration of the arrival and service process is presented in Figure 44. Transition mode ends after service of request "11" because it enters the queue while service of request "10" and must wait for server release. Shaded boxes mark free server, which is associated with an established stationary mode. Based on the scheme, the duration of the transient mode is $11T$.

Figure 44. Illustration of the service in Situation (2)

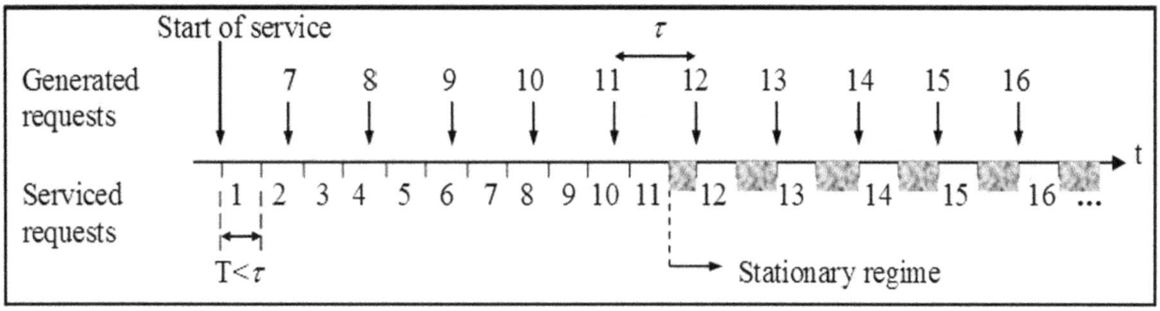

To study the transition mode, during which, in addition to the initial N requests, N^* new requests are served, the following analytical dependencies can be defined.

Number of additional requests received. For the analytical determination, $T<\tau \Rightarrow \delta=\tau-T$ and

$T_{tr}=(Q-1).T=(N+N^*).T \leq (N^*+1)\tau$, are used, from where it can be determined:

$N.T+N^*.T \leq (N^*+1)\tau \Rightarrow N.T-\tau \leq N^*(\tau-T)$, i.e. $N^* \leq \dfrac{N.T-\tau}{\tau-T} \Rightarrow N^*=\dfrac{N.T-\tau}{\tau-T}+1$.

For integer values of the parameters, $N^*=(N-1).T/(\tau-T)$ is obtained.

Total number of requests served. For the transient mode $Q-1=N+N^*$ and from the last expression for N^* it can be determined that for the non-stationary (transient) mode the total number of requests is

$Q-1=N+N^*=N+(N-1).T/(\tau-T)$

After conversion of the last expression is determined:

$Q-1=[(N\tau-T)/(\tau-T)]=[(N-1)\tau/(\tau-T)+1]$

Duration of the transitional regime. From $T_{tr}=(Q-1).T=(N+N^*).T \leq (N^*+1)\tau$ it follows that the time for the transition mode is

$$T_{tr}=\left(N+N^*\right).T=\left(N+\frac{N.T-\tau}{\tau-T}+1\right)T \Rightarrow T_{tr}=\frac{N.\tau-T}{\tau-T}.T\,.$$

Number of pending requests at time $0 \leq t \leq t_Q$. It is defined as the difference between requests received with intensity $\lambda=1/\tau$ up to time t and requests already served with intensity $\mu=1/T$ up to that moment and characterizes a non-stationary mode:

$$N_w(t) = \left[(N-1)+t.(1/\tau)\right]-\left[t.(1/T)\right] = N-1+\frac{t}{\tau}-\frac{t}{T} = N-1+t\left(\frac{T-\tau}{\tau T}\right).$$

Waiting time for a request generated at time t until it is fully serviced. It is determined depending on the mode by:

$$w(t)=\begin{cases}(i-1).T;?@\otimes t=0;i=1,2,...,N\\ w(t-\tau)+T;?@\otimes 0<t\leq t_Q\\ 0;?@\otimes t\geq t_O\end{cases}$$

For non-stationary mode $0<t<t_Q$ is valid the expression:

$$w(t) = N_w(t).T = \left[N-1+t\left(\frac{T-\tau}{\tau T}\right)\right].T.$$

REFERENCES

Andries, V. D., Goras, L., Buzo, A., & Pelz, G. (2017, July). Automatic tuning for a DC-DC buck converter with adaptive controller. In *2017 International Symposium on Signals, Circuits and Systems (ISSCS)* (pp. 1-5). IEEE. 10.1109/ISSCS.2017.8034938

Antonelli, F., Cortellessa, V., Gribaudo, M., Pinciroli, R., Trivedi, K. S., & Trubiani, C. (2020). Analytical modeling of performance indices under epistemic uncertainty applied to cloud computing systems. *Future Generation Computer Systems*, *102*, 746–761. doi:10.1016/j.future.2019.09.006

Baccelli, F., Foss, S., & Gaujal, B. (1996). Free-choice Petri nets-an algebraic approach. *IEEE Transactions on Automatic Control*, *41*(12), 1751–1778. doi:10.1109/9.545714

Bożek, A., Rak, T., & Rzonca, D. (2022). Timed Colored Petri Net-based event generators for Web systems simulation. *Applied Sciences (Basel, Switzerland)*, *12*(23), 12385. doi:10.3390/app122312385

Brezovan, M., Precup, R. E., Selişteanu, D., & Stănescu, L. (2022). Colored Petri nets-based control and experimental validation on three-tank system level control. *International Journal of General Systems*, 1–47.

Burke, A., Leemans, S. J., & Wynn, M. T. (2021). Stochastic process discovery by weight estimation. In *Process Mining Workshops: ICPM 2020 International Workshops, Padua, Italy, October 5–8, 2020, Revised Selected Papers 2* (pp. 260-272). Springer International Publishing 10.1007/978-3-030-72693-5_20

Cassandras, C. G., Lafortune, S., Cassandras, C. G., & Lafortune, S. (2021). Petri nets. *Introduction to discrete event systems*, 259-302.

Desel, J., & Esparza, J. (1995). Free choice Petri nets (40). Cambridge university press.

Digalovski, M., & Rafajlovski, G. (2020). Distribution transformer mathematical model for power losses minimization. *International Journal on Information Technologies and Security*, *12*(2), 57–68.

Dubois, D., & Prade, H. (2006). *Possibility theory and its applications: a retrospective and prospective view*. Springer Vienna.

Giua, A., & Silva, M. (2018). Petri nets and automatic control: A historical perspective. *Annual Reviews in Control*, *45*, 223–239. doi:10.1016/j.arcontrol.2018.04.006

Gochev, V. P., Gocheva, P. V., & Hinov, N. L. (2021, March). NET implementation of electronic circuit design. In. AIP Conference Proceedings, (p. 070016). AIP Publishing LLC.

Hazır, Ö. (2015). A review of analytical models, approaches and decision support tools in project monitoring and control. *International Journal of Project Management*, *33*(4), 808–815. doi:10.1016/j.ijproman.2014.09.005

He, Z., Zhang, R., Ran, N., & Gu, C. (2022). Path planning of multi-type robot systems with time windows based on timed colored Petri nets. *Applied Sciences (Basel, Switzerland)*, *12*(14), 6878. doi:10.3390/app12146878

Hinov, N., Gocheva, P., & Gochev, V. (2022b). Index matrices-based software implementation of power electronic circuit design. *Electronics (Basel)*, *11*(5), 675. doi:10.3390/electronics11050675

Hinov, N. L., Gocheva, P. V., & Gochev, V. P. (2022a). Fuzzy reasoning on buck DC-DC power converter parameters. *International Journal on Information Technologies & Security*, *14*(4), 33–44.

Huang, B., Zhou, M., Lu, X. S., & Abusorrah, A. (2023). Scheduling of resource allocation systems with timed Petri nets: A Survey. *ACM Computing Surveys*, *55*(11), 1–27. doi:10.1145/3570326

Jitmit, C., & Vatanawood, W. (2021, April). Simulating Artificial Neural Network using hierarchical Colored Petri Nets. In *2021 6th International Conference on Machine Learning Technologies* (pp. 127-131).

Karaseva, T., & Semenkin, E. (2021). On the automatic identification of differential equations using a hybrid evolutionary approach. In *2021 IEEE International Conference on Information Technologies (InfoTech-2021)*, (pp. 128-133). IEEE. 10.1109/InfoTech52438.2021.9548643

Kravets, O.Ja., Aksenov, I.A. & Rahman, P.A. (2022). Algorithmization of analytical methods for finding motion vectors when processing image series. *International Journal on Information Technologies and Security*, *14*(2), 39–50.

Kuchárik, M., & Balogh, Z. (2019). Modeling of uncertainty with petri nets. In *Intelligent Information and Database Systems: 11th Asian Conference*, (pp. 499–509). Springer.

Kumar, N. N., Patil, S. R., Srikanth, S., & Ravichandran, S. (2021, September). Feedback controller design for a DC-DC buck converter. In *2021 Fourth International Conference on Electrical, Computer and Communication Technologies (ICECCT)* (pp. 1-5). IEEE.

Kung, S. Y. (2020). Wavefront array processors. In *Systolic Signal Processing Systems* (pp. 97–160). CRC Press.

Kurniawan, N. B., Bandung, Y., & Yustianto, P. (2019). Services computing systems engineering framework: A proposition and evaluation through soa principles and analysis model. *IEEE Systems Journal*, *14*(3), 3105–3116. doi:10.1109/JSYST.2019.2939433

Liu, H. C., You, J. X., Li, Z., & Tian, G. (2017). Fuzzy Petri nets for knowledge representation and reasoning: A literature review. *Engineering Applications of Artificial Intelligence*, *60*, 45–56. doi:10.1016/j.engappai.2017.01.012

Liu, X., Zhao, M., Wei, Z., & Lu, M. (2022). The energy management and economic optimization scheduling of microgrid based on Colored Petri net and Quantum-PSO algorithm. *Sustainable Energy Technologies and Assessments*, *53*, 102670. doi:10.1016/j.seta.2022.102670

Nai-Zhi, G., Ming-Ming, Z., & Bo, L. (2022). A data-driven analytical model for wind turbine wakes using machine learning method. *Energy Conversion and Management*, *252*, 115130. doi:10.1016/j.enconman.2021.115130

Parrot, R., Briday, M., & Roux, O. H. (2023). Design and verification of pipelined circuits with Timed Petri Nets. *Discrete Event Dynamic Systems*, *33*(1), 1–24. doi:10.100710626-022-00371-7

Pereira, P., Araujo, J., Melo, C., Santos, V., & Maciel, P. (2021). Analytical models for availability evaluation of edge and fog computing nodes. *The Journal of Supercomputing*, *77*(9), 9905–9933. doi:10.100711227-021-03672-0

Romansky, R., & Noninska, I. (2016). Discrete formalization and investigation of secure access to corporative resources. *International Journal of Engineering Research and Management*, *13*(5), 97–101.

Romansky, R., & Noninska, I. (2021b). Deterministic modelling of a management system with protected access to virtual resources. AIP Conference Proceedings (American Institute of Physics), 2333, art.090003, 090003-1-090003-10 () doi:10.1063/5.0041760

Silva, J. R., & Del Foyo, P. M. (2012). Timed Petri nets. In Petri Nets: Manufacturing and Computer Science (pp. 359-378). InTech.

Su, J., & Schön, S. (2022, April). Deterministic approaches for bounding GNSS uncertainty: A comparative analysis. In *2022 10th workshop on satellite navigation technology (NAVITEC)* (pp. 1-8). IEEE. 10.1109/NAVITEC53682.2022.9847545

Teren, V., Cortadella, J., & Villa, T. (2022, August). Decomposition of transition systems into sets of synchronizing Free-choice Petri Nets. In *2022 25th Euromicro Conference on Digital System Design (DSD)* (pp. 165-173). IEEE. 10.1109/DSD57027.2022.00031

Ventisei, A., Yakovlev, A., & Pacheco-Peña, V. (2022). Exploiting Petri nets for graphical modelling of electromagnetic pulse switching operations. *Advanced Theory and Simulations*, *5*(3), 2100429. doi:10.1002/adts.202100429

Wang, X., Zhang, P., Du, Y., & Qi, M. (2020). Trust routing protocol based on cloud-based fuzzy Petri net and trust entropy for mobile ad hoc networks. *IEEE Access : Practical Innovations, Open Solutions*, *8*, 47675–47693. doi:10.1109/ACCESS.2020.2978143

Zhou, K. Q., & Zain, A. M. (2016). Fuzzy Petri nets and industrial applications: A review. *Artificial Intelligence Review*, *45*(4), 405–446. doi:10.100710462-015-9451-9

Chapter 8
Features and Application of Stochastic Analytical Modeling

ABSTRACT

The chapter deals with the main features of stochastic modeling and its application in the investigation of probabilistic processes in computer processing. At the beginning, a brief overview of the possibilities offered by the probabilistic analytical modeling was made with the definition of basic characteristics and short comments. The relationship of discrete and continuous random variables to stochastic modeling is discussed in the first part, focusing on Markov processes with application examples presented. The second part is devoted to the stochastic analytical approximation. The third part is directed to the discussion of the stochastic modeling by means of the theory of Markov processes, concentrating on random processes with discrete states and discrete time (Markov Chain) and on analogous processes, but with continuous time. Each of these two sub-parts is illustrated with numerous examples of sample computer processing situations with analysis of the results and appropriate tabular and graphical illustration of the resulting estimates.

1. BASIC FEATURES OF THE PROBABILISTIC ANALYTICAL MODELING

For most of the studied systems and processes, the presence of random factors is characteristic, which determines the effectiveness when applying the probabilistic modeling approach (Efatmaneshnik et al., 2019). It is related to the use of various approaches and means of probability theory (Durrett, 2019; Gnedenko, 2018). Probability theory (PT) has been applied to various studies over the centuries, but as a branch of mathematics it was established in the mid-seventeenth century. A consistent historical overview of the development and use of probabilities is given in (Pandit et al., 2022). The applied aspect of probabilities is directly related to random processes (RPs), random sequences and statistics (Girardin & Limnios, 2022). Various publications analyze the relationship of PT with statistical reasoning about the degree of uncertainty in process development, as done in (Huang, 2022). The goal of PT is determined as a task "*to express uncertain phenomena using a set of axioms*" or as it is added in the article, probabilities for the realization of events and scenarios are calculated and analyzed when there is no

DOI: 10.4018/978-1-6684-8947-5.ch008

necessary certainty for their realization. In order to confirm the stated thesis, the article presents basic mathematical concepts and theorems. The relationship of mathematical statistics to PT is discussed in (Stoffels & Hohmann, 2022), where the two theories are defined as fundamental parts of stochastics. It is stated that the theory of statistics is relevant to empirical sciences because it reflects numerical values for measured parameters of various processes. This allows description of datasets, hypothesis building and testing by applying statistical methods.

The probability $P(A)$ is a number in the range [0,1] that indicates how likely it is that event A occurs when an experiment is run. All possible elementary events in the realization of a given experience form the space of events $\Omega = \{\omega_i / i=1, 2, \ldots, n\}$. Then, if $A = \{\omega_1, \omega_2, \ldots, \omega_m\} \subset \Omega$, the classical definition of probability is $P(A) = m/n$. Or in other words – the ratio of the number of occurrences m of event A to all possible equally likely outcomes n. The following axioms apply:

1. $0 \leq P(A) \leq 1$.
2. $P(\Omega) = 1$.
3. $P(A \cup B) = P(A) + P(B)$, if events A and B are mutually exclusive.
   Example: Let the sides of a coin be denoted by h (face) and t (back). Then for an experiment "tossing a coin three times" the space of elementary events will be Ω = {hhh, hht, hth, htt, thh, tht, tth, ttt}, with úΩú = 8. Let A be the event "one face and two backs", i.e. A = {htt, tht, tth} ⇒úAú = 3, and the probability of its occurrence will be $P(A) = 3/8$.

In modeling, it is often necessary to examine the occurrence of an event A, conditional on the occurrence of another event B. For this purpose, a conditional probability $P(A/B)$ is applied. The new event A/B will have a space formed by the intersection of A and B (requiring that B is not a null event), and the formula is:

$$P(A/B) = \frac{P(A \cap B)}{P(B)}.$$

In the general case, independent events do not necessarily have non-intersecting spaces A and B. For their union, the formula $P(A \cup B) = P(A) + P(B) - P(A \cap B)$, where $P(A \cap B) = P(A).P(B)$ is the intersection of the events, is valid. If the two events have no point in common, their intersection is the empty set ∅, and the probability is $P(\emptyset)=0$.

When creating a probabilistic analytical model (PAM), random variables (RVs) are most often used, which take on different values at each trial and they cannot be predicted in advance. Thus, the random variable X at each trial gives a result satisfying the conditions:

1. The set $\{X \leq x\}$ represents an event for each value of x.
2. $P(X = +\infty) = P(X = -\infty) = 0$.

A random variable can be discrete (takes on a finite or countable number of values) or continuous (has an infinite number of values).

***Continuous RV** X* can be described by:

Distribution Function:	$F(x)=P(X\leq x) F(-\infty)=0$ and $F(\infty)=1$
Density Function:	$f(x)=\frac{dF(x)}{dx}$; $\int_{-\infty}^{t} f(x).dx=1$; $F(x)=\int_{-\infty}^{t} f(x).dx$

The application of continuous probability distributions for modeling conditional modulation parameters in semantic image synthesis is presented in (Tan et al., 2021). A new framework for the synthesis in terms of different distributions of semantic classes, replacing discrete values with continuous probability distributions is proposed. This is implemented by sampling modulation parameters for each instance applied to the entire network. As an additional step during testing, noise pre-remapping is applied, which facilitates supervised learning and instance style control.

Discrete RV is described by function $f(x)$:$\{<x_i, p_i> / i=1, 2, \ldots, n\}$, which is defined analytically, by table of graphically, and the function $F(x)$ – by equations:

$$p_i=P(X=x_i);\ i=1,2,\ldots,n \text{ and } \sum_{i=1}^{n} p_i = 1$$

Information theory mainly considers two types of theoretical models known as entropy and divergence. To the first group of entropy models, many standard models based on discrete probability distributions have been developed. Confirming the above two new parametric entropy models with discrete probability distributions are defined in (Om Parkash, 2021). The discrete probability distribution was also used in (Bhati et al., 2020) to develop a model for counting symmetric and asymmetric data using the "*discrete skew logistic distribution*". In addition to discussing the basic properties of the distribution, an assessment of the parameters of the method of proportions and the effectiveness of the maximum likelihood method has been made.

Stochastic modeling should be perceived as part of the possibilities of probability theory and its main features are presented in (Lanchier, 2017). Random variables and processes based on standard limit theorems are used in describing the studied objects and real-world problems. The article reviews the tools used, such as martingales, discrete-time Markov chains, Poisson processes, and continuous-time Markov chains. Various examples of application of stochastic modeling in the study of processes in physics, biology and sociology are presented.

The basic characteristics of random variables briefly presented below provide a summary of the tools used in stochastic modeling.

- *Frequency Ratio* – the ratio of the number of occurrences $N_m(x)$ of a given value x from the probability space to the total number m of the performed attempts:

$$P_m(x)=\frac{N_m(x)}{m};$$

- *Expectation* or average value (mean):

$$E[X]=\mu_X=\int_{-\infty}^{\infty}x.f(x).dx \text{ – for continuous RV;}$$

$$E[X]=\mu_X=\frac{1}{n}\sum_{i=1}^{n}x_i.p_i \text{ – for discrete RV;}$$

- *Variance*:

$$V[X]=\sigma^2=E[X-\mu_X)^2]=E[X^2]-\mu_X^2];$$

Standard Deviation: $\sigma_x=\sqrt{V[X]}$.

As noted above, stochastic modeling is based on the so-called *stochastic/random process (RP)*, representing a family of random variables $\{X(t), t\in T\}$ as a function of time. Depending on the accepted values for the two parameters "time" and "states" (discrete, continuous), different types of RPs are defined. Usually, in modeling, two basic types are used in a generalized sense - discrete RP and continuous RP, and the conditions for approaching (falling into) *steady state (stationary RP)* are most often sought. It is such a process that its probabilistic characteristics do not change with time. One of the typical definitions of such a process is related to a constant mathematical expectation, i.e. $E[X(t)]=\mu=const.$ The description and study of an RP is based on:

- Transition probabilities $p_{ij}(t)=P(s_i\rightarrow s_j)$ from state $s_i=S(t)$ to another permitted state $s_j=S(t+1)$;
- Initial probabilities $p_i(0)$ for the initial state s_i ($i = 1, ..., n$) of the process;
- Probabilities for transition in time $p_j(t)/j = 1, ..., n$, which are constant values for each given time moment $t = 1, 2, ...$ and define the probability for transition to a specific process state s_j in the time t:

$$p_j(t)=\sum_{i=0}^{n}p_i(0).p_{ij}(t).$$

More important discrete RPs are Markov chains, random flow of discrete type (Poisson, Erlang, Bernoulli, etc.), random sequences, etc. Modeling a continuous RP boil down to obtaining a discrete random sequence whose properties approximate the real process with sufficient accuracy. A comparison of both discrete-time and continuous-time approaches to network data analysis is made in (Ryan & Hamaker, 2022). A disadvantage of discrete-time models in defining the network structure with measures of centrality is their dependence on time intervals. As an alternative, continuous-time modeling is indicated, but it is specified that this requires some modification of the initial concept based on the assumption that the parameters reflect general rather than direct effects. In this respect, a new decision regarding the problem is proposed in the article.

An example of the application of discrete processes in defining a deterministic stochastic model for the study of renewable energy and large-scale utility networks is presented in (Hwangbo et al., 2022). The structure of the developed model includes two parts – a deterministic model and a stochastic model. The first is applied in the optimization of utility networks, while stochastic modeling of the processes of building clean electric networks with renewable energy is the task of the second part. To conduct the experiments, different scenarios have been developed under uncertain parameters, and the goal is to find the best of them.

Another direction in discrete RPs and the application of nonlinear ordinary differential equations based on the concept of partial differentiation discussed in (Sweilam et al., 2022). The article is oriented towards chaotic systems, modeling of two deterministic-stochastic chaotic systems using hybrid fractional derivatives was done. The conducted study of the behavior of the models was done under defined different possible cases and numerical methods were introduced: the non-standard method of Milstein for solving the stochastic models and the non-standard finite difference of Grünwald–Letnikov for solving the deterministic models of the hybrid fractional order.

Continuous RPs are characterized by continuous time and are the basis of the corresponding type of stochastic models. A probabilistic framework for modeling stochastic dynamics is proposed in (Park et al., 2022). A model called neural controlled stochastic differential Markov equation was developed. Conducted experiments show that the model efficiently generates complex time series in the data space without additional meshes with performance comparable to other methods. Stochastic continuous-time modeling has also been applied in (Laomettachit et al., 2022) to investigate a cell division cycle. The specific in this case is that to develop a continuous-time stochastic model, they used seven Boolean variables to represent the activities of the main cell cycle regulators and one continuous variable to represent cell growth. A continuous update of the time and cell size of the tested yeast is performed with the calculation of statistical properties of the cell cycle progression.

An example of an application of stochastic modeling in the study of stack cache processes is presented below. The purpose of the model is to investigate the processes of monitoring and predicting the usability of the stack data and to optimize the structural implementation. When defining a model, it is based on the traditional stack organization of access through specialized registers 'Stack Pointer' (SP) and 'Base Pointer' (BP), as well as on the peculiarities of local data in a stack memory:

a) They are valid only during the execution of the attached procedure;
b) Implementation of control, activation and deactivation is automatic;
c) They are grouped in a very compact area, the location of which is easy to predict.

In its implementation, an application works with the so-called procedure window formed as a continuous array with boundaries defined by *SP* and *BP* (Figure 1). The context of an active procedure (the set of parameters) is formed on the basis of the stack allocated for it and includes: ✓ starting address from the stack (associated with register *BP*); ✓ end address from stack (associated with register SP); ✓ address of the control transition to enter the procedure. It is assumed that the execution of an application can be viewed as an RP of procedure calls $\{A, B, C, \dots\}$. Each RP $X(t)$ at a time $t = t_i$ takes a specific value $X(t_i)$ representing RVs with values $\omega \in \Omega$. This allows RP to be represented as the function $X(t)=\varphi(t,\omega)$, for $\omega \in \Omega$, $t \in T$, $X(t) \in \Sigma$,, where: ω – value of RV; Ω – space of possible values of RV; T – set of values of the time argument; Σ – set of possible values of $X(t)$.

Figure 1. Procedure window in the stack

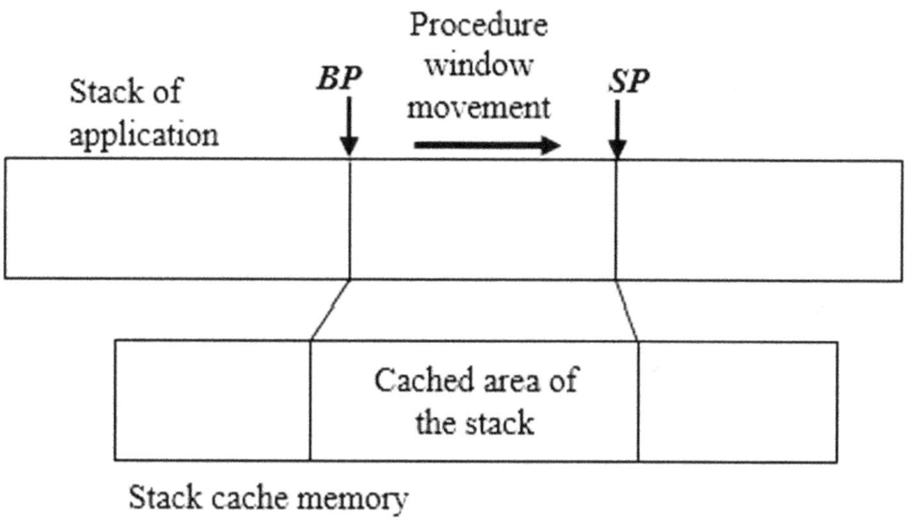

The initial hypothesis is that the execution process of an application is a discrete RP with discrete RP with discrete states and it can be descripted by sequence of system states <$S(0)$, $S(1)$, $S(2)$,..., S(k),...> for the separate time moments t=0,1,...,k,... . Each transition $[S(k\text{-}1)=S_i]\rightarrow[S(k)=S_j]$ is probability and is determined by the formula:

$$p_j(k)=\sum_{i=1}^{n} p_i(k-1).p_{ij};\ j=1,2,...n$$

For the purposes of the study, a stochastic model is defined to analytically represent the process of prefetching and saving stack data as a function of prefetch time. It is assumed that the execution process of an application is a discrete Markov RP with a finite set of states $S=\{S_1,...,S_n\}$, that it enters at discrete time instants $t_i \in T$ (procedure execution steps). An additional assumption is that the stack of windows is divided into equal size blocks in memory and the time to download each of them is equal to M processor cycles. It is assumed that the start of the prefetch process is initiated during entry into a procedure, respectively return from a procedure, as defined for the stack cache operation algorithm. When implementing a given procedure, the processor can initiate K accesses to the stack, and after the first M cycles the first block will be fetched.

Any access by the processor to a block fetched from the stack memory is considered as elementary event A, and access to a non-fetched block as elementary event B. Therefore, the access process by the processor can be considered as a series of random events for many $\Omega = \{A, B\}$, and the process itself – as RP with function $F(\Omega,t)$. The following edge cases apply to this random process:

- At the initial time $t_0 \leq (1...M)$ all addresses to the stack memory result in misses because the first block of the procedure window has not yet been pulled into the stack.

- At time $t_n \geq (N_* M)$ all blocks from the procedure window have been pushed onto the stack and all CPU calls result in successful lookups.

Based on the defined conceptual model, two base states $S = (A, B)$ are defined with transition probabilities reflected in the state graph of Figure 2.

Figure 2. Graph of the states for the stochastic model

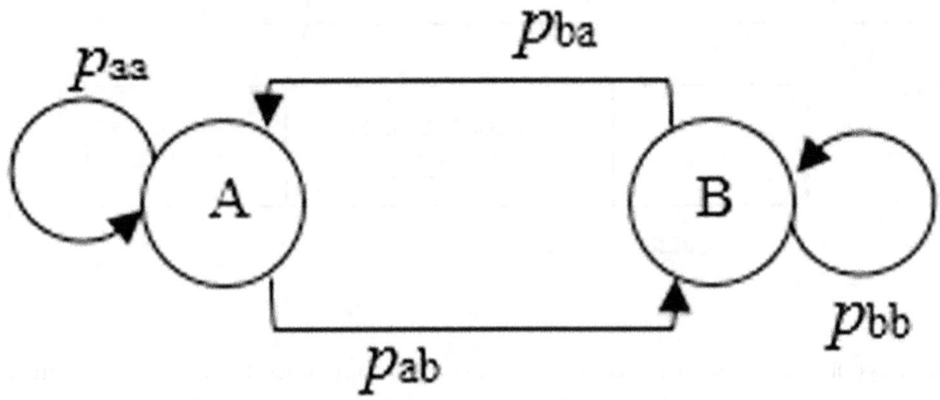

For the first M cycles, the process can go through K events. Based on a transition matrix P and assuming that the primary vector is $\mu^{(0)}=(1,0)$, for the first K steps the probability vector $\mu^{(k)}$ is calculated by the formula, where is the K-th product of the matrix P, i.e.

$$P_0^k = \prod_{i=0}^{k} P^i = P^0 * P^1 * \ldots * P^k$$

After applying the full likelihood formula for $K \leq M$, the following analytic model is defined, and its analytic solution and the resulting probabilities p_a and p_b are presented below.

$$\begin{cases} p_a = p_{aa} \cdot p_a + p_{ba} \cdot p_b \\ p_b = p_{ab} \cdot p_a + p_{bb} \cdot p_b \\ p_a + p_b = 1 \end{cases}$$

$$p_a - p_{aa} \cdot p_a = p_{ba} \cdot p_b \Rightarrow p_a = \frac{p_{ba}}{(1 - p_{aa})} \cdot p_b;$$

$$\frac{p_{ba}}{(1 - p_{aa})} \cdot p_b + p_b = 1 \Rightarrow p_b \frac{(p_{ba} + 1 - p_{aa})}{(1 - p_{aa})} = 1;$$

$$p_a = \frac{p_{ba}}{\left(1 + p_{ba} - p_{aa}\right)} \text{ and } p_b = \frac{\left(1 - p_{aa}\right)}{\left(1 + p_{ba} - p_{aa}\right)}$$

The resulting expressions for p_a and p_b allow estimates of the final probabilities for the process of ($K \leq M$) consecutive cycles to be computed. For specific values of transition probabilities, numerical values can also be calculated for them. For example, with a matrix of transition probabilities:

P[1,*]:= [0.8; 0.2] and P[2,*]:= [0.9;0.1],

quantitative estimates are determined for the final probabilities:

$p_a = 0.818181... \approx 0.8182$ and $p_b = 0.181818... \approx 0.1818$.

2. STOCHASTIC ANALYTICAL APROXIMATION

The stochastic approximation approach is analogous to the functional approximation in deterministic modeling, but uses probabilistic variables. Usually, a suitable law is sought to approximate a given real RP or RV sufficiently accurately. Most often, assumptions of stationarity, ordinariness and absence of consequence are made for the studied real stochastic process, which allows it to be approximated by a simple Poisson flow or another known probability law, the study of which is relatively easy. For example, if the movement of packets in a network is modeled by a stochastic flow, then the number of packets arriving in a given time interval has a Poisson distribution:

$$P(\text{arrival of } k \text{ elements in time } T) = \frac{(\lambda.T)^k}{k!} e^{-\lambda T}$$

The definition of a Poisson counting process $\{N(t), t \geq 0\}$) is based on the following:

1. $N(t)$ has stationary independent increments.
2. $N(0) = 0$.
3. For $0 < t_1 < t_2$, the value $N(t_2) - N(t_1)$ is equal to the number of points in the interval (t_1, t_2) and it has a Poisson distribution with average value $\lambda(t_2 - t_1)$.

In this case for $N(t)$ is obtained: $P\left[N(t) = k\right] = \frac{(.t)^k}{k!} e^{-t}$ and $E[N(t)]=V[N(t)]=t$.

The process $N(t)$ is non-stationary because the average value depends on time and its behavior in time has a step-like character (Figure 3a). The stationary process associated with $N(t)$ is called a Poisson increment process and for the case (at mean value λt and constant $L>0$) it has the following definition – Figure 3(b):

$$X(t)=\frac{N(t+L)-N(t)}{L}.$$

The mean of $X(t)$ is k/L, where k is the number of points in the interval $(t, t+L)$. The equal $E[X(t)]=\lambda$ is valid for this process.

Figure 3. Poisson processes

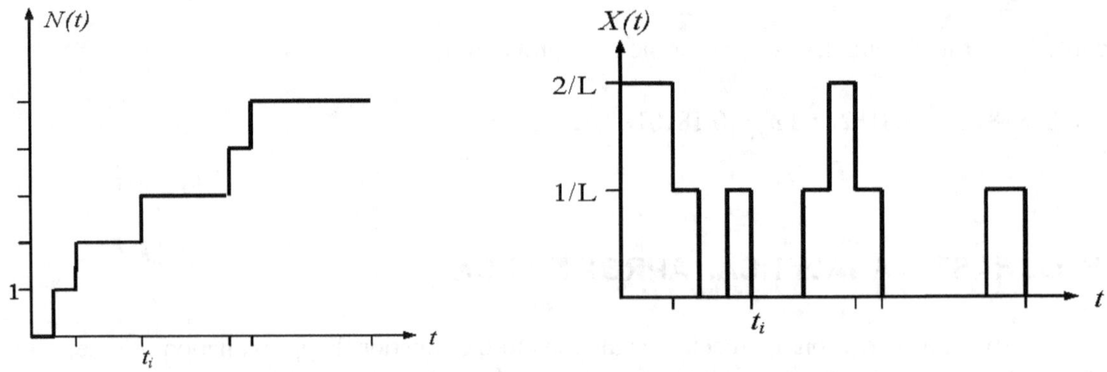

a) counting Poisson process *b) incremental Poisson process*

Stochastic analytical models are mainly used in the investigation and evaluation of operational and functional characteristics of CS, e.g. operability, reliability, readiness for work, productivity, etc. For example, the reliability of a system over time can be described by $R(t)=e^{-\lambda t}$, which is an exponential function with parameter λ (intensity of failures under normal operating conditions conditions). If U is a random variable uniformly distributed in the range [0, 1], the system is said to fail by time t if $R(t)\leq U$.

Analytical approximation is often applied in various research projects where stochastic analysis of processes is required (Ghadimi et al., 2020). An example area is the stochastic evolution of chemical reaction networks, which are analyzed by solving the underlying chemical equation or performing extensive simulations, but as stated in (Cardelli et al., 2016), both approaches are inapplicable to complex systems. In this respect, probabilistic logic for stochastic analysis of the evolution of molecular species populations is presented in the paper. A model checking algorithm based on the linear noise approximation was developed, the computational complexity of which does not depend on the population size of each species and is polynomial in the number of different species. In its implementation, polynomial differential equations of the first order are solved.

Problems in applying the stochastic approximation for the investigation of financial risk measures are discussed in (Bourgey, 2020), emphasizing that "*the need for fast, efficient, and reliable analytic approximation formulas is of primal importance to financial institutions*". In order to fulfill the stated requirement, some multi-level approximation methods are studied in these dissertations and applied to solve specific practical problems. A general framework for meta-modeling large sums of weighted Bernoulli RP variables that are conditionally independent of a common factor X is proposed. In general, the approach is based on a polynomial chaos expansion on the common factor together with some

Gaussian approximations and are calculated estimates for error in the factor X associated with classical orthogonal polynomials.

Non-Markovian models are often applied to investigate stochastic biochemical kinetics, but there are certain difficulties in their analysis, including the inference of their parameters from data (Jiang et al., 2021). The main reason is the dependence of the dynamics on the history of the system. To overcome this shortcoming, the paper proposes the use of an artificial neural network to approximate the distributions of non-Markovian models that are time-dependent. It is shown that the approximation does not increase the size of the model, as the training of the neural network uses a relatively small set of measurements from experimental data or stochastic simulations of the non-Markovian model.

3. STOCHASTIC MODELING BY USING MARKOV CHAIN APPROACH

The application of the theory of Markov processes allows the creation of adequate SAMs in the description and study of computer systems and processes. The basis for this statement is the fact that the nature of computer processes is very close to that of Markov processes with discrete states (Markov chain – MC). Depending on the nature of the time, both varieties are applied – with discrete and with continuous time. Markov models are based on the concept that an RP $f(t)$ gives rise to a sequence $f_0, f_1, f_2, \ldots$ of states corresponding to times $t = 0, 1, 2, \ldots$, with the initial state $f_0 = s_i$ is determined by the initial probability P_0, and the subsequent states for depend on the probabilistic transitions $[S(k\text{-}1)=s_i]\rightarrow[S(k)=s_j]$ based on the matrix of transition probabilities. If in the matrix of transition probabilities there exists a probability $p_{ii} = 1$ for a given state, then this state is absorbing (it cannot be left upon entering it) and such a MC is called absorbing.

3.1. Markov Chains with Discrete Time

An introduction to the features of MC has been presented in the Chapter 5. Investigation of a MC model is based on the two formulas – for the full probabilities and the condition for normality:

$$p_j(k)=\sum\nolimits_{i=1}^{n} p_i(k-1).p_{ij};\ j=1,2,\ldots,n;$$

$$\sum\nolimits_{j=1}^{n} p_{ij}(k)=1; i=1,2,\ldots,n.$$

The traditional form of MC with discrete time is suitable for model investigation of different computer and network processes, as the routing efficiency in network spice (Kravets et al., 2021), hierarchical data investigation (Wanduku, 2022), study of the presence of noise in the spectrum (Sinha & Trivedi, 2022), evaluation of hydropower plant reliability (Dimishkovska & Iliev, 2019), etc.

To clarify the features and possibilities of modeling with MC, an addition is made below to the probabilistic analytical modeling of a stack cache memory presented at the end of part 8.1. Here we discuss the issue of data prefetching, defined as the asynchronous process of moving data from the low levels to the high levels of the stack cache before it is real used. It is possible that some of the pre-fetched data will not be used and this will reduce the performance of the system. Various hardware and software mechanisms are used to solve this problem. All these solutions have one drawback - data access is considered as a single set. An efficient approach is to prefetch the stack and trace procedure calls.

Data for each application can be defined as global and local (stack) data. Local data allows efficient use of prefetching because it is grouped in a limited memory environment. The main problem with data prefetching is to predict the probability of stack moves during application execution. The stack memory is part of the system memory and, as mentioned above, the two specialized registers – Stack Pointer (SP) and Base Pointer (BP) – are used for its management, and the functioning is based on the procedure window (PW). It can be assumed that the stack data form a continuous data set defined by PW, and the execution of an application could be described as a random process of procedure calls {*A*, *B*, *C*, …} – Figure 4.

For a concrete example it is assumed that an application consists of procedures {*A*, *B*, *C*} and each execution could be presented as a sequence of procedure calls. The MC model is presented in Figure 5 as an initial definition and graph of states. The values for the transition probabilities (arcs in the graph) are determined on the base of empirical data obtained by using SPECint benchmarking.

Figure 4. An example for graph description of application execution

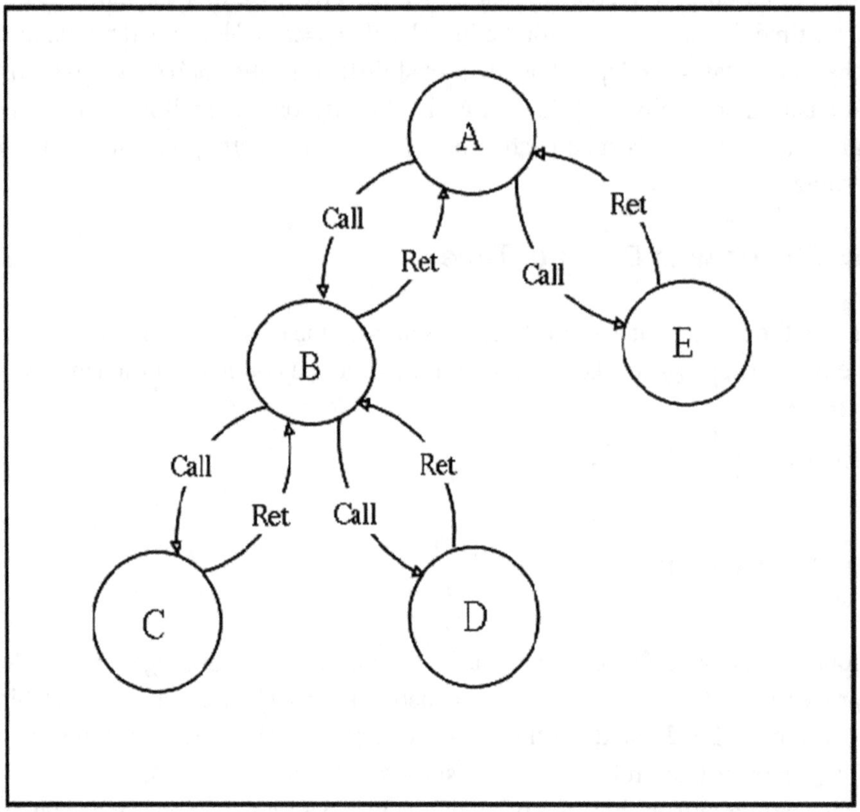

Figure 5. Definition of MC model

Discrete set of states and Vector of initial probabilities:

$S = \{S_A, S_B, S_C\} = \{A, B, C\}$

$P_0 = (1,0,0)$

Transition matrix:

$P(A \to *) := [0.1, 0.8, 0.1]$

$P(B \to *) := [0.2, 0, 0.8]$

$P(C \to *) := [0.1, 0.9, 0]$

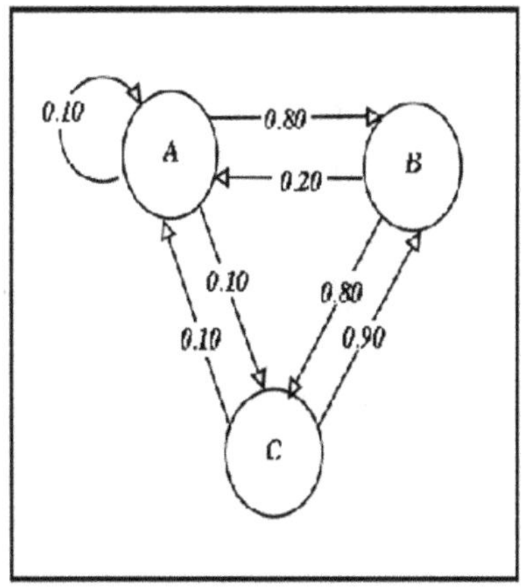

The analytical MC model as system of equations is presented below.

$$p_A = 0.1, p_A + 0.2, p_B + 0.1, p_C$$
$$p_B = 0.8, p_A + 0.9, p_C$$
$$p_C = 0.1, p_A + 0.8, p_B$$
$$p_A + p_B + p_C = 1$$

Evaluation of the probabilities for steady state regime is connected with the final probabilities' determination. A solution of the analytical model (system of equations) is made by the following transformation started from the state A:

p_A=0.1.p_A+0.2(0.8.p_A+0.9.p_C)+0.1.p_C $\Rightarrow$ p_A=0.1.p_A+0.16.p_A+0.18.p_C+0.1.p_C $\Rightarrow$ p_A=0.1.p_A–0.16. p_A=(0.18+0.1).p_C $\Rightarrow$ $p_C = \dfrac{74}{28} p_A$

The obtained assessments for final probabilities are the following: $p_A \approx 0.1466$; $p_B \approx 0.466$; $p_C \approx 0.3874$.

The sequence of procedure calls forms a chain of states for the steps k=1,2,3,…, and each sequence could be formatted with any probability. For example, the investigation of sequence <*ABAA*> is based on:

$S(k=0) = A \to S(k=1) = B \to S(k=2) = A \to S(k=3) = A$;

$P[S(k=3) = A/S(k=1) = B$ and $S(k=2) = A]$;

$P_{ABAA} = p_{AB} \cdot p_{BA}$ = 0.8,0.2,0.1 = 0.016.

In this way it is possible to determine each probability, for example:

$$p_{ABCABC} = \prod_{k=1}^{5} p_{AB} \cdot p_{BC} \cdot p_{CA} \cdot p_{AB} \cdot p_{BC} = 0.8, 0.8, 0.1, 0.8, 0.8 = 0.04096\,.$$

The number of possible sequences for the set {A, B, C} without any limit is $N \to \infty$. In this reason for our investigation is assumed that $N \leq 5$ and some results are shown in Table 1.

Table 1. Example for calculated freciences

Sequences	*AAAAA*	…	*ABAAA*	…	*ABCAB*	…
Frequencies	0.01%		0.16%		5.12%	

For investigation of all possible transitions between duple of states in the stochastic model request to define the probability of next procedure that will be called. On this base it will be possible to determine the new PW using the parameters of the previously PW. The problem decision is based on condition probability $p_{AB}(k=2)= P[S(k=2)=B/S(k=0)=A]$. The analysis gives the assessment $p_{AB}(k=2) = 0{,}17$ that determinates frequency of 17% for realization of sequences type '*AxB*', where $x \in \{A, B, C\}$. By this way it is possible to determine the estimations for each calls sequence '*Axx...xA*', '*Axx...xB*' or '*Axx...xC*'.

Generalized assessments for random parameters of data prefetching in several cases of the transition matrix are presented in Table 2. Results of sequences distribution invitation at the procedure calls based on finite discrete set {A, B, C} is show in Figure 6.

Table 2. Determination of probabilities for state visiting at the application execution

	A	B	C	A	B	C	A	B	C	A	B	C
A	0.1	0.8	0.1	0.3	0,5	0,2	0,5	0,2	0,3	0,7	0,3	0
B	0.2	0	0.8	0.2	0,7	0,1	0,1	0,6	0,3	0,7	0,2	0,1
C	0.1	0.9	0	0.1	0,1	0,8	0,2	0,3	0,5	0,2	0,5	0,3
	$P_A = 0.1466$; $P_B = 0.466$; $P_C = 0.3874$			$P_A = 0.1786$ $P_B = 0.4286$; $P_C = 0.3929$			$P_A = 0.2292$; $P_B = 0.3958$; $P_C = 0.3750$			$P_A = 0.680$; $P_B = 0.280$; $P_C = 0.040$		

Figure 6. Distribution of double transitions

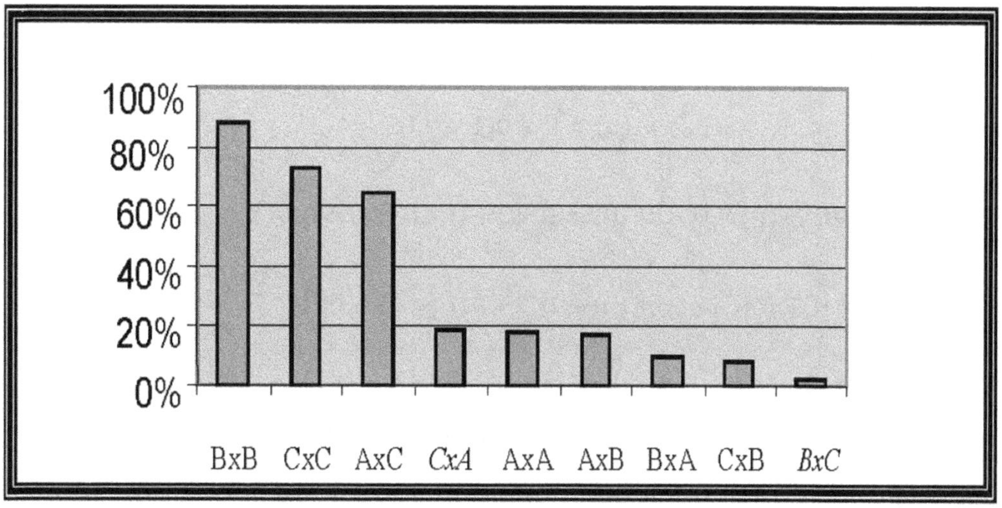

The study of probabilistic processes by MC is carried out for the non-stationary regime and for the conditions for reaching stationarity, which are discussed below.

When studying a ***non-stationary mode***, the probabilities of the states $p_j(k)=P[S(k)=s_j]$ for step k are determined, which are united in the vector $P(k)=\{p_1(k), p_2(k),...,p_n(k)\}$. In homogeneous MC, the normality condition is satisfied by every row of the transition probability matrix for which $p_{ij}(k) = p_{ij} = const.$ The probabilities of the process falling into a given state at a sequence of steps ($k = 1, 2, \ldots$) or of falling into a state $S(k) = s_j$ under the condition that $S(k\text{-}1) = s_i$, for which the probability is $p_i(k\text{-}1)$. An example study of a non-steady state is shown below. An example of creating an MC model and determining the probabilities of successively falling into the first 3 states is given in Figure 7.

Figure 7. An example for MC model definition

$S = \{s_1, s_2, s_3\}$

$P_0 = \{0{,}7;\ 0{,}2;\ 0{,}1\}$

P	s_1	s_2	s_3
s_1	0	0,6	0,4
s_2	0,5	0	0,5
s_3	0,7	0,3	0

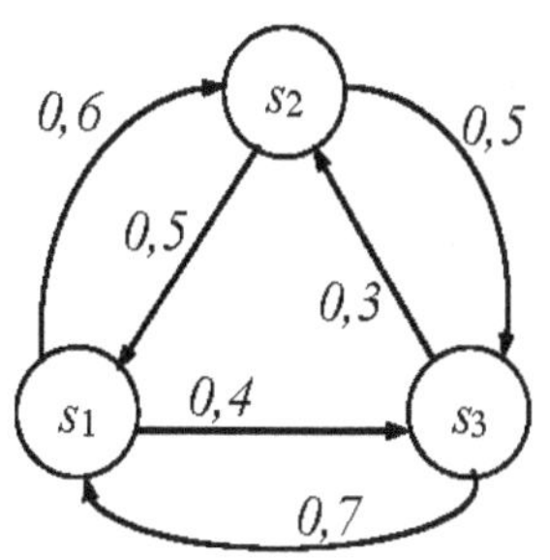

Markovian model:

$p_1 = 0{,}5.p_2 + 0{,}7.p_3$

$p_2 = 0{,}6.p_1 + 0{,}3.p_3$

$p_3 = 0{,}4.p_1 + 0{,}5.p_2$

$p_1 + p_2 + p_3 = 1$

A solution is sought by the full probability formula for successive steps $k = 1, 2, 3$.

- $k = 0 \Rightarrow$ the initial state is determined by the vector of initial probabilities, i.e. $p_1(0)$=0.7; $p_2(0)$=0.2; $p_3(0)$=0.1.
- $k = 1 \Rightarrow$ the probabilities are calculated according to equations:

$$p_j(1) = p_{1j} \cdot p_1(0) + p_{2j} \cdot p_2(0) + p_{3j} \cdot p_3(0)\text{; for } j = 1, 2, 3,$$

and the solution is as follows:

$$p_1(1) = 0.p_1(0) + 0{,}5.p_2(0) + 0{,}7.p_3(0) = 0 + 0{,}1 + 0{,}07 = 0{,}17$$

$$p_2(1) = 0{,}6.p_1(0) + 0.p_2(0) + 0{,}3.p_3(0) = 0{,}42 + 0 + 0{,}03 = 0{,}45$$

$$p_3(1) = 0{,}4.p_1(0) + 0{,}5.p_2(0) + 0.p_3(0) = 0{,}28 + 0{,}1 + 0 = 0{,}38$$

- $k = 2 \Rightarrow$ the probability $p_j(2)$ is calculated based on the equations:

$$p_j(2) = p_{1j} \cdot p_1(1) + p_{2j} \cdot p_2(1) + p_{3j} \cdot p_3(1)\text{; for } j = 1, 2, 3,$$

where the probabilities $p_j(1)$ are those determined in the previous step, and the calculated values are summarized in Table 3.

- $k = 3 \Rightarrow$ probabilities $p_j(3)$ are calculated analogous to the previous one (the values are presented in Table 3):

$$p_j(3) = p_{1j} \cdot p_1(2) + p_{2j} \cdot p_2(2) + p_{3j} \cdot p_3(2)\text{; for } j = 1, 2, 3,$$

Table 3. Values for the state's probability

k	$p_1(k)$	$p_2(k)$	$p_3(k)$
1	0.17	0.491	0.3131
2	0.45	0.216	0.3825
3	0.38	0.293	0.3044

Figure 8 illustrates obtaining an answer to the task of determining after how many steps the process presented by the MC model from Figure 7 will reach state s_3 if it started from state s_1. The solution is to calculate the probability $p_{13}(k=3)$. Analytically, this can be done after analyzing the transition probability matrix by analyzing row (1) and its associated columns according to receiver probabilities.

Figure 8. Graphical illustration of the task solution

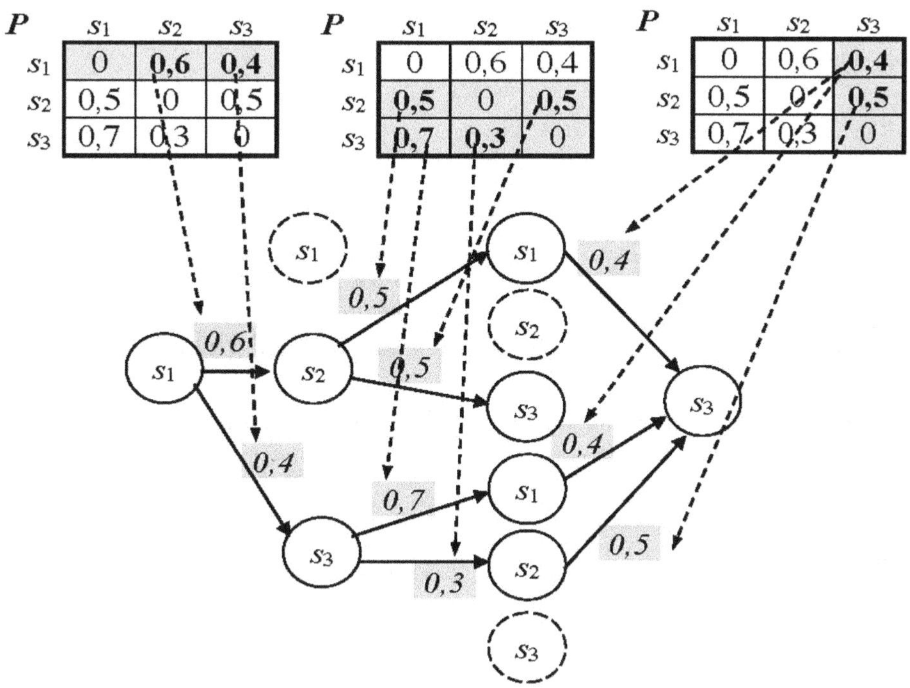

The successive transitions for steps 1, 2, and 3 are determined by evaluating the zero probabilities (the Markov model has no reflexive nodes). Three paths are formed to reach state s_3 starting from state s_1. The total probability is defined as follows:

$p_{13}(k=3)= p_{12}{\cdot}p_{21}{\cdot}p_{13} + p_{13}{\cdot}p_{31}{\cdot}p_{13} + p_{13}{\cdot}p_{32}{\cdot}p_{23}$;

$p_{13}(k=3)$= 0.6,0.5,0.4 + 0.4,0.8,0.4 + 0.4,0.3,0.5 = 0.12+0.112+0.06 = 0.292

In the ***stationary mode*** of the MC (at $\kappa\rightarrow\infty$) the transition probabilities $p_{ij}(k)$ and the state probabilities $p_j(k)$ do not depend on the step k and tend to the same constant value $p_j=const$:

$$\lim_{k\to\infty} p_{ij}(k) = p_j\,;\ i,j=q,2,\ldots,n$$

$$\lim_{k\to\infty} p_{ij}(k) = p_j\,;\ i,j=q,2,\ldots,n$$

$$\lim_{k\to\infty} p_j(k) = p_j\,;\ j=1,2,\ldots,n$$

The determination of these final probabilities is carried out based on constructing a system of probability equations, which, together with the normality condition, define the Markov analytical model:

$$\begin{cases} p_1 = p_{11} \cdot p_1 + p_{21} \cdot p_2 + \dots + p_{n1} \cdot p_n \\ p_2 = p_{12} \cdot p_1 + p_{22} \cdot p_2 + \dots + p_{n2} \cdot p_n \\ \vdots \\ p_n = p_{1n} \cdot p_1 + p_{2n} \cdot p_2 + \dots + p_{nn} \cdot p_n \\ p_1 + p_2 + \dots + p_n = 1 \end{cases}$$

Solving the model allows determining the final probabilities for the individual states upon reaching a steady state.

Continuing with the investigation of the MC-model defined in Figure 7, it can determine the final probabilities for the states in stationary regime. The solution can be done by the following substitutions in the system of equations.

$p_1 = 0.5,p_2 + 0.7,p_3 = 0.5,p_2 + 0.7(0.4,p_1 + 0.5,p_2) = 0.5,p_2 + 0.28,p_1 + 0.35,p_2(1 - 0.28),p_1 = (0.5+0.35),p_2$

$$\Rightarrow p_2 = \frac{0.72}{0.85} p_1;$$

$$p_2 = 0.6, p_1 + 0.3, p_3 \Rightarrow \frac{0.72}{0.85} p_1 = 0.6, p_1 + 0.3, p_3 \Rightarrow p_3 = \left(\frac{0.72}{0.85} - 0.6 \right) . p_1 = 0.3, p_3$$

$$p_3 = \frac{0.21}{0.255} p_1;$$

$$p_1 + \frac{0.72}{0.85} p_1 + \frac{0.21}{0.255} p_1 = 1 \Rightarrow \left(1 + \frac{0.72}{0.85} + \frac{0.21}{0.255} \right) p_1 = 1 \Rightarrow p_1 \approx 0.375;$$

$$p_2 = \frac{0.72}{0.85} p_1 \Rightarrow p_2 \approx 0.317 \text{ and } p_3 = \frac{0.21}{0.255} p_1 \Rightarrow p_3 \approx 0.308.$$

The investigation of algorithmic structures presented in Chapter 6.2 can be extended by introducing the stochastic approach. For this purpose, it is assumed that the development of the processes from the initial start of the computer processing to its completion along one of the possible paths has a probabilistic character, as it is. Since the individual algorithmic parts of the processing (programs or tasks) are discrete, and the transitions from one structure to the next occur at discrete moments in time, this allows the MC apparatus to be applied. In this case, the graph structure (GSA) of Figure 7 from Chapter 6 can be viewed as an MC state graph with the following definition.

a) Discrete set of states $S = \{A_1, A_2, A_3, A_4, A_5, A_6, A_7, A_8\}$;

b) Vector of initial probabilities $P_0 = \{1, 0, 0, 0, 0, 0, 0, 0\}$;
c) Matrix of the transition probabilities $P = \{p_{ij} \,/\, i,j = 1\div8\}$.

P	A_1	A_2	A_3	A_4	A_5	A_6	A_7	A_8
A_1		0.7					0.3	
A_2			0.4	0.1	0.2	0.1		0.2
A_3						1		
A_4								
A_5								
A_6								
A_7								
A_8					1			

To calculate the stochastic characteristics of a given MC, a program function "MARKOV" was developed in the APL2 language environment from the analytic APL-space (Romansky, 2022f). It allows to determine the vector of probabilities for the states $P(k)=\{p_1(k), \ldots, p_n(k)\}$ for successive steps, the number of which is set by the user. After starting, it requires defining the main characteristics of the Markov chain: N – number of states; P[I,J] – the elements of the matrix of transition probabilities; PO[1÷N] – the elements of the vector of initial probabilities. The default MC definition when starting the function "MAROV" is shown in Figure 9, the program code is given in Figure 10, and the implementation – in Figure 11.

Figure 9. Initial conditions for carried out an experiment

```
      SETUP P
      NODES
1 2 3 4 5 6 7 8
      SIZE
8
      NETWORK
0    0.7  0    0    0    0    0.3  0
0    0    0.4  0.1  0.2  0.1  0    0.2
0    0    0    0    0    1    0    0
0    0    0    0    0    0    0    0
0    0    0    0    0    0    0    0
0    0    0    0    0    0    0    0
0    0    0    0    0    0    0    0
0    0    0    0    1    0    0    0
```

Figure 10. APL-function MARKOV for MC solution

```
MARKOV
'1. NUMBER OF STATES N:'
N←□
P←(N N)ρ0
'2. MATRIX OF PROBABILITIES P(I,J):'
I←1
ETI:J←1
ETJ: 'P(',I,',',',',J,')='
P[I;J]←□
→(N≥J←J+1)/ETJ
→(N≥I←I+1)/ETI
'3. INITIAL VECTOR P0(1÷',N,'):'
PSNEW←P0←□
K←0
LOOP:
'STEP = ',K
'   VECTOR OF STATES PROBABILITIES:',PSNEW
'--------------------------------------------------'
'NEXT STEP: 1 (FOR YES),  0  (FOR NO)'
YES←□
→(YES=0)/OUT
PSOLD←PSNEW
J←1
K←K+1
LABJ:I←1
Q←0
LABI:Q←Q+(PSOLD[I]×P[I;J])
→(N≥I←I+1)/LABI
PSNEW[J]←Q
→(N≥J←J+1)/LABJ
→ LOOP
OUT: 'END OF MODEL'
```

Figure 11. APL-function MARKOV execution for the defined MC

```
STEP=0
   VECTOR OF STATES PROBABILITIES:  1  0  0  0  0  0  0  0
---------------------------------------------------------------------------
NEXT STEP:  1 (FOR YES),  0  (FOR NO)
□:
   1
STEP=1
   VECTOR OF STATES PROBABILITIES:  0  0.7  0  0  0  0  0.3  0
---------------------------------------------------------------------------
NEXT STEP:  1 (FOR YES),  0  (FOR NO)
□:
   1
STEP=2
   VECTOR OF STATES PROBABILITIES:  0  0  0.28  0.07  0.14  0.07  0  0.14
---------------------------------------------------------------------------
NEXT STEP:  1 (FOR YES),  0  (FOR NO)
□:
   1
STEP=3
   VECTOR OF STATES PROBABILITIES:  0  0  0  0  0.14  0.28  0  0
---------------------------------------------------------------------------
NEXT STEP:  1 (FOR YES),  0  (FOR NO)
□:
   1
STEP=4
   VECTOR OF STATES PROBABILITIES:  0  0  0  0  0  0  0  0
---------------------------------------------------------------------------
NEXT STEP:  1 (FOR YES),  0  (FOR NO)
□:
   0
END OF MODEL
```

The initial result at STEP=0 reflects the starting position represented by the vector of initial probabilities. The next steps reflect the development of the computer processing, and obtaining a zero vector for the distributed probabilities of all states determines the end of the experiment (reaching the final task) – in the case of the execution of STEP=4.

To apply the defined function "MARKOV" is needed the investigated graph of states to have only one input (start) node and only one final (input) node. In principle, this does not contradict the vector of initial probabilities, and if necessary, one passive node can be introduced for the start and end of the processing process. An example study under the same initial conditions for MC is shown in Figure 12 (with additional comments).

Figure 12. Stochastic investigation of defined MC by using APL-spice "OR"

```
    PATHSFROM  1                                            list of determined paths
1 2 3 6   1 2 4   1 2 5   1 2 6   1 2 8 5   1 7
    ρPATH←PATHSFROM  1                                      number of determined paths
6
    ↑PATH                                                   first path in the list
1 2 3 6
    ARCS  ↑PATH                                             probability weights for the first path
0.7  0.4  1
    ×/ARCS  ↑PATH                                           probability for the first path realization
0.28
    ×/¨ARCS¨PATH                                            probabilities for each path realization
0.28  0.07  0.14  0.07  0.14  0.3
    VALUE¨PATH                                              sum of the weight on the each path
2.1  0.8  0.9  0.8  1.9  0.3
    ⎕←MIN←⌈/V                                               minimal probability for MC realization
0.07
    ⎕←MAX←⌊/V                                               maximal probability for MC realization
0.3
    (V=⌈/V)/PATH                                            path with maximal probability
1  7
    (V=⌊/V)/PATH                                            path with minimal probability
1 2 4   1 2 6
```

3.2. Markov Chains with Continuous Time

Some authors call this kind of RP "continuous MC" based on the discrete type of the states. This approach is quite often applicable in the investigation of computational processes due to their probabilistic nature (Tsaregorodtsev et al., 2028; Hassani & Berang, 2018). The development of the process is similar to discrete MC, but in continuous time, and the transitions between states are carried out on the basis of intensities - when defining a continuous Markov model, instead of a matrix of transition probabilities, a matrix of transition intensities $\{\lambda_{ij}\}$ is set. For a system of states $S = \{s_1, ..., s_n\}$, the development of the process starts at the zero moment from a given state $S(0)$, and the probabilities of the states are $p_j(t)$; $t \geq 0; j = 1, ..., n$. If at any moment $t \geq 0$ the process is in state $S(t)=s_i$,, then at the next moment $t+\Delta t$ the process can fall into state $S(t+\Delta t) = s_j$ with transition probability $p_{ij}(t,\Delta t) = P[S(t+\Delta t)=s_j / S(t) = s_i]$; $i,j = 1, 2, ..., n$ which depends on the previous state s_i and possibly on $t+\Delta t$. Transitions are carried out with intensities:

$$\lambda_{ij}(t) = \lim_{\Delta t \to 0} \frac{p_{ij}(t, \Delta t)}{\Delta t} \; ; \; j \neq i .$$

If $S(t)=s_j$ and under the influence of a Poisson flow, the system passes into another state $S(t+\Delta t)=s_i$ with intensity λ_{ij}, then the transition takes place instantly upon the occurrence of the first event from the flow. If all the flows of events that bring S to different states are Poisson (stationary or non-stationary), then the random process running in S will be continuous-time Markovian (the reverse is also true). The residence time of the system in a given state is a random variable determined only by the fluxes causing the transitions.

For the Markovian RP with continuous time the formula for full probability is:

$$p_j(t+\text{”}\,t)=\sum_{i=1}^{n} p_i(t).p_{ij}(\text{”}\,t)=p_j(t).p_{jj}(\text{”}\,t)+\sum_{i=1(i\neq j)}^{n} p_i(t).p_{ij}(\text{”}\,t).$$

It is used for determining final probabilities based on:

$$\lim_{t\to\infty} p_j(t)=p_j;\, j=1\div n,$$

where it is taken into account that for each state the total output flow is equal to the total input flow and based on the normality condition the system of linear equations of the form is constructed and solved:

$$\sum_{\substack{j=1\\ j\neq i}}^{n} {}_{ij}\cdot p_i=\sum_{\substack{k=1\\ k\neq i}}^{n} {}_{ki}\cdot p_k,\ for\ i=1\div n$$

$$\sum_{j=1}^{n} p_j=1$$

Illustration of the application of continuous MC in modeling the performance of a dual-processor computer system when failures occur. It is assumed that for the overall performance it is enough that at least one of the processors is in work. Each processor is characterized by a failure intensity λ and a recovery intensity μ (transitions are assumed to obey a simple flow). The following basic states are defined in which the CS can fall: s_0 – both processors are working; s_1 – one processor is up and the other is under repair; s_2 – both processors are under repair. For the study, it is assumed that the system will start from state s_0 (fully operational).

Since the system is homogeneous, the intensities are the same for the type transitions. In state s_0, both processors are healthy, each subject to its own failure flow—the total intensity is 2λ. In state s_1, one of the processors is down (it is subjected to a recovery flow of intensity μ), and the second, which is healthy, is subjected to a failure flow of intensity λ. In state s_2, both processors are in a repair state, i.e., are subjected to their own recovery currents – total intensity 2μ. The definition of the Markov model is presented in Figure 13, and the state graph is of the “death and reproduction” type.

Figure 13. Definition of Markov model with continuous time

$$S = \{s_0, s_1, s_2\}$$

$$P_0 = \{1; 0; 0\}$$

P

	s_0	s_1	s_2
s_0	0	2λ	0
s_1	μ	0	λ
s_2	0	2μ	0

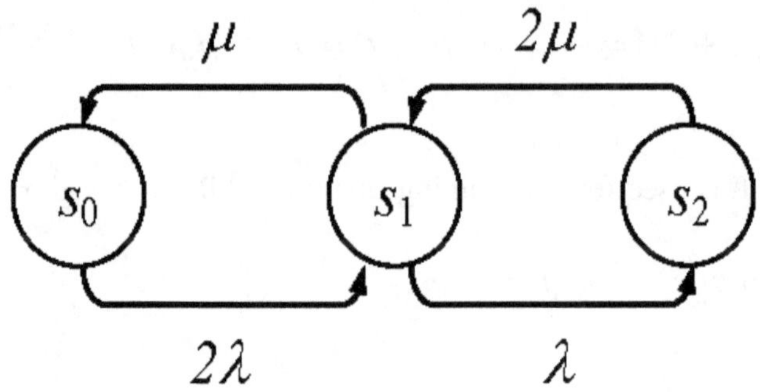

Analytical model for stationary regime has 4 equations or the 3 states – for each state the sum of input flows must be equal to the sum of output flows.

s_0: (2), p_0=μ,p*1*

s_1: (μ+), p*1*₌(2), p*0*₊(2μ),p2

s2: (2μ),p2=ₚ1

p*0* + p*1* + p2 = 1

The ***analytical solution***: From "s_0" → $p_1 = \left(\frac{2}{\mu}\right).p_0$ and after substituting into equation "s_2" →

$p_2 = \left(\frac{}{\mu}\right)^2 .p_0$ and in the last equation:

$$p_0 + p_1 + p_2 = 1 \Rightarrow p_0 + \frac{2»}{¼} p_0 + \frac{»^2}{½¼} p_0 = 1 \Rightarrow p_0 = \frac{1}{\left(1 + \frac{»}{¼}\right)^2} .$$

The solution for the steady-state probabilities is:

$$p_0 = \frac{½¼}{(» + ¼)^2};\ p_1 = \frac{2»¼}{(» + ¼)^2};\ p_2 = \frac{»^2}{(» + ¼)^2} .$$

If $\rho=\lambda/\mu$ is set specific calculations can be made for different values of ρ. Example results are presented in Figure 14.

Figure 14. Assessments of the model solution

ρ	0,5	1	2	3	4
	$2\lambda=\mu$	$\lambda=\mu$	$\lambda=2\mu$	$\lambda=3\mu$	$\lambda=4\mu$

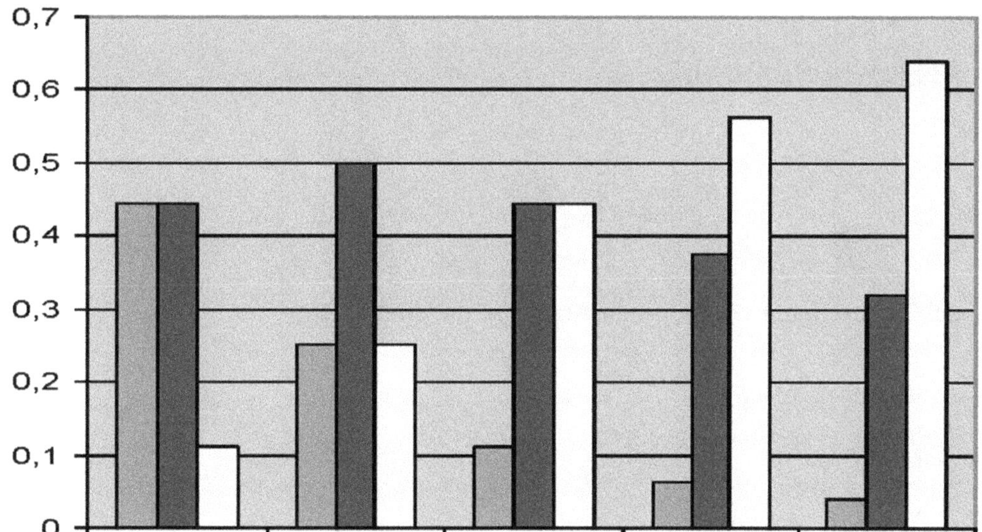

	0,5	1	2	3	4
p0	0,4445	0,25	0,111	0,0625	0,04
p1	0,4445	0,5	0,4445	0,375	0,32
p2	0,111	0,25	0,4445	0,5625	0,64

Another case that can be discussed is the reliable operation of a computer structure with serially connected processing devices U_i ($i=1\div N$) to process a workflow passing through them. Let us assume the hypothesis that at any given time no more than one device may have failed and is under repair. Then for each device U_i two states can be defined - working state (($\sigma_i=0$) and failure state (($\sigma_i=1$), which allows the system readiness to be described by the states $s_j=<(\sigma_i\ /\ i=1\div N>$ for $j=0\div 2^N-1$. The basic state $s_0=<\sigma_1\ \sigma_2\ \sigma_3\ \ldots\ \sigma_N>=<000\ \ldots 0>$ presents fully worked system and all other states present a system with failure in a unit.

Presented conceptual situation permits to use Markov process with continuous time with $N+1$ states $s_j=S(t)\in S$ for stochastic modeling of the reliability work of a sequential processing structure. The model investigation must starts from the basic state s_0 and each other transition can be determined by failure intensity $\lambda_j(t){:}s_0\rightarrow s_j$ or restoration intensity $\mu_j(t){:}s_j\rightarrow s_0$. Each fault make transition from s_0 to the state s_j defined by the number j of the defected unit U_j in the structure. After repair and return to working condition, the transition is reversed. The Markov model investigation starts from initial condition – vector of initial probabilities $P_0 = (1, 0, 0, \ldots, 0)$, ú$P_0$ú=$N$+1. Two analytical stochastic Markov models can be determined.

- Model for sequential processing describing:

$$\frac{dp_{0(t)}}{dt} = \left[\sum_{i=1}^{n} \mu_{iP_i}(t)\right] - \left[\sum_{j-1}^{N}{}_{j} P_0(t)\right]$$

$$\frac{dP_i(t)}{dt} =_i P_0(t) - \mu_i P_i(t); i = 1, 2, \ldots, N$$

$$\sum_{i=0}^{N} P_i(t) = 1$$

- Model for description of study-state regime of the processes:

$$\left[\sum_{i=1}^{N} \mu_i P_i\right] - \left[\sum_{j-1}^{N}{}_{j} P_0\right] = 0$$

$${}_i P_0 = \mu_i P_i; i = 1, 2, \ldots, N$$

$$\sum_{i=0}^{N} P_i = 1$$

The solving of the second model gives the next result:

$$p_i = \frac{i}{\mu_i} p_0; i = 1, 2, \ldots, N$$

$$p_0 + \frac{1}{\mu_1} p_0 + \frac{2}{\mu_2} p_0 + \ldots + \frac{N}{\mu_N} p_0 = 1 \rightarrow p_0 = \frac{1}{1 + \sum_{j=1}^{N} \frac{j}{\mu_j}}$$

$$p_i = \frac{1}{\mu_i \left[\sum_{j=1}^{N} \frac{j}{\mu_j} \right]}; i = 1, 2, \ldots, N$$

The probability for reliable failure-free system is $P(t) = e^{-\int_0^t (t).dt}$.

An example for 4-units system is presented in Figure 15 and the Markov model can be modified based on this structure for N=4. Two cases can be evaluated:

(1) If is assumed that the two type intensities are not equivalent ($\lambda_j \neq \mu_j$) and the values are for example $\lambda_1=\lambda_4=\eta$; $\lambda_2=\lambda_3=2\eta$; $\mu_1=\mu_4=2\eta$; $\mu_2=\mu_3=3\eta$, it can to obtain the next probabilities: p_0=0,3; $p_1=p_4$=0,15; $p_2=p_3$=0,2; $P(t)$=0,3.

(2) Equivalent intensities is assumed ($\lambda=\mu$) – in this case all probabilities are equivalent, i.e. p_j=0,2 for j=0,1,2,3,4.

Figure 15. Example of sequential processing structure

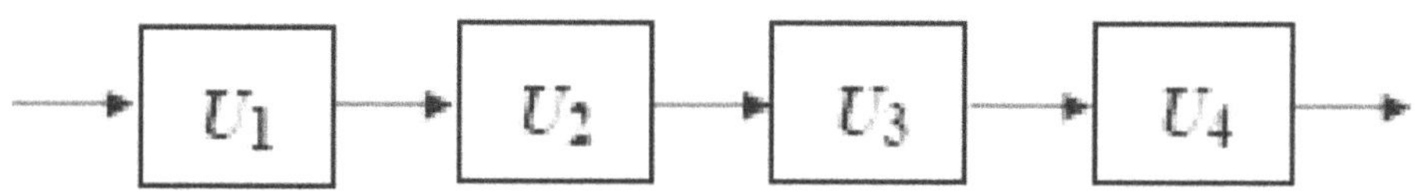

4. STOCHASTIC MODEL INVESTIGATION OF PROCESSES IN COMPUTING

Investigation by using MC with Continuous Time

An extension of the above case is a model study of a two-processor CS with shared memory and identical service parameters (intensity μ) for both processors (Figure 16). Each of the processors PU1 and PU2 has a buffer local memory (B1 and B2), which for simplicity is assumed to store only the current task. Workload is formed based on homogeneous stochastic workflow with constant intensity λ. On the basis of the Markovian model with continuous time, the falling of the system into a stationary mode is investigated, and the following processing states are defined.

s_0 – both processors are not working on a task;
s_1 – only one task is processed (only one processor is loaded);
s_2 – two tasks are processed (both processors work in parallel).

Figure 16. Two-processors parallel system type Shared Memory MIMD with bus connection

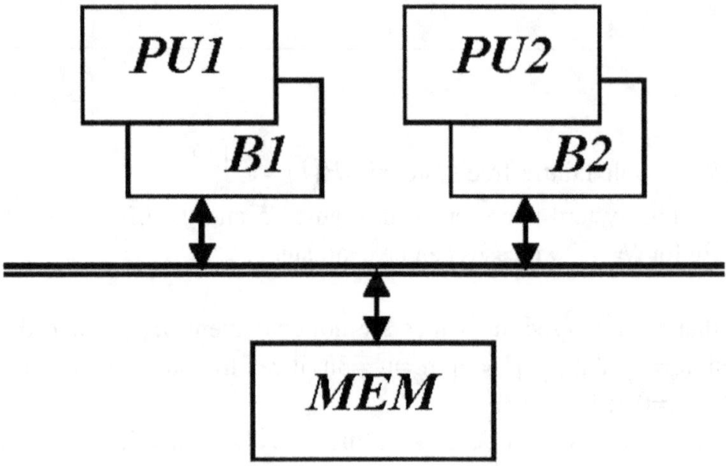

The defined Markov model with continuous time and graph of the states is presented in Figure 17. The arcs describe the transitions between states based on the two intensities λ and μ. It is assumed that the probability system to be in the state s_i in time moment t is $\lim_{t\to\infty} p_i(t) = p_i, for\ 1 \le i \le n$.

Figure 17. Markov model of the processing

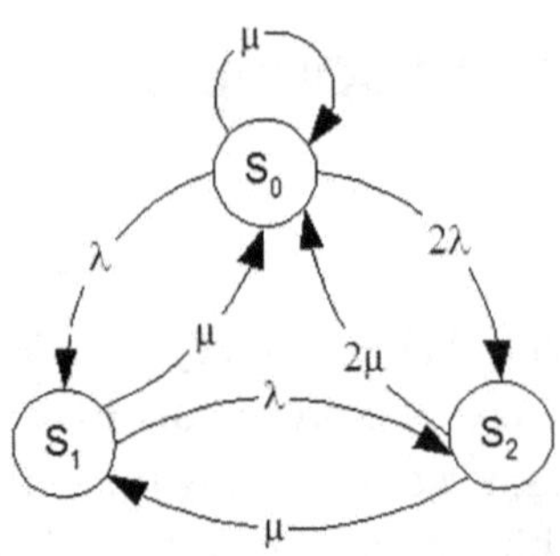

$$s_0:\ \mu.p_0+\lambda.p_0+2.\lambda.p_0= \mu.p_0 + \mu.p_1 + 2.\mu.p_2$$
$$s_1:\ (\mu + \lambda).p_1 = \lambda.p_0 + \mu.p_2$$
$$s_2:\ 3.\mu.p_2 = 2.\lambda.p_0 + \lambda.p_1$$
$$p_0(s_0)+p_1(s_1)+p_2(s_2)=1$$

Analytical solution of the stochastic Markov model is presented below.

$$p_0 = \frac{1}{3»}(¼p_1 + 2¼p_2);\ p_1 = \frac{».p_0 + ¼p_2}{» + ¼};\ p_2 = \frac{2».p_0 + ».p_1}{3¼}$$

The final probabilities are formed based on the substitutions:

$$p_0 = \frac{1}{3\lambda}\left[\mu p_1 + 2.\mu\left(\frac{2.\lambda.p_0+\lambda.p_1}{3\mu}\right)\right] = \frac{1}{3\lambda}\left[\mu p_1 + \frac{2}{3}(2\lambda.p_0+\lambda.p_1)\right] = \frac{1}{3\lambda}\mu p_1 + \frac{1}{3\lambda}\frac{2}{3}2\lambda.p_0 + \frac{1}{3\lambda}\frac{2}{3}\lambda.p_1$$

$$\Rightarrow p_0\cdot\left(1-\frac{4}{9}\right) = p_1\cdot\left(\frac{\mu}{3.\lambda}+\frac{2}{9}\right) \Rightarrow \frac{5}{9}\cdot p_0 = p_1\cdot\left(\frac{3.\mu}{9.\lambda}+\frac{2.\lambda}{9.\lambda}\right) \Rightarrow p_0 = \frac{3.\mu+2.\lambda}{5.\lambda}\cdot p_1$$

$$p_1 = \frac{\lambda.p_0+\mu p_2}{\lambda+\mu} = \frac{\lambda.\frac{3.\mu+2.\lambda}{5.\lambda}\cdot p_1+\mu p_2}{\lambda+\mu} = \frac{(3.\mu+2.\lambda).p_1+5.\mu p_2}{5.(\lambda+\mu)}$$

$$\Rightarrow p_1 = \frac{3\mu+2\lambda}{5(\lambda+\mu)}p_1 + \frac{\mu p_2}{(\lambda+\mu)} \Rightarrow p_2\frac{\mu}{\lambda+\mu} = p_1\frac{3\lambda+2\mu}{5(\lambda+\mu)} \Rightarrow p_2 = \frac{3\lambda+2\mu}{5\mu}\cdot p_1$$

And finally based on the question of normality the analytical expressions for the final probabilities are received:

$$\frac{3\mu+2\lambda}{5\lambda}\cdot p_1 + p_1 + \frac{3\lambda+2\mu}{5\mu}\cdot p_1 = 1$$

$$p_0 = \frac{3.\mu^2+2.\lambda.\mu}{3.(\mu^2+3.\lambda.\mu+\lambda^2)};\; p_1 = \frac{5.\mu\lambda}{3.(\mu^2+3.\lambda.\mu+\lambda^2)};\; p_2 = \frac{3.\lambda^2+2.\lambda.\mu}{3.(\mu^2+3.\lambda.\mu+\lambda^2)}.$$

The condition for reaching a steady state is $\lambda<\mu$, i.e. $\rho=\lambda/\mu<1$. An estimate of the full workload of the system can be obtained by determining the probability that the system will enter state s_2 (loading both processors). If $\pi=3\mu^2+9\lambda\mu+2\lambda^2$, the following analytical expressions are obtained for the final probabilities:

$$p_0 = \frac{3\mu^2+2\lambda.\mu}{\pi};\; p_1 = \frac{5\mu\lambda}{\pi};\; p_2 = \frac{3\lambda^2+2\lambda.\mu}{\pi}.$$

The calculated quantitative estimates based on the last expressions are presented in Table 4.

Table 4. Assessments for different values of ρ

λ	μ	ρ	π	P_0	P_1	P_2
0.5	0.4	1.25	3.03	0,2904	0,3300	0,3795
0.5	0.5	1	3.75	0,3333	0,3333	0,3333
0.4	0.5	0.8	3.03	0.3795	0.3300	0.2904
0.25	0.5	0.5	2.0625	0.4848	0.3030	0.2121
0.2	0.6	0.33	2.28	0.579	0.2632	0.1579
0.12	0.6	0.2	1.7712	0.6911	0.2033	0.1057

The graphical interpretation of the obtained estimates is presented in Figure 18, with $\rho = 1$ the three states are equally likely (the condition for ensuring a steady state is $\rho < 1$). The diagram in Figure 19 shows the dependency $p_2(\rho=\lambda/\mu)$ which is the probability for parallel operation of the two processors. As can be seen from the presented results, as the service time increases (decreasing μ) and the intensity of the input flow of tasks λ increases, the trends lead to saturation.

If we consider the fact that two processors are running at the same time and the input flow can be 2 times more intensive, then the condition for uniform load will be reduced to $\rho = 2\lambda/\mu$. This leads to an increase in the stationarity threshold, by about 16.65% compared to single-stream service.

Figure 18. Graphical interpretation of the assessment for final probabilities as function of ρ

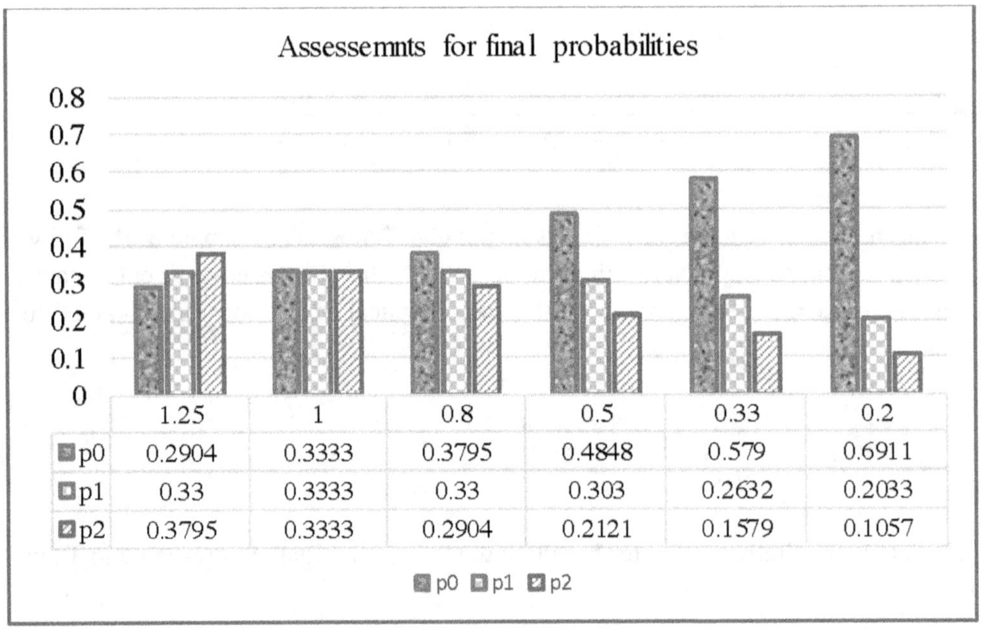

Figure 19. Dependency p2(ρ)

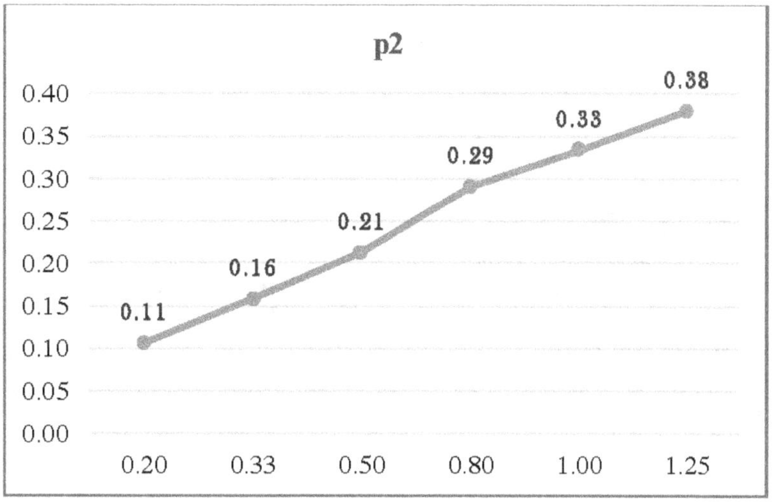

Examples for Investigation by using MC with Discrete Time

Distributed learning environments (DLE). A formalization of a distributed space for distance learning is presented in Chapter 5 (see Figure 11), and a more detailed representation is given Figure 20.

The following base component groups are defined:

- Users $U = \{U_i \,/\, i=1 \div N\}$ $(U \neq \varnothing)$ which access different remote information resources in Internet based on lines of the global network based on a request $req: U_i \rightarrow R_j$, for $\forall U_i \in U$ & $\forall R_j \in R$.
- Resources $R = \{R_j \,/\, j=1 \div M\}$ $(R \neq \varnothing)$, which presents learning contents as an information objects for common use with multiple access $Inf: R_j \rightarrow U_i$, for $\forall U_i \in U$ & $\forall R_j \in R$.
- Network communication means (Transmitters) $T = \{T_q \,/\, q=1 \div K\}$ $(T \neq \varnothing)$ for transferring information objects (requests and information blocks) between different DLE nodes.
- Distributor (D) of information objects which unites hardware and software means for objects routing in the network space.

Figure 20. Abstract model for relationships in distributed learning environment (DLE)

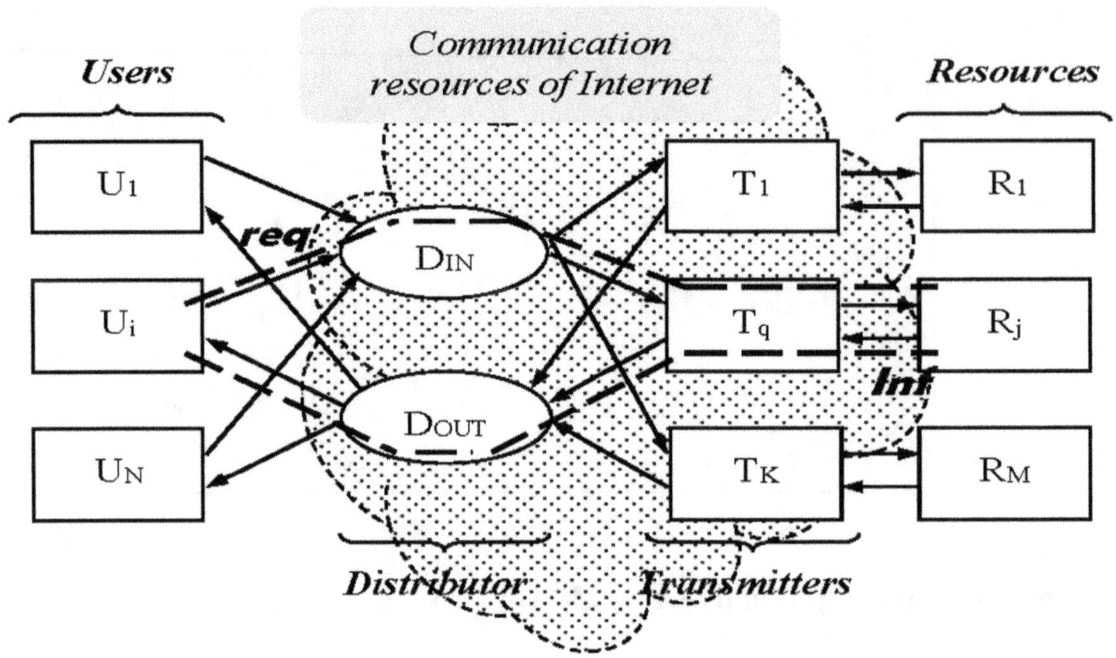

This formalization permits to describe the DLE as a ordered set *DLE*={*U, R, T, D*} with two types of relations between components (*req* & *Inf*). The stochastic nature of the DLE permits to use the MC theory for modeling and investigation of processes. Definition of such model has been presented in Figure 27 of Chapter 5 – finite set of states *S*, vector of initial probabilities P_0, matrix of transition probabilities P_{ij} and the graph of the states. The stochastic MC-model is presented below.

$$p_1 = 0.3p_1 + 0.1p_2 + 0.3p_4 + 0.8p_7$$
$$p_2 = 0.7p_1 + 0.2p_3$$
$$p_3 = 0.9p_2$$
$$p_4 = 0.8p_3$$
$$p_5 = 0.7p_4 + 0.3p_6 + 0.2p_7$$
$$p_6 = p_5$$
$$p_7 = 0.7p_6$$
$$p_1 + p_2 + p_3 + p_4 + p_5 + p_6 + p_7 = 1$$

One possible solution to the analytical model starting from equation 2 is as.

$$p_2 = 0.7p_1 + 0.2p_3 = 0.7p_1 + 0.2(0.9p_2) \Rightarrow p_2 = \frac{0.7}{0.82}p_1$$

$$p_3 = 0.9\frac{0.7}{0.82}p_1 \Rightarrow p_3 = \frac{0.63}{0.82}p_1$$

$$p_4 = 0.8(0.9p_2) = 0.8(0.9\frac{0.7}{0.82}p_1) \Rightarrow p_4 = \frac{0.504}{0.82}p_1 .$$

$$p_5 = 0.7\frac{0.504}{0.82}p_1 + 0.3p_5 + 0.2(0.7p_5) \Rightarrow p_5 = 0.7683p_1$$

$$p_6 = p_5 = 0.7683p_1$$

$$p_7 = 0.7p_6 = 0.7(0.7683p_1) = 0.53781p_1$$

The determination of the final probabilities is performed after substitution in the last equation of the system, and the obtained estimates are presented in Figure 21.

Figure 21. Final probabilities at stationarity of processes

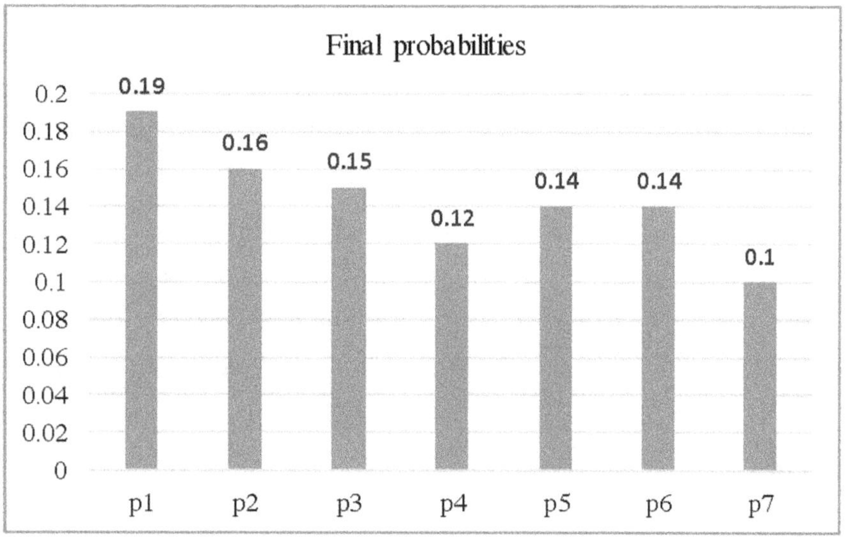

To investigate the influence of the volume of the content of the exchanged information objects (Case 2), a new MC-model (Figure 22) was defined with the following base states: s_1 – waiting state; s_2 – generation of a user request for access to an information resource; s_3 – transfer of a request through the communication medium; s_4 – service of received content access request; s_5 – work with provided information resource. The initial probability vector specifies starting from the first state $P_0 = (1, 0, 0, 0, 0)$, and the transition probability matrix can be derived from the state graph. To create the model, it was accepted:

- The process starts from initial state s_1.

- Transitions from the "waiting" state to the next state are equally likely.
- If you get into the "request generation" state with a probability of 0.2, you can refuse to work with the system.
- A value of 0.2 was adopted for both probabilities – for the occurrence of an error (failure) during transmission through the communication medium and for the occurrence of multiple transmission (chain of requests).
- When servicing an incoming request, the probability of permitted access is 0.8.
- When working with the provided information resource, the following situations are possible: connection breakdown (probability 0.1); continuous work depending on the volume of content (probability π); completion of work with provided resource (with probability 0.9-π).

Figure 22. Definition of the MC-model for the case 2

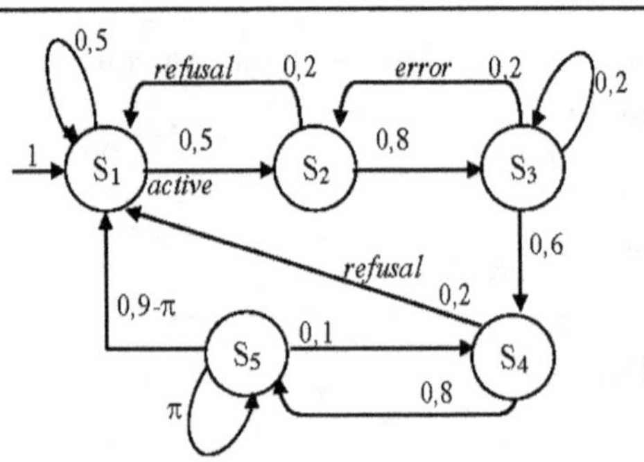

$$\begin{cases} p_1 = 0{,}5p_1 + 0{,}2p_2 + 0{,}2p_4 + (0{,}9-\pi)p_5 \\ p_2 = 0{,}5p_1 + 0{,}2p_3 \\ p_3 = 0{,}8p_2 + 0{,}2p_3 \\ p_4 = 0{,}6p_3 + 0{,}1p_5 \\ p_5 = 0{,}8p_4 + \pi.p_5 \\ p_1 + p_2 + p_3 + p_4 + p_5 = 1 \end{cases}$$

Analogous to the approach from the last case 1, the model is solved by determining the following dependencies:

$$p_2 = p_3 = 0.625, p_1; p_4 = \frac{0.375,(1-\pi)}{1-\pi-0.08}P_l;\ P_5 = \frac{0.8}{1-\pi}P_4 \rightarrow P_5 = \frac{0.3}{1-\pi-0.08}P_l$$

$$p_1 = \frac{1-\grave{A}-0.08}{2.745-2.625\grave{A}}$$

To assess the influence of the volume of content on the service processes of accessing and using information resources in DLE, the probability value π was varied, and the results are summarized in Table 5.

Table 5. Assessment for the final probabilities based on the parameter π

Volume	No	π	p_1	p_2	p_3	p_4	p_5
Small	1	0,1	0,33	0,21	0,21	0,13	0,12
	2	0,3	0,32	0,20	0,20	0,13	0,15
Average	3	0,5	0,30	0,18	0,18	0,13	0,21
Big	4	0,6	0,27	0,17	0,17	0,13	0,26
	5	0,8	0,19	0,11	0,11	0,12	0,47

In the stochastic investigation presented above and based on the obtained model results, the following conclusions can be drawn:

- The only functional dependence that has an increasing trend with the increase of the information volume of the used resources is the time to work with them. The remaining time parameters – for generation, for transfer and for processing requests show decreasing trends.
- The graph of the functional dependence of the working time on the size of the learning content becomes significantly steeper for large volumes of information resources.
- The relative share of time to service a request in the server of the information resource is kept almost constant.
- The sensitivity of the relative times taken to generate a request from a user and to transfer it to the communication environment is manifested in information blocks with a volume above the average.

Investigation of virtual environments. This example is related to the design of virtual research environment (VRE) with multiple access to resources. A conceptual model has been defined for the organization of the structure and the implementation of the processes (Figure 23) including two main subsystems – Front office (regulation and access management) and Back office (intelligent search engine and specialized virtual environments). Graph formalization is presented in Figure 24.

Basic components, included in the environment, are Digital Rights Management System (DRMS – support high level of information security and management of access to resources) and Distributed Systems for Access management (DAMS – for control access to internal and external resources).

Figure 23. Generalized conceptual model of virtual research environment (VRE)

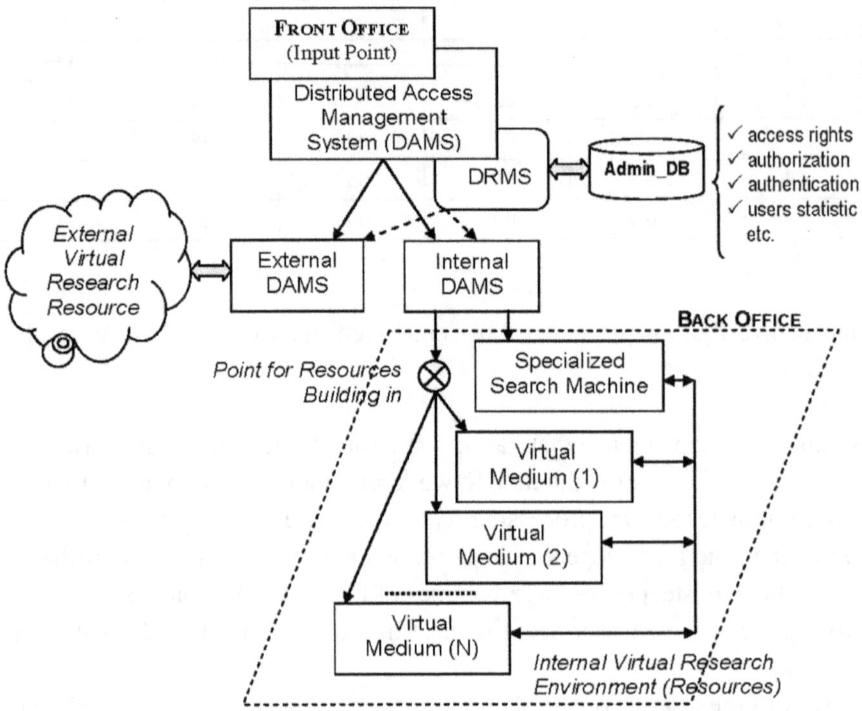

Figure 24. Graph formalization of processes in VRE

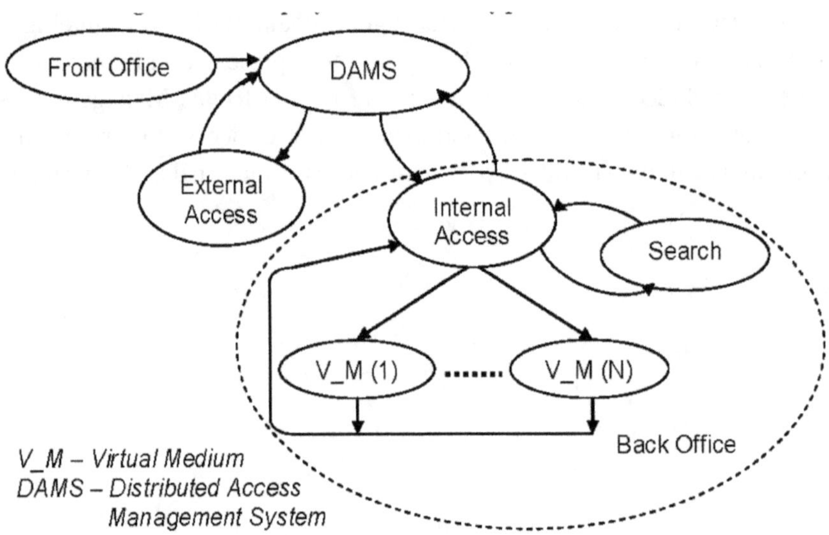

The formalization is a transition to developing a model of the information service, and due to the probabilistic nature of the processes, the stochastic approach was chosen. The developed MC-model is presented below.

The graph of the states and the analytical definition of the MC-model are presented in Figure 25 with the states which corresponds of the nodes in the graph formalization upper: s_1 – activity of the front office; s_2 – using DAMS possibilities for management; s_3 & s_4 – management of the access to the external & internal resources respectively; s_5 – search machine usage; s_6 – work in selected V_M(*j*) / j=1,2,…,*N*.

Figure 25. Stochastic MC-model of the VRE processes

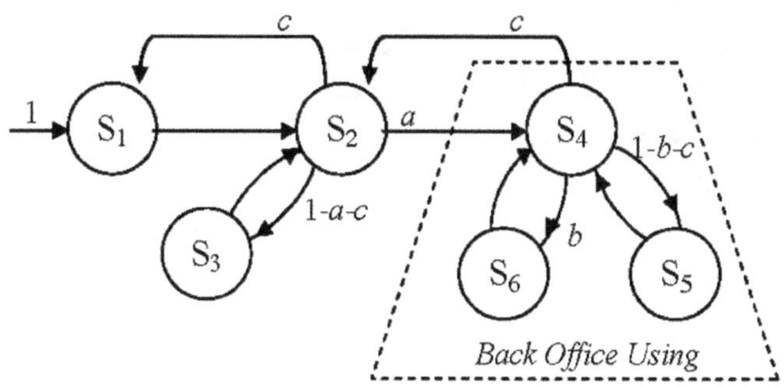

(a) Graph of states

$$1)\ p_1 = c.p_2$$
$$2)\ p_2 = p_1 + p_3 + c.p_4$$
$$3)\ p_3 = (1 - a - c).p_2$$
$$4)\ p_4 = a.p_2 + p_5 + p_6$$
$$5)\ p_5 = (1 - b - c).p_4$$
$$6)\ p_6 = b.p_4$$
$$7)\ p_1 + p_2 + p_3 + p_4 + p_5 + p_6 = 1$$

(b) Analytical model

After solving the model, the following expressions for the final probabilities are obtained (substitution $\pi=2(a + c + ac)$ is used):

$$p_1 = \frac{c^2}{\pi};\ p_2 = \frac{c}{\pi};\ p_3 = \frac{c.(1-a-c)}{\pi};\ p_4 = \frac{a}{\pi};\ p_5 = \frac{a.(1-b-c)}{\pi};\ p_6 = \frac{a.b}{\pi}.$$

Two experimental plans are presented below. The first case is a single random plan for a concrete combination of main model parameters ($a = 0.6$; $b = 0.7$; $c = 0.1$) and results are shown in Figure 26. The results shows that the states s_4 (access to internal resources) and s_6 (work with virtual medium) have maximal loading.

Figure 26. Results of the first case: Single random plan

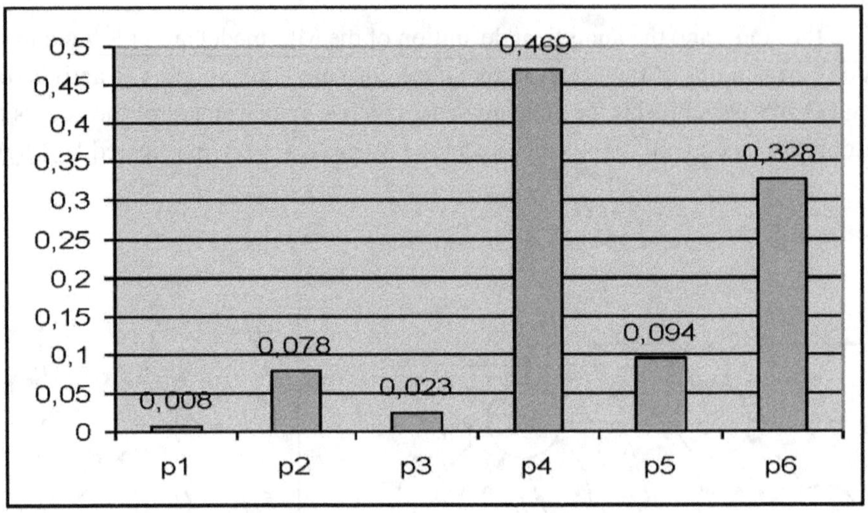

The second case is organization of full factors plan for obtaining more precise assessments. It is based on selected spice of the controlled factors – $a,b \in \{0.4; 0.5; 0.6; 0.7; 0.8\}$ and $c \in \{0.1; 0.2; 0.3\}$. This is made based on the hypotheses:

- Probability for the factor c is relatively low.
- The factors a and b are related to the states s_2 and s_4 and their sum $(a+b)$ should be 1.

Graphical interpretation of selected assessment is presented in Figure 27 (for first 4 final probabilities and for a segment of dependence $p_4=F(a)$. The dependence of final probabilities p_5 and p_6 as a function of factors b & c (the value of the factor 'a' is fixed) is presented in Figure 28.

Figure 27. Assessments for pi=F(a); i=1,2,3,4 and for p_4

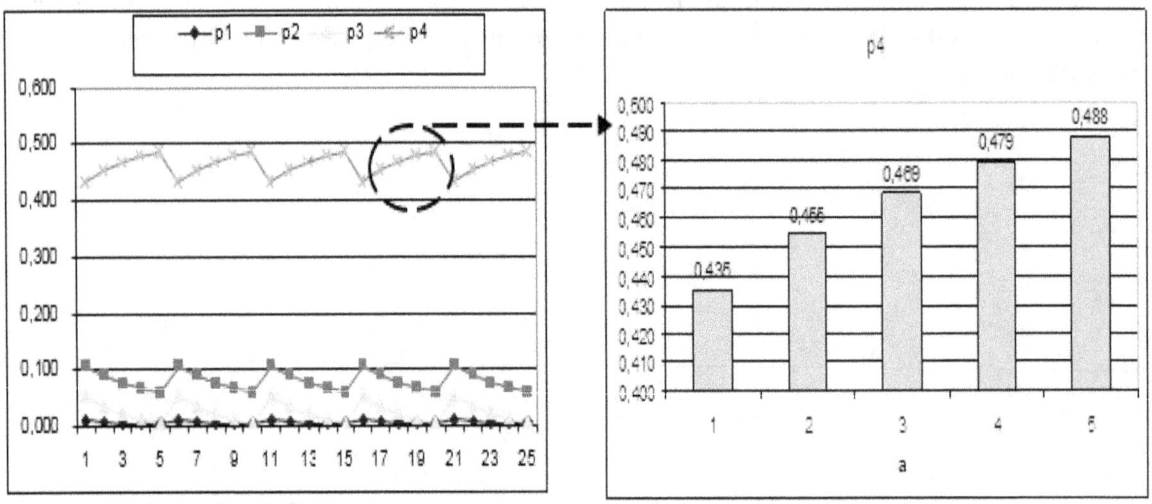

Figure 28. Assessments for p5 & p6 as functions F(b,c / a=fix)

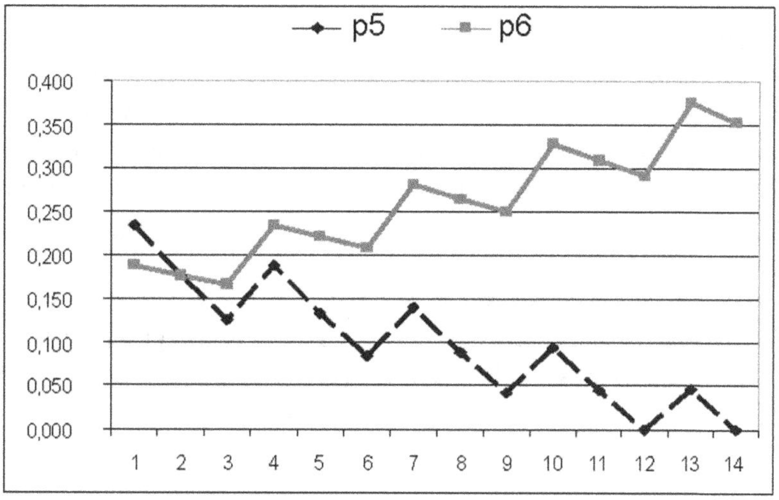

Investigation of access to system resources. The formal description of a combined environment based on internal and external (in cloud for example) resources has been presented in Figure 6 (Chapter 6). This graph formalization is based on the finite number of nodes in the structure and each observed even has constant behavior determined by fixed procedure. The main sub-systems of the environment are presented in the end of part 6.1 (Chapter 6). This stricture can be presented as a discrete graph of states, but the processes have stochastic nature and this permits to use MC with discrete time for model investigation organization. It is assumed that a state transition from the current time t_i to the next time t_{i+1} is independent on the time (discrete time) and the final probabilities for steady state regime can be determined by MC-model solution. The defined discrete state of MC are: S_1 – Access to the front office; S_2 – Registration procedure; S_3 – Authentication procedure; S_4 – Request analysis; S_5 – Access to external resources; S_6 – Authorization procedure for access to internal resource; S_7 – Using educational resource; S_8 – Using system resource & personal data processing. Matrix of transition probabilities is given in Table 6, and the vector of initial probabilities is $P_0 = \{1, 0, 0, 0, 0, 0, 0\}$.

Table 6. Matrix of transition probabilities $P = \{p_{ij}\}$

	S_1	S_2	S_3	S_4	S_5	S_6	S_7	S_8
S_1	0	*a*	1-*a*	0	0	0	0	0
S_2	0	1	0	0	0	0	0	0
S_3	1-*b*	0	0	*b*	0	0	0	0
S_4	1-*c*-*d*	0	0	0	*c*	*d*	0	0
S_5	1	0	0	0	0	0	0	0
S_6	0	0	0	1-*e*-*f*	0	0	*e*	*f*
S_7	0	0	0	1	0	0	0	0
S_8	0	0	0	1	0	0	0	0

The following parameters are adopted, which are involved in determining the transition probabilities:

a – Probability for unregistered (new) user;
b – Probability for correct authorization of registered user;
c – Probability for a request to access and using external educational resources;
d – Probability for authorization (determining the right) for using an internal resource;
e – Probability for an internal educational resource using (after successful authorization);
f – Probability for using system resources and personal data processing.

To make a preliminary stochastic investigation a MC with discrete time is designed (see Figure 29) and it assumed that the probability of access to the system by new unregistered user is no more than 0.3 ($a \leq 0.3$) and the unauthorized access to the resources (including attacks from external nodes) is in the frame [10%, 30%], i.e. $0,1 \leq b \leq 0,3$. Another assumption is that refusal of service is no more than 10% of all cases mines $(1 - c - d) \leq 0.1 \Rightarrow (c + d) \leq 0.9$ and $(1 - e - f) \leq 0.1 \Rightarrow (e + f) \leq 0.9$.

Figure 29. MC model – analytical definition and graph of the states

$$\begin{cases} 1) p_1 = p_2 + (1-b).p3 + (1-c-d).p_4 + p_5 \\ 2) p_2 = a.p_1 \\ 3) p_3 = (1-a).p_2 \\ 4) p_4 = b.p_3 + (1-e-f).p_6 + p_7 + p_8 \\ 5) p_5 = c.p_4 \\ 6) p_6 = d.p_4 \\ 7) p_7 = e.p_6 \\ 8) p_8 = f.p_6 \\ 9) p_1 + p_2 + p_3 + p_4 + p_5 + p_6 + p_7 + p_8 = 1 \end{cases}$$

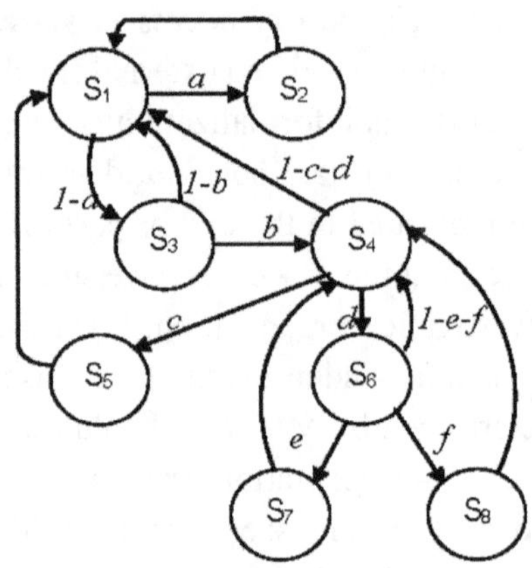

Solution of the analytical model can be made by determining probabilities as expressions on the base of the probability p_1 and forming the last equation of the system from Figure 29:

$$p_1 + ap_1 + (1-a)p_1$$
$$+\frac{b(1-a)}{(1-d)}p_1 + \frac{bc(1-c)}{(1-d)}p_1 + \frac{bd(1-a)}{(1-d)}p_1 + \frac{bde(1-a)}{(1-d)}p_1 + \frac{bdf(1-a)}{(1-d)}p_1 = 1$$

If it assumed the substitution π = [2-2d+b(1-a)(1+c+d+de+df)] the final probabilities for steady state regime can be determined by following analytical expressions:

$$p_1 = \frac{(1-d)}{\grave{A}};\ p_2 = \frac{a(1-d)}{\grave{A}};\ p_3 = \frac{(1-a)(1-d)}{\grave{A}};\ p_4 = \frac{b(1-a)}{\grave{A}};$$

$$p_5 = \frac{bc(1-a)}{\grave{A}};\ p_6 = \frac{bd(1-a)}{\grave{A}};\ p_7 = \frac{bde(1-a)}{\grave{A}};\ p_8 = \frac{bdf(1-a)}{\grave{A}}$$

Experimental investigation of the processes in steady state regime is made based on the selected values for the control parameters, for example: $a \in \{0.2; 0.25; 0.3\}$; $b \in \{0.7; 0.75; 0.8; 0.85; 0.9\}$; $c \in \{0.35; 0.4; 0.45\}$; $d \in \{0.45; 0.5; 0.55; 0.6\}$ and $(1\text{-}e\text{-}f)=0{,}1 \Rightarrow e = f = 0.45$ (equal probabilities for access after authorization to educational or system resource). Two factor plans can be realized and the results are presented below.

Partial multi-factor experimental plan is realized on the base of all combinations of accepted probabilities' values. The statistical realization of this plan which has been made by software "Develve" is shown in Figure 30, and the assessment are presented in the Table 7.

Figure 30. Results of multi-factor experimental plan by using "Develve" software

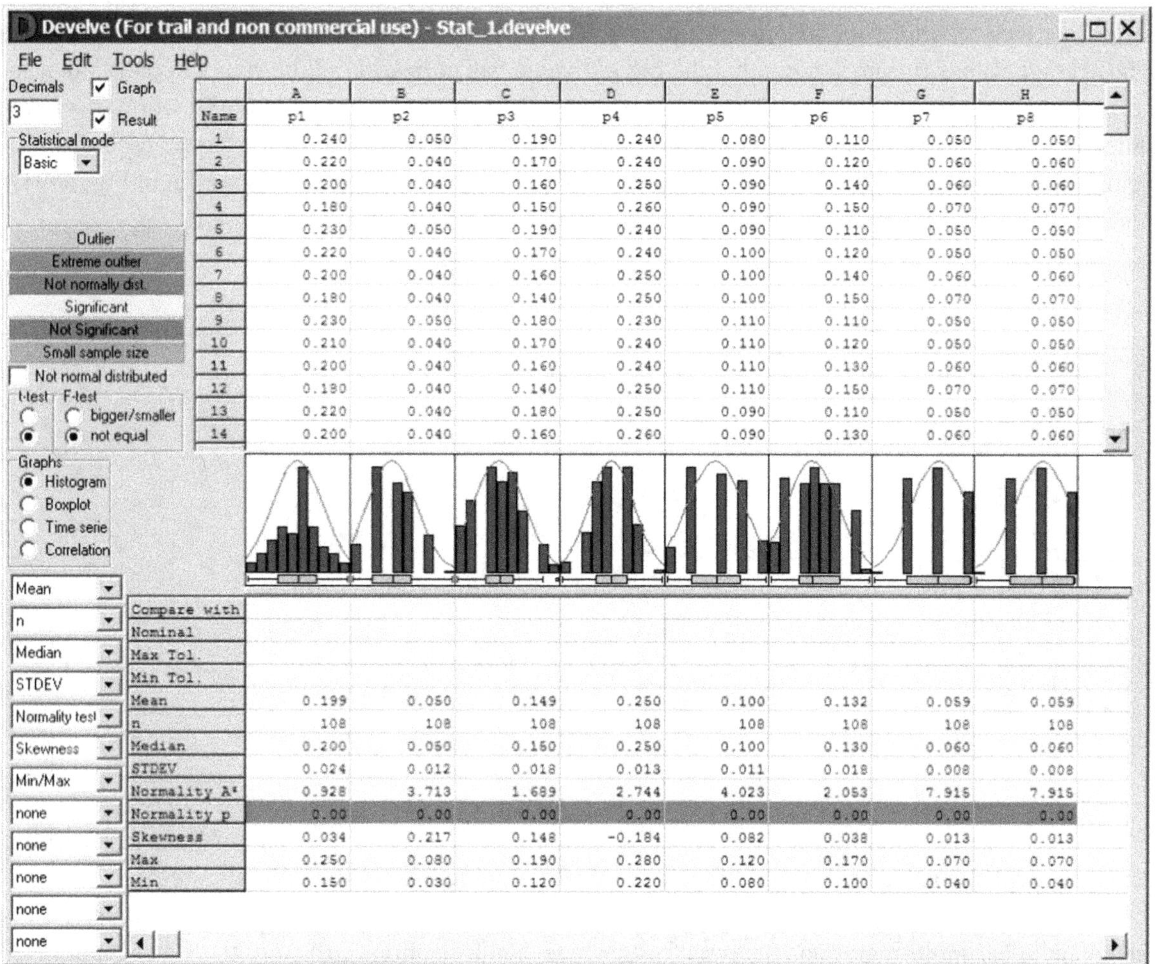

Table 7. Generalization of the statistical assessments for the probabilities

	p1	p2	p3	p4	p5	p6	p7	p8
Mean	0.1995	0.0501	0.1493	0.2505	0.1001	0.1318	0.0593	0.0593
Median	0.1992	0.0492	0.1490	0.2512	0.1003	0.1313	0.0591	0.0591
MIN	0.1501	0.0300	0.1151	0.2198	0.0787	0.0989	0.0445	0.0445
MAX	0.2522	0.0757	0.1885	0.2777	0.1216	0.1666	0.0750	0.0750
Δ=[max-min]	0.1021	0.0457	0.0733	0.0579	0.0429	0.0677	0.0305	0.0305
Variance	0.0006	0.0001	0.0003	0.0002	0.0001	0.0003	0.0001	0.0001
St.Dev.	0.02391	0.01139	0.01757	0.01293	0.01051	0.01778	0.008	0.008
Conf.Int. (T)	0.005	0.002	0.003	0.002	0.002	0.003	0.002	0.002
Conf. Int. (N)	0.005	0.002	0.003	0.002	0.002	0.003	0.002	0.002

Experimental results are obtained for n=108 combinations of factor's values and the abbreviation for calculated assessments are follows – standard deviation (St.Dev.) and confidence intervals (Conf. Int.) by using the Student's T-distribution (T) and by using normal distribution (N). The difference Δ is largest for the probability p_1 that presents loading front office portal by access of different remote users. The average values (means) shows that the highest value has utilization of the state "Request analysis" (p_4) and the assessments for two security procedures – authentication (p_3) and authorization (p_6) have a quiet little difference (about 0.017) that is determined by the access to external resources. A comparative statistical analysis of using these two procedures which is made by "Develve" is shown in Figure 31.

Figure 31. Comparative analysis of procedures presented by p_3 and p_6

Compare with	A	A
Nominal		
Max Tol.		
Min Tol.		
Difference	-0.050	-0.068
t-test t	-17.355	-23.117
t-test DF	194	197
t-test p	0.00	0.00
F test F	1.937	1.837
Min Samples	70	83
F test p	0.00	0.00
Correl r	0.869	-0.930
Correl pairs	108	108
Correl p	0.00	0.00
Regression	1.21X+0.02	-1.26X+0.37

"Develve" software permits to carry out a correlation analysis between procedures (final probabilities) and the calculated all correlation coefficients in steady state regime is shown in Figure 32.

One-way Anova analysis for states '1', '3'and '4' is shown in Figure 33(a) and calculated assessments by linear multi-regression analysis for the state "authorization" (p_6) from the states "input" (p_1) and "authentication" (p_3) is presented in Figure 33(b).

Figure 32. Results of correlation analysis

	p2	p3	p4	p5	p6	p7	p8
p1	0.721	0.869	-0.813	-0.478	-0.930	-0.863	-0.863
	0.00	0.00	0.00	0.00	0.00	0.00	0.00
p2		0.345	-0.720	-0.418	-0.603	-0.561	-0.561
		0.00	0.00	0.00	0.00	0.00	0.00
p3			-0.636	-0.369	-0.877	-0.816	-0.816
			0.00	0.00	0.00	0.00	0.00
p4				0.270	0.678	0.598	0.598
				0.00	0.00	0.00	0.00
p5					0.270	0.218	0.218
					0.00	0.02	0.02
p6						0.933	0.933
						0.00	0.00
p7							1.000
							0.00

Figure 33. Dependency analysis by Develve

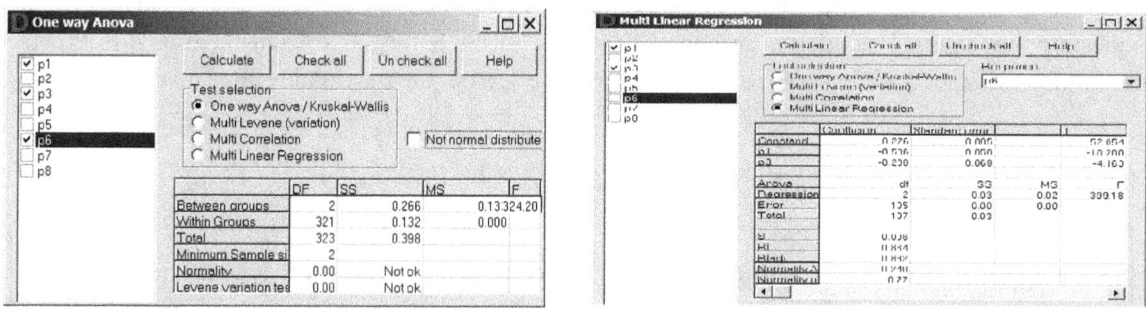

a) One way ANOVA analysis

b) Multi linear regression

One factor experimental plan has been made to evaluate functional dependencies between activities of procedures presented by final probabilities. For the experiments investigation of the process of registration is accepted consecutively increasing factor "*a*" from 0,02 to 0,70 by step 0,02 (number of registrations *n*=36). For other factors (*b, c, d, e, f*) its average value from Table 7 is used. The graphical interpretation of obtained results is summarized in Figure 34.

Figure 34. Security procedures using with increasing new user access

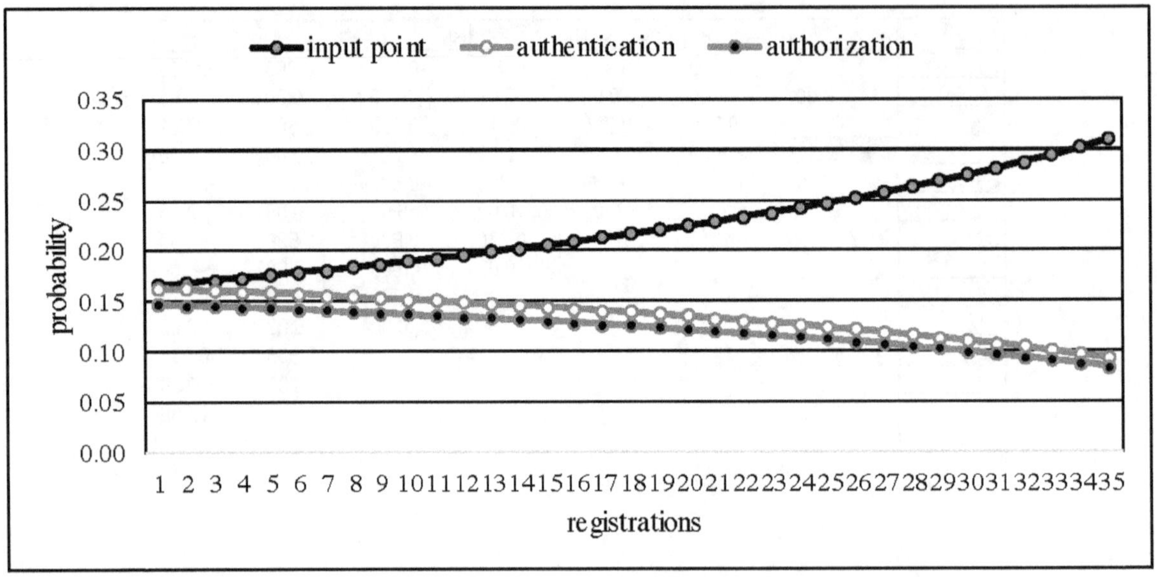

Diagram in Figure 35 presents a comparison between utilization of "registration" procedure and requesting any external or internal resource. Factor "*a*" accepts the same values and as it is expected registration utilization increases in the large range of the set, but the maximal level is no more than 0,22.

Figure 35. Relation between registration and access to the resources

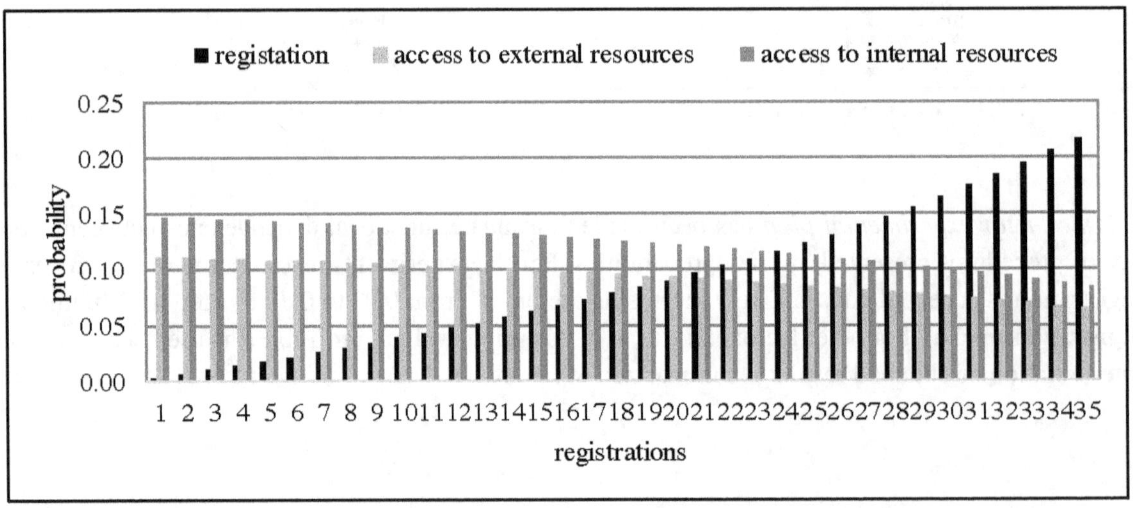

Stochastic investigation of secure access in corporative system. This example presents an investigation of the processes in a corporative system which general functional structure has been shown in Figure 3 of Chapter 7. The general structure organized in three sub-parts (Front office, back office and

Virtual space of resources) has been discussed with presentation of the algorithm for system functionality modeling (Figure 4 of Chapter 7). Here, an extension of modeling capabilities will be made based on applying discrete-time MC and conducting research by analytically solving the stochastic model. The goal is to obtain assessments for activity of the Security Information System (SIS) in the structure of the studied corporative system. In this reason, as an initial step the algorithm for SIS procedures is presented in Figure 36. The used abbreviations are PR – access to public resource and CR – access to private corporative resource.

Figure 36. Algorithm for processes supported by Security Information System

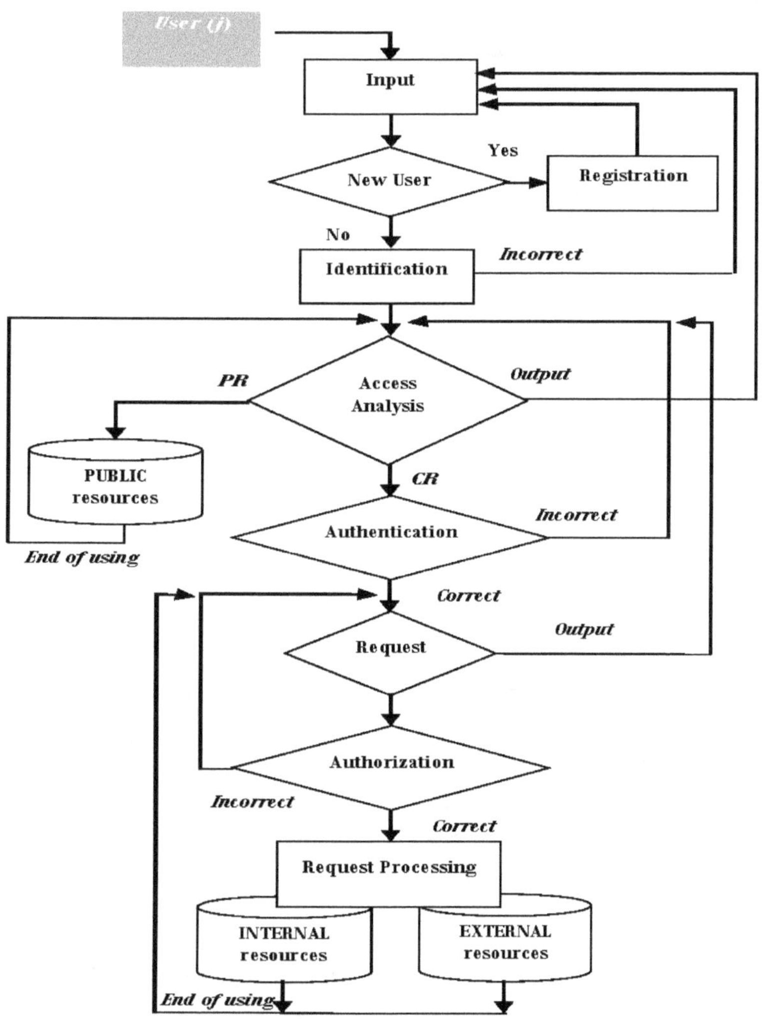

The model by using discrete-time MC, presented below, consists of 11 states with the next definitions: s_1 – input point for remote access of external users; s_2 – procedure for preliminary registration; s_3 – identification procedure for registered access; s_4 – procedure for analysis (what is the required access); s_5 – free access and using public resource; s_6 –user (access) authentication; s_7 –analyses of request for access to private resources; s_8 – authorization for the access rights; s_9 – execution of a resolved request into one of the following two directions; s_{10} – access and using internal resources; s_{11} – access and using external resources. The last three states are united based on relations between them – transitions ($s_9 \rightarrow s_{10}$) and ($s_9 \rightarrow s_{11}$) with probabilities ($p_{9\rightarrow 10}$) + ($p_{9\rightarrow 11}$) = 1. This allows to accept as a main node state s_9 for this triple presentation, which is reflected in the analytical model below. Graph of the states is shown in Figure 37. Concrete values of the transition probabilities are visible in the arcs of the graph.

$$(1): p_1 = p_2 + (1-a), p_3 + 0.1, p_4$$
$$(2): p_2 = 0.2, p_1$$
$$(3): p_3 = 0.8, p_1$$
$$(4): p_4 = a, p_3 + p_5 + (1-b), p_6 + 0.2, p_7$$
$$(5): p_5 = 0.1, p_4$$
$$(6): p_6 = 0.8, p_4$$
$$(7): p_7 = b, p_6 + (1-c), p_8 + p_9$$
$$(8): p_8 = 0.8, p_7$$
$$(9): p_9 = c, p_8$$
$$(10): \sum_{j=1}^{9} p_j = 1$$

Figure 37. Graph of the states of the designed MC model

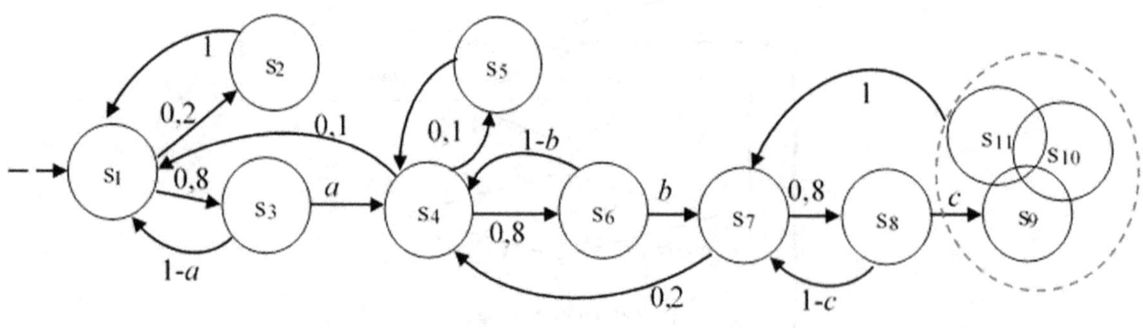

The analytical solution of the system of equations allows obtaining the following expressions for the probabilities:

$$p_1;\ p_2 = 0.2, p_1;\ p_3 = 0.8, p_1;\ p_4 = 8, a, p_1;\ p_5 = 0.8, a, p_1;\ p_6 = 0.8(8, a, p_1) = 6.4, a, p_1;$$
$$p_7 = 4, b, (8, a, p_1) = 32, a, b, p_1;\ p_8 = 0.8, (32, a, b, p_1) = 25.6, a, b, p_1;\ p_9 = 25.6, a, b, c, p_1$$

And after substitution in the last equation (10) it is received:

$$p_1 = \frac{1}{(2+15.2a+57.6ab+25.6abc)} = \frac{1}{\grave{A}},$$

which permits to receive

$$p_2 = \frac{0.2}{\grave{A}};\ p_3 = \frac{0.8}{\grave{A}};\ p_4 = \frac{8a}{\grave{A}};\ p_5 = \frac{0.8a}{\grave{A}};\ p_6 = \frac{6.4a}{\grave{A}};\ p_7 = \frac{32ab}{\grave{A}};\ p_8 = \frac{25.6ab}{\grave{A}};\ p_9 = \frac{25.6abc}{\grave{A}}$$

The assumption a, b, $c \in [0{,}7 \div 1]$ is introduced for experimental investigation organizing and for minimization number of combinations of values for final probabilities is determined experimental factor plan with fixed discrete values for these control parameters {0,7; 0,8; 0,9; 1,0}. Some experimental results are shown in Figure 38 and in Figure 39.

Figure 38. Assessments for final probabilities received based on full experimental plan for control factors

a	b	c	p_1	p_2	p_3	p_4	p_5	p_6	p_7	p_8	p_9	total	p_{10}	p_{11}
0,85	0,85	0,85	0,0138	0,0028	0,0111	0,0941	0,0094	0,0753	0,3200	0,2560	0,2176	1,0000	0,1305	0,0870
1,0	1,0	1,0	0,0100	0,0020	0,0080	0,0797	0,0080	0,0637	0,3187	0,2550	0,2550	1,0000	0,1530	0,1020
0,9	1,0	1,0	0,0110	0,0022	0,0088	0,0795	0,0080	0,0636	0,3180	0,2544	0,2544	1,0000	0,1527	0,1018
0,8	1,0	1,0	0,0124	0,0025	0,0099	0,0793	0,0079	0,0634	0,3171	0,2537	0,2537	1,0000	0,1522	0,1015
0,7	1,0	1,0	0,0141	0,0028	0,0113	0,0790	0,0079	0,0632	0,3160	0,2528	0,2528	1,0000	0,1517	0,1011
1,0	0,9	1,0	0,0109	0,0022	0,0087	0,0869	0,0087	0,0695	0,3128	0,2502	0,2502	1,0000	0,1501	0,1001
0,9	0,9	1,0	0,0120	0,0024	0,0096	0,0867	0,0087	0,0693	0,3120	0,2496	0,2496	1,0000	0,1498	0,0998
0,8	0,9	1,0	0,0135	0,0027	0,0108	0,0864	0,0086	0,0691	0,3111	0,2489	0,2489	1,0000	0,1493	0,0995
0,7	0,9	1,0	0,0154	0,0031	0,0123	0,0861	0,0086	0,0689	0,3099	0,2479	0,2479	1,0000	0,1487	0,0992
....														
1,0	0,8	0,7	0,0129	0,0026	0,0103	0,1031	0,0103	0,0825	0,3298	0,2639	0,1847	1,0000	0,1108	0,0739
0,9	0,8	0,7	0,0143	0,0029	0,0114	0,1028	0,0103	0,0822	0,3289	0,2631	0,1842	1,0000	0,1105	0,0737
0,8	0,8	0,7	0,0160	0,0032	0,0128	0,1024	0,0102	0,0819	0,3277	0,2622	0,1835	1,0000	0,1101	0,0734
0,7	0,8	0,7	0,0182	0,0036	0,0146	0,1019	0,0102	0,0816	0,3262	0,2610	0,1827	1,0000	0,1096	0,0731
1,0	0,7	0,7	0,0143	0,0029	0,0114	0,1142	0,0114	0,0913	0,3197	0,2558	0,1790	1,0000	0,1074	0,0716
0,9	0,7	0,7	0,0158	0,0032	0,0126	0,1138	0,0114	0,0911	0,3187	0,2550	0,1785	1,0000	0,1071	0,0714
0,8	0,7	0,7	0,0177	0,0035	0,0142	0,1134	0,0113	0,0907	0,3174	0,2540	0,1778	1,0000	0,1067	0,0711
0,7	0,7	0,7	0,0201	0,0040	0,0161	0,1128	0,0113	0,0902	0,3158	0,2527	0,1769	1,0000	0,1061	0,0707

Figure 39. Assessments for average case a=b=c=0.85

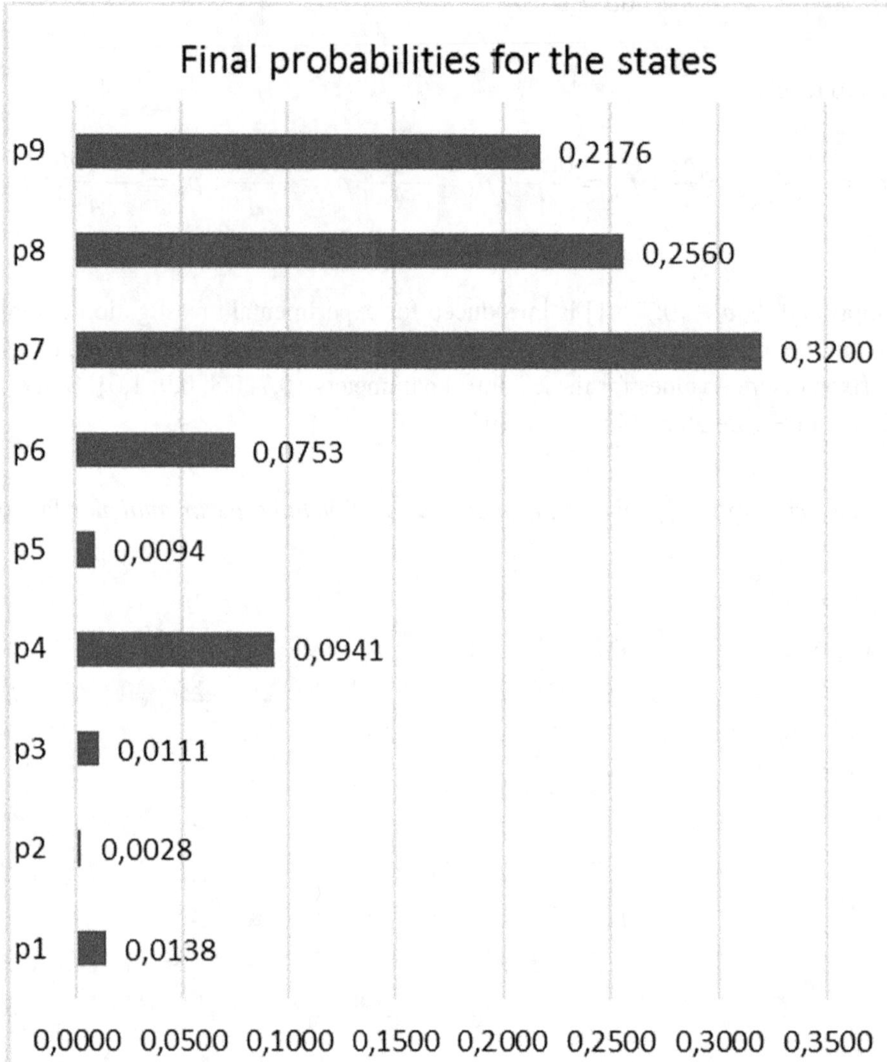

Additional assessment calculated by using software packet "Develve" are summarized in Figure 40 and Figure 41.

Figure 40. Assessments obtained by using "Develve" software (Histograms)

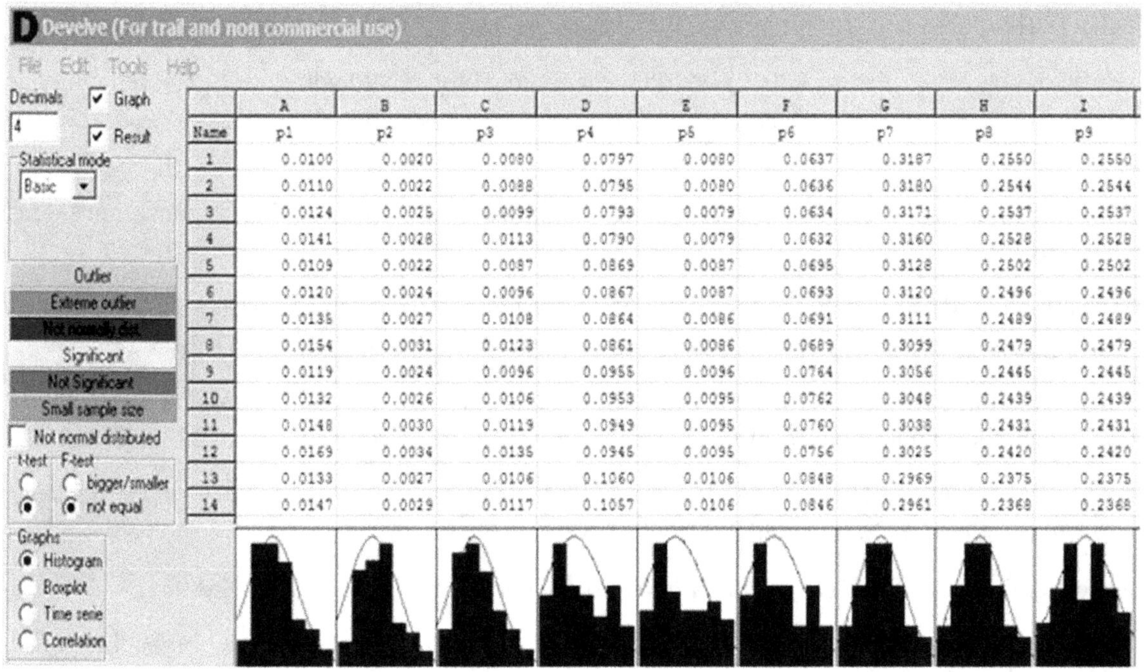

	A	B	C	D	E	F	G	H	I
Name	p1	p2	p3	p4	p5	p6	p7	p8	p9
1	0.0100	0.0020	0.0080	0.0797	0.0080	0.0637	0.3187	0.2550	0.2550
2	0.0110	0.0022	0.0088	0.0795	0.0080	0.0636	0.3180	0.2544	0.2544
3	0.0124	0.0025	0.0099	0.0793	0.0079	0.0634	0.3171	0.2537	0.2537
4	0.0141	0.0028	0.0113	0.0790	0.0079	0.0632	0.3160	0.2528	0.2528
5	0.0109	0.0022	0.0087	0.0869	0.0087	0.0695	0.3128	0.2502	0.2502
6	0.0120	0.0024	0.0096	0.0867	0.0087	0.0693	0.3120	0.2496	0.2496
7	0.0135	0.0027	0.0108	0.0864	0.0086	0.0691	0.3111	0.2489	0.2489
8	0.0154	0.0031	0.0123	0.0861	0.0086	0.0689	0.3099	0.2479	0.2479
9	0.0119	0.0024	0.0096	0.0955	0.0096	0.0764	0.3056	0.2445	0.2445
10	0.0132	0.0026	0.0106	0.0953	0.0095	0.0762	0.3048	0.2439	0.2439
11	0.0148	0.0030	0.0119	0.0949	0.0095	0.0760	0.3038	0.2431	0.2431
12	0.0169	0.0034	0.0135	0.0945	0.0095	0.0756	0.3025	0.2420	0.2420
13	0.0133	0.0027	0.0106	0.1060	0.0106	0.0848	0.2969	0.2375	0.2375
14	0.0147	0.0029	0.0117	0.1057	0.0106	0.0846	0.2961	0.2368	0.2368

Figure 41. Assessments obtained by using "Develve" software (Boxplots and time series for the final probabilities)

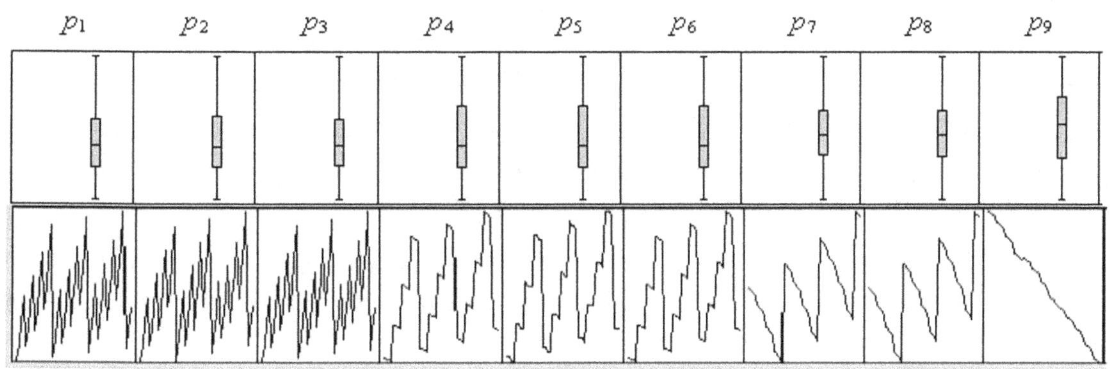

REFERENCES

Bhati, D., Chakraborty, S., & Lateef, S. G. (2020). A discrete probability model suitable for both symmetric and asymmetric count data. *Filomat*, *34*(8), 2559–2572. doi:10.2298/FIL2008559B

Bourgey, F. (2020). *Stochastic approximations for financial risk computations* [Doctoral dissertation, Institut polytechnique de Paris].

Cardelli, L., Kwiatkowska, M., & Laurenti, L. (2016). Stochastic analysis of chemical reaction networks using linear noise approximation. *Bio Systems*, *149*, 26–33. doi:10.1016/j.biosystems.2016.09.004 PMID:27816736

Dimishkovska, N., & Iliev, A. (2019). Markov chain model for small hydropower plant reliability and operation evaluation. In *33rd International Conference on Information Technologies (InfoTech-2019)* (pp. 98-109).

Durrett, R. (2019). *Probability: theory and examples* (Vol. 49). Cambridge university press. doi:10.1017/9781108591034

Efatmaneshnik, M., Shoval, S., & Joiner, K. (2019). System test architecture evaluation: A probabilistic modeling approach. *IEEE Systems Journal*, *13*(4), 3651–3662. doi:10.1109/JSYST.2019.2899697

Ghadimi, S., Ruszczynski, A., & Wang, M. (2020). A single timescale stochastic approximation method for nested stochastic optimization. *SIAM Journal on Optimization*, *30*(1), 960–979. doi:10.1137/18M1230542

Girardin, V., & Limnios, N. (2022). *Applied probability: from random experiments to random sequences and statistics*. Springer Nature. doi:10.1007/978-3-030-97963-8

Gnedenko, B. V. (2018). *Theory of probability*. Routledge. doi:10.1201/9780203718964

Hassani, M. M., & Berangi, R. (2018). An analytical model to calculate blocking probability of secondary user in cognitive radio sensor networks. *International Journal on Information Technologies & Security*, *10*(2), 3–12.

Huang, J. (2022, May). Fundamental theorems in the probability theory. In *2nd International Conference on Applied Mathematics, Modelling, and Intelligent Computing (CAMMIC 2022)* (Vol. 12259, pp. 126-132). SPIE.

Hwangbo, S., Heo, S., & Yoo, C. (2022). Development of deterministic-stochastic model to integrate variable renewable energy-driven electricity and large-scale utility networks: Towards decarbonization petrochemical industry. *Energy*, *238*, 122006. doi:10.1016/j.energy.2021.122006

Jiang, Q., Fu, X., Yan, S., Li, R., Du, W., Cao, Z., Qian, F., & Grima, R. (2021). Neural network aided approximation and parameter inference of non-Markovian models of gene expression. *Nature Communications*, *12*(1), 2618. doi:10.103841467-021-22919-1 PMID:33976195

Kravets, O. J., Atlasov, I. V., Aksenov, I. A., Molchan, A. S., Frantsisko, O. Y., & Rahman, P. A. (2021). Increasing efficiency of routing in transient modes of computer network operation. *International Journal on Information Technologies & Security*, *13*(2), 3–14.

Lanchier, N. (2017). *Stochastic modeling*. Springer. doi:10.1007/978-3-319-50038-6

Laomettachit, T., Kraikivski, P., & Tyson, J. J. (2022). A continuous-time stochastic Boolean model provides a quantitative description of the budding yeast cell cycle. *Scientific Reports*, *12*(1), 20302. doi:10.103841598-022-24302-6 PMID:36434030

Om Parkash, M. (2021). Two new parametric entropic models for discrete probability distributions. [TURCOMAT]. *Turkish Journal of Computer and Mathematics Education*, *12*(6), 2949–2954.

Pandit, R. A., Waghmare, S. A., & Bhagat, P. M. (2022). History of Probability Theory. *Journal of Social Sciences and Humanities*, *4*(5), 140–142. doi:10.53469/jssh.2022.4(05).29

Park, S. W., Lee, K., & Kwon, J. (2022). Neural markov controlled SDE: Stochastic optimization for continuous-time data. In *International Conference on Learning Representations (ICLR 2022)*. IEEE.

Romansky, R. (2022f). An approach for program investigation of computer processes presented by Markov models. *International Journal on Information Technologies and Security*, *14*(4), 45–54.

Ryan, O., & Hamaker, E. L. (2022). Time to intervene: A continuous-time approach to network analysis and centrality. *Psychometrika*, *87*(1), 214–252. doi:10.100711336-021-09767-0 PMID:34165691

Sinha, K., & Trivedi, Y. N. (2022). Spectrum sensing based on two state discrete time Markov chain in additive Laplacian noise. *Wireless Networks*, *28*(6), 2393–2402. doi:10.100711276-022-02979-x

Stoffels, G., & Hohmann, S. (2022). Comparison: Stochastics with a focus on Probability Theory. In *Comparison of Mathematics and Physics Education II: Examples of Interdisciplinary Teaching at School* (pp. 277–297). Springer Fachmedien Wiesbaden. doi:10.1007/978-3-658-36415-1_20

Sweilam, N. H., Al-Mekhlafi, S. M., Hassan, S. M., Alsenaideh, N. R., & Radwan, A. E. (2022). Numerical treatments for some stochastic–deterministic chaotic systems. *Results in Physics*, *38*, 105628. doi:10.1016/j.rinp.2022.105628

Tan, Z., Chai, M., Chen, D., Liao, J., Chu, Q., Liu, B., Hua, G., & Yu, N. (2021). Diverse semantic image synthesis via probability distribution modeling. In *Proceedings of the IEEE/CVF Conference on Computer Vision and Pattern Recognition* (pp. 7962-7971). IEEE. 10.1109/CVPR46437.2021.00787

Tsaregorodtsev, A. V., Kravets, O. J., Choporov, O. N., & Zelenina, A. N. (2018). Information security risk estimation for cloud infrastructure. *International Journal on Information Technologies & Security*, *10*(4), 67–76.

Wanduku, D. (2022). The multilevel hierarchical data EM-algorithm. Applications to discrete-time Markov chain epidemic models. *Heliyon*, *8*(12), e12622. doi:10.1016/j.heliyon.2022.e12622 PMID:36643325

Chapter 9
Stochastic Analytical Modeling Based on Queuing Systems

ABSTRACT

The subject of discussion in this chapter is the stochastic approach to model investigation of queuing systems (QS) with waiting buffer (queue) and server (service unit). Two types of QS are presented: single-channel and multi-channel QS. Single-channel QS is discussed in the first part of the chapter with presentation of main characteristics such as workload of the resource, queuing length, total number of requests in system, waiting time and time for presence in the system. The second part deals with organization of analytical investigation of QS and in particular presentation of discrete and continuous time Markov chains. The two possibilities for investigation were considered – during the development of a transient regime and the conditions for reaching a steady state regime. Three versions of QS are presented in QS – with an infinite buffer, with a limited buffer, and QS with request rejections in the absence of a buffer. Some examples for stochastic model investigation are presented in the last part.

1. MAIN CHARACTERISTICS OF SINGLE-CHANNEL QUEUING SYSTEM

Mathematical modeling based on networks of queuing systems (QSs) is widespread in various fields such as optimization of various technical, physical, economic, industrial, and administrative systems. Confirming this, the paper (Dudin et al., 2020) discusses investigation by using stochastic approach various complex multi-server QS or single-server queues with random distribution of service time or semi-Markov service. A similar confirmation of the applicability of the QSs was made in (Atlasov et al., 2020) for minimizing the total time for searching and processing information throughout the system, with remote servers divided into groups of equal capacity.

One of the most frequently studied QS is single channel with different types of synchronous or asynchronous workflow (Gortsev & Nezhelskaya, 2022). The research conducted is of different orientation, and the basis of each is QS. One example is the improvement of services provided by a government institution, discussed in (Darmayunata et al., 2023). The purpose of the research is to establish effective organization of the queue for ordering customers based on registration through a website with previ-

DOI: 10.4018/978-1-6684-8947-5.ch009

ously expected service time (T_s). To model the process, Single Channel Multi Steps QS was used with the requirement to minimize waiting time (T_w) and the number of waiting requests (L).

The application of QT and, in particular, the model researched on the basis of single-channel QS, is relevant to processes of different nature, including in the investigation of the traffic characteristics of multi-service communication networks. An example of such an application is the random variable-based traffic analysis of incoming requests with constant time intervals, which is presented in (Likhttsinder & Bakai, 2022). Based on the generalized Hinchin-Pollachek formula for average queue values, queue parameters such as dispersion, correlation properties and load factor are defined. To represent the workflow of requests, a group Poisson flow is chosen, which has no correlation component, and the average value of the queue is determined entirely by the number of incoming requests and their service intervals. As a result of the research, it is shown that the requests are processed cyclically and in the last interval of each cycle the queue is empty.

The main components of a classical single-channel QS and the related parameters of the request service process were briefly discussed in Chapter 5 (part 5.3.2, Figure 28). A presentation of the parameters of single-channel QS related to a stochastic model investigation of computer processes is presented below.

Workload of the resource (R) – for single-channel QS it is the ratio of the real busy time of a server unit (SU) to the total busy time T, while for multi-channel QS it is the average number of SUs busy serving the requests for an interval T. The parameter is commensurate with utilization factor ρ and allows to determine a general utilization factor. A case study of low intensity of the input flow of requests λ and high speed of service time is discussed in (Zyulkov et al., 2022). The object of a model study is QS type M/M/1, and the study was conducted for a very short observation time and approximate dependencies for the system utilization factor were obtained. A simple probabilistic model in the form of a probabilistic mixture was developed to conduct the experiments. An analogous study of QS with high loading and small tail is done in (Tsitsiashvili, 2021), where two alternatives are proposed. The first one is uniting many single channel QSs in a new multi-channel QS. The second variant is based on the model of single-channel QS in which random fluctuations are determined depending on the load factor. The exponent of this degree has a critical value above which the tail tends to zero and below which it tends to infinity.

Another article states that the implementation of the known models for virtualization of resources through a virtual server does not allow to make a comprehensive analysis of the efficiency of their allocation (Martyshkin et al., 2020). The objective is to conduct an experimental investigation of mathematical models of a CS for virtualization, proposing a closed QS. It is indicated that the determination of the most efficient option for optimal resource allocation depends on a specific field and task of use. The study of various structural solutions for CS with resource virtualization was carried out with developed simple QS models, monitoring, and adjusting the workload of the service channel under specified conditions.

The QS with distributer workload are widely used in contemporary computing environments because they offer good parameters for stability and scalability (Staffolani et al., 2023). In addition, they allow simple allocation of requests to individual queues on a "least used" basis or apply manually formulated requirements for resources or execution time. To solve the problem with the correct distribution of tasks in the system and applied Markovian approach using synthetic and real forms of the workload. A comparative analysis of the two forms of workload was made, and results were obtained for better service by reducing waiting and service times.

The workload of a QS depends on the structure and type of resources in it, for example, what is the level of homogeneity, what is the type of queues, and the number of service devices. Usually, real systems have a heterogeneous nature of structural organization and different strategies for serving requests. In this direction, a study of a heterogeneous resource queuing system with a two-node flexible request-response facility is conducted in the paper (Moiseeva et al., 2022). It is assumed that each node has a certain service resource capacity (buffer space) and therefore certain resource load capabilities. When conducting the research, input flows of requests with a Poisson distribution and a steady state with different intensities were defined.

Queue length (*QL*) – current number of requests waiting at any given time in the buffer (queue) for service. Mathematical theory of probability workflows defines the formulas for mathematical expectation and variance of steady flow. The QL has a maximum value where all input requests are concentrated in the beginning of the served cycle, and the average QL decreases if the requests generation is further away from the start of the cycle (Likhttsinder & Bakai, 2022). The paper discusses the development of the interval method for queuing analysis and examines the number of requests in the time intervals leaving the queue for transition to service. It is concluded that if the conditions for stationarity, ergodicity and convergence of the covariance function are met, the conditional average values of QL in a single-channel QS are determined entirely by the variance and the sum of the covariance of the requests number.

The *QL* parameter changes with the frame depending on the intensities of the input flow and the service flow, as the dependency of *QL* as a function of workload R is shown in Figure 1.

Figure 1. Functional dependency QL(R)

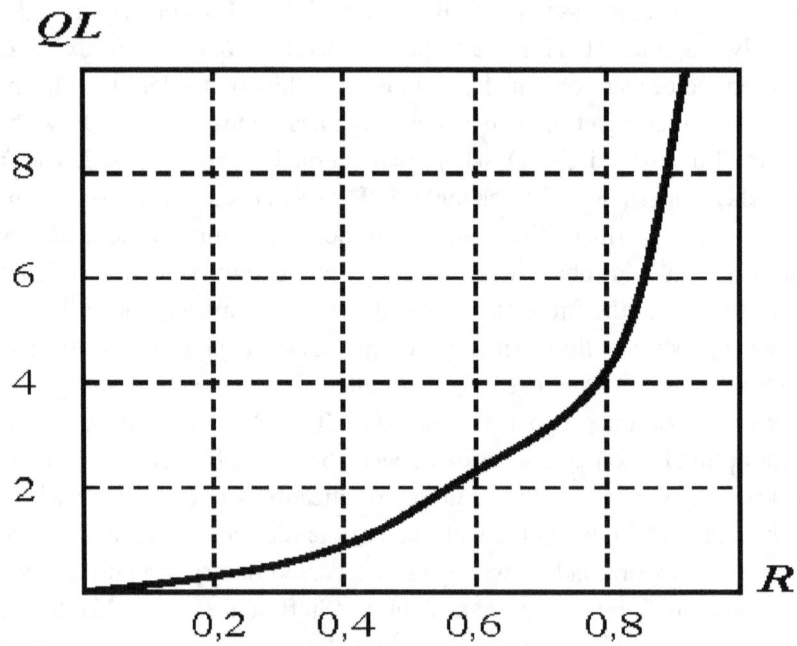

An investigation to obtain estimates of a mean QL for single-channel QS is presented in (Likhttsinder et al., 2022). The object of model study is a simple QS with deterministic service time and random input workflow and various generalizations of the Pollaczek–Khinchine formula for the average QL have been obtained. In this investigation, too, an interval model of the arrival flow was applied, and an expression for the average QL was formed in terms of statistically unconditional second-order moments to build the model. All results are obtained under very general assumptions of ergodicity and stationarity. Numerical experiments were conducted to confirm the theoretical results of the modeling.

The case of multi-server QS with exponential distributed time between arriving time and servicing of request (Markovian process) is discussed in (Cruz et al., 2017). A Bayesian technique was used to estimate QS parameters of this type. The choice was made because this approach allows the determination of performance measures essential in determining the evaluation of important practical applications in various fields, including in the computer field and telecommunication networks. The obtained numerical results confirm the accuracy and efficiency of the applied approach, but also determine some limitations.

The contemporary multiservice telecommunication systems are characterized by packetized request flows, and a model study with such defined QS applying the Pollaczek–Khinchine formula for a general type of flows is done in (Likhttsinder, 2017). Based on an analysis of the dependencies of the average queue size at a low fill factor, a concept of conditional average queue size was defined, which reflected the total number of requests in QS. For this purpose, the assumption is made that there are no idle intervals of the processor. An algorithm is proposed for approximating the size dependencies of QL queues in QS with batched query flows, considering the time delay in the queues. To ensure the QL size of the queues, bounding values are set for the final resource load levels (R) and time delays.

Total number of requests in the QS (n) – all present in QS queries at any given time. The time dependence $n(t)$ of this parameter is a step function, as an example of single-channel QS is shown in Figure 2. The figure shows the times T_R and T_S for the first two requests (with their respective indexes) arriving at the QS. At the beginning, the SU is not yet busy, allowing the first request to pass without waiting in the queue and its service to start immediately. The second request must wait for the SU to be released, therefore $T_{R2}>T_{S2}$. This case is at a higher intensity λ for the incoming request flow relative to the service intensity μ in the SU, which forces the second request to wait a certain amount of time in the queue. When n=0 QS is idle; for n=1 in QS there is one request that is served in SU; the situation of pending requests is illustrated by n>1.

Figure 2. An example for function n(t) in QS

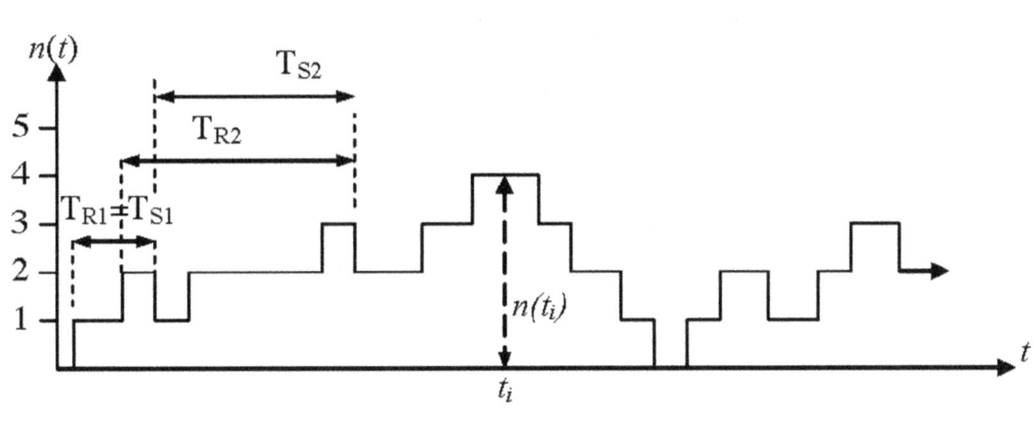

It is obvious that the total number of requests at any moment is different and is determined by a number of factors, including whether one is investigating single-channel QS, multi-channel QS or QS with one SU and multiple queues (for example priority servicing or service of different type of requests). A study of the last-mentioned case is presented in (Uryvsky & Martynova, 2019), where the object of the study is a unified service model for servicing priority requests. The model reflects the performance of a cloud server when processing multiple service requests, which affects the total number of requests. The analytical model investigation was made based on universal mathematical ratios used to determine the service indicators. As a result, a comparative characterization of the service quality indicators when organizing priority access using absolute and relative priorities is presented.

The application of multi-channel QS (multiple service units and common queue) is typical for different systems in computing, for example multiprocessor systems, synchronous parallel processing, networks, etc. What these cases have in common is that the input stream of processing tasks (input workflow) is generated from a single node, for example shared memory. This allows defining a common queue for incoming requests which will be routed to different SUs in the QS. Such model investigation is made in (Matyushenko & Ermolayeva, 2021), where a QS with several parallel devices and common storage with limited capacity is discussed. The input WF is Poisson type, and the QS is requesting denial system (Figure 3), because if all SUs are occupied, the incoming request is lost and does not affect the general functioning of the system.

Figure 3. Multi-channel QS with failure

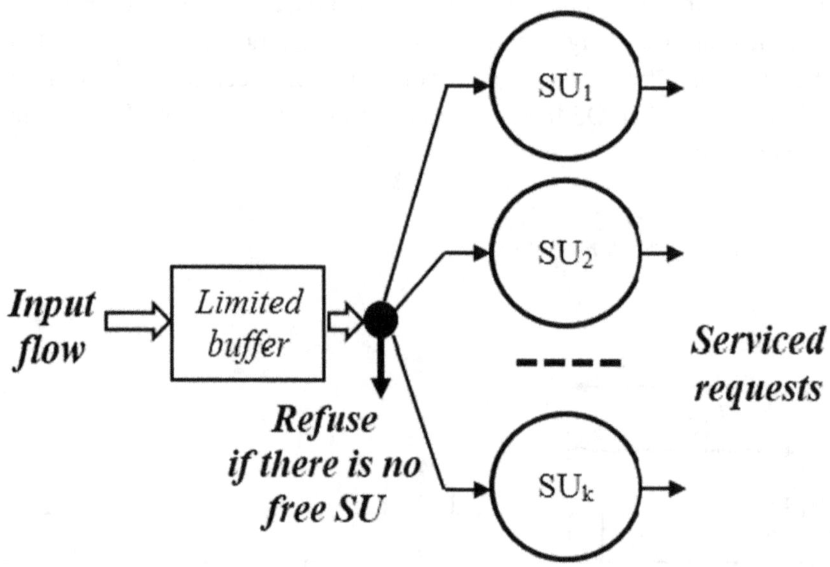

The requests servicing is random and obeys an exponential law. When developing the model and implemented a service strategy based on selecting from each request a free SU from the available ones that has the highest intensity μ. The order in which the requests are completed corresponds to the order in which they are received in the QS. A buffer is added at the system output for additional reordering of queries that have violated the entered order (QS with reordering of queries). Modeling has made it possible to obtain analytical expressions for various stationary characteristics of the behavior of the object under study.

Modeling using QS allows us to study the load of a given transport infrastructure, as done in (Bychkov et al., 2021). It is stated that the efficiency of the transport industry depends on the quality of functioning of the units involved in it. The basis for applying the stochastic approach with QS is the complex multi-level structure and probabilistic nature of the traffic flow. This justifies the application of mathematical modeling based on a network of QS and input flow as a Markov process. The paper presents a complex queuing model for describing the routing of requests in an object with a non-linear hierarchical structure. Simulation software was used to implement the developed models during numerical experiments. The obtained numerical results allowed to determine the total number $n(t)$ of service requests at different moments of time, some "bottlenecks" in the structure, as well as an assessment of the risk when switching to an irregular mode of operation.

Waiting time (T_W) – this time-parameter is determined from the moment of entry into the buffer (queue) until the start of its actual service. In stationary mode, the requests do not have to wait in the buffer, but immediately pass from it to the SU. This is because the condition $\mu \geq \lambda$ is satisfied. If the last condition is not fulfilled, it is necessary to wait for the requests in the buffer and even its continuous filling. The reason for this non-stationary mode is that servicing a single request in the SU requires more time than the input request arrival intervals. In a QS with an infinite FIFO buffer, this leads to a continuous increase in the number of pending requests, while with a limited buffer, when it fills up, the QS becomes a system with denial.

One solution to the above problem is multi-channel QS with a common queue. This was used in a model investigation of a radio monitoring complex with a large amount of communication signals presented in (Romanov et al., 2022). Increasing the number of channels leads to higher performance, but the processing of requests (signals and protocols) combined in layers is sequential and after the completion of the previous one. The consequence is that individual components of the system are in processing mode while others are idle. To solve the problem, the possibility of simultaneous operation of all complex processing units from different layers is discussed in the article, by developing a three-phase QS model. The first phase is a QS without a queue (buffer), while the second and third are QSs with an infinite buffer. Each phase is provided with a mathematical definition of formulas for calculating the probabilities of the state and characteristics of the systems. An analytical comparison of the probability of service and average waiting time ($T_{W\text{-}av}$) of a request in the radio monitoring complex was made. Experimental dependencies for the processes were formed. Based on the processed experimental results, it has been proven that better working conditions are offered by the three-phase complex compared to the single-phase one.

Another comparative analysis of the two types of QSs – single-channel and multi-channel, was conducted in (Epishkin, 2023). A mathematical model with reliable channels for parallel car service is developed, and an equation for the throughput of multi-channel QS is defined. The functional dependency accounts for the number of channels, the intensity λ of the input flow of requests, service time T_S, as well as channel failures and their service time recovery. Based on calculations, it has been proven that the waiting time T_W (throughput) depends mainly on the intensities of the input flow λ and of the service μ. The influence of other parameters on QS throughput was additionally investigated.

Time of presence in the QS (T_R) – includes the entire time of existence of a request in the QS (from entering the buffer to leaving the QS). The probabilistic service process in QS through the flows of incoming requests $a(t)$ and outgoing requests $b(t)$ is illustrated in Figure 4. The two functional dependencies are step-like, but for significant values of time and queue length they can be approximated by continuous functions. The two stochastic flows can be used to determine: ✓ the current number of requests at a given time $n(t_i) = a(t_i) - b(t_i)$; ✓ average time for presence of a request in the QS $T_{R\text{-}av}$.

Figure 4. Stochastic flows of incoming and outgoing flows of requests

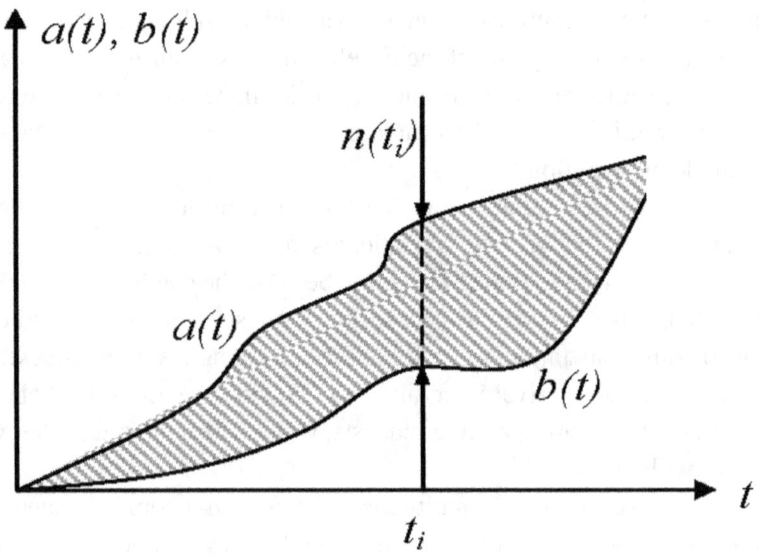

In most model investigations and analytical solutions based on QS theory, the assumption is made of a stable and unchanging system structure and Poisson input flows, Erlang flow with limited consequence, flow as a Markov process with finite set of discrete states, and flow presented as a semi-Markov process. In (Porshnev & Korelin, 2017) it is stated that there is a smaller number of QS studies conducted on stochastic input flow of requests with intensity $\lambda = \lambda(t)$ as a dependence of time. It is further stated that this is the situation that corresponds to real service systems, and conducting this type of research will assist in the design or improvement of various systems. To solve the problem, the article presents the results of a model study of a non-stationary QS, with an intensity λ of the input flow set similar to that in a real object access control system. The model, based on statistical modeling, offers analytical formulas to describe basic QS characteristics at maximum values for $\lambda(t)$ and average value of service intensity μ in the system. Deterministic functions are applied, allowing to make the connection between a real input workflow and average service frequency. The obtained results allow determining the average time for presence of a request ($T_{R\text{-}av}$) in the QS, as well as the dependence of the number of served requests on time.

2. ORGANIZATION OF ANALYTICAL INVESTIGATION OF QUEUING SYSTEMS

The analytical study of a stochastic service system is simplest when the processes taking place in it are Markov (usually the processes in computer systems are like this). This is the case when the events driving the QS into successive states obey Poisson flows, and especially if these flows are simple. For a uniform simple input flow of requests with intensity λ and a simple service flow with intensity μ and mathematical expectation $1/\mu$ (mean service time) for single-channel QS and k-channel QS, the notations M/M/1 and M/M/k are used respectively. An example of the use of QS type M/M/1 is the study of an incoming packet stream with an intensity with Poisson distribution presented in (Ahmed & Ibrahim, 2018). The packets are generated by independent nodes in the sensor network, with an intensity of arrival λ. It is also assumed that the service is the same for all nodes and the processing speed is determined by a probability flow with intensity μ. Below is presented a brief overview of the features of the two main types of Markov chains (MCs), using the classification presented in (Sukhorukova & Chistyakova, 2020) – discrete MC and continuous MC.

When studying QS by using ***discrete MC (with discrete time)*** the transitions between service states s_j $(j=0\div n)$ are described by the transition probabilities, as:

$$\sum_{i=1}^{n} p_{ij} = \sum_{m=1}^{n} p_{jm} = 1.$$

A study of the features, possibilities and rules when using discrete time MC is presented in (Privault, 2018). The discussion is focused on the properties of MC and on the role of matrixes of transition probability. In addition, several simple examples are presented. As noted in (Zoppo et al., 2021), "*problems involving discrete Markov Chains are solved mathematically using matrix methods.*" In addition, it is specified that matrix multiplication has recently been demonstrated analytically within a single time step using electronic circuit called a "memristor". This demonstrates the possibility of solving systems of linear algebraic equations in one step using similar hardware. The paper concludes that this approach can be used in the analysis of discrete MCs to determine the successive states of the process (by matrix multiplication) and in determining the steady state (by solving the system of equations). To confirm this possibility, the paper presents MC-models for open-loop and feedback configurations. The analysis made includes possible memristor programming errors, the impact of external noise in operation, and problematic elements. An adequate mathematical toolkit for a formal description of MC is provided, and the corresponding comparison between the two approaches is made to determine the correspondence of the obtained analog results with similar calculations by the digital method.

Discrete MC has been used in developing a stochastic model of the movement of a set of robots and the communications between them in a common environment (Shirsat et al., 2020). The research object is a probabilistic consensus strategy for processes in disaster-affected areas that are resilient to communication link failures. Modeling with a Markov process allows the exchange of information with neighboring robots, which changes over time the communication environment. The effect and size of the robot population and the level of information uncertainty on the weather statistics is investigated through numerical experiments. The results showed reaching a consensus between the robots in a finite time. Another problem solved by applying discrete MC is a model study of air pollution and more precisely "*to predict the air quality index (AQI) and identify the prime air pollutants in a specific area*", presented in (Chen & Wu, 2020). Collected air quality data obtained from direct monitoring over different time periods were used to develop the MC-model.

When studying QS by using ***continuous MC (with continuous time)***, the transitions are described by the corresponding intensities, and the probabilities for the stationary mode (if there is a condition for existence) are determined after solving $[dp_{ij}(t)/dt] = 0$ and are presented below, where p_0 is determined based on the condition of normality $p_0 + p_1 + ... + p_n = 1$.

$$p_j = p_0 \prod_{m=1}^{j} \frac{\lambda_{m-1}}{\mu_m}$$

Continuous-time MCs are mathematical models which are used for investigation of dynamic systems with stochastic behavior. To ease the mathematical calculation of MC parameters, certain assumptions are most often introduced, which can lead to inaccuracies in the model description (Krak et al., 2017). Therefore, when formulating specific assumptions, it is necessary to specify how well they correspond to the field of application for which the model is being developed. In this respect, the article discusses what are the tolerable limits for imprecision in constructing continuous-time Markov chains so as to preserve the stability of the stochastic investigation and the computational tractability of the processes. From a technical point of view, continuous-time MCs are a set of stochastic processes with a limited set of states and continuous time in the transitions between them. The article emphasizes that, unlike Bayesian methods, the elements of this set are perceived on equal grounds, which is why it is possible to intuitively interpret the formulated limits in both directions – to represent best and worst scenarios for the considered set of stochastic processes. Adopting this formulation, a formal description of stochastic processes for continuous-time MC is made without the above-mentioned simplifying assumptions. Based on this formalism, MCs are defined and their properties are investigated. The purpose of the conducted research is to define easy-to-use algorithms (with polynomial complexity of the execution time compared to the maximum numerical error) for calculating a lower mathematical expectation of the functions.

A review of opportunities of stochastic model investigation by using MC with continuous time is made in (Zeifman et al., 2020). Basic approaches to study the stability of homogeneous and inhomogeneous MCs with respect to changes in their intensities as well as infinitesimal features are discussed. An approach is presented for formulating disturbance estimates for queue service parameters, with bounds calculated within process-acceptable limits. In the conducted research, it was assumed that the evolution of the discussed object is described by a known space of states with an infinitesimally small matrix, which is defined with some inaccuracy. This allows different classes of tolerable disturbances to be analyzed by discussing various small deviations from the original matrix.

Another application of continuous-time MCs concerning the model investigation of the time evolution of individual sets of genetic mutations is presented in (Gotovos et al., 2021). It is pointed out that the problem defined for research most often does not have a sufficient volume of necessary data, therefore the article adds (as stated "a surprising means") the inclusion of additional independent elements. The reason is the possibility to help determine the time series by supplementing the insufficient specification. It is appreciated that this is perhaps somewhat at odds with the usual practice of limiting analysis to a small subset of items. In order to confirm the relevance of the chosen approach, an approximate method for maximizing the likelihood for the analysis of continuous-time MCs, allowing scaling to hundreds of elements, has been developed. The effectiveness of the proposed approach has been confirmed by applying it to real data, and it has been proven to be faster than other known methods.

The most common presentation of the analytical definition of QS based on continuous-time MC is by using graph of states type "*death and reproduction*" – Figure 5. Each node S_j of the graph corresponds to the presence of *j* requests in the QS at a given time. The transitions between the states are carried out under the influence of the two flows – of admission (with intensities λ) and of service (with intensities μ).

Figure 5. Graph of states "death and reproduction"

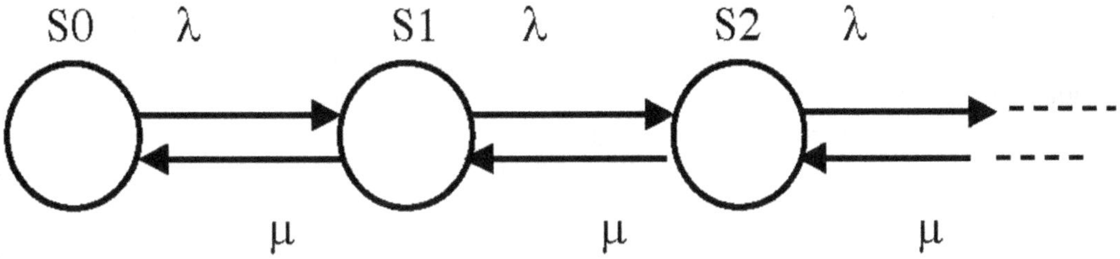

An investigation using the graph of states "death and reproduction" is presented in (Chkalova & Pavlidis, 2020). The developed stochastic model is of a real technological process (feed mills) and is based on the theory of correlated random functions, theory of automatic control, Markov processes and operational calculus. The conducted experimental investigations have confirmed a high adequacy of the modeled processes to the real ones at optimal system load. The modeling considered the variety of factors influencing the efficiency, and the random processes of the "death and reproduction" type were defined as Markovian. Correspondences for both states to real processes are defined, with the assumption of continuous operation in stationary mode being accepted. The individual states in which the model passes are indicated by consecutive nodes from the graph, for which the corresponding transition probabilities for "death" and "reproduction" are defined. The probabilities of falling into a given state $p_i(t)$ ($i = 1,2...$) have been calculated, and regardless of a continuous change of QS states in a stationary mode, these probabilities remain constant $p_i(t) = P\{S(t) = s_i\}$, where $S(t)$ represents one of the possible states at time t.

For a QS of type M/M/1 with service discipline FCFS and starting the service at an empty queue, the condition for the existence of a ***steady state regime*** is $\rho=\lambda/\mu<1$ (or $\lambda<\mu$), and the analytical model is:

$$\begin{cases} (\lambda+\mu).p_j = \lambda.p_{j-1} + \mu p_{j+1} \;;\; j \geq 1 \\ \lambda.p_0 = \mu p_1 \end{cases}.$$

Determining the values for the probability parameters in the stationary mode for MC with continuous time requires solving a system of probability equations, and their number depends on the number of states in the defined graph. In a certain case, this number can be infinite or too large, and in both cases solving them analytically is problematic. One way to overcome this problem is to use approximations and truncate the set of states, a review on the subject is presented in (Kuntz et al., 2021). As a basis, a theoretical justification of the stationary distributions and their relation to the general behavior of the MC in a period of time is given. Possible state-reduction approximation schemes are discussed, emphasizing the convergence of the schemes and the errors they may introduce. At the end of the review, certain computational trade-offs related to error control and several open questions are suggested.

A study of the service process in QS using hyper-exponential and Erlang approximations to represent the flows is presented in (Zhernovyi, 2018). The calculation of stationary characteristics of QS of type G/G/1/m was carried out by the fictitious phase method, and a simulation approach was used to verify the obtained results from the model experiments.

The stationarity conditions are the subject of research in (Gortsev & Nezhelskaya, 2022), where a single-channel QS with an input asynchronous double stochastic flow, represented by the authors as a Markov modulated Poisson process, is chosen as the main object. Analytical formulas for the probabilistic distribution of states in the steady state are presented. In addition, clear analytical expressions for QS numerical characteristics are given: ✓ average queue length (QL_{avr}); ✓ average number of requests (n_{av}) in QS; ✓ probability of system in idle state. The case of a special two-state asynchronous input stream (switched Poisson process) is also discussed.

For λ≥μ (ρ=λ/μ>1), a solution for a ***transient regime*** is sought, since there is no condition for establishing a steady state situation for the QS with continuous MC. Analytically solving the above model involves the following transformations:

$$\left| \begin{array}{l} (1+\acute{A}).p_j = p_{j+1} + \acute{A}.p_{j-1} \;\; ; \; j \geq 1 \\ p_1 = \acute{A}.p_0 \end{array} \right. ,$$

and after sequentially solving the system for the probabilities (j=1, 2, ...) we get:

$$p_1 = \rho.p_0$$
$$p_2 = (1+\rho).p_1 - \rho.p_0 = (1+\rho).(\rho.p_0) - \rho.p_0 = \rho^2 . p_0 \Rightarrow p_j = \rho.p_{j-1} = \rho^j . p_0 \text{ (for } j>0)$$

3. QUEUING SYSTEMS WITH INFINITE AND LIMITED BUFFERS

QS with infinite buffer. For the normality condition, the following transformation is obtained:

$$\sum_{j=0}^{\infty} p_j = 1 \Rightarrow \left[\sum_{j=0}^{\infty} \left(\acute{A}^j p_0 \right) \right] = \left[p_0 \left(\sum_{j=0}^{\infty} \acute{A}^j \right) \right] = p_0 \left(\frac{1}{1-\acute{A}} \right) = 1,$$

from where, after calculation, the following final probabilities are determined:

$$p_0 = 1 - \rho$$
$$p_j = \rho^j (1 - \rho) \text{ ; for } j>0.$$

The analytical solution of the stochastic QS-model allows to calculate stochastic parameter estimates as presented in Table 1.

Table 1. Analytical estimates for service parameters

mathematical expectation of the total number n of requests in QS	$E[n]=N=\frac{\rho}{1-\rho}=\frac{\lambda}{\mu-\lambda}$
variance of the total number of requests n in QS	$V[n]=\left[\frac{\rho}{1-\rho}+\frac{2\rho^2}{(1-\rho)^2}\right]-\frac{\rho^2}{(1-\rho)^2}=[N+2N^2]-N^2=N+N^2$
mathematical expectation of the number L of requests waiting in the buffer	$E[L]=L_{av}=\frac{\rho^2}{1-\rho}$
mathematical expectation of the time w to wait for a request	$E[T_w]=T_{w-av}=\frac{1}{\lambda}.L_{av}=\frac{\rho^2}{\lambda(1-\rho)}$ $E[T_w]=T_{w-av}=\frac{1}{\mu}.N=\frac{\rho}{\mu(1-\rho)}$

The main stages in the organization of the investigation of CS and the processes taking place in them through the means of mass service are the following:

(1) Determining the type of QS, defining the states and conditions for the transitions, and constructing a state graph.
(2) Checking the condition of stationarity $\rho = \lambda/\mu < 1$ (ρ is the utilization coefficient of the QS) and determining the final probabilities for the states depending on the type of the MC.
(3) The load factor $R = \rho/k$ is determined for QS with k serving channels and service probabilities and service failure probabilities for which *Pserv* + *Pfail* = *1*.

QS with limited buffer. For this case, there is some simplification of the computation expressions, insofar as the capacity of the buffer (queue) is known in advance ($0<q\leq k$). Keeping the notations already accepted for the two intensities λ and μ, the analytical model of continuous-time MC presented below is defined. The corresponding graph of states type "death and reproduction" will have a finite number q+2 states $\{S_0, S_1, S_2, \ldots, S_q, S_{q+1}\}$, which represent the currently available requests in the QS.

$$\left|\begin{array}{l}\lambda.p_0=\mu p_1\\(\lambda+\mu).p_j=\lambda.p_{j-1}+\mu p_{j+1}\;;\;1\leq j\leq q\\\mu p_{q+1}=\lambda.p_q\\p_0+p_1+\ldots+p_{q+1}=1\end{array}\right.$$

State S_0 reflects absence of request in QS, i.e., $n(t)$=0, while state S_{q+1} (with corresponding probability p_{q+1}) reflects a busy serving device and a full buffer. If the system is in this state and a new service request arrives, it will be refused, with the probability of refusal $P_{REF} = p_{q+1}$. An illustration of this case can be made by a QS model with a two-place buffer and example parameters λ=0,3; μ=0,6 (ρ = λ/μ = 0,5) – Figure 6.

Figure 6. Continuous-time MC-model with limited buffer

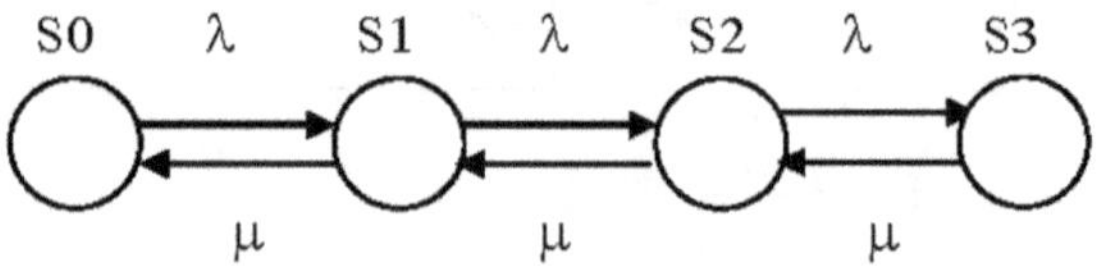

$$\begin{cases} \lambda.p_0 = \mu.p_1 \\ (\lambda+\mu).p_i = \lambda.p_{i-1} + \mu.p_{i+1}; i = 1,2 \\ \mu.p_3 = \lambda.p_2 \\ p_0 + p_1 + p_2 + p_3 = 1 \end{cases}$$

The expressions p_0=1-ρ and p_j=ρ^j.p_0, obtained above can also be used for the calculations, and the final solution is found by substitution in the last equation in the system:

$$\sum_{j=0}^{q+1} p_j = 1 \Rightarrow p_0(\acute{A}^0 + \acute{A}^1 + \acute{A}^2 + \acute{A}^3 + ... + \acute{A}^q + \acute{A}^{q+1}) = 1.$$

An example of using continuous-time MC for model presentation and analytical investigation of the processes for access to remote distributed information resources is presented below. The initial formalization and the abstract model of this object has been discussed in Chapter 5 (Figures 10 and 11). In addition, the detailed formal model and preliminary modeling by using discrete-time MC has been presented in Chapter 8 (Figure 13). Modeling in this part is oriented towards application of a MC with state graph type "death and reproduction" with preserving definitions for abstract objects (set of independent users $U = \{U_i / i=1 \div N\}$; set of distributed learning resources in different nodes $R = \{R_j / j=1 \div M\}$; set of transmitters $T = \{T_q / q=1 \div K\}$; distributor D) and for the interactions between defined abstract objects (requests $req: U_i \rightarrow R_j$ and received information $Inf: R_j \rightarrow U_i$). Two possible forms for communication between abstract defined objects is shown in Figure 7, which permits to define the stochastic model.

Figure 7. Abstract presentation of communications between participants in the system

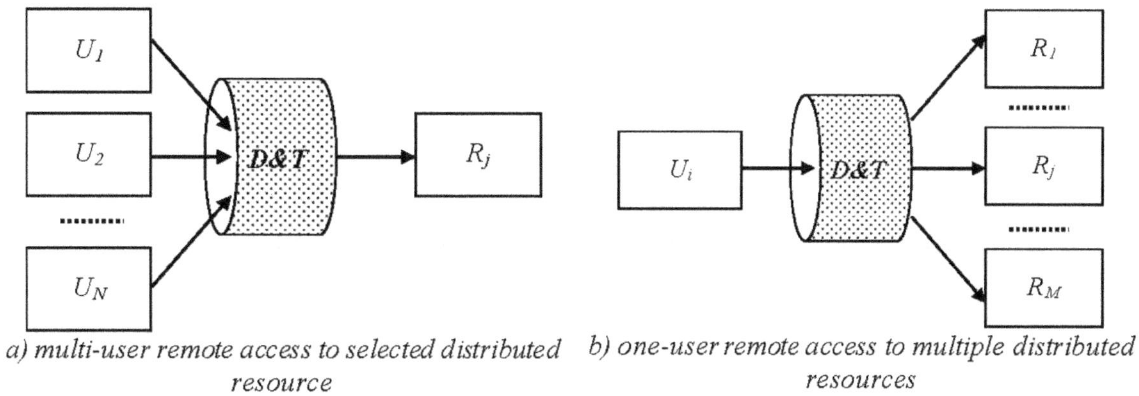

The stochastic nature of supported process is determined by the traditional intensities – intensity of the input flow of requests (λ) and intensity of the flow of servicing in the node of the requested resource (μ). The formal description from Figure 7 determines two possible types of access to the information resource R_j in the environments. Each resource can be requested by a different number of users from 0 (minimum) to N (maximum). Each new user will generate request with the intensity λ and after servicing the Markov's process will be returned to the previous state by the intensity μ. This is illustrated by using the state graph "death and reproduction", which is shown in Figure 8.

Figure 8. Graph of the states for the continuous-time MC model

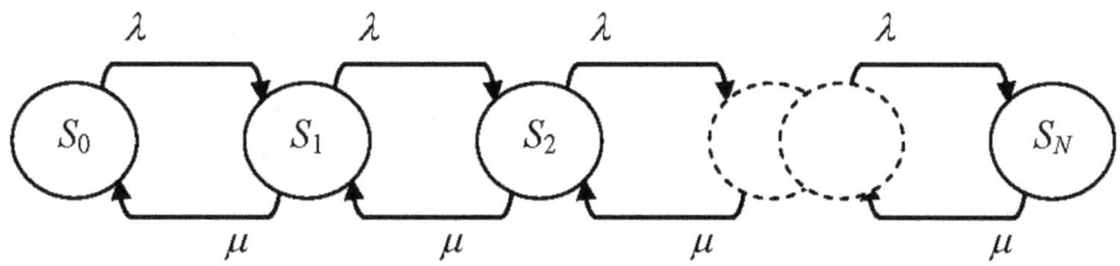

The general analytical definition of the MC-model is a system of probabilistic equations estimating the relationship between individual states and the transitions between them. To ensure entry into the steady-state regime it is necessary to by realized the condition $\rho = \lambda/\mu < 1$, i.e. $\lambda < \mu$.

$$\begin{cases} \text{»}.p_0 = \text{¼}p_1 \\ (\text{»}+\text{¼}).p_i = \text{»}.p_{i-1} + \text{¼}p_{i+1};\ for 1 \le i \le N-1 \\ \text{¼}p_N = \text{»}.p_{N-1} \\ \sum_{i=0}^{N} p_i = 1 \end{cases}$$

After some transformations of the (1) we define the next form of the Markov's model:

$$\begin{cases} p_1 = \frac{\text{»}}{\text{¼}}.p_0 \Rightarrow p_1 = \text{Á}.p_0 \\ (\text{»}+\text{¼}).p_i = \text{»}.p_{i-1} + \text{¼}p_{i+1} \Rightarrow \left(\frac{\text{»}+\text{¼}}{\text{¼}}\right).p_i = \frac{\text{»}}{\text{¼}}.p_{i-1} + \frac{\text{¼}}{\text{¼}}.p_{i+1} \Rightarrow \\ \quad \Rightarrow (\text{Á}+1).p_i = \text{Á}.p_{i-1} + p_{i+1} \Rightarrow p_i = \text{Á}^i.p_0;\ \ for 1 \le i \le N-1. \\ \text{»}.p_{N-1} = \text{¼}p_N \Rightarrow p_N = \text{Á}.p_{N-1} = \text{Á}^N.p_0 \\ \sum_{i=0}^{N} p_i = 1 \Rightarrow p_0 + \text{Á}^1.p_0 + \text{Á}2.p_0 + \ldots + \text{Á}^N.p_0 = 1 \Rightarrow \\ \quad \Rightarrow p_0(1+\text{Á}+\text{Á}^2+\ldots+\text{Á}^N) = 1 \end{cases}$$

Based on the theory of probabilities can be determined for discussed case some assessments for the main stochastic parameters if $N \rightarrow \infty$:

- Expectation of the number of the active users $E[N] = N_{av} = \rho/(1-\rho) = \lambda/(\mu-\lambda)$;
- Variance of the number of the active users $V[N] = N_{av} + N_{av}^2$;
- Expectation of the number of the users waiting servicing $E[W] = W_{av} = \rho^2/(1-\rho)$;
- Expectation of the waiting time for servicing $E[T_W] = N/\mu = \rho/[\mu\ .(1-\rho)]$.

A preliminary solution of the discussed MC-model with continuous time can be made for its verification and defining the model sensitivity. For this purpose, a QS with limited buffer will be regarded and concrete values for the basic model parameters can be seen, for example $N = 3$, $\lambda = 0.3$ (the average time between user's requests is 3s) and $\mu = 0.6$ (the average time for a request servicing is 1.5s) ⇨ $\rho = 1/2 < 1$. This permits to reorganizing the MC-model in the form:

$$\begin{cases} 0.3, p_0 = 0.6, p_1 \\ (0.3+0.6), p_1 = 0.3, p_0 + 0.6, p_2 \\ (0.3+0.6), p_2 = 0.3, p_1 + 0.6, p_3 \\ \sum_{i=0}^{3} p_i = 1 \end{cases} \Rightarrow \begin{cases} p_1 = 0.5, p_0 \\ p_2 = 0.5^2, p_0 \\ p_3 = 0.5^3, p_0 \\ \sum_{i=0}^{3} p_i = 1 \end{cases}$$

At ρ = 1, we gat p_0=1/(q+2), and at $\rho \neq 1$, after summing a geometric progression with first term 1 and parameter ρ, we get:

$$p_0 = \frac{1}{\left(1-\rho^{q+2}\right)/\left(1-\rho\right)} = \frac{1-\rho}{1-\rho^{q+2}} \Rightarrow p_0 = \frac{0.5}{1-0.5^4} = 0.5333\,.$$

From $p_j = \rho^j.p_0$ (j=1,2,3) can be calculated the rest final probabilities, including the probability for refusal $P_{REF}=p_3$.

Diagrams to illustrate the results obtained based on the analytical solution of the model are presented in Figure 9. Each value for the p_i (i=0, 1, 2, 3) presents the probability of the number of active users in the steady-state regime. The measures of the stochastic parameters' values are: 'number of active users' for $E[N]$ and $V[N]$; 'time in seconds' for $E[W]$ and $E[Tw]$.

Figure 9. Results from the preliminary model solution

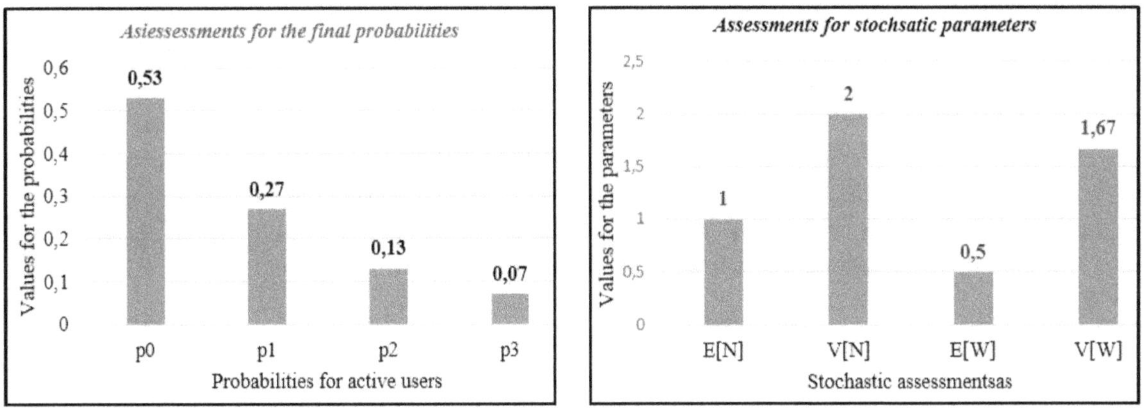

4. BASIC PARAMETERS AND ESTIMATES OF SERVICING

The study of basic parameters of QSs by performing analysis of real systems with service flow of requests is a typical task in various technological fields. One such problem is estimating the state of traffic generated by heterogeneous sources and determining important parameters of the model, for example, frequency of queue release from a consecutive request, intensity of request arrival in the queue, length of stay in the queue, etc. Research in this direction was conducted in (Cheng et al., 2022) to determine traffic bottlenecks using queue models. A review of the applied numerical methods for describing characteristics of QSs is made and a spatial queuing model is developed for systems with heavy traffic, depending on the intensity of the arrival of requests. Analytical formulations with a polynomial functional approximation for the size of the virtual flow are applied to describe the dynamics of the loaded system. This has made it possible to formulate a framework for modeling and evaluating important parameters of the state of the queue formed by traffic and to define bottlenecks in its development.

Proper queue management in request service systems is an important design task, especially in network environments. It is clear that the generation of packets and their movement along communication lines is a probabilistic process, and in many places (for example, in intermediate routers) it is necessary to queue and wait for sending to the next point. To study the behavior of the service system in (Mehta, 2022) a discrete-time single-channel model with FIFO service strategy is proposed. The object of research is service parameters such as queue length, transfer delay, loading of the server device, and the time the packets stay in the system. In addition, a conventional multi-objective genetic algorithm was used to model the relationship between key system parameters. It allowed additional estimates to be calculated for parameters such as queuing delays, network traffic intensity, packet transmission rate, and loss rate. Unlike the previous paper, in (Elliriki et al., 2022) the network service problems in a multi-server environment are modeled using multi-channel QS. The goal is to obtain more correct service evaluations when conducting complex calculations with internal .and external communication analysis. It is recognized that, in principle, available capacity is limited, with multiple servers subject to a probabilistic flow of failures. This necessitates a modeling study of QS transitions from one state to another depending on the impact of the probability flows of failure (failure of a given server) and repair (returning a failed server to a working state). To conduct the model experiments, a workflow with a non-uniform distribution of the arrival times of the requests (packages) was used.

Expressions for calculating the estimates for basic parameters for three types of single-channel QS are given in Table 2, assuming the following assumptions:

- The input flow of service requests has a Poisson distribution.
- Scheduling of requests for service is done according to the FCFS (First Come First Served) discipline, also known as FIFO (First In First Out).
- All requests are equal and there is no removal of requests from the queue.

Table 2. Estimates for parameters of single channel QS

Basic parameters	Type of single-channel QS		
	M/G/1*	**M/M/1****	**M/D/1*****
Total number of requests in QS (n)	$\rho+\frac{\rho^2 A}{1-\rho}$	$\frac{\rho}{1-\rho}$	$\frac{\rho^2}{2(1-\rho)}+\rho$
Number of requests waiting in queue (L)	$\frac{\rho^2 A}{1-\rho}$	$\frac{\rho^2}{1-\rho}$	$\frac{\rho^2}{2(1-\rho)}$
Time of presence in the QS (T_R)	$T_S+\frac{\rho T_S A}{1-\rho}$	$\frac{T_S}{1-\rho}$	$\frac{T_S(2-\rho)}{2(1-\rho)}$
Waiting time of a request in queue (T_w)	$\frac{\rho T_S A}{1-\rho}$	$\frac{\rho T_S}{1-\rho}$	$\frac{\rho T_S}{2(1-\rho)}$

Remarks:

* QS type **M/G/1** has Poisson distributed input flow and random distribution service. To simplify the definition of the expressions, a scaling factor A is introduced, depending on the root-mean-square deviation of the service time to its mean value:

$$A=\frac{1}{2}\left[1+\left(\frac{\sigma_{T_S}}{T_S}\right)^2\right].$$

** QS type **M/M/1** is the traditionally used system when modeling computer objects because it defines an exponential distribution for both intensities – the input flow and the service. This is the simplest case and the calculation of the scores is the easiest.

*** QS type **M/D/1** is characterized by deterministic service (regular service flow), i.e. the root mean square deviation of the service time is zero (T_s=const).

The occurrence of each service event, including the generation of the input workflow, is random and determines the connection time in a given period T, as was indicated in Chapter 5 (Figure 23). It is known that the individual generated units have a discrete character, but the formed flow is stochastic, the simplest structure being a regular stream with constant intervals of incoming requests (τ=*const*). Such a flow has a discrete nature, while the representation by a simple Poisson flow without consequences reflects the probabilistic nature of the modeled object. The definition is by the Poisson distribution function:

$$P_m(\ddot{A})=\frac{a^m}{m!}e^{-a}(m=0,1,2,...)$$

It is proved that for the Poisson flow the intervals τ_j between events has exponential distribution based on the distribution function $F(t)=P(T<t)=1-e^{-\lambda\tau}$. It determines the probability of obtaining value less than t for the parameter T. In addition the theory of probability define the parameters: intensity $\lambda=1/\tau$ (for $t\geq 0$), mathematical expectation $E[t]=\mu=1/\lambda=\tau$ and variance $V[t]=\sigma^2=(1/\lambda)^2=\tau^2$.

Let us assume that a specific set of tasks (procedures) must be performed in a computer environment. Graph theory makes it possible to make a complete connection graph and, accordingly, to move to an ordered graph structure and to determine all the paths from the beginning to the end of the set. This was discussed in detail in Chapter 6 (part 6.2). This formalization allows to define structure with discrete nodes (states of the processing), where the transitions between them in a real case have probability nature (probability transactions in discrete moment of time). This allows us to use discrete-time MC for organization of stochastic investigation and determining assessments for parameters of the behavior. Determining parameters for making the transitions are the correlation coefficients between the tasks a_{ij}: $A_i \rightarrow A_j$, which in general have a probabilistic nature. In this respect, the implementation of each path in the graph structure defines a specific sequence of tasks directed for execution at probabilistic moments of time, and the times for completing each task depend on the internal implementation parameters (Figure 10). The definition of each execution sequence depends on the scheduling strategy of the system dispatcher, and the selection of tasks and their ordering, in addition to their interdependence, should ensure the optimization of time parameters and performance of processor processing.

Figure 10. An example for sequence of procedures realization

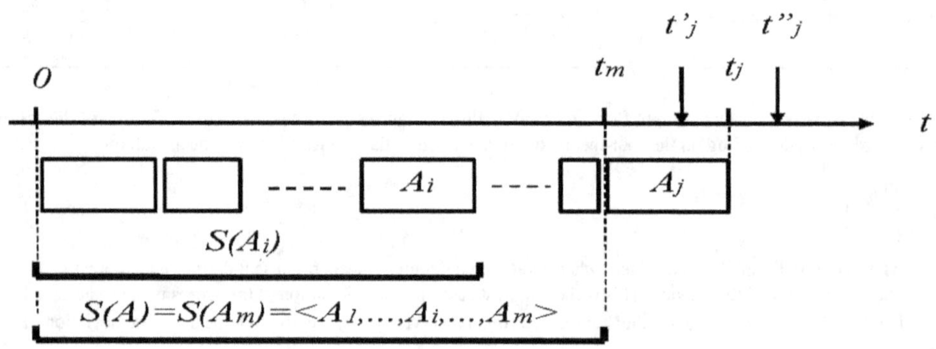

To determine a concrete limited sequence of procedures for execution $S(A)=<A_1,...,A_i,...,A_m>$ is accepted that each of them has probability personal time for realization $T(A_i)$, $i=1\div m$, which permits to determine the total time $T(S)$ for the realization of this sequence of tasks. This time has a normal distribution with average time $E[T(S)]=\sum E[T(A_i)]$ and variance $D[T(S)]=\sum D[T(A_i)]$. It can be seen from the figure that the execution of a given sequential task A_j starts at time t_m after the completion of the previous sequence. The total execution time $T(A_j)$ is a random quantity and depends on the specific execution conditions of the current calculation (data, processes, management, etc.). In this reason, the mathematical expectation for the moment $t(A_j)=t_j$ to complete the task is also a random variable. Limit values can be calculated for it – the earliest permissible moment ($t_j^{'}$) and the latest permissible moment ($t_j^{''}$). To evaluate the service, the characteristics presented below can be defined.

Earliest allowable completion time $t_j^{'}$ – it can be estimated by the variance and the standard deviation:

$$D\left[t_j^{'}\right]=D\left[t_m\right]+D\left[t_j\right] \text{ and } \sigma\left[t_j^{'}\right]=\{D\left[t_j^{'}\right]\}^{1/2}.$$

Latest allowable completion time ($t_j^{''}$) – the mathematical expectation permits to determine the assessment:

$$\left[t_j^{''}\right]=E\left[t''(A_F)\right]-E\left[t\left(S_{j\rightarrow F}\right)\right],$$

where A_F is the final task in a sequence <A_j, ..., A_F>. The variance of ($t_j^{"}$)is estimated analogously by:

$$D\left[t_j^{"}\right] = D\left[t''(A_F)\right] - D\left[t\left(S_{j\to F}\right)\right].$$

The probability $P(t_j)$ for completion of task A_j in a given time can be determined by expression:

$$P\left(t_j\right) \le t_j^{*} = \int_{-\infty}^{t^{*}} \varphi_j\left(x\right).dx$$

where $\varphi_j(x)$ is the density of the probability distribution of the random variable t_j. If $\varphi_j(x)$ is a normal distribution law with parameter $\mu = E[t_j]$ and variance $\sigma^2 = D[t_j]$, then:

$$\AE_j(x) = \frac{1}{\tilde{A}\sqrt{2\acute{A}}} e^{-\left[\frac{(x-¼)^2}{2\tilde{A}^2}\right]}.$$

The analytical investigation of servicing requires to be obtained additional assessments of stochastic characteristics. For example, f the labor intensity $\tau(A_j)$ of the task $A_j \in S(A)$ is known, then the time $t(A_j)$ will depend on the speed B of the computer system and can be calculated by $E[t(A_j)] = E[\tau(A_j) / B]$. The same is valid for the variance D.

If the complexity of the tasks is known, it is possible to determine the probabilistic assessments as follows:

- Assessments of the total labor intensity of the implementation of a selected realization of sequence in a structure $S(A)$:

$$E[\ddot{A}(S_i)] = \sum_{A_j \in S_i} E[\ddot{A}(A_j)];$$

$$D[\ddot{A}(S_i)] = \sum_{A_j \in S_i} D[\ddot{A}(A_j)] = \tilde{A}^2[\ddot{A}(S_i)];$$

- Assessment of the realization of a certain task included in a given path $S_i \in S(A)$ of the structure:

$$E[t(S_i)] = E[\ddot{A}(S_i)] / B;$$

$$D[t(S_i)] = D[\ddot{A}(S_i)] / B = \tilde{A}^2[t(S_i)];$$

- Estimation of the correlation between S_i and S_j:

$$r_{ij} = \frac{\sum_{k=1}^{n} D[t(A_k^{ij})]}{\tilde{A}[t(S_i)].\tilde{A}[t(S_j)]}$$

where $D\left[t\left(A_k^{ij}\right)\right]$ is the time variance for solving the problem A_k belonging to both structures (tasks sequences) S_i and S_j simultaneously.

REFERENCES

Ahmed, A. E., & Ibrahim, M. E. (2018). Colored Petri net models for clustered and tree-based data aggregation in wireless sensor networks. *International Journal on Information Technologies & Security*, *10*(3), 25–36.

Atlasov, I. V., Bolnokin, V. E., Kravets, O. J., Mutin, D. I., & Nurutdinov, G. N. (2020). Statistical models for minimizing the number of serch queries. *International Journal on Information Technologies & Security*, *12*(3), 3–12.

Bychkov, I., Kazakov, A., Lempert, A., & Zharkov, M. (2021). Modeling of railway stations based on queuing networks. *Applied Sciences (Basel, Switzerland)*, *11*(5), 2425. doi:10.3390/app11052425

Chen, J. C., & Wu, Y. J. (2020). Discrete-time Markov chain for prediction of air quality index. *Journal of Ambient Intelligence and Humanized Computing*, 1–10. doi:10.100712652-020-02036-5

Cheng, Q., Liu, Z., Guo, J., Wu, X., Pendyala, R., Belezamo, B., & Zhou, X. S. (2022). Estimating key traffic state parameters through parsimonious spatial queue models. *Transportation Research Part C, Emerging Technologies*, *137*, 103596. doi:10.1016/j.trc.2022.103596

Chkalova, M., & Pavlidis, V. (2020). Modeling of grain raw stuff grinding process. *Engineering for Rural Development*, *19*, 341–348. doi:10.22616/ERDev.2020.19.TF080

Cruz, F. R., Quinino, R. D. C., & Ho, L. L. (2017). Bayesian estimation of traffic intensity based on queue length in a multi-server M/M/s queue. *Communications in Statistics. Simulation and Computation*, *46*(9), 7319–7331. doi:10.1080/03610918.2016.1236953

Darmayunata, Y., Devega, M., & Yuhelmi, Y. (2023). Development of web-based single channel multi steps online queuing system with model view controller. *Sinkron: jurnal dan penelitian teknik informatika*, *8*(1), 390-397.

Dudin, A. N., Klimenok, V. I., & Vishnevsky, V. M. (2020). *The theory of queuing systems with correlated flows*. Springer. doi:10.1007/978-3-030-32072-0

Elliriki, M., Reddy, C. S., Anand, K., & Saritha, S. (2022). Multi server queuing system with crashes and alternative repair strategies. *Communications in Statistics. Theory and Methods*, *51*(23), 8173–8185. doi:10.1080/03610926.2021.1889603

Epishkin, V. E. (2023). Mathematical model of technological equipment while analyzing the capacity of car server stations. *Transportation Research Procedia*, *68*, 622–629. doi:10.1016/j.trpro.2023.02.085

Gortsev, A. M., & Nezhelskaya, L. A. (2022). Analytical investigation of a single-channel QS with incoming asynchronous event flow. *Automation and Remote Control*, *83*(8), 1200–1212. doi:10.1134/S0005117922080045

Gotovos, A., Burkholz, R., Quackenbush, J., & Jegelka, S. (2021). Scaling up continuous-time Markov chains helps resolve underspecification. *Advances in Neural Information Processing Systems*, *34*, 14580–14592.

Krak, T., De Bock, J., & Siebes, A. (2017). Imprecise continuous-time Markov chains. *International Journal of Approximate Reasoning*, *88*, 452–528. doi:10.1016/j.ijar.2017.06.012

Kuntz, J., Thomas, P., Stan, G. B., & Barahona, M. (2021). Stationary distributions of continuous-time Markov chains: A review of theory and truncation-based approximations. *SIAM Review*, *63*(1), 3–64. doi:10.1137/19M1289625

Likhttsinder, B. (2017, October). Conditional average value of queues in queuing systems with bath request flows. In *2017 4th International Scientific-Practical Conference Problems of Infocommunications. Science and Technology (PIC S&T)*, IEEE. 10.1109/INFOCOMMST.2017.8246347

Likhttsinder, B. Y., & Bakai, Y. O. (2022, March). *Development of an interval method for queue analysis in queueing systems. In 2022 Systems of Signals Generating and Processing in the Field of on Board Communications*. IEEE.

Likhttsinder, B. Y., Blatov, I. A., & Kitaeva, E. V. (2022). On estimates of the mean queue length for single-channel queuing systems in terms of statistical unconditional second-order moments of the modified arrival flow. *Automation and Remote Control*, *83*(1), 92–105. doi:10.1134/S0005117922010076

Martyshkin, A. I., Pashchenko, D. V., Trokoz, D. A., Sinev, M. P., & Svistunov, B. L. (2020). Using queuing theory to describe adaptive mathematical models of computing systems with resource virtualization and its verification using a virtual server with a configuration similar to the configuration of a given model. *Bulletin of Electrical Engineering and Informatics*, *9*(3), 1106–1120. doi:10.11591/eei.v9i3.1714

Matyushenko, S., & Ermolayeva, A. (2021). On stationary characteristics of a multiserver exponential queuing system with reordering of requests. In *2021 13th International Congress on Ultra Modern Telecommunications and Control Systems and Workshops (ICUMT)*, Brno, Czech Republic, pp. 98-103. 709.10.1109/ICUMT54235.2021.9631709

Mehta, R. (2022). Discrete-time simulation for performance modelling of FIFO single-server queuing system. *International Journal of Systems. Control and Communications*, *13*(2), 112–132.

Moiseeva, S. P., Bushkova, T. V., Pankratova, E. V., Farkhadov, M. P., & Imomov, A. A. (2022). Asymptotic analysis of resource heterogeneous QS under equivalently increasing service time. *Automation and Remote Control*, *83*(8), 1213–1227. doi:10.1134/S0005117922080057

Porshnev, S., & Korelin, I. (2017). Non-stationary single-channel queuing system features research in context of number of served queries. *ITM Web of Conferences (2017 Seminar on Systems Analysis) 10*, 03006. DOI: 10.1051/itmconf/20171003006

Privault, N. (2018). *Discrete-time Markov chains. Understanding Markov Chains: Examples and Applications*. Springer. doi:10.1007/978-981-13-0659-4

Romanov, O., Nikolaev, S., & Orliuk, Y. (2022). Radio monitoring complex model as multi-phase queuing system. *Radioelectronics and Communications Systems*, *65*(3), 155–164. doi:10.3103/S0735272722030050

Shirsat, A., Elamvazhuthi, K., & Berman, S. (2020, November). Multi-robot target search using probabilistic consensus on discrete Markov chains. In *2020 IEEE International Symposium on Safety, Security, and Rescue Robotics (SSRR)* (pp. 108-115). IEEE. 10.1109/SSRR50563.2020.9292589

Staffolani, A., Darvariu, V. A., Bellavista, P., & Musolesi, M. (2023). RLQ: Workload allocation with reinforcement learning in distributed queues. *IEEE Transactions on Parallel and Distributed Systems*, *34*(3), 856–868. doi:10.1109/TPDS.2022.3231981

Sukhorukova, I., & Chistyakova, N. (2020). Methodology for the formation of a special course on applications of Markov processes. *Revista ESPACIOS, 41*(09).

Tsitsiashvili, G. (2021). Alternative designs of high load queuing systems with small queue. In *Informatics and Cybernetics in Intelligent Systems: Proceedings of 10th Computer Science On-line Conference 2021,* Vol. 3 (pp. 69-76). Springer International Publishing. 10.1007/978-3-030-77448-6_8

Uryvsky, L., & Martynova, K. (2019). Complex analytical model of priority requires service on cloud server. In *2019 International Conference on Information and Telecommunication Technologies and Radio Electronics (UkrMiCo)*, Odessa, Ukraine. 10.1109/UkrMiCo47782.2019.9165323

Zeifman, A., Korolev, V., & Satin, Y. (2020). Two approaches to the construction of perturbation bounds for continuous-time Markov chains. *Mathematics*, *8*(2), 253. doi:10.3390/math8020253

Zhernovyi, Y. V. (2018). Calculating steady-state characteristics of single-channel queuing systems using phase-type distributions. *Cybernetics and Systems Analysis*, *54*(5), 824–832. doi:10.100710559-018-0084-2

Zoppo, G., Korkmaz, A., Marrone, F., Palermo, S., Corinto, F., & Williams, R. S. (2021). Analog solutions of discrete Markov chains via memristor crossbars. *IEEE Transactions on Circuits and Systems. I, Regular Papers*, *68*(12), 4910–4923. doi:10.1109/TCSI.2021.3126477

Zyulkov, A., Kutoyants, Y., Perelevskiy, S., & Korableva, L. (2022, December). Single channel queuing system utilization factor model. []. IOP Publishing.]. *Journal of Physics: Conference Series*, *2388*(1), 012043. doi:10.1088/1742-6596/2388/1/012043

Chapter 10
Organization of Queuing Analysist of Complex Systems

ABSTRACT

The purpose of this chapter is to present approaches and solutions in organizing the study of complex systems using queuing systems (QSs) and networks (QNs). For the realization of this, specific problems are presented and their resolution is proposed. In the first part, the object of the model study is a multi-computer system with a cube architecture. A graph formalization of the inter-process communications, realized through message exchange between processor nodes, is presented, which is used for development of a queuing model of processor node. The second part is dedicated to priority QSs whit discussion of various applications of queuing modeling and presentation of concrete examples for single-channel priory QS and for priority multiprocessor system with two processors modeled by queuing multi-channel QS and investigated by using Markov chain with continuous time. The third part deals with QNs as combinations of different QSs. The stochastic workload modeling is the object of the last part.

1. MODEL INVESTIGATION OF CUBE MULRICOMPUTER ARCHITECTURE

Multiple information processing is associated with the Multiple Instruction streams Multiple Data streams (MIMD) architectural form, where multiple processors independently and relatively independently process different tasks or subtasks (Massini, 2017). In this type of architecture, coarse-grained parallelism is mainly supported, which introduces the essential difference compared to other architectures (Wang, 2021), such as the synchronous parallel environments of type SIMD (Single Instruction stream Multiple Data streams), and pipeline systems type MISD (Multiple Instruction streams Single data stream). The main components of the MIMD-architecture are the following.

- A set of independent processors $P = \{P_i / i = 1 \div n\}$ – a set of homogeneous or heterogeneous processor cores, determining with their hardware and system parameters the time for processor calculation (Processing Time – *PT*).

DOI: 10.4018/978-1-6684-8947-5.ch010

- Multiple memory modules $M = \{M_j / j = 1 \div m\}$, making up the system memory – individual memories can be physically connected to the respective processors, but are logically treated either as common (Shared Memory MIMD) or as distributed memory (Distributed Memory MIMD).
- Set of Input-Output devices $I/O = \{I/O_k/k = 1 \div q\}$ – make up the system periphery, including distributed devices to individual computing nodes (local) or generally available devices with collective access.
- Interconnection network (IN) – based on the organized physical connections, the possibilities for logical connection between the processes are defined, which affects the communication time (Communication Time – *CT*).

The total execution time (*ET*) of a set of tasks in a MIMD-environment is determined by the computing power of the processors, the strategy for allocating tasks to processors, the level of possible parallelism for individual tasks or the presence of information dependence between them, as and from a number of parameters of the connection network (bandwidth, the type of connection of individual components and the way of organization of connection channels, management of communication processes, etc.). In the general case, $ET = PT + CT$, and when developing MIMD systems, minimization of the total time *ET* is sought regarding the average value of the set of processed tasks $min\{ET(k)\}$. There are two main approaches to reducing communication losses presented below.

- At the expense of optimal distribution of the solved task (code and data) on the processors and/or on the memory modules (object of research in the construction of the parallel algorithms).
- At the expense of increasing the speed of operation of the IN itself (it is realized either through hardware means with the use of faster-acting logic, or with an appropriate structural organization of the communication network).

To build an effective parallel architecture, the second approach is of greater interest, but in practice a common balance between the two approaches is sought for building a suitable interconnection network.

Depending on the organization of interactions between processes (inter-process interactions) and the type of coupling from IN, two types of MIMD architectures are defined below (Stone, 2015).

- Shared Memory MIMD Architectures (SM MIMD), also known as multiprocessor architectures. They are characterized by a highly connected topology, logical separation of processors from memory modules and synchronization of processes through common variables, semaphores, monitors (UKEssays, 2018).
- Distributed Memory MIMD Architectures (DM MIMD), considered as multicomputer systems because the individual processor nodes are a typical computer configuration (processor, memory, peripherals and communication processor unit). The topology of IN is loosely connected (relative independence of connections), with the processes in individual nodes being independent and communicating with each other by exchanging messages according to a relevant protocol (Habib et al., 2017) – Figure 1.

Figure 1. Base model of communications in DM MIMD architecture

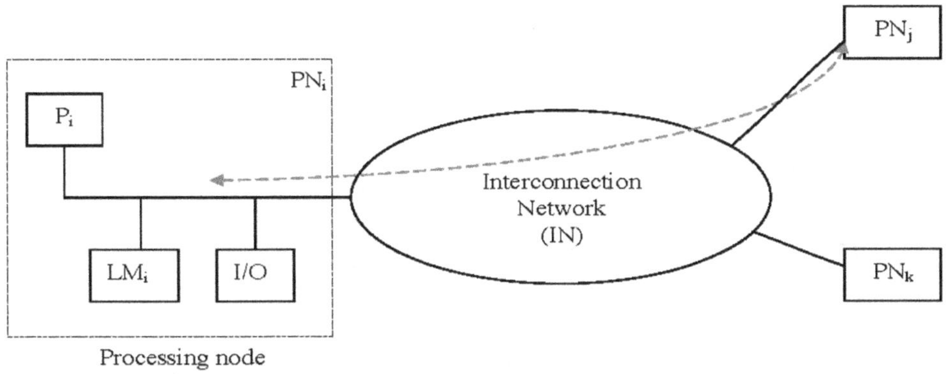

The main difference between the two architectural models lies in the way communication is organized in the set of active processes and the memory access possibilities. With SM MIMD, it is possible for multiple processes to access shared memory, which requires ensuring precise synchronization of shared objects. In DM MIMD, each processor has its own memory, and communication between different processes is reduced to exchanging messages through system management software (Message Passing System), which supports a corresponding exchange protocol, most often 'handshaking'. In the DM MIMD architecture, each processor has its own physical address space, in contrast to the global address space of SM MIMD, as a graphical representation of communications between two processes (in different processor nodes) is presented in Figure 2. The states are as follows: **E** (execute) – execution of a task, allowing the activation of several processes (E^1, E^2) depending on the interruptions; **S** (send) – send a message; **W** (wait) – waiting even realization; **R** (respond) – forming and sending a response. The communication protocol supports the following sequence.

(1) Process (1) needs additional information from another process and notifies the system kernel, which interrupts it and from state E^1 is passed to state S (start of communication).
(2) The MPS establishes a connection over the physical lines and sends a message *msg*(1) over the formed channel to the receiving node's buffer.
(3) The system kernel (dispatcher) activates a new task (starts another process), with which control is transferred to state E^2.
(4) The message *msg*(1) has arrived in the input buffer of a node executing process (2) and is in state W (waiting) until it becomes possible to interrupt the process and process the received message - transition to state E.
(5) As a result of the processing, a response *msg*(2) is formed – state R, which is returned to the sender (enters its buffer – state W).
(6) If possible, process (1) is restored and continues its execution based on the received information.

Figure 2. Graph formalization of communication between processes

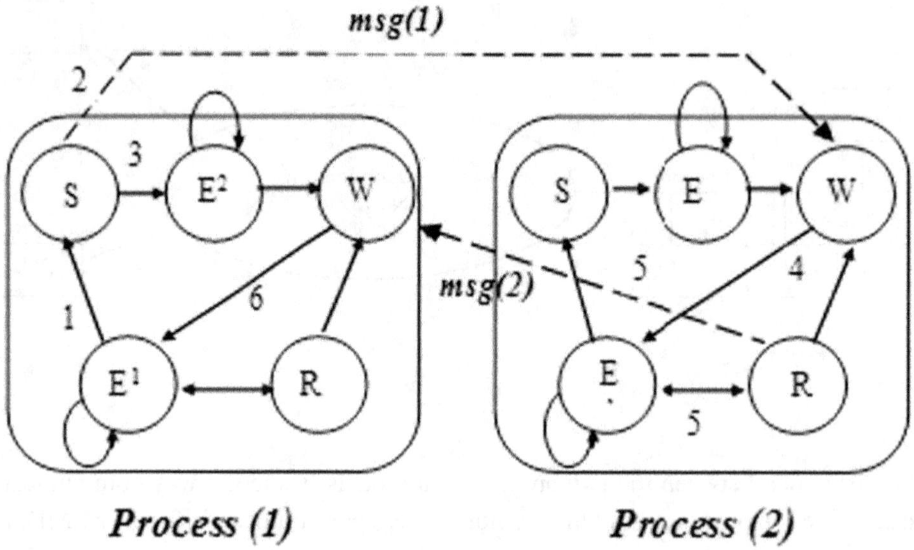

Cube multicomputer architecture is DM MIMD type and support N-dimensional parallelism and connection of $n=2^N$ processor nodes. Theoretically, the number of nodes can vary depending on the dimension ($N = 0, 1, 2, 3, \ldots$), which is illustrated by the variants of Figure 3. Actually technically feasible for the topologies at N=3 (Boolean cube) and N=4 (Hypercube).

Figure 3. Modification based on n-2^N nodes (N=0÷3)

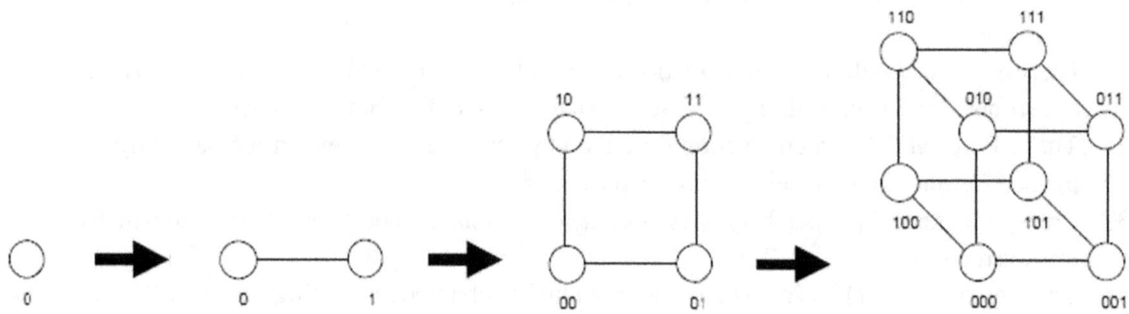

The object of study is an architecture with a cube/hypercube topology, providing connection between processor nodes through $N.2^{N-1}$ connection lines (arcs) at a maximum communication distance $\chi=F(n)=N$. Each processor node is assigned a unique index (binary address), and in a binary sense all neighbors have a Hamming distance of 1 – Figure 4. This also determines the message routing strategy – routing in an order determined by the number of distinct bits in the binary address. As the number of nodes increases, the number of links for each node grows proportionally to N, with the average distance being $(N.2^{N-1})/(2^{N-1})$.

Figure 4. Communications in Boolean cube

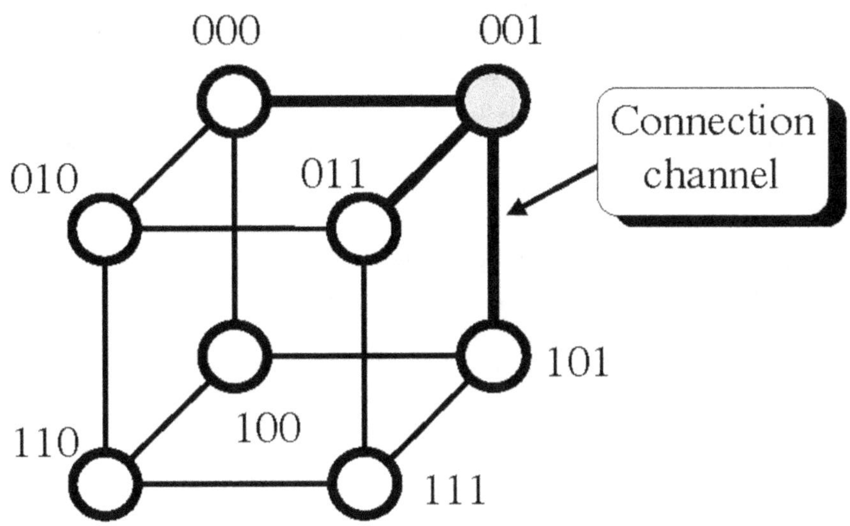

Processor nodes for a multicomputer architecture with a Boolean cube (bCube) topology have individual numbers from 0 to 7, with communication management determined by the Hamming difference of their binary codes. The communication algorithm can be formalized as follows:

FOR j=0 TO N
IF $s_j \neq d_j$ THEN <transmission $PN_S \rightarrow PN_D$>

When organizing the model study, the following working variables were defined:

- $N = 3, 4, \ldots$ – dimension of the cube architecture.
- $n = 2^N$ – total number of processor nodes.
- $I = \sum_{q=0}^{N-1} \tilde{A}_q.2^q$ – Boolean identifier number of a processor node $I=<\sigma_{N-1},\ldots,\sigma_1,\sigma_0>$; $\sigma_q \in \{0,1\}$.
- $i \in \{1,2,\ldots,n\}$ – decimal processor node number equivalent to I ($\rho:I \rightarrow i$). It is assumed that 'i'='I+1' to preserve the situation 'i=0' for indication inside a processor node.

For the formalization of the processes in the research object when building the mathematical model, model parameters are defined as follows:

S – Decimal source node number (PrN_S) to direct a generated request from the input stream to a corresponding node $\Rightarrow S \in \{1, 2, \ldots, n\}$. After initial processing in PrN_S, the parameter is reset (S=0) for the next indication on re-entry to the same node.

CV – A control vector of length N bits, which is formed as an EXOR between the source number (I_S) and the receiver number (I_D), i.e. $CV=I_S \oplus I_D$. It is good to maintain N such attributes $\{CV_1, CV_{2,\ldots,}$

CV_N}, which at startup receive random values $CV_j \in \{0,1\}$, with the situation $CV_1 = CV_2 = \ldots = CV_N = 0$ marking reaching a sink node (PrN_D).

D – Auxiliary parameter specifying current transmission request (when *D*=1) or current request reaching receiver (*D*=0) then subject to destruction $D \in \{0,1\}$, and $D = CV_1 \vee CV_2 \vee \ldots \vee CV_N$).

T – Management of the internal routing of requests in a node, where *T*=1 the request is directed to the execution processor (EXECUTE) and when *T*=0 it is directed to the communication processor (TRANSMIT). It is determined by the expression $T = (S \neq 0) \vee (D = 0)$.

Formal flow model of the processes in designed queuing model is presented in Figure 5 and the main process object are summarized in Table 1.

Figure 5. Process flow model

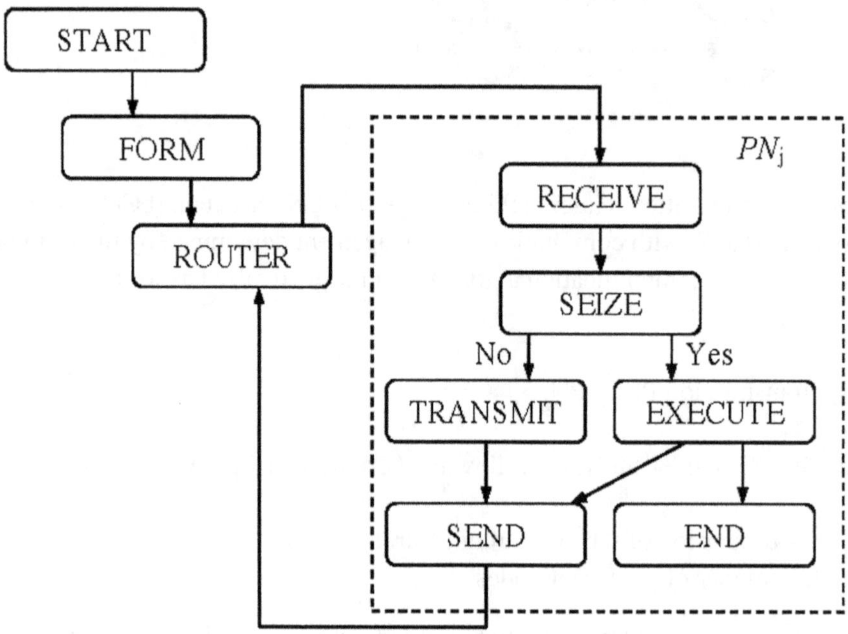

Table 1. Main model objects

START	Generate an input flow of requests that is routed (Message Allocation) to PrN_i (i=1,2,…,n);); each request is assigned an attribute $S\in\{1,2,\ldots,n\}$and a parameter D=0 is defined.
FORM	Formation of control vector $CV=<CV_1, \ldots, CV_N>$; $\forall CV_q\in\{0,1\}$.
ROUTER	It implements the operations: (a) selection (Message Allocation) – by directing the input stream depending on S; (b) communications (Message Transferring) between PrN_S and PrN_D after a basic request received in PrN_S and already processed.
RECEIVE	Incoming a request into a PrN and forming internal routing attributes: $D= CV_1 \vee CV_2 \vee \ldots \vee CV_N$ and $T = (S\neq0) \vee (D=0)$.
SEIZE	Allocates the flow of requests execution or their redirection in communication network.
EXECUTE	Current request processing.
TRANSMIT	The current request transferring through the buffer of PrN.
SEND	Selection of output ports (by CV analysis) for transmitting the request to a neighboring node.
END	Serviced request

A simple algorithm for realization of the operation 'SEND' is as follows:

```
procedure SEND(CV, q)          q:=1;          while CV[q]=0    do
q:=q+1;           CV[q]=0;<transfer by using OUT[I,Jq], q=1÷N>;end;
```

The model of a selected processor node as a queuing service system is presented in Figure 6. It support the functionality from Figure 5 and uses the object presented in the Table 1. It can be regarded as a basic single-channel QS for construction of queuing network described servicing in a cube architecture regardless of the dimension N and number n of processor nodes.

Figure 6. Queuing model of a processor node

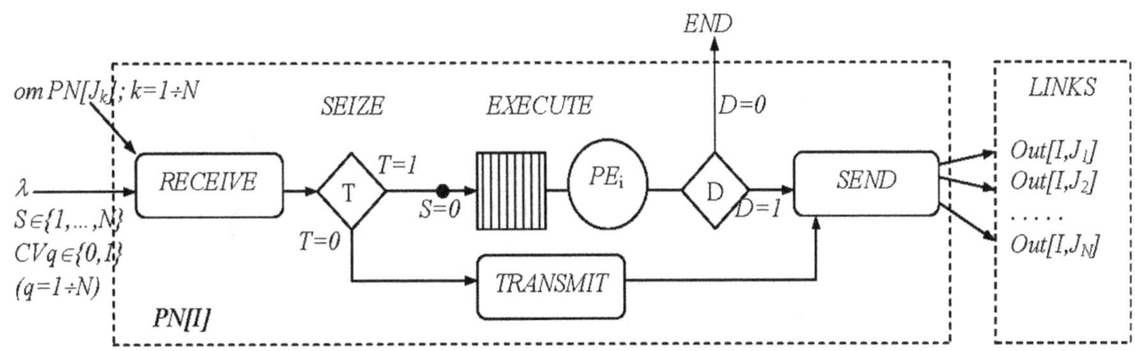

The real processing of each task in the multicomputer system, uses processor time (*PT*) and additional communication time (*CT*) are spent to transfer the messages between the nodes. To account for *CT*, *LINKS* the communication channels *LINKS* are defined as resources with a specific message transfer time. This ensures that at a given time only one request will be able to use the communication channel on a corresponding $OUT[I,J_q]$ port.

The workload model is based on generated input workflow which is a probability flow of requests. Work parameters are set for individual requests, which are analyzed when adjusting the flow of service in the processing nodes and turning to the basic resources (processors and communication channels).

Processing takes place in the processing node to which the request is directed and after traversing a possible route (if communication is required). In this situation, a transaction received for processing is assigned the lower priority, which will allow processing to be interrupted when a redirect (routing) request is received on the same processor. After the final processing of a request, the transaction is removed from the system (destroyed).

In order to avoid a possible endless waiting ("deadlock"), it is possible to assign priorities to requests. A higher priority (1) is assigned to newly generated queries to be analyzed and routed to the processor that will process them, and a lower priority (0) to queries computed in the respective processor. This sets QS as priority, which is discussed in the next section.

2. PRIORITY QUEUING SYSTEMS

Traditional QSs have FCFS-type queuing service discipline. In many cases, however (network service, operating systems, etc.), priorities are applied when servicing the real objects. They can be different (for example, depending on the type of network traffic). In the field of computer networks, an important point is the determination of priorities depending on the average service time. Often requests with less expected service time are given higher priority. For example, a router may assign a higher priority to a stream of digital speech packets than to a stream of data packets. With such a dispatching scheme, higher network traffic performance is achieved. When modeling such objects, QS with priorities is applied when servicing requests from the queue (buffer).

A priority QS for modeling a processor that serves k different classes of requests is presented in (Xu & Gautam, 2020). Workflow is a random flow with Poisson distribution and exponential flow of service. Object of investigation are two scenarios where the size of the buffer for each queue is unlimited, but the strategies for service are different – FCFS and LCFS (Last Come First Served). The purpose of the investigation is to determine the effect of prioritization in the two service disciplines for the different priority classes of queues. The use of priority QS has been implemented in (Yadavalli et al., 2021) in modeling the distribution of ships in a port when searching for a suitable docking or berthing location. Different policies and priority scenarios in docking management (incoming different service requests) are considered. The objective is to optimize the management with maximum allocation to the berths and to evaluate the efficiency of the system by evaluating the service parameters. The model considers a system with integrated congestion management and steady-state behavior of the expected service time. Experiments were conducted to evaluate waiting time, queue length, workload of the service unit and to determine the optimal number of units needed for servicing different priority levels.

Another aspect of priority service is discussed in (Ritha & Yasodai, 2021), where a priority QS with "*Preemptive-Resume priority of a fuzzy queuing model with Erlang service distribution*" is investigated. The parameters of the incoming request and service flows are represented by triangular fuzzy numbers, and the main idea is to use a suitable ranking algorithm to represent the fuzzy values as clear numbers.

Expressions for priority single-channel QS with two classes of flows with different service times are presented below. The following assumptions are made for their determination.

- The input flow of service requests has intensity with a Poisson distribution.
- Requests with priority "1" are served before those with priority "2".
- In case of equal priorities, the FCFS discipline is applied.
- No interruption of started service of a request.
- No requests are removed from the queue.

$\lambda=\lambda_1+\lambda_2 \rightarrow \rho=\rho_1+\rho_2$, where $\rho_1=\lambda_1.T_{S1}$ and $\rho_2=\lambda_2.T_{S2}$;

$$T_S = \frac{\lambda_1}{\lambda} T_{S1} + \frac{\lambda_2}{\lambda} T_{S2};\ T_R = \frac{\lambda_1}{\lambda} T_{R1} + \frac{\lambda_2}{\lambda} T_{R2}.$$

For exponentially distributed service flow intensity, the following expressions are valid:

$$T_{R1} = T_{S1} + \frac{\rho_1.T_{R1} + \rho_2.T_{R2}}{1-\rho_1} \text{ and } T_{R2} = T_{S2} + \frac{T_{R1} - T_{S1}}{1-\rho}.$$

Example: A network stream of short and long data packets is transmitted between the nodes of a packet-switched network. The intensities for the arrival of packets of both types are the same. The lengths for the two types of packets are assumed to be exponentially distributed with mean lengths of 80B (short packets) and 800B (long packets), respectively, and the link bandwidth is 64 Kb/s. Let the average service times of short and long packets be 0.01s and 0.1s, respectively, and the arrival rate of each type be 8 packets per second.

For the priority service system, a higher priority is assigned to short packets. The ratings are as follows:

$\lambda=\lambda 1+_{\lambda}2=8_{+}8=16$; $\rho 1=8._{0},01=0.08$; $\rho 2=8.0\,1=0.8$; $\rho=0.88$;

$$T_S = \frac{8}{16},0.01+\frac{8}{16},0.1 = 0.005+0.05 = 0.055;$$

$$T_{R1} = 0.01+\frac{0.08,0.01+0.8,0.1}{1-0.08} = 0.098;$$

$$T_{R2} = 0.1+\frac{0.098-0.01}{1-0.88} = 0.833;$$

$$T_R = \frac{8}{16}0.098+\frac{8}{16}0.833 = 0.4655s.$$

The application of priority service embraces various technology areas. For example, Wireless Fidelity (Wi-Fi) broadband network technology has seen major developments in wireless networks, mainly regarding the requirements to increase transmission speed and coverage. As stated in (Najim et al., 2022), various Wi-Fi hotspots are available everywhere through the global network, but the Internet cannot provide the required quality of service. In order to determine the level of service efficiency, the article examines the three main mechanisms – First-In-First-Out (FIFO), Priority Queuing (PQ) and Weighted Fair Queuing (WFQ) (service strategies PQ and WFQ use priority service at traffic classification). The following different scenarios were applied in the study: (1) application of the strategies without service discipline; (2) application of FIFO; (3) application of PQ; (4) implementation of WFQ. Experimental results have shown that priority service (and especially with WFQ) gives better queuing estimates than others.

Another study considers query servicing in a single-channel QS with a queue supporting both M/G/1 and M/D/1 principles (Madan, 2011). The formed input workflow contains two independent Poisson-distributed streams containing priority and non-priority requests, respectively. The rule of priority service without preference is applied. For priority requests the server applies general type service, while for non-priority requests – deterministic service. For the second type of tasks (requests), it is assumed that servicing can be suspended indefinitely until it has finished servicing all queued priority tasks. The purpose of the study is to determine the steady state parameters, including the probability distribution of the queue size.

Priority multi-channel QS with two types of requests is the object of investigation in (Lee et al., 2022). Each of the two request types has a designated server to serve, and QS also includes additional servers for general use. Requests of type 1 have service priority in flexible usage depending on the current number of requests of this type in the queue (infinite buffer). Type 2 requests have no input buffer and leave the QS unserved if there is no free server when they arrive. On the basis of the formulated QS, an analytical model was developed as a multidimensional MC to study the dynamics of service with a fixed total number of servers and a fixed number of reserved servers. For modeling the service process, a correlated model of input flows and mechanisms of servicing and rules for denying type 2 requests was chosen.

An analogous case of priority multi-channel QS with different types of requests is discussed in (Klimenok et al., 2020). The input flow is Markovian and includes two types of registers which differ in priority and service time. The structuring of QS is consistent with the requirements for modeling information transmission processes in telecommunication networks, where the general information flow is a complex of several types of flows with correlation dependence of arrival times. Analytical modeling of the QS is done using a multivariate MC with an external query generator, for which the steady state is investigated and the system performance estimates and the loss probabilities of different query types are calculated. The QS outflow is also analyzed, examining the importance of correlation in the arrival process modeled with a Poisson flow. A continuation of the study is presented in (D'Apice et al., 2022), considering a "*priority queuing model with many types of requests and restricted processor sharing*". A new discipline of service with limitation of the capacity of the service unit and the number of simultaneously served requests in QS is proposed, which the authors define as "*hybrid of the traditional discipline of service in a multi-server system and the discipline of the limited processor sharing*". Requests with a higher priority are served with priority over those with a lower one, and to determine the number of concurrently served requests, an MC model is defined and analyzed at any fixed set of QS parameters.

Priority servicing based on multi-channel QS with Poisson input flow is discussed in (Tatashev et al., 2022). Request service time is randomly distributed depending on priorities, but the special thing is that requests with the same priority are served according to the LIFO (Last In First Out) strategy. If, upon the arrival of a new request, all servers are busy, the priorities of the requests served on them are analyzed, and if at least one request with a lower priority than the new one is found, its service is interrupted in order to activate the service of the new request with a higher priority. The service of the interrupted request is restored at a later time, and one of the evaluations in analyzing the model is the dwell times of the requests in the QS.

An example for reliability investigation of parallel dual processor structure with shared memory is discussed below. It is assumed that the tasks (requests) have two priorities 'Pr1' and 'Pr2' for determining the number of processor which must process the current task (Figure 7).

Figure 7. Structure of the investigated system

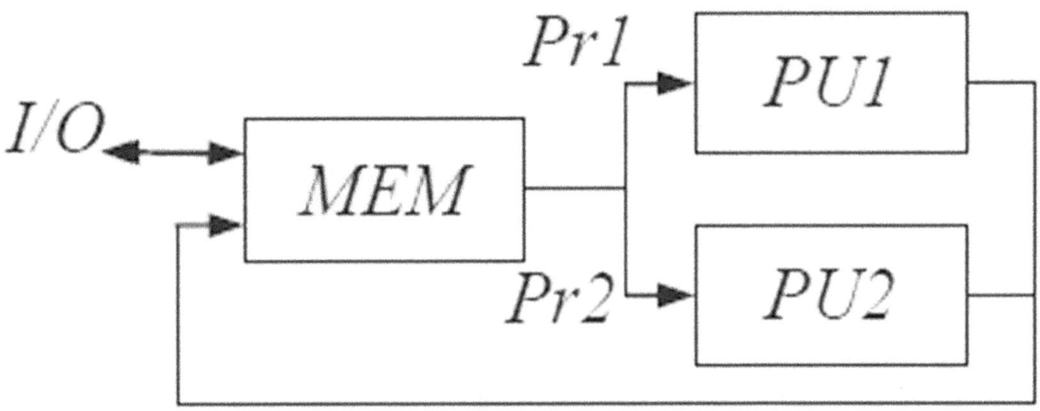

In case of failure of one of the two processors, the servicing of the requests is taken over by the working one, but under the following conditions:

- If the PU1 is faulted the other processor PU2 will process only task with priority 'Pr1' and other task with priority 'Pr2' will not be processed.
- If the PU2 is faulted the other processor PU1 will continues processing of own type tasks with 'Pr1' and any task with 'Pr2' will be processed if this is needed for execution of high priority task.

The model investigation is made based on multi-channel QS presented as a MC with input workflow formed by two type requests. The stochastic model has three states S = {A, B, C}, which describe system availability for processing (A), task execution (B) and failure in the system (C). The following intensities are determined which permit definition of the transition matrix: χ:A→B (input request and processing start); ν:B→A (task processing is completed); λ:A→C (intensity of failure stream); β:B→C (fatal failure of processing); μ:C→A (restoration of the system). The vector of initial probabilities is $P_0 = (1, 0, 0)$ and the matrix of transition probabilities (intensities) and the analytical MC-model are presented in Figure 8.

Figure 8. Analytical definition of MC-model

P	A	B	C
A	0	χ	λ
B	ν	0	β
C	μ	0	0

$$\nu . p_A + \mu . p_C - (\chi + \lambda) . p_A = 0$$
$$\chi . p_A - (\nu + \beta) . p_B = 0$$
$$\lambda . p_A + \beta . p_B - \mu . p_C = 0$$
$$p_A + p_B + p_C = 1$$

The analytical solution of the model provides the final probabilities for the steady state regime in the sequence below.

$$p_B = \frac{\cdot p_A}{(+\beta)}$$

$$p_C = \frac{(+\beta)+\beta}{\mu(+\beta)} \cdot p_A$$

$$p_A = \frac{\mu(+\beta)}{(+\beta),(\mu+)+(\mu+\beta)}$$

$$p_B = \frac{}{(+\beta),(\mu+)+(\mu+\beta)}$$

$$p_C = \frac{(+\beta)+\beta}{(+\beta),(\mu+)+(\mu+\beta)}$$

Let the p_1 and p_2 are the probabilities for trouble-free work of processors PU1 and PU2 and based on that the states A and B are working states the probability for correct worked structure can be defined as follows:

$$p_1 = p_2 = p_A + p_B = \frac{\mu(+\beta+)}{(+\beta),(\mu+)+(\mu+\beta)}$$

An example for numerical solution of the model is presented below based on the following assumptions:

- The relation between states “availability” and “execution” as 4:3.

- The failure do not depend on the current state of processing in the structure and their intensity is 1/100 part of the execution intensity.
- The value of restoration time is 10 times less than the time for execution of task.

These assumptions permit to determine values $\chi = 75$, $\nu = 100$, $\lambda = \beta = 1$, $\mu = 10$, and to calculate the assessments: $p_A = 0{,}522$; $p_B = 0{,}387$; $p_C = 0{,}091$; $p_1 = p_2 = 0{,}909$.

3. QUEUING NETWORK (NETWORK OF QS)

Queuing networks (network of queuing systems) are combination of different QSs with separate service of input flows of requests which are transmitted between nodes. Each QS has own stochastic parameters of service and structure of the buffer (servicing discipline – FIFO, priority, etc.). This apparatus is very suitable for model investigation of real systems like computer network configurations, parallel computer systems, multiprocessor or multicomputer structures, etc. Such a model is a collection of interconnected individual QSs (single-channel or multi-channel), which represent separate resources in the investigated object. IGI Global gives the follows definition of Queuing network[1]: "*A modeling technique for representing resource sharing systems, which consists of a collection of service centers that represent the resources for providing services to a collection of customers that represent tasks*".

Supporting the above are numerous publications that define QNs as a suitable apparatus for studying various processes of a stochastic nature developing in a discrete environment. Such are the telecommunication mechanisms, which can be reduced to connections between two data exchange points. Possibilities for conducting research in this area based on queuing theory are discussed in (Afolalu et al,, 2021). The review examines the practical application of the stochastic queuing approach for network communication optimization. As a basis, it is stated that modern network structures represent clusters of nodes for processing information and directing the transmitted units (packets) along the formed route (virtual channel). To carry out these communications, initial access is usually made to link-maintaining units. The survey seeks an answer to the question of what is the optimal number of these units, given four defined indicators for an efficiency index. The research itself confirms the importance of queuing theory and QNs in particular as a tool for optimizing telecommunication systems.

It is possible to define several types of QN.

- Open QN where requests come from an external source and after service leave the network. The input stream may be directed to the first QS in the QN, but external requests may also be made to other QSs from the network. Each request arriving at a QN can visit several different QSs, as well as return for re-service in an already visited queue (see Chapter 5, Figure 29).
- Closed QN serves a fixed number of requests that move between individual QSs with no external sink and no external source. This determines the internal generation of requests that form the workload of the QN. If it generates a new request, so the total number of requests remains constant (see Chapter 5, Figure 29).
- QN with losses and request limitations – a variant of open QN with a shared pool of united queues to which requests from an external source are routed, then a server form QN "chooses" a request for servicing. If the system is full, an incoming request is lost. An example is presented in Chapter

9, Figure 3 where denial of service (request loss) occurs on busy servers. Another possibility is that the point of failure (request loss) is in front of the buffer (if it is full).

- Mixed QN, which is any combination of previous types. For example closed QN designed by using separate open QS – Figure 9.

Figure 9. An example for mixed QN

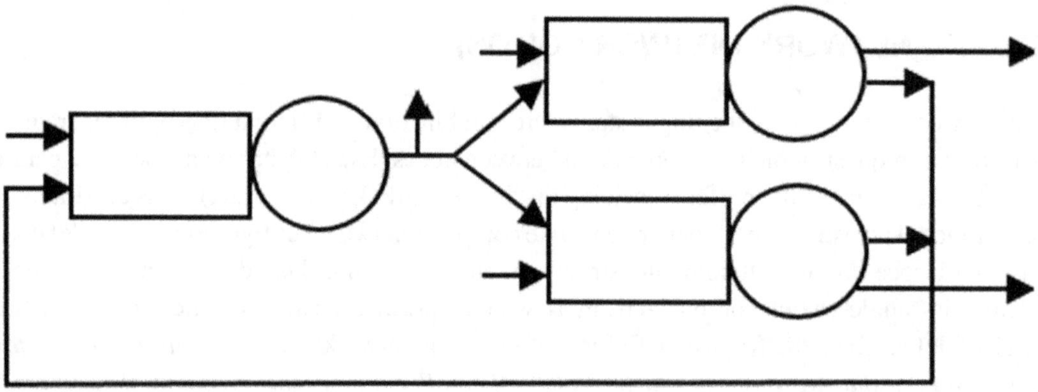

The traditional ***open QN*** finds quite a large application in the organization of researches, especially if they are related to network communication processes and virtualization of resources. The latter greatly facilitates the scaling of software services, which is related to determining the necessary amount of resources (resource sizing) to achieve the desired performance under a given load. An effective solution to this problem is modeling the performance of the system as an analytical model based on open QN with notification G/G/m is proposed in (Prados-Garzon et al., 2019). It is intended to evaluate the response time of software network service, and the created model is validated experimentally in a test platform equivalent to a typical real-world virtualization environment of a data center. The QN Analyzer and mean value analysis methods were used to solve the developed QN, which allowed the estimation error to be determined.

QN with losses is used for model investigation of multi-server systems in (Naumov & Samuilov, 2018). It is accepted that a request occupies a resource with a given distribution function in dependence of their class. The service of a request at a given node (QS) may be interrupted by another request arriving with exponential time from the start of processing, and it is possible that the interruptions may be more than one. The incoming request streams for the individual QSs are Poisson distributed. Estimates were calculated for key features of the model, including the joint distribution functions of the number of requests at individual nodes and the level of occupied resources.

One of the discussed QN problems in systems with incoming requests is getting overflow which leads to denial of service. This is a significant problem for high-priority queues, where service may be available on an idle (free) server, while on fully busy servers, an incoming request will not be joined to the main queue. In this direction is the study of two heterogeneous queuing networks presented in (Saritha et al., 2022). Incoming requests are Poisson distributed with exponential service time in the server nodes, which are ranked by levels – from lower to higher. A model based on the queuing network to investigate processes in a multi-airport system is discussed in (Zhao et al., 2022). Point stationary queues are used, analyzing the changes of non-stationary queues based on the utilization level for the request flow. Additionally, regression analyzes were performed to determine the relationship between arrival and departure performance for the various airports (inflows and outflows for the individual QSs of the network). Another aspect of the queuing network is discussed in (Stankevich et al., 2022), where the modeling is done with an open QN with a Poisson input flow, an infinite queue for each server, and an exponential service law. What is different about this model is that the requests are served in batches, and an algorithm is designed to determine the size of the batches with minimum average network response time. An analytical expression for the average time for presence in each node (T_R) is defined for the operation of the algorithm as initial data.

Some authors define ***semi-open QN*** as a special type of QN consisting of two separate parts: "*an inner network with a population constraint and an external queue to accommodate jobs whose entrance is delayed*" (Jia & Heragu, 2009). The research presented in the article describes two stages. The first is related to the study of semi-open QN for a class of tasks with tandem configuration. The second makes an extension for multiclass configurations of general and special type. The proposed method combines the matrix-geometric method with the decomposition-aggregation approach, and the obtained numerical results have shown that the desired accuracy and efficiency have been reached. The applicability of the semi-open QN apparatus is also confirmed in (Roy, 2016), where it was used in the investigation of resource size and the minimization of request waiting time probabilities. It is specified that the determination of the waiting time in an external queue to a corresponding resource is important for the evaluation of the parameters of the service of the requests and the performance of the system. Solving this task is supported by the capabilities of semi-open QN to separate incoming requests (transactions) from network resources with the help of a synchronizing station (semaphore queue). This allows for more precise determination of transaction latency with better definition of the network model. As a result of the study, several stochastic models for real systems using semi-open QN have been developed.

An extension of the latter study was done in (Dhingra et al., 2018) and (Kumawat & Roy, 2021). In the first paper, it is pointed out that the methods developed for semi-open QN with constant demand flow intensity λ cannot provide a fully efficient evaluation of the system performance, and an efficient solution with changing input flow intensity is proposed. A Markov modulated Poisson process was used to characterize the variability of λ and a QN solution approach based on a matrix-geometric method was developed. Numerical experiments were performed to validate the method. The second article examines the application of multi-stage semi-open QN and the possibility to decompose it into single-stage semi-open QNs. This will allow to make evaluation of individual single-level parts and a two-moment approximation approach is proposed to estimate the coefficient of variation of time between individual service terminations. The output flow of a single-stage network can be served as the input flow to connect multi-stage semi-open QNs. The robustness of the approach is tested through numerical experiments, setting different settings of input parameters for single and multi-class problems. The results showed good estimates, especially with a coefficient of variation for admission times (intensity λ) less than 2.

An example for network of QSs is shown in Figure 10. Three types of flows are defined: input flows (Φ_{IN}) of requests entering service, internal flows (Φ_{IJ}) of requests circulating between individual QSs, and output flows (Φ_{OUT}) of requests leaving after final service in the network. The complexity of the model is determined by the presence of splitting and merging of request streams, as well as the possibility of creating chains of consecutive queues (with sequentially connected QSs).

Figure 10. An example for queuing network

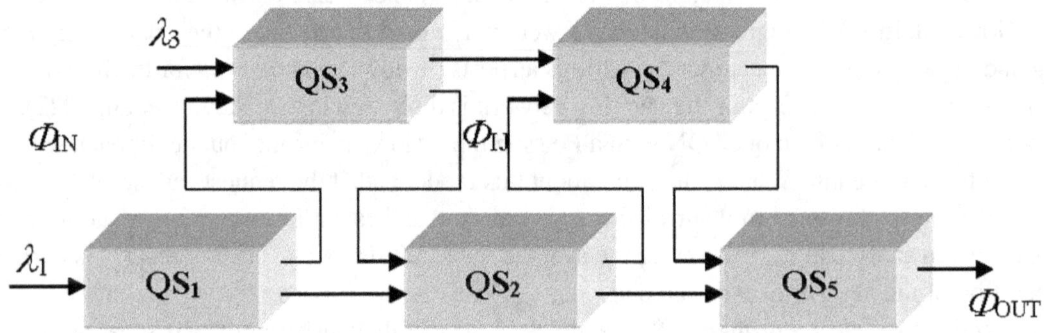

A queuing network is defined by the following parameters:

- Intensities of input flows (of request sources);
- Structure of connections in the network;
- Transition probabilities between individual QSs;
- Individual parameters for each QS of the network composition (number of channels, service discipline, service flow intensity, priority, etc.).

There is no general methodology for analyzing queuing networks, but if the traffic in the network obeys a Poisson distribution law and the service time is exponentially distributed, there is a relatively simple solution. Analytical modeling of computer systems and processes by using QN has some disadvantages:

- Difficult to use simple analytical expressions.
- In a number of cases it is necessary to use numerical methods.
- Difficult to reflect memory usage, switches and non-uniform flow of requests.

QN process management usually relies on the precise definition of the values for the model parameters, mainly paying attention to the arrival (λ) and service (μ) intensities. These parameters can easily be transformed into time values for the two stochastic flows. It is a problem if such data cannot be easily determined, and it is possible that they change over time. One solution to this problem is given in (Xie & Jin, 2022), where control policies that are independent of the data model are reviewed. Data model control rules are defined, which are oriented towards multi-class Markovian QNs under centralized and decentralized policies. Management is focused on the routing processes, sequence of requests and delay (hold) times in the system. Based on developed partial-linear test functions, criteria are defined for checking the stability of QN under a certain data management policy. They are used to confirm the existence of a centralizing policy to stabilize the data model in multi-class QNs. For single-class QNs, research has shown that a decentralized data model management policy can be created. The set goal is to create specific policies to achieve maximum feasibility of the system, which has been confirmed by a numerical method.

What is important in conducting QNs averaging experiments is the possibility of automated calculation of probabilistic parameters of behavior over time, of the level of performance and assessment of system reliability. A software package for analysis of QN behavior and experimental work with it is presented in (Marzola, 2022). This queuing package is based on numerical algorithms methods which permits to calculate probability parameters for the two main regimes (transient and steady state) for discrete-time MC and continuous-time MC. It is applicable for investigation of single-channel MCs and different QNs and the main features of the package with giving examples of its use are described.

4. STOCHSTIC WORKLOAD MODELING

The workload (WL) of a CS is determined by the incoming external information units (input workflow) and the set of internal program objects that are transmitted between individual components and subsystems. These service request flows are probabilistic in nature (stochastic flows subject to probability laws) and load the service devices. This generates workload of the system resources with probability parameters described by stochastic flow with intensity and average time for servicing (Feitelson, 2015).

Today's technological development and, in particular, the growing possibilities of ICT pose the question of optimizing the workload for different structures, such as in the cloud, for IoT devices, for CPS realization, etc. For example, workload forecasting in the cloud requires elastic scaling, where optimal configuration choices must be made based on available resources. In principle, various techniques can be used to solve this task, such as architecture modeling, machine learning, environment validation, tuning, etc. Each of them has its own peculiarities and challenges, as noted in (Daradkeh et al., 2021), and an analytical mathematical model is proposed to describe the load behavior in the cloud. It is assumed that workloads are generated from various sources (user applications, information flows, registrations from sensors, IoT data for control of physical objects, etc.). It is stated that in real-time CPS operation, ensuring high accuracy for resource delivery is very important to maintain a good level of performance. For this reason, the workload modeling can help to make suitable prognosis of the cloud processes and to increase performance. The developed workload model provides excellent predictive values at a reduced cost compared to other methods. In the same direction is the study of the workload in service delivery, formed by the incoming calls, presented in (Gonçalves et al., 2016). A hierarchical model is proposed for the parameterization of which measurements made in four different networks were used. The proposed model is applied to further refine a synthetic workload generator in simulating network traffic in a cloud environment under various scenarios. To confirm the effectiveness of the model study, the obtained results were compared with actual data from re-working networks.

A broader view of the importance and role of workload in the technical field is presented in (Calzarossa et al., 2016). The overview of the basic characteristics of work flow establishes it as an important part of performance engineering research in managed productivity. This also defines the significant role of workload models and the paper offers a comprehensive survey of the state of workload characterization in popular application areas (conventional web workloads, online social networks, video services, mobile applications and infrastructures for cloud computing). Methodological approaches and workload modeling techniques for different scenarios (performance assessment, capacity planning, content distribution, resource provisioning) are analyzed.

The Internet of Things (IoT) provides multiple services for the cloud, and depending on the increasing load, delays and interruptions due to cloud resource management technology are possible. Two options for scaling cloud systems are presented in (Razzaq et al., 2021) – vertical scaling (adding more resources to a single cloud server) and horizontal scaling (increasing the number of active servers). The peculiarities and mainly negatives of the choice of scaling are analyzed, indicating that the increase in requests deteriorates the quality of work in vertical scaling when the maximum caps of the server are reached. On the other hand, horizontal scaling leads to economic inefficiencies as servers grow. To resolve this contradiction, the paper proposes an auto-scaling approach with detection of peak moments of dynamic workloads and their prediction to reduce response time. The approach is based on a hybrid cloud structure model with auto-scaling, enabling optimal convergence of vertically and horizontally scalable systems. The model uses an ensemble algorithm to pre-scaling the vertical system based on workload prediction and its estimation. Experimental results have confirmed efficiency in managing packages and services through a balanced workload.

The above shows the importance of the workload in ensuring good performance of the processing system, as well as the need to correctly define the indicators in its modeling. Some main features can be pointed out.

- The composition of the WL depends on the selected boundaries of the particular study. Typically, in a computing or computing environment, it is formed by external input streams of requests and internal processes related to active user tasks. System software also participates in computer processing and generates load impacts.
- WL depends on the reporting period against which subsets of the full WL (over the entire operating period) can be determined.
- Input workflow comes from an external source, but also has a stochastic nature and indicates an influence on the formation of WL, especially in computer processing (real-time work, dialog systems, multiprocessor processing, etc.). Due to the difficulty of prediction, insensitivity from the study object is usually allowed in WL modeling.

In analytical modeling of WL, two tasks are solved - specification of the type and its representation by appropriate stochastic means.

Specifying the type of WL is important, because usually one is looking for such a load that leads to a stationary mode of the investigated system or, in some optimization tasks and determination of the maximum throughput, to saturation of the system.

It is accepted to consider the WL as a stochastic flow of service requests of a certain type depending on the distribution function $F(x)$ or the density $f(x)$. Depending on the specific application and the features of the WL, stochastic flows with a probability distribution of discrete or continuous type are applied. A simple flow (stationary Poisson flow) is easiest to study. It is related to the Poisson distribution defined by the formula

$$P_m(\tau) = \frac{a^m}{m!} e^{-a} \left(m = 0, 1, 2, \ldots\right).$$

For such a flow, the intervals τ_j between consecutive requests are distributed according to an exponential law $f(t)=\lambda.e^{-\lambda t,}$ where $\lambda=1/\tau$ (за $t\geq 0$), with mathematical expectation $E[t]=\mu=1/\lambda=\tau$ and variance $V[t]=\sigma 2 = (1/\lambda)2 = \tau 2$. The distribution function $F(t)=P(T<t)$ expresses the probability that the random

variable T *t*akes a value smaller than t, which is determined by the opposite event, i.e. no event occurs in the interval t *(*m=*0)*:

$$P(T < t) = 1 - P_0(t) = 1 - \frac{a^0}{0!} e^{-a} = 1 - e^{-a} = 1 - e^{-\lambda\tau}$$

From the above, we derive the property that if the individual events in the random stream are distributed according to the Poisson law (discrete distribution), then the intervals between consecutive occurrences of events are exponentially distributed $F(t)=1-e^{-\lambda\tau}$ (continuous distribution). The opposite is also true. A random flow with an exponential distribution is without consequences.

Palm flow is also a steady flow of limited consequence. For him, the intervals τ_j between consecutive events are independent uniformly distributed random variables. The simple flow is a special case of the Palm flow because the intervals between events are distributed according to an exponential law.

An Erlang flow is formed by diluting a simple event flow, keeping every r^{th} event and discarding the remaining (*r*-1) events. The distribution of intervals between events obeys Erlang's law:

$$f_r(t) = \frac{\lambda(\lambda t)^{r-1}}{(r-1)!} e^{-\lambda t} (70t \geq 0),$$

as at *r*=1 the Erlang flow becomes a simple stochastic flow.

Other continuous distributions used as an analytical model of WL are the uniform (rectangular) – Figure 11, triangular, normal (Gaussian), Student distribution, etc. From the discrete distributions, the binomial distribution (Bernoulli), geometric, etc. are used.

Figure 11. Types of basic probability distributions

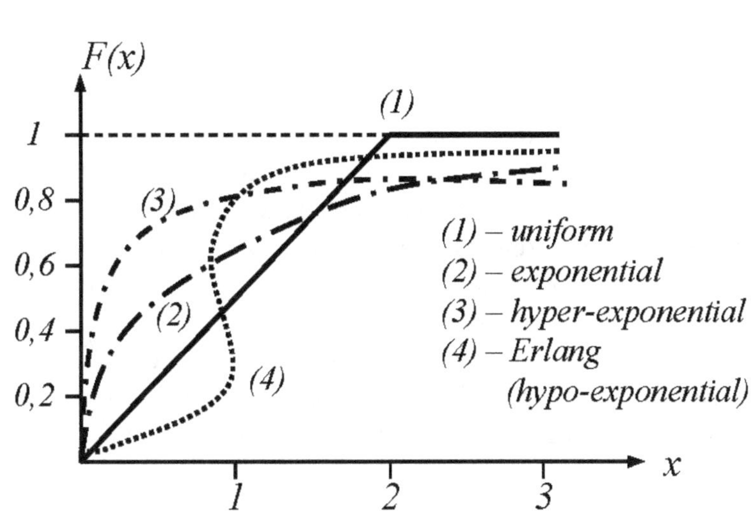

The presentation of WL can be done with a different degree of detail, and two ways are possible.

- Uniform presentation – a flow of requests of the same type, which can be described by uniform probability parameters;
- Non-uniform representation – WL is represented as a collection of different classes of queries, each class characterized by its own probability parameters.

In real systems, WL is mostly inhomogeneous and in modeling can be represented as a collection of several types of uniform flows. An example of such a representation is shown in Figure 12. It is multi-modal and can be viewed as a mixture of three different type distributions with values of the parameter x distributed around the modes x_i and bounds $[a_{i-1}, a_i]$ for each individual class i.

Figure 12. Multimodal presentation of WL

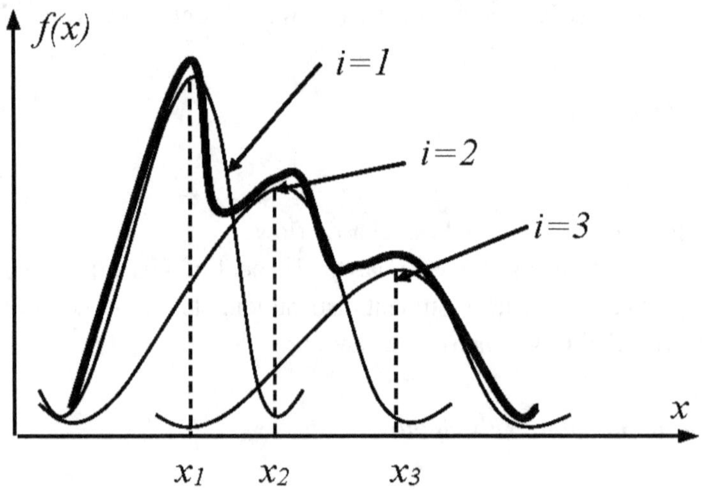

The exponential distribution is the most common continuous distribution, which is characterized by a coefficient of variation C_V=1. Such are the flow of program tasks in CS, the flow of data transmission and exchange of messages, the service of dispatching strategies such as FCFS (First Come First Served), LCFS (last Come First Served), RR (Round Robin), PS (Processor Sharing), SJF (Shortest Jobs First), etc. Other classes of continuous distributions that differ in their coefficient of variation from the exponential are the hyper-exponential (C_V>1) and the hypo-exponential (C_V<1), also known as the Erlang distribution (see Figure 11). The hyper-exponential can be used to represent WL under good justification of model adequacy. The hypo-exponential (Erlang) distribution is used to analytically represent WL as a regular flow that is stationary and ordinary, with intervals obeying a uniform probability distribution.

REFERENCES

Afolalu, S. A., Ikumapayi, O. M., Abdulkareem, A., Emetere, M. E., & Adejumo, O. (2021). A short review on queuing theory as a deterministic tool in sustainable telecommunication system. *Materials Today: Proceedings*, *44*, 2884–2888. doi:10.1016/j.matpr.2021.01.092

Calzarossa, M. C., Massari, L., & Tessera, D. (2016). Workload characterization: A survey revisited. *ACM Computing Surveys*, *48*(3), 1–43. doi:10.1145/2856127

D'Apice, C., Dudin, A., Dudin, S., & Manzo, R. (2022). Priority queueing system with many types of requests and restricted processor sharing. *Journal of Ambient Intelligence and Humanized Computing*, 1–12. doi:10.100712652-022-04233-w

Daradkeh, T., Agarwal, A., Zaman, M., & Manzano, R. (2021, June). Analytical modeling and prediction of cloud workload. In *2021 IEEE International Conference on Communications Workshops (ICC Workshops)* (pp. 1-6). IEEE. 10.1109/ICCWorkshops50388.2021.9473619

Dhingra, V., Kumawat, G. L., Roy, D., & de Koster, R. (2018). Solving semi-open queuing networks with time-varying arrivals: An application in container terminal landside operations. *European Journal of Operational Research*, *267*(3), 855–876. doi:10.1016/j.ejor.2017.12.020

Feitelson, D. G. (2015). *Workload modeling for computer systems performance evaluation*. Cambridge University Press. doi:10.1017/CBO9781139939690

Gonçalves, G. D., Drago, I., Vieira, A. B., da Silva, A. P. C., Almeida, J. M., & Mellia, M. (2016). Workload models and performance evaluation of cloud storage services. *Computer Networks*, *109*, 183–199. doi:10.1016/j.comnet.2016.03.024

Habib, H., Khanam, S. H., Azhar, A., & Khalid, Z. (2017). *Distributed Memory Architecture (Non Shared MIMD Architecture)*. Slideshare. (https://www.slideshare.net/HBukhary/distributed-memory-architecture-non-shared-mimd-architecture)

Jia, J., & Heragu, S. S. (2009). Solving semi-open queuing networks. *Operations Research*, *57*(2), 391–401. doi:10.1287/opre.1080.0627

Klimenok, V., Dudin, A., & Vishnevsky, V. (2020). Priority multi-server queueing system with heterogeneous customers. *Mathematics*, *8*(9), 1501. doi:10.3390/math8091501

Kumawat, G. L., & Roy, D. (2021). A new solution approach for multi-stage semi-open queuing networks: An application in shuttle-based compact storage systems. *Computers & Operations Research*, *125*, 105086. doi:10.1016/j.cor.2020.105086

Lee, S., Dudin, A., Dudina, O., & Kim, C. (2022). Analysis of a priority queueing system with the enhanced fairness of servers scheduling. *Journal of Ambient Intelligence and Humanized Computing*, 1–13. doi:10.100712652-022-03903-z

Madan, K. C. (2011). A non-preemptive priority queueing system with a single server serving two queues M/G/1 and M/D/1 with optional server vacations based on exhaustive service of the priority units. *Applied Mathematics*, 2(06), 791–799. doi:10.4236/am.2011.26106

Marzolla, M. (2022). A Software package for queueing networks and Markov chains analysis. *arXiv preprint arXiv:2209.04220*.

Massini, A. (2017). *Advanced parallel architectures*. Uni Romal. (https://twiki.di.uniroma1.it/pub/AAP/WebHome/2017-lesson14-Multiproc-InterconnectionNetworks.pdf)

Najim, A. H., Mansour, H. S., & Abbas, A. H. (2022). Characteristic analysis of queue theory in Wi-Fi applications using OPNET 14.5 modeler. *Eastern-European Journal of Enterprise Technologies, 2*(9), 116, 35-46. doi:10.15587/1729-4061.2022.255520

Garzon, J., Ameigeiras, P., Ramos-Munoz, J. J., Navarro-Ortiz, J., Andres-Maldonado, P., & Lopez-Soler, J. M. (2019). Performance modeling of softwarized network services based on queuing theory with experimental validation. *IEEE Transactions on Mobile Computing*, *20*(4), 1558–1573.

Razzaq, M. A., Mahar, J. A., Ahmad, M., Saher, N., Mehmood, A., & Choi, G. S. (2021). Hybrid auto-scaled service-cloud-based predictive workload modeling and analysis for smart campus system. *IEEE Access : Practical Innovations, Open Solutions*, *9*, 42081–42089. doi:10.1109/ACCESS.2021.3065597

Ritha, W., & Yasodai, P. (2021). Exploration of fuzzy preemptive-resume priority queuing system using robust ranking method. *International Journal of Aquatic Science*, *12*(2), 3224–3230.

Roy, D. (2016). Semi-open queuing networks: A review of stochastic models, solution methods and new research areas. *International Journal of Production Research*, *54*(6), 1735–1752. doi:10.1080/00207543.2015.1056316

Saritha, S., Mamatha, E., Reddy, C. S., & Rajadurai, P. (2022). A model for overflow queuing network with two-station heterogeneous system. *International Journal of Process Management and Benchmarking*, *12*(2), 147–158. doi:10.1504/IJPMB.2022.121592

Stankevich, E., Tananko, I., & Pagano, M. (2022). Optimization of open queuing networks with batch services. *Mathematics*, *10*(16), 3027. doi:10.3390/math10163027

Stone, D. (2015). *The Difference between Multiprocessors & Multicomputer Systems*, Techwalla (https://www.techwalla.com/articles/the-difference-between-multiprocessor-multicomputer-systems)

Tatashev, A. G., Seleznjev, O. V., & Yashina, M. V. (2022). Approximate Formulas for Characteristics of Multichannel LIFO Preemptive-Resume Priority Queueing System. arXiv preprint arXiv:2206.09263 (5 p.).

UKEssays. (2018). *Shared Memory MIMD Architectures*. UK Essays. (https://www.ukessays.com/essays/architecture/shared-memory-mimd-architecture.php)

Wang, S. P. (2021). Advanced computer architecture. In *Computer Architecture and Organization: Fundamentals and Architecture Security* (pp. 163–212). Springer., doi:10.1007/978-981-16-5662-0_7

Xie, Q., & Jin, L. (2022). Stabilizing queuing networks with model data-independent control. *IEEE Transactions on Control of Network Systems*, *9*(3), 1317–1326. doi:10.1109/TCNS.2022.3145752

Xu, J., & Gautam, N. (2020). Peak age of information in priority queuing systems. *IEEE Transactions on Information Theory*, *67*(1), 373–390. doi:10.1109/TIT.2020.3033501

Yadavalli, V. S. S., Adetunji, O., & Alrikabi, R. (2021). Optimization of the berth allocation problem to the vessels using priority queuing systems. In *Soft Computing in Inventory Management* (pp. 41–68). Springer., doi:10.1007/978-981-16-2156-7_3

Zhao, X., Wang, Y., Li, L., & Delahaye, D. (2022). A queuing network model of a multi-airport system based on point-wise stationary approximation. *Aerospace (Basel, Switzerland)*, *9*(7), 390. doi:10.3390/aerospace9070390

ENDNOTE

1 https://www.igi-global.com/dictionary/queuing-networks-for-designing-shared-services/40141

Chapter 11
Technology and Application of Computer Simulation

ABSTRACT

The chapter is devoted to the technological aspects of computer simulation and its application in solving specific problems. In the first part, a comparison of the features and possibilities of simulation and emulation is made. In addition, the main advantages of simulation and possible disadvantages are discussed, the mine phases in simulation model developing are presented and clarifying its adequacy is made. In the second section, the possible simulation approaches determined by the type of events and time are presented, which are discrete-even simulation with discrete time intervals between request and continuous simulation based on continuous sequences in time. A combination of these methods is presented too. The third section discusses the two mine structural components of a simulation model which can be named system model and workload model. Different examples for application of computer simulation are presented in the last section, which are united in the directions – simulation of service systems, simulation of parallel computer architecture and remote access simulation.

1. TECHNOLOGICAL ASSPECTS OF COMPUTER SIMULATION

Simulation and Emulation

Computer simulation is a numerical method for conducting computer-based experiments with mathematical models describing the behavior of a real object over a period. The approach is applicable in various fields, including virtual reality research (Nassar & Tekian, 2020), the social sphere (Huttar & BrintzenhofeSzoc, 2020), learning (Ben Ouahi et al., 2021), etc. The simulation model is a computer program designed for iterative execution by solving a dynamic system for the behavior of a given object, having developed a certain algorithm. The created model simulates real situations (with a certain scale and approximation) to study the behavior of the real object under the influence of external factors. In this reason, the simulation model must register the occurrence of certain events in time and accumulate the necessary data for conducting the analysis. In its operation, the simulator works with a model

DOI: 10.4018/978-1-6684-8947-5.ch011

of the input flow of requests. When conducting a computer simulation, the created program (program simulation model) is executed in a computer environment, starting with the assignment of initial values for basic parameters (definition of initial conditions). During the execution of the program, values for the observed parameters are calculated, and at the end, corresponding estimates are formed. Typically, sequentially calculated scores are stored in a suitable manner so that they can be interpreted or analyzed together at the end. The main difference of computer simulation from other types of models (for example, analytical models) is that in it the corresponding program is executed in time, i.e., time is an important and sometimes critical factor.

Emulation is a special type of simulation to complement the functionality of a given system. An emulator is a program that replaces a missing hardware node and program code in the operation of a computer system. Typical examples of emulators are programs for supplementing system commands when ensuring compatibility between different versions of the operating system, emulating an arithmetic coprocessor, etc. The main feature of the emulating program is that it processes incoming real input data or programs, i.e., works with a real workload.

The benefit of emulation is that it supports scientific research using computer experiments based on computer models. In these, it is possible to change the input effects, mostly by modulating the input stream, to examine the output results. Flows with Gaussian intensity are one of the popular tools in computational means because it allows interpolation of input settings (Katzfuss et al., 2022). The paper argues that their use on large datasets does not lead to useful results, and therefore proposes an extended and powerful model for large-scale analysis and emulation of computer experiments. The emulation uses an ordered conditional approximation when transforming the input space, for which parameters in the covariance function of the Gaussian process are pre-estimated. The result is the ability to obtain an estimate of joint prediction and simulation in nearly linear time in the number of model's runs.

Emulation as an approach is applied in various studies where real input workflow data is available. For example, in (Gregor et al., 2022) emulation was used in a model investigation of the processes in a business logistics system, with the possibility of flexible response to possible changes. The emulation is implemented in a mixed real-world and virtual-world environment to verify the functionality of the entire process in route optimization of automated guided vehicles. The model proposed in the article was developed in the Tecnomatix Plant Simulation software environment, and the results provided estimates for routing capacity research in internal logistics processes. In general, the achieved effect is a shortening of the time for preliminary design and optimization of operations, including the time for testing the logistics system. The technological aspect of application of emulation is also reflected in (Ram et al., 2018) in the analysis of energy resources and the search for alternative sources of energy. Solar energy is one of the leading sources of "new" energy, allowing it to replace traditional ones, but the real-time study of photovoltaic systems is a difficult task. The reason is that such experiments require an accurate emulator to reproduce the nonlinear characteristics of the photovoltaic cell. In order to offer researchers, the opportunity to make an adequate choice, the article reviews various emulators, uniting them under general indicators of comparability, for example, costs and accuracy of emulation, complexity of development, level of sensitivity to external influences, etc.

Quantum technologies are new to the computing field, with standard primitives of quantum computing involving deterministic unitary gates distinct from traditional operations (Bartolucci et al., 2023). The article presents a model for fault-tolerant quantum computing built from physical primitives. It is stated that the model offers direct error handling from the quantum correction protocol, and the proposed architecture has a modular organization and reduced requirements compared to other quantum architectures. Another paper developed a class of emulators to study the quantum scattering problem (Zhang & Furnstahl, 2022). A combination of the variational method of scattering the observable parameters and the concept of eigenvector continuation is used. The emulator is pre-trained by a flow Hamiltonian for a small number of points in space, and as a next step, the necessary interpolations and extrapolations are performed in that space. Emulation computation time estimates are on the millisecond scale with highly minimized errors and low memory usage.

Computer simulation is a reliable approach to studying systems and processes if it provides sufficiently correct results in the relevant domain. It also depends on the correctness of the initial conditions and the set values for the input controllable parameters. It is desirable that the simulation gives very accurate results, but in certain cases a certain tolerance and approximation of the estimates is permissible. However, the accuracy of the analysis also depends on the task at hand and the size of the research area – the wider the research domain, the more difficult it is to obtain precise simulation results.

Computer simulation is a suitable approach for modeling systems in which random (probabilistic) processes develop, as well as in situations such as the following.

- Inaccessible research object, impossible or very expensive research in real conditions of a system or a real process.
- Tasks allowing the formulation of an acceptable analytical or functional model, which, however, would be difficult to solve with known mathematical means or due to peculiarities of the used modeling apparatus, the necessary solution cannot be found.
- Serious difficulties or impossibility to correctly validate an established mathematical model of a system or process due to unclear operating conditions or insufficient behavioral data.

Main Advantages and Disadvantages of Simulation

Advantages of computer simulation.

- Provides the opportunity to study non-existing systems or those in the process of preliminary design.
- Enables the exploration of systems of arbitrary complexity with the ability to test under different resource occupancy scenarios and under varying input impacts.
- It does not limit the level of detail, as by expanding or reducing the time simulation, it is possible to investigate different situations of object behavior.
- Gives an unlimited amount of data about the behavior of the system under study. With a properly developed and validated simulation model, it is possible to examine different policies, procedures, and situations without increasing model development costs. The reason is that the changes are made in the computer environment and not in the real object.
- Simulation makes it easier to study the relationships between individual components in a complex system, which is done in a programming environment, rather than in the real one. In addition, it helps to specify the impact of components on the overall performance of a modeled system, as well as the specific specification of each individual part by defining specific requirements for it. The effect is to clarify the functioning of a given system and possibly uncover problematic parts of the behavior as well as initially invisible design flaws.

- The simulation allows to build on an ongoing study with another apparatus, for example an additional simulation investigation of an analytical stochastic queuing model or a Markov chain model. Another possibility is to combine simulation experiments with subsequent statistical processing of obtained results.
- There are many software packages and simulation environments, most of which have a specific orientation to a given area or type of problem. This allows, when choosing suitable software, to ensure high accuracy and correctness of the results and evaluations. In addition, the continuous improvement of the software allows shortening the time to run the simulation experiments. One reason is that specialized simulation languages contain macros that represent entire parts of a traditional program.

The development of quantum computers and their application in some technologies raises the question of 'quantum advantages', including the advantages of quantum simulation (Daley et al., 2022). This field includes the modeling of quantum properties of microscopic particles, processes in quantum physics and quantum chemistry, etc. The article reviews the technological possibilities of quantum simulation, its advantages, and prospects for future development. The application of quantum simulation in specialized applications of analog devices and the possibilities of use in digital devices are discussed. Another direction in which simulation is used is for the evaluation of health technologies, the features of which are summarized in (Caro & Möller, 2016). Discrete event simulation has been shown to be applicable in the field by investigating operations with competing entities and the formation of queues on demand. Key advantages of this application are highlighted, such as the ability to solve many current problems in health technology, the traceability of the time involved in each condition, the appropriate specification of an event and its attributes, compactness in model representation, and the study of competing risks.

Main disadvantages of computer simulation.

- Requires significant machine time to run the simulation regardless of the time scale used. Depending on the parameters defined for the simulation experiment, the running time may be shorter or longer. In both cases, there may be a negative effect on the effect - with a small execution time, the results may be of reduced accuracy, while a very long simulation duration will increase the value of the experiment.
- The results obtained in a specific simulation experiment are most often of a private nature, which necessitates conducting multiple experiments.
- When setting large requirements and more tasks to the simulation, it is possible that the developed simulation model will become very complex, which will affect its preliminary validation.
- To create an adequate simulation model, a good knowledge of the technology and the methods used is necessary, which is obtained with experience. With different applications of the approach by two developers, it is possible that the two models differ and give different results, possibly and contradictory in certain aspects.
- To obtain correct data, a good knowledge of the used simulation software is necessary. Simplistic development of the model or its incorrect programming leads to inaccurate conclusions, and this is fatal for the designed system.

Phases of Simulation Model Development

The development of a simulation model follows the general methodology for modeling, presented in Chapter 3, and the features of the individual stages are summarized below.

Phase (1). Preliminary clarifications. Analogously to the generalized procedure, it is necessary to make a clear formulation of the problem and the purpose of the planned research during the simulation. Correct formulation is reflected in subsequent phases and lays the foundation for adequate research. At this stage, assumptions are accepted that should not significantly disturb the effect of the simulation and the correct description of the original object. For complex systems, it is necessary to make a correct decomposition and accept approximations based on defined initial requirements. A clearly defined goal defines the main tasks before the simulation and formulates initial conditions for its organization. In addition to the tasks of this phase, a plan of future experimentation can be developed by defining controllable and observable variables (factors). This experiment plan can be updated after the completion of some of the following phases to adapt to the already created simulation model and its software implementation.

Phase (2). Conceptual model defining. The main task of this phase is to develop the first "working" model, which will allow the subsequent creation of the mathematical and program models. In essence, this is a formalization of the object of research, which is done as a result of the formulation of the main problems and principles for building the simulation model made in the previous phase. The defined initial conditions and approximations allow us to formulate the main requirements, limitations, and criteria for evaluating model performance. One of the requirements is the specification of primary factors and the selection of controllable factors, which will determine the "direction" of research and the purpose of the obtained estimates from the experimental results. The decomposition made is the basis of detailing the functional algorithm and an adequate description of the external environment. In essence, a conceptual model is a set of mathematical and logical functional dependencies representing the components of the original object and their interaction within the entire structure. Finally, specific hypotheses for the research should be defined in the phase.

Phase (3). Development of mathematical model and functional algorithm. Usually, an appropriate analytical apparatus, most often stochastic in nature, is used to formulate the mathematical description of the conceptual model. For example, developing an analytical model such as a queuing network for serving streams of input requests and routing them along different paths in the network. In complex analytical descriptions, it is possible to create separate sub-systems that are relatively independent and communicate only by input and output. This will allow easier analytical description of each sub-system and their subsequent analytical combination (connection) in the final mathematical model. Since the computer simulation implies a subsequent program implementation of the model, it is necessary, after composing the mathematical model, to develop a functional algorithm as a transition from the mathematical model to its program implementation. It is necessary to solve the following tasks:

- the synchronization of model components;
- the transfer of request flows between components;
- solving possible conflict situations in the model;
- preliminary determination of the type of output information;
- choice of simulation management method and its duration.

Phase (4). Determination of a priori information. A basic requirement is to make a reasoned choice of values (levels) for the input variables and for the controllable factors in experimentation. One example is the determination of the intensities of the input flows and of the service flows, and due to subsequent program implementation it is good to define time parameters based on an average value and tolerance ($t_{av} \pm \Delta t$) for the time of arrival of a current request and for the time for serving a request received in the server. Another requirement is to determine the range of change of the controllable factors (MIN÷MAX) and the step of change to determine specific values for conducting the simulation experiments. It is known that a very small step change will lead to a significant increase in simulation time and the number of simulation experiments. Major differences between successive levels of a controllable factor may produce results leading to inaccurate conclusions.

Phase (5). Program model implementation. This includes a program description of the developed mathematical model in a selected language environment based on the functional algorithm (phase 3) and when reporting the selected values for input and controlled model variables (phase 4). As specified in Chapter 4, general-purpose or specialized software can be used, the latter allowing for higher efficiency. This is the model that will be applied in simulation experimentation and its correctness must be ensured by testing and validating the program code. Testing can be done for each of the separate parts (if the model has such a structure) and after the completion of all, test the correctness of the connections between them. If testing confirms the correctness of the program code, then validation is a procedure of proving that the developed model corresponds to the real system in the sense of the specific research task set. Symbolically, it can be represented as the principal equivalence of the model to the original object ($\Omega_M \approx \Omega_O$). This step should provide an answer to the question of whether the results of a model analysis can be applied to the real object. This can easily be specified if some variant of the real object exists or if there is any real data on the relationship of input parameters to output results in known situations. However, this is not possible when designing new systems.

Phase (6). Realization of simulation experiments. Before starting the actual execution of the program simulation model, it is necessary to check and update the implementation plan (if already created) or to develop a correct plan for the experiment. This includes the time for a single simulation, the number of runs of the model at different values of the controllable parameters, and the scheduling of those values. The planning of the experimental work itself must comply with the set task and goal, allowing to ensure high efficiency and usefulness of the results. The execution of the experimental plan will allow the expected results to be obtained, which can be attributed to the behavior of the object-original Ω_O.

Adequacy of Computer Simulation and Application

A key point is ensuring the plausibility of the simulation model, for which it undergoes verification at various stages of development. Adequacy verification rules are analogous to the general rules in a technological aspect and are applied at the individual stages of development. Some of the verification approaches are presented below.

- Testing parts or the whole model for inputs for which results are known.
- Study of the behavior of the output parameters (observable factors) with atypical input data (borderline, incorrect, prohibited, etc.).
- Assessment of the accuracy and sensitivity of the model when changing the levels of the controllable parameters within wide limits.

- Comparing model results with known data on the behavior of the original or with data obtained from another type of modeling or monitoring.

In case of a low rating or lack of adequacy, calibration of the model is required. Possible reasons for low adequacy are:

- Incorrect assumptions or initial hypotheses.
- Too much simplification of the functional behavior and working conditions of the object.
- Incorrect approximations, ignoring significant factors, etc.

To confirm the reliability of the simulation, it is recommended to ensure the following characteristics of the model:

- Accuracy of the simulation – depends on the used stochastic variables and their impact on the functional algorithm.
- Stability of the model results – characterized by the convergence of the observed parameters to certain values when the total modeling time increases (it is associated with the establishment of a stationary mode).
- Sensitivity of the model – it is estimated by the range of a selected observable parameter when varying the levels for the controllable parameters in the range of permissible values. The absence of change determines the insensitivity of the model to some or all controllable parameters and makes it pointless to conduct experiments.

An analysis of basic concepts that make a simulation adequate is done in (Swarup, 2019). Validation of adequacy makes the simulation model good for the specific application, stating that it is a "*nuanced concept that has to be applied at every stage of the development of the simulation, from initial conceptualization to final use*". To confirm the latter, the article proposes specific practices for developing adequate models.

Computer simulation is used in various fields because it allows us to obtain results for an object that is currently not available or needs improvement. When experimentation in a real environment is very difficult, simulation modeling comes to the rescue. The growing importance of cloud services in practice necessitates the evaluation of different cloud resource drivers, one solution being cloud simulation (Mansouri et al., 2020). The article states that the cloud is a large and complex system with a heterogeneous composition and participants, which requires the proper formation of the appropriate configuration of resources to overcome possible limitations. Such an assessment is very difficult to make in a real cloud environment, which makes computer simulation a useful tool for analyzing system behavior by examining a specific component under different scenarios. To address this challenge, the paper provides an in-depth analysis of various cloud problem-oriented simulation tools. The reason for this review is the fact that traditional means of computer simulation of the processes in distributed environments can provide the necessary accurate results in the modeling study of cloud computing. On the other hand, the available cloud simulators must be precisely analyzed to make the right choice for conducting a specific study. The simulation must match the research objectives and provide correct results for several parameters, such as access time, load balancing, scalability, elasticity of heterogeneous data, latency, energy saving, and last but not least requirements for reliable storage and security of the information (Romansky &

Noninska, 2019b). These requirements are related to the need to develop complex modules and tools for managing cloud resources.

Computer simulation can lead to cost reductions and process efficiency gains if conducted when the specific implementation is made. The utility of the simulator is that it allows a preliminary observation of the behavior of the system. Some advantages of cloud simulation can be mentioned which complement the general advantages presented above.

- They do not require specialized installation and maintenance, which saves additional costs.
- Provides an opportunity to pre-assess the risk associated with the design of the individual components and parameters of the cloud.
- Easily change scenarios as well as input impacts and cloud workload in operation.
- With a good knowledge of the programming language, modifications to the simulation model can be made relatively easily.

Another discussion of the application of simulation in a cloud environment is made in (Magalhães et al., 2015), where the object of the study is the workload of cloud data centers. The presence of heterogeneous hardware, virtualization, and the complexity of the workload with different characteristics are indicated as the basis for the need for modeling and simulating the processes. This determines the need to develop a model to capture the behavioral characteristics of user profiles in the simulation of cloud resource usage. To solve the defined problems in the paper, a model suitable for several different domains of behavior with the possibility of dynamically scaling resources and supporting different user profiles is proposed. Statistical distributions for realistic parameters in a dynamic environment are applied to define a workload model. The developed model is proposed as a supplement to the cloud simulator CloudSim, and graphical and analytical methods are applied to prove the effectiveness and validate the model. The CloudSim simulator enables easy simulation and experimentation in a cloud computing environment. It is a toolkit for system and behavioral modeling of cloud components (data centers, virtual machines, etc.). Basically, it is not a working simulation framework because it does not provide a ready-to-use environment but requires the development of a cloud scenario in advance on specific inputs. In this sense, the model proposed in the paper is taken as an extension of the cloud simulation environment.

2. MINE DIRECTIONS IN SIMULATION RESEARCH

The main directions in the organization and conduct of computer simulations are related to model time, which can be discrete or continuous. In the first group, the model time is represented as discrete time intervals for the registrations of the occurring model events, while in the second group, the model processes are represented by continuous functions of time. In both variants, the simulation paradigm can be combined with analytical queuing networks, through which stochastic processes are modeled. When using the simulation approach, the probabilistic parameters are represented by discrete values such as minimum, maximum, average value, etc. The two regimes in the analytical stochastic models (transient and steady state regimes) can be simulated by organizing simulation experiments at specific values for the parameters and conditions for the conduct. Possible simulation options are presented below.

Discrete-Event Simulation

It is characterized by changing the model parameters in discrete time intervals. During these intervals, the state of the objects is preserved unchanged. The implementation supports a model clock for timing events that change the state of model objects. This is how model time is managed, with two basic approaches illustrated in Figure 1.

(a) *Event-driven (asynchronous) simulation* – model time is governed by the occurrence of events in the model. It determines an event with a minimum time (earliest event) from the set of generated requests and changes the model timer with this time. After that, the necessary actions are taken for its realization (occupying a device, moving requests, determining next events, etc.). Simultaneous execution of several events becomes pseudo-parallel without changing the model time.

(b) *Timer simulation (synchronous)* – the model time is varied by a constant interval Δt during which the registrations are made. At the beginning, the model time $T_m = t_0 = 0$ is established and the system parameters $S_m(t_0) = s_0$ characterizing the object at the beginning of the simulation are determined. Cyclic execution of the simulation follows during the intervals Δt, with the vector of current events determining the new states $s_{i+1} = S_m(t_i + \Delta t)$, becoming current.

Figure 1. Model time management in simulation

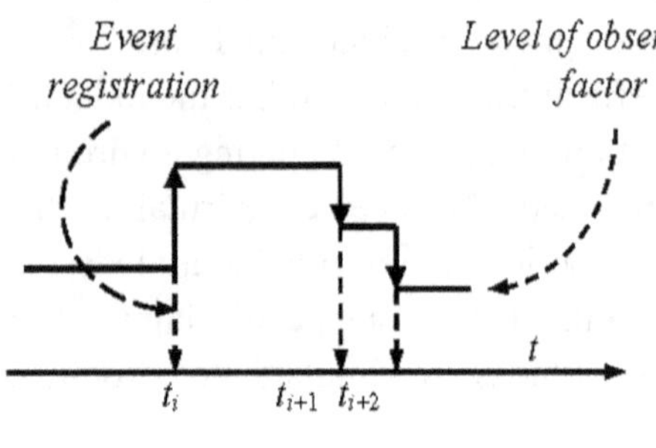

(a) Event simulation

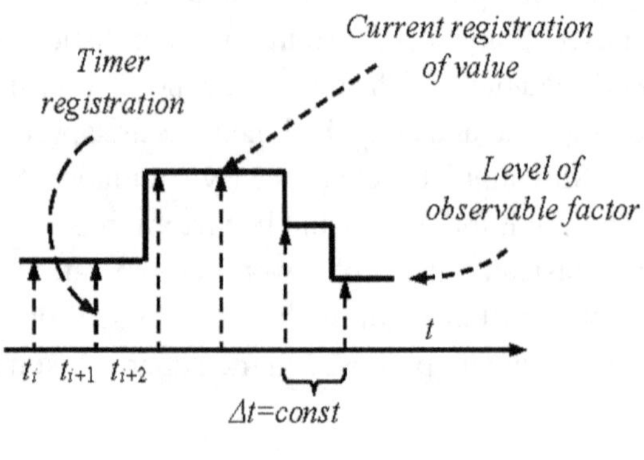

(b) Timer simulation

Discrete-event simulation (DES) is a dynamic process for studying a stochastic process but with discrete events, with changes occurring only at discrete moments in time. The input effects can be represented by one or more workflows, which affects the output results, which should be treated as approximations of the true model results. The reason for this is that it is necessary to conduct multiple experiments to obtain adequate statistical estimates. The simulation times themselves are discrete events. Conducting experiments at different values for the controllable factors allows specific estimates to be determined, which should be aggregated into a common array for analysis and determination of final statistical estimates for the observable factors.

A specific solution to a problem using discrete-event simulation is presented in (Rotunno et al., 2023). Initially, the studied system is described using Unified Modeling Language (UML) and then simulated in a corresponding programming environment. The simulation was applied to a specific real case, and the obtained results showed the influence of internal variables on the behavior of the processes. The relationship between simulation and queuing network analytical models is the subject of experiments in (Frichi et al., 2022), where two queuing models were developed and compared, then investigated through discrete event simulation. Experiments were conducted for different scenarios with different numbers of participants, reflecting real cases of service processes.

Continuous Simulation

Unlike the previous type, this simulation implements the processes and variables as continuous dependencies in time, described by appropriate differential equations. If these equations are simple, they can be solved analytically, otherwise numerical methods are applied. In terms of description, these models are close to analytical models, but their solutions provide insight into the development of the modeled processes over time. It is usually applied to unknown future behavior of a system that can be estimated by a continuous-time probabilistic approach, with continuous simulation requiring an appropriate input stream of impacts. This can be achieved by collecting statistical data. Continuous simulation uses a queuing system or queuing network with continuous time, assuming that the number of requests and the number of busy servers in the system take an arbitrary continuous value.

At the heart of continuous simulation is the definition of an adequate description of the state of a system over time. If the current state of the investigated system is presented as a variable x, then the main goal is to define a functional dependence $x=f(t,\lambda,x_0)$, where t is time, λ represents basic parameters of the model and x_0 sets the initial conditions of the system . This dependence is defined as the functional state equation of the system. In complex systems, it is usually difficult to make a direct formulation of the equations of state, but it is possible to find a relationship between the rate of change of a parameter x with respect to time, i.e., to determine the derivative of x. A continuous simulation model may contain one or more equations of state to describe the behavior of the system. If the model contains all possible equations of state, then the simulation covers all states of the time factor given initial conditions and model parameters. Continuous simulation is usually applied in cases where it is possible to represent each studied process as a function of the state in time, and the rate of its change can be determined based on the differentiating function of the state.

Continuous simulation was used in power flow modeling of AC-DC hybrid microgrids operating for different generation and consumption scenarios in (López-Santos et al., 2022). The initial analytical description of the object is as a multiple-input-multiple-output model, and the simulation is performed in a virtual instrument in the LabVIEW platform to obtain the continuous property. When creating the model and running the simulation, the experimenter performs an initial configuration of the parameters and sets the coefficients of the performance profiles of the components and algorithms. In addition, there is the possibility of online changes of environmental variables and energy consumption. It is possible by using global variables that the running simulation interacts with other algorithms.

Combined Discrete-Continuous Simulation

The organization of simulation research combines both approaches presented above to achieve more accurate and flexible modeling of random processes. The combined approach has certain advantages over others. For example, it provides higher efficiency than the deterministic approach in the investigation of service delays and in most cases provides more accurate results. One possibility for combined simulation is to initially develop a discrete simulation model based on an analytical description and then combine with continuous time variation. The main interactions between events and variables when conducting a combined discrete0continuous simulation can be represented as follows.

- A discrete event can cause a discrete change in the value of a continuous variable.
- A discrete event affects a continuous variable to change individual time.
- A continuous variable can reach a certain value to cause a discrete event to occur.

Simulation modeling is used in various fields, with discrete-even simulation being the most used. One example is a study of the possibilities of effective prevention in providing reliable cyber-protection conducted in (Ivanova, 2022). The paper discusses a simulation of a Denial-of-Service (DoS) countermeasure technique by investigating the impact of attacks on control centers where the goal is to disrupt communications via a high-range signal. Such a vulnerability exists in Wireless Sensor Networks (WSNs) due to the possibility of blocking legitimate signals that cause a denial of service. It is indicated that for successful simulation it is necessary to choose an appropriate programming environment. In this case, Riverbed Modeler Academic Edition 17.5 software was used to organize the simulation, effective in studying complex functional models with a high level of reality. The product uses multiple real network devices with advanced settings on each, as well as sets of security protocols, encryption algorithms, etc. for network security reliability analysis purposes. The open-source CST Studio Suite 2021, which is applicable for simulation modeling of jamming and anti-jamming processes, was used to simulate interference antennas separately.

3. STRUCTURAL COMPONENTS OF THE SIMULATION MODEL

It is clear that the model is any presentation of system or process by using means and tools for describing and analysis. In the case of simulation, the time is the important part of execution and experiments realization. This was discussed in the previous sections. In addition, the most used discrete-event simulation model is the one where changes are made only at discrete points in time, as opposed to a continuous model in which time and change are subject to continuous flows.

A simulation model includes various structural components that may make mathematical or logical sense as well as be discrete or continuous in nature. Discrete-event model is based on the concepts of subjects and resources, realization of processes in a system, separate events generated in the process realization, system activities and status. The typical component of a simulation model is the time because it determines the realization of events during the experimental execution. The realization of each event can generate other new events and processes in the investigated system. A brief description of some other components of the simulation model is presented below.

- The structure of the model is built by using different subjects which can be static or dynamic. The first type determines the system model as a collection of servers, queues, and transactions between them. Each of them has its own attributes and parameters that define it functionally. Static entities have a constant structure and set behavior parameters. Dynamic objects define the movement of flows through the static structure and, in a general sense, characterize workflow and workload. A more detailed discussion of these structural components is presented below.
- The resource is a structural object that provides a service to dynamic objects. In this group can be included the static entities that provide services such as waiting for the release of the server, the server itself, providing the possibility of service under probabilistic parameters, etc. If the infinite buffer is excluded, the resources are characterized by a fixed capacity, predetermined when structuring the model. In some models, specific resource characteristics may be defined, such as in the priority service in the simulation of priority QNs.
- The process is a component which is related to the activity of the simulation model. It is connected to the realization of the requests processing and their transfer to other subjects. Practically, each generated process is characterized by its own attributes and parameters which are preliminary determined or are formed during the realization in the model. The process is combination of events, or it generates sequence of events during the simulation.
- The event is a situation of change status of the investigated system. It is realized in a concrete time moment of the activity of the simulation model (model execution). Such events are the new request arriving, input in the server, server refusing, etc. The events change the status of the simulation model. Two types of events can be defined. The first group are the initial (basic) events, which are managed by data as the arriving time for requests, time for service finishing, etc. These events are planned during the simulation model design. The secondary events are generated during the simulation realization and are determined by internal logical behavior of the model.
- An activity is a length of time to do something specific in the model. It starts in an initial moment and ends after some time, determined by the situation. An example is the request servicing in one of the servers included in the simulation model. The event "request input in a server" initiates the activity "service time" which will continue to the event "end of service" and output of the current request with the occupied server release. Activities are usually governed by probabilistic processes, for example, the formation of an input flow of requests, the occupation of a given resource for a certain time, the redirection to other resources, etc.
- The current status of a simulation model is determined by a list of values for the main structural components. This can be presented as an ordered vector or a set of numerical or logical values for the selected important parameters. These values must be available to define fully the system each time.

The simulation model must accurately reflect the structure of the studied system and its functional algorithm (behavior). In this respect, a simulation model is designed by two main parts as a structural component with static and dynamic subjects presented below – system model and workload model.

The ***system model*** describes the static structure of the modeled object and creates the environment for the implementation of the functional algorithm. It represents a set of interacting model elements, each describing a given functional node (device) of the overall structure. Behavior rules are set for elements, which includes:

- Discipline for selecting a service request.
- Reaction of a busy service device when a new request is received.

- Reaction of the server unit in the event of a malfunction or denial of service.

Most often, a system model is created as a queuing network with certain functional connections between individual nodes. Requests are propagated from the input to the output, with multiple servicing performed on them. There are two types of primitives for such service which are presented below.

Type (1). Parallel multi-service, where several servers manage the incoming flow of requests. Allocation can be done based on a common input queue or by pre-allocating the flow (separate queues) – Figure 2.

Figure 2. Parallel multiservice

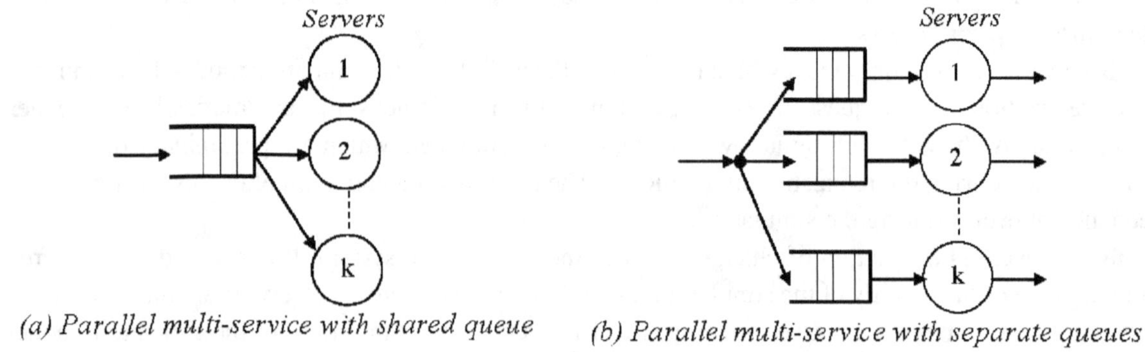

Type (2). Sequential multiple serving, where a given request passes through several sequentially chained servers, each performing a specific processing of the request. Two types are also allowed depending on the organization of the queues – Figure 3.

Figure 3. Consecutive multiple service

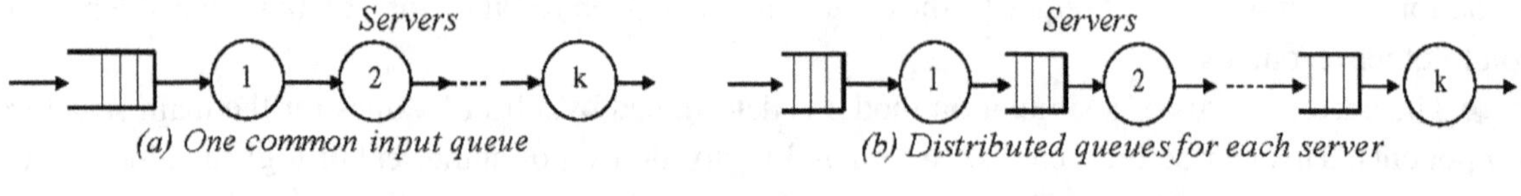

The choice of system model type should be tailored to the task set before the simulation. The reason is that each simulation model includes basic structural components and parameters to determine their functionality, the main ones are presented below.

- System objects (queues and servers).
- Functional relations between them.
- Input flow with personal parameters.
- Measures for efficiency.

An example for describing simulation model is M/M/1 where this determines the type of the two main system objects – queue (with exponential distribution of the input workflow) and one server (service channel) with exponential intensity of service. The parameter "total time of a request in the system" as a sum of the time for waiting and time for service can be determined as a functional relation between system objects. And finally, the parameters as percentage of arriving, percent of service, average time for waiting, maximum queue length are defined as measures for efficiency.

The ***workload model*** reflects the dynamic features and external influences on the behavior of the internal system variables and the values of the output reactions. It describes the rules for determining the moments of arrival of requests from the input stream, their routing in the network of structural elements depending on the logical connections, defines the priorities of individual classes of requests during service, as well as the allowable times for requests to stay in the system model. In this respect, the workload model manages the work of the system model in a logical sense and is defined with it. In the implementation, stochastic or deterministic flows of units (requests) can be used, which is associated with the corresponding probabilistic or regular distribution law. Standard distributions are usually used to represent the input workflow and of the workload, for example exponential flow, flow with Poisson distribution, normal, hyper-exponential. The reason is that flows are simple for realization in a simulation model. Most simulation packages offer options for choosing certain standard distributions when modeling workflows. To implement other non-standard distributions, or when the distributions offered are quite limited in type, it is necessary to use random number generators. In this way, random sequences can be generated that exactly match the modeled processes and the flows maintained in the investigated system.

The dynamic entity simulation approach has been applied in (Jodayree et al., 2019) to the study of dynamic clouds and the processes of load balancing and uniform distribution of resources. The article presents a claim that simulation allows determining the conditions for effective management of cloud resources. For this purpose, an algorithm based on point-to-point system predictions is proposed, and the simulation was performed using the CloudSim simulator. The effectiveness of the proposed algorithm is confirmed by the results obtained from the conducted simulation experiments for predicting the dynamic processes in cloud services.

4. APPLICATION OF COMPUTER SIMULATION

Examples for Simulation of Service Systems

An example of the application of computer simulation is the model study of a single-processor system with external disk memory, described as a queuing system which is presented in Figure 4. The system model is realized by two servers (CPU and DISK) with their queues. The workload model is determined by input flow of request (task for calculation) with uniform distributed times between them in the range [5, 15], and the time for service of a separate requests is $t_{SERV} = 8 \pm 2$ model time units (MTUs). The servicing is non-priority based on the strategy FCFS, with a probability of 0.6, requests require turning to external memory (DISK) with a service time in it $t_{DISK}=30 \pm 10$, then they are again directed to the processor (CPU) for recovery (continued) of service.

Figure 4. Queuing model for one-processor computer system with disk memory

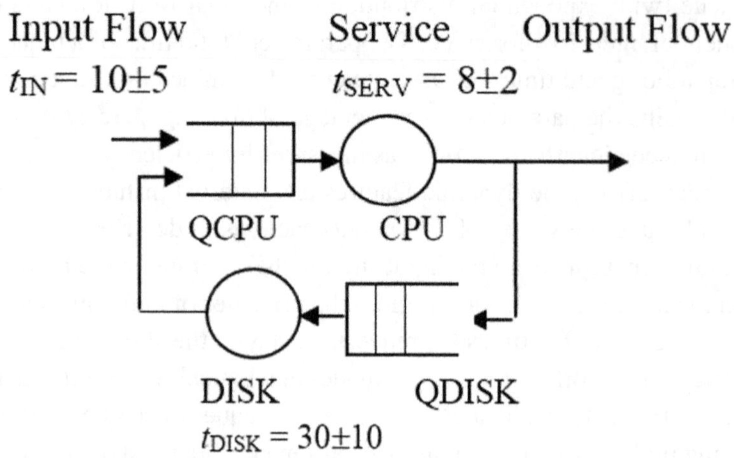

The program simulation model developed in the GPSS environment is shown in Figure 5, and results of its execution for 300 MTUs are summarized in Table 1(a & b).

Figure 5. Discrete-even simulation model based on QS from Figure 4

```
--      Model segment (1)
GEN     GENERATE     10,5        -- request generation
BASK    QUEUE        QCPU        -- arrival in queue
        SEIZE        CPU         -- input in CPU
        DEPART       QCPU        -- leaving the queue
        ADVANCE      8,2         -- service in CPU
        RELEASE      CPU         -- CPU release
--
        TRANSFER     .6, ,OUT    -- probability transfer
--
        QUEUE        QDISK       -- input in DISK memory
        SEIZE        DISK
        DEPART       QDISK
        ADVANCE      30,10
        RELEASE      DISK
--
        TRANSFER     ,BACK       -- return to CPU
--
OUT     TERMINATE                -- leaving the system
--
--      Model segment (2)        -- timer
        GENERATE     300         -- 300 time units for execution
        TERMINATE    1
--
        START        1
        END
```

Table 1(a). Results of the simulation model execution (for servers)

Parameters of servers	*CPU*	*DISK*
Medium usage	0,93	0,78
Number of inputs	36	9
Average time for a request	7,81	26

Table 1(b). Results of the simulation model execution (for queues)

Parameters of queues	*QCPU*	*QDISK*
Maximum content	4	4
Average number of pending requests	1,27	1,41
Number of inputs	40	14
Average time for a request	9,57	30,29

The queuing model of car wash with a for waiting area for two vehicles (SPACE) and node for service (WASH) is presented in Figure 6. Cars arrive with intensity λ with exponential distribution with parameter 0.2 and mean time 5 min. If there is no free space in the parking lot, the car (another request) leaves the car wash (QS with limited buffer). A time *t* uniformly distributed in the range [4, 8] minutes is required to service one car. The simulation program model in GPSS is presented in Figure 7.

Figure 6. Queuing model for simulation organization

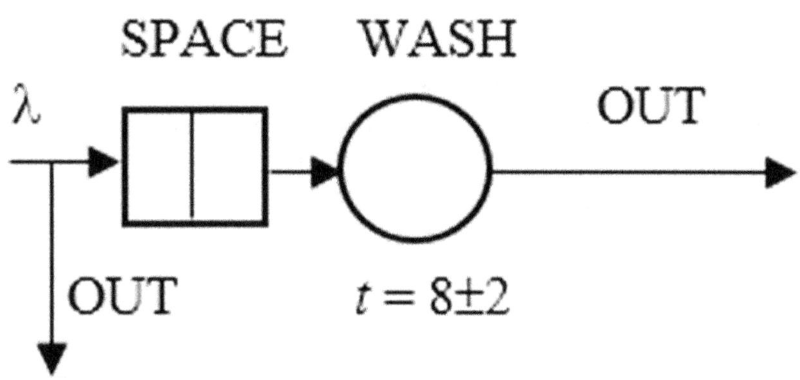

Figure 7.GPSS simulation model of car wash

```
EXPO     FUNCTION        RN1,C24              -- function definition
0, 0 / .1, .104 / -2, .222 / .3, .355 / .4, .509 / .5, .69 / .6, .915/
.7, .1,2 / .75, 1.38 / .8, 1.6 / .84, 1.83 / .84, 1.83 / .88, 2.12 / .9, 2.3 /
.99, 4.6 / .995, 5.3 / .998, 6.2 / .999, 7 / .9998, 8
--
--       Definition of two-places space
         STORAGE         S$SPACE,2
--       Segment (1)
         GENERATE        500,FN$EXPO
         TRANSFER        BOTH, , OUT
         ENTER           SPACE
         SEIZE           WASH
         LEAVE           SPACE
         ADVANCE         600,200
         RELEASE         WASH
OUT      TERMINATE
--
--       Segment (2)
         GENERATE        12000                -- time 2 hours
         TERMINATE       1
--
         START           1
         END
```

The simulation is planned to investigate the work of the wash for a time of 2 hours (120 minutes) and with a scale of 1:100 (when creating the model, the MTU was chosen to be 0.01 minutes). An EXPO function with an exponential distribution is defined to model the input flow. The statistical results of a simulation experiment are summarized in Table 2.

Table 2. Results of simulation

Statistics for WASH			
Name	Average load	Number of entries	Average time/transact
WASH	0,97	20	580,55

Statistics for SPACE					
Name	Capacity	Average content	Average load	Number of entries	Average time/ transact
SPACE	2	1,73	0,8630	20	1035,68

The next example of a simulation investigation (Figure 8) is for a manufacturing system with two workplaces, where additional finishing is performed on parts arriving from a molding shop at uniformly distributed time intervals in the range [*a*, *b*] minutes. Processing each detail takes from 3 to 9 minutes and from 4 to 12 minutes, respectively. A controller checks the details for 2 to 4 minutes, and if they do not meet the standard, they are returned for further processing. With probability 0.35, the parts are within the standard and leave the workshop.

Figure 8. Simulation model of a production system

```
GEN      GENERATE     10,5
LAB1     QUEUE        QWORK           -- first worker
         TRANSFER     BOTH, , LAB2
         SEIZE        WORK1
         DEPART       QWORK1
         ADVANCE      6,3
         RELEASE      WORK1
         TRANSFER     ,LAB3
LAB2     SEIZE        WORK2           -- second worker
         DEPART       QWORK2
         ADVANCE      8,4
         RELEASE      WORK2
         TRANSFER     ,LAB2
LAB3     QUEUE        QTEST           -- check for standard
         SEIZE        TEST
         DEPART       QTEST
         ADVANCE      3,1
         RELEASE      TEST
         TRANSFER     .35, ,LAB1      -- redirection of detail after inspection
         TERMINATE
--
         GENERATE     60              -- execution for 1 hour
         TERMINATE    1
--
         START        1               -- first model execution
--
         CLEAR                        -- second model execution
GEN      GENERATE     4,2
         START        1
--
         CLEAR                        -- third model execution
GEN      GENERATE     20,5
         START        1
         END
```

The experimental plan envisages a simulation run of the model for one working hour in three consecutive runs for parameter values [*a*, *b*], respectively [5, 15], [2, 6] and [15, 25]. A simulation time to real time scale of 1:1 is selected, which sets the total simulation time to 60 MTUs. The results for the first two executions are summarized in Table 3.

Table 3. Experimental result of production system simulation

Name	Average load	Number of entries	Average time/transact
Statistics for servers – EXPERIMET (1)			
WORK1	0,80	10	4,90
WORK2	0,30	3	6,00
TEST	0,46	11	2,55
Statistics for queues – EXPERIMENT (2)			
WORK1	0,91	11	5,09
WORK2	0,30	3	6,00
TEST	0,46	11	2,55

A simulation model of a system with priority service of each request is presented in Figure 9. Two priority levels are defined – absolute priority and relative priority. With absolute priority, a request with a higher priority interrupts a started service of a request with a lower priority. With relative priority, a service that has already started is waiting.

Figure 9. Simulation model of a priority system

```
EXPO    FUNCTION      RN1,C24              -- function definition
0, 0 / .1, .104 / -2, .222 / .3, .355 / .4, .509 / .5, .69 / .6, .915/
.7, .1,2 / .75, 1.38 / .8, 1.6 / .84, 1.83 / .84, 1.83 / .88, 2.12 / .9, 2.3 /
.99, 4.6 / .995, 5.3 / .998, 6.2 / .999, 7 / .9998, 8
--
        GENERATE      100,FN$EXPO, , ,3    -- flow type 3
        TRANSFER      ,INPUT
        GENERATE      25,FN$EXPO, , ,2     -- flow type 2
        TRANSFER      ,INPUT
        GENERATE      20,FN$EXPO, , ,1     -- flow type 1
INPUT   QUEUE         QLINE
        SEIZE         LINE
        DEPART        QLINE
        ADVANCE       5, FN$EXPO
        RELEASE       LINE
        TERMINATE     1
--
        START         100
        END
```

With absolute priority, interrupted requests can be serviced in one of the following ways.

- An interrupted request can be further serviced after completion of all requests with higher, considering the time from the start to the moment of service.
- An aborted request is destroyed.
- the service of an interrupted request starts again from the beginning without considering the time of previous attempts.

For the purpose of the investigation, 3 types of input flows of requests with Poisson distribution are generated, each with relative priority (set with the fifth attribute for the GENERATE statement) and with parameters λ_1=0,01 [1/MTUs], λ_2=0,04 [1/MTUs] and λ_3=0,05 [1/MTUs]. The service time follows an exponential law with parameter λ=0.2 [1/MTUs] (parameter 5 in the first attribute of the operator ADVANCE). The model was run for 100 requests, and the values for the absolute and relative timers and the standard statistics of the LINE server are presented in Table 4.

Table 4. Results of a priority system simulation

Values for the timers			
Relative timer 944	Absolute timer 944		
Statistics for the server			
Name	Average load	Number of entries	Average time/transact
LINE	0,52	100	4,93

Figure 10 shows an example GPSS program model of a lossless multi-line system that includes two parallel processing lines with identical instrument servers. The system is powered by a common input stream, with the generated requests waiting in a common queue, from where they are directed to a free service container. If there is no free device, the requests wait according to the FCFS strategy.

Figure 10. Simulation model of two-linear system

```
SYST        STORAGE          1
EXPO        FUNCTION         RN1,C24
0, 0 / .1, .104 / -2, .222 / .3, .355 / .4, .509 / .5, .69 / .6, .915/
.7, .1,2 / .75, 1.38 / .8, 1.6 / .84, 1.83 / .84, 1.83 / .88, 2.12 / .9, 2.3 /
.99, 4.6 / .995, 5.3 / .998, 6.2 / .999, 7 / .9998, 8
--
            GENERATE         10, FN$EXPO
            ENTER            SYST
            QUEUE            1
            SEIZE            1
            DEPART           1
            ADVANCE          5, FN$EXPO
            RELEASE          1
            QUEUE            2
            SEIZE            2
            DEPART           2
            ADVANCE          2, FN$EXPO
            REELASE          2
            LEAVE            SYST
            TERMINATE        1
--
            START            50
            END
```

The input stream of requests has a Poisson distributed intensity with parameter 0.1 (time parameter 1/0.1 in GENERATE). The service intensity in the two devices 1 and 2 has an exponential distribution with a parameter of 0.05 (time parameter 1/0.5 in ADVENCE). Generated requests enter the SYST common queue, accumulating wait statistics before being directed for service. The results of a conducted simulation experiment for 50 generates requests are summarized in Table 5.

Table 5. Simulation model of a priority system

Statistics for the servers			
Name	Average load	Number of entries	Average time/transact
1	0,77	7	11,06
2	0,02	1	1,53

Simulation of MIMD Computer Architecture

The object of simulation research is the organization of processes in Distributed Memory MIMD architecture of multicomputer processing nodes with cube topology of the connections. Preliminary organization of modeling has been discussed in Section 10.1 of Chapter 10, where the queuing model of a separate processor node is presented. This section could be considered as a continuation of the study by applying a simulation based on the GPSS World language. As an addition to the one presented in Chapter 10, the architecture of hypercube multicomputer with dimension $N=4$ is the object of investigation here. The supported N-dimensional parallelism is implemented based on a set of binary-ordered $n=2^N$ nodes connected by $N.2^{N-1}$ arcs. Each processor node (PN) is assigned a unique index, and in a binary sense all neighbors have a Hamming distance of 1. Indexing uses N-bit binary code for the corresponding architectural implementation – Figure 11.

Figure 11. Architecture of hypercube multicomputer whit communications

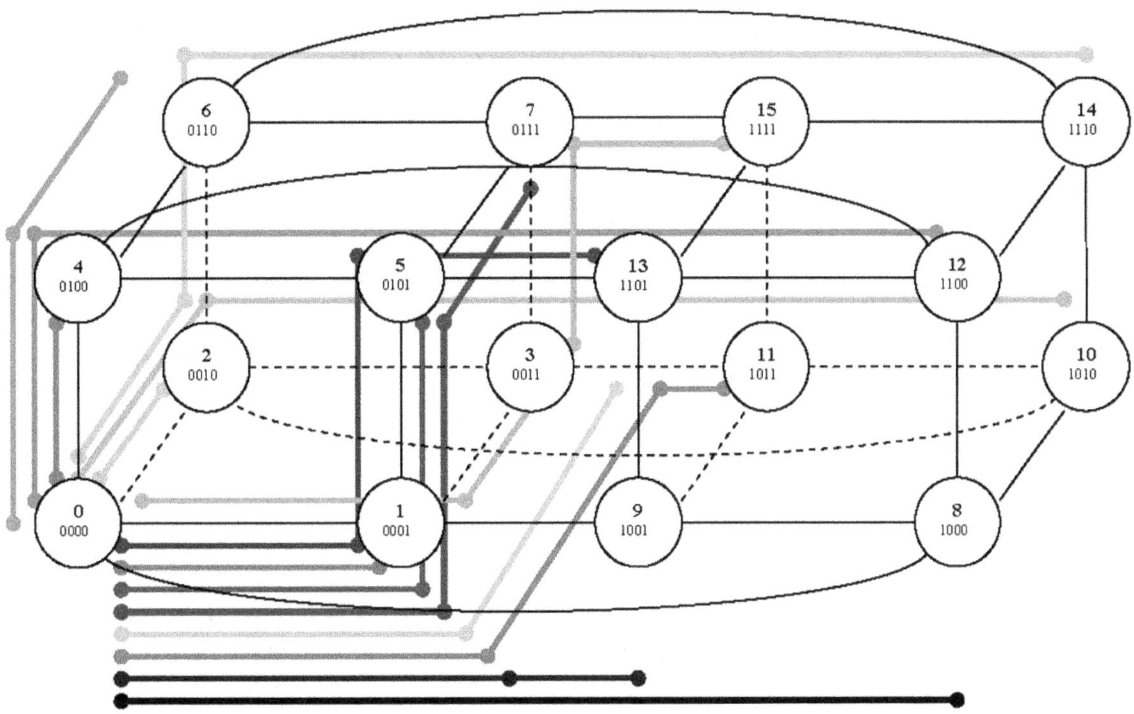

The defined binary indices (addresses) of the individual nodes determine the routing strategy when performing inter-process communications, which is based on a Hamming comparison. A neighboring PN with a difference in the least significant bit in a binary address comparison is selected for message transmission. This strategy is embedded in the mathematical formalization of communications and the organization of modeling based on the created QN from Figure 12.

Figure 12. Queueing network of Hypercube for simulation model creation

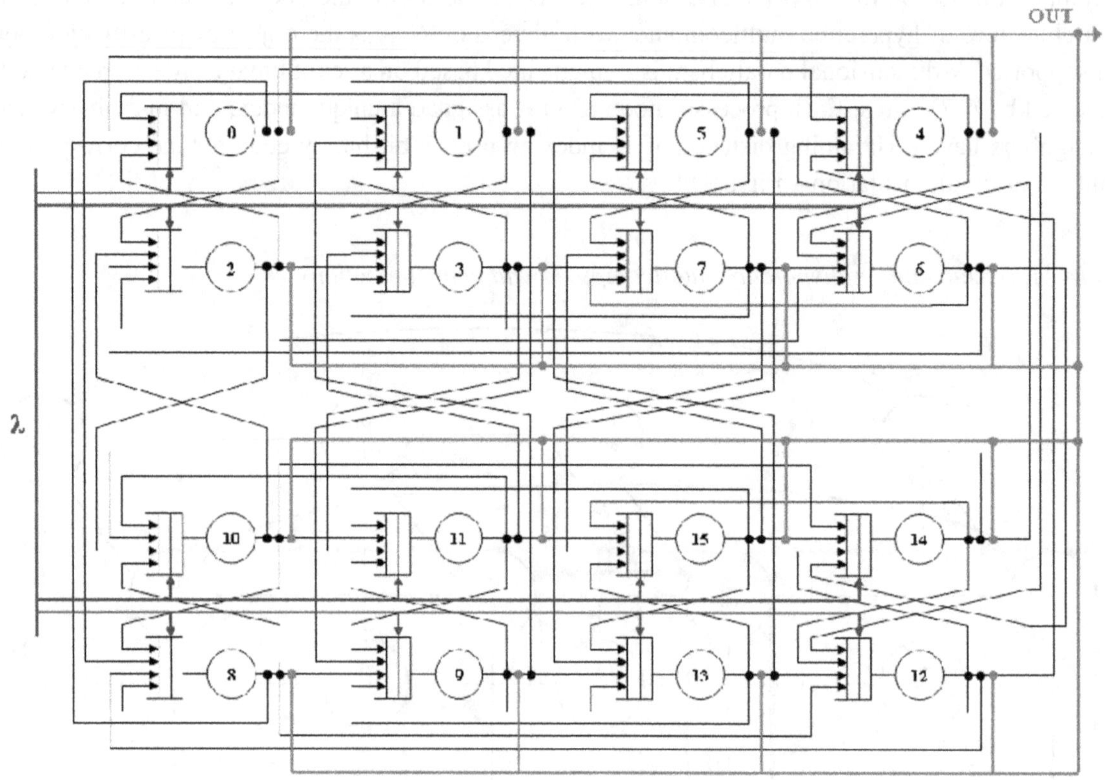

When designing the simulation models, working variables are provided for managing the input flow of requests (setting different conditions during generation, subject to a selected probability law), as well as for regulating the service flow in the processor nodes. To refer to the base resources (processors and communication channels) and queues in the model, a global variable is used to define them with a personal number. All transactions carry a certain number of parameters (12 for a hypercube) that take random values assigned to them by a work function FNC. These parameters, in addition to the attributes described above, also set additional values, such as a binary index for directing a request to a processor node, a binary index of a processing processor, an auxiliary code for finding a route in the topology diagram, etc. The processing (analysis) of these binary structures is always in ascending direction starting from bit "0" of the binary value.

Processing takes place in the PN to which the request is directed and after traversing a possible route (if communication is required). In this situation, a transaction received for processing is assigned the lower priority, which will allow processing to be interrupted when a redirect (routing) request is received on the same processor. After the final processing of a request, the transaction is removed from the system (destroyed).

In the organization of the simulation modeling, the possibility of applying different probability distributions when generating the input flow of requests is provided, and in the planning of the experiments, the variation of the total model time and the control parameters (number of generated transactions, control of the model timer, conditions for end of experiment etc.). The main model results and estimates obtained from the simulation experiments are summarized below.

An evaluation of the processing acceleration compared to the sequential method is presented in Figure 13. The formula $SPEED = TIME_{SEC}/TIME_{PAR}$ is applied, and the time for sequential processing is determined based on $TIME_{SEC}=N.C$ (where N is the number of serviced tasks/requests, and C is the complexity of task). It can be seen from the graph that as the complexity of the tasks increases, saturation of the acceleration is reached.

Figure 13. Changing the acceleration depending on the complexity of the tasks

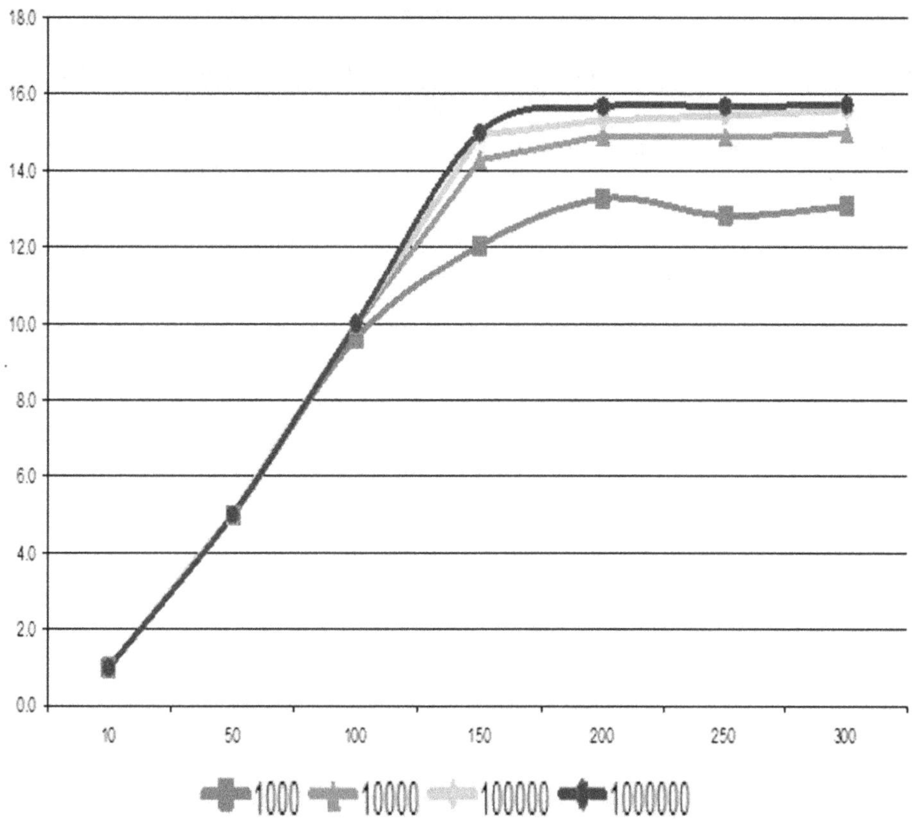

The estimated complexity scores are presented in Table 6. The trend of this estimate is clearly visible in the graphical representation (Figure 13). The results show that for very short and small number of tasks, the acceleration is less, while as the complexity of the tasks increases, the acceleration relative to their sequential processing grows sharply until a saturation point is reached. The conclusion is that the acceleration increases with the increase in the number of tasks processed in the architecture.

Table 6. Assessments for complexity

C	N			
	1000	**10000**	**100000**	**1000000**
10	1.0	1.0	1.0	1.0
50	5.0	5.0	5.0	5.0
100	9.6	9.9	10.0	10.0
150	12.0	14.3	14.9	15.0
200	13.3	14.9	15.3	15.7
250	12.8	14.9	15.4	15.7

The assessment of communication time (Figure 14) as a relative characteristic is carried out based on the ratio of communication time (CT) to task processing time (PT). The graphical representation shows that in case increasing number of tasks, the CT in general rapidly decreases, which reflects an increase in the efficiency of parallel processing.

Figure 14. Relative communication time (in %) to process 1000 transactions depending on complexity

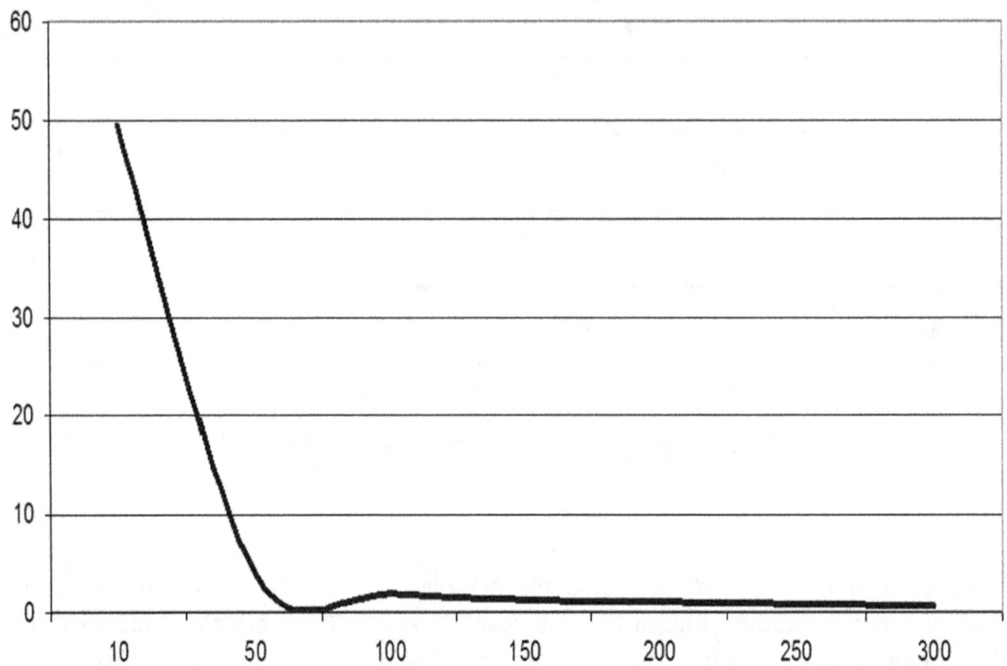

Additional simulation modeling is done for a Boolean cube architecture (N=3), keeping the basic parameter requirements. The goal is to compare important features. For example, Figure 15 presents a graphical comparison of the total processing time for tasks of different complexity (ranging from 10

to 300) under an exponential input workflow distribution and a fixed simulation time corresponding to 100000 transactions. A constant processing time is observed for the hypercube for tasks of low and medium complexity, and compared to the Boolean cube, the simulation shows significantly lower values, especially for tasks of higher complexity.

Figure 15. Total processing time

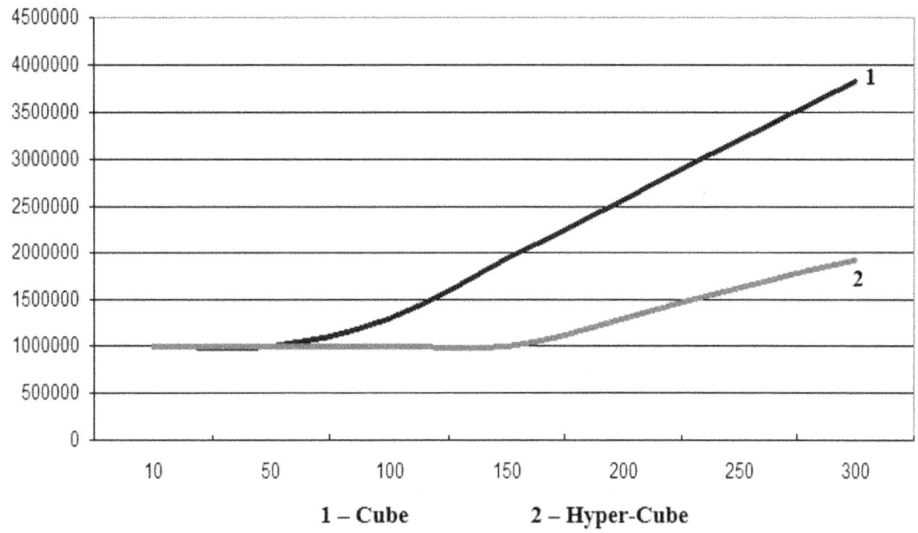

A summary of the estimates regarding the average load (number of transacts) of the two architectures against the complexity of the tasks (defined by the parameters of ADVANCE) is presented in Figure 16. A steep increase in saturation dependence is observed in the second half of the studied range, with the hypercube architecture giving better scores than the Boolean cube.

Figure 16. Average workload

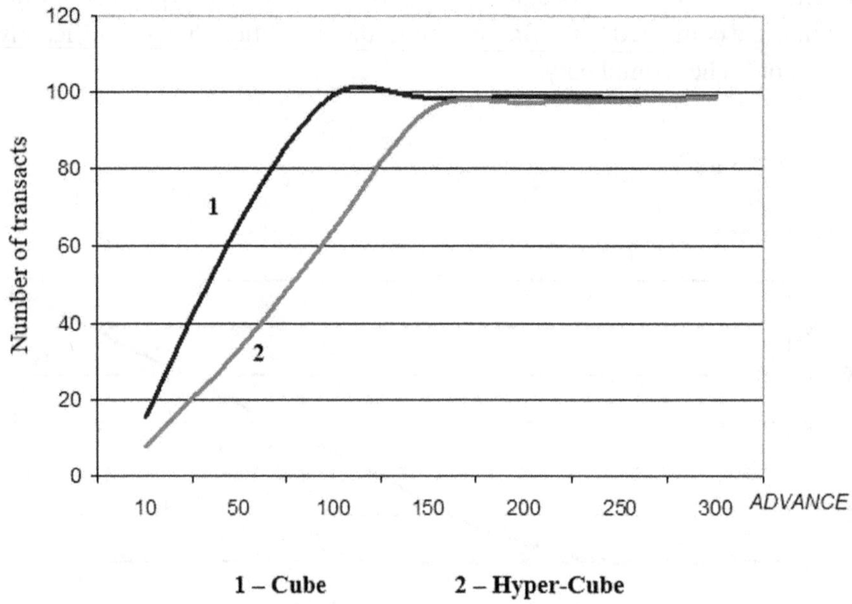

Simulation of Remote Access to Distributed Resources

This part deals with some problems of organization of remote access and using distributed resources in Distributed Learning Environment (DLE) which was presented as an abstract formalization in sub-section 5.2.2 of Chapter 5 (see Figures 10 and 11). In addition, an investigation of this virtual environment by using Markov chain with discrete time was presented in Chapter 8 (sub-charter 8.4 and Figure 20). Mine goal here is to present an example for using discrete simulation to investigate the organization and performance indexes of remote access and information services. Preliminary formalization of communications in the environment for simulation organization is made and the abstract model is shown in Figure 17. This is an extension of the formalization from Figure 11 (Chapter 5) and is used to develop a queueing network presented in (Figure 18). The abbreviations used in both figures have the following meaning: U – users; R – resources; T – transmitters; D – distributors.

Figure 17. Abstract model for simulation model development

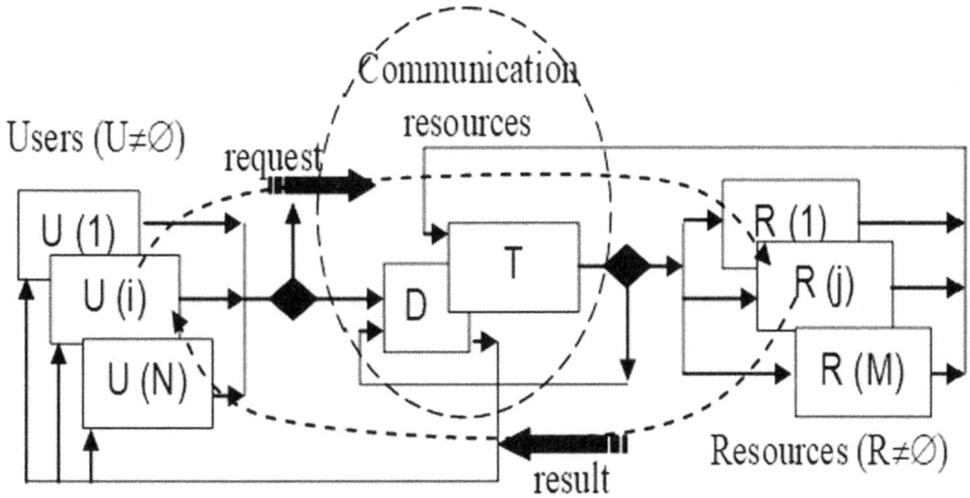

Figure 18. Queueing network for simulation model design

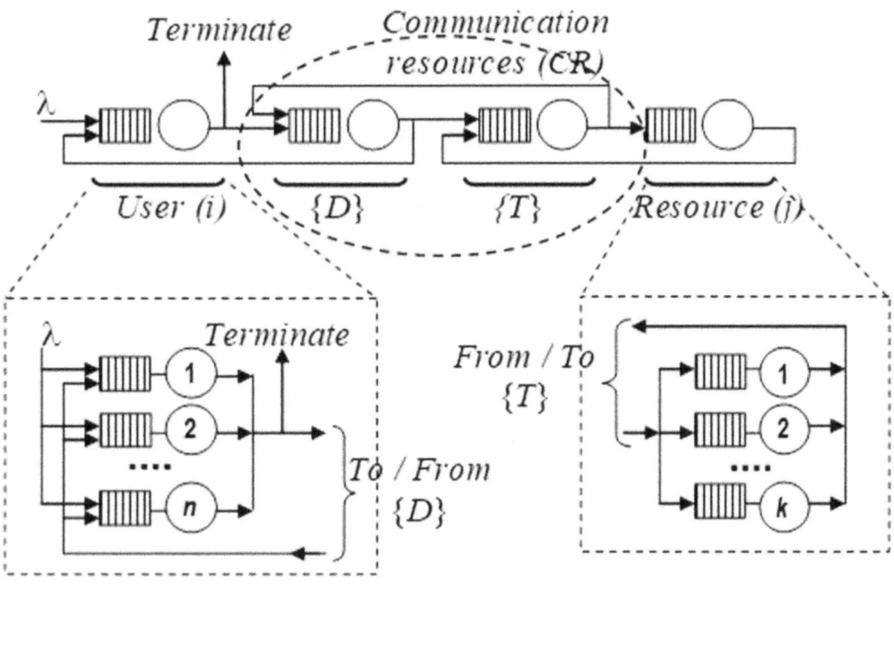

A simulation model based on WebGPSS program environment (Stahl, 2009) is designed and after validation some experiments are carried out. The simulation is organized in two parts – investigation of processes for information servicing and investigation of remote access to resources. Selected experimental results received during information servicing simulation are presented in Figure 19. The graphic illustrates the dependency of the utilization of the administrative components of the front office (INT) and virtual modules (MOD) of the back office (MOD_{AV} is average value) from the intensity of the stochastic input workflow of user's requests.

Figure 19. Simulation result for utilization study

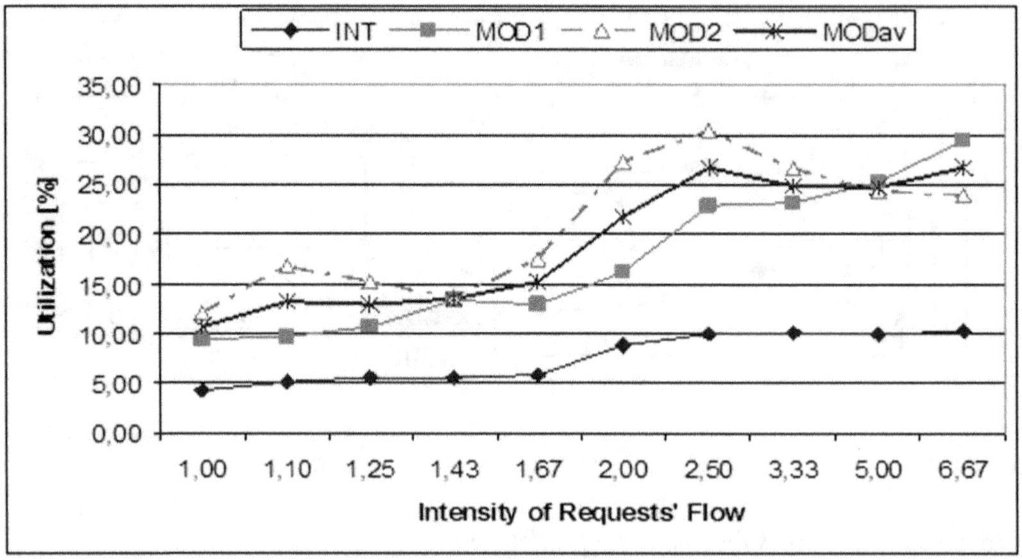

The simulation investigation of remote access is organized in two directions: ✓ single-user access to single resource; ✓ multi-user access to multiple resources. The simulation is carried out by using one factor experimental plan with variation of the intensity λ of the input flow of users' requests. The connections between users and resources are supported by the communication space (router D and transmitter T) and request could be terminated only by the user after returning the needed information.

Single-user access to single resource. Some experimental results after single execution of the simulation model are summarized in Table 7 (assessments for the model parameters of facilities and their buffers) and in Table 8 (assessments for workload and waiting time).

Table 7. Assessments for model components

Facilities	*Average load*	*Number of entries*	*Average time/ transact*		
USER	98.61	164	30.06		
DISP	33.48	174	9,62		
TRANS	68.0	175	19.45		
RES	50.06	88	28.44		
Buffers	*Maximum contents*	*Average contents*	*Total entries*	*Zero entries*	*Percent zeros*
BUSER	22	10.91	184	3	1.63
BDISP	2	0.04	174	142	81.61
BTRANS	3	0.30	175	87	49.71
BRES	1	0.04	88	71	80.68

Table 8. Simulation assessments for workload (WL) and waiting time (WT)

$\lambda.10^{-2}$	U		D		T		R	
	WL	WT	WL	WT	WL	WT	WL	WT
1,00	57,27	5,68	17,80	0,69	41,19	1,29	28,47	0,00
1,10	64,30	5,48	22,05	0,93	44,70	1,63	34,18	0,00
1,25	74,78	8,62	23,31	0,97	51,02	3,11	37,25	0,89
1,43	83,87	13,17	28,50	1,06	58,33	3,98	42,59	1,59
1,67	92,80	23,12	32,67	1,21	63,82	5,77	44,91	1,57
2,00	98,61	296,55	33,48	1,01	68,08	8,65	50,06	2,12
2,50	98,74	588,72	34,93	1,94	75,10	9,75	58,31	4,24
3,33	99,04	995,58	39,60	1,24	76,98	13,01	55,19	4,96
5,00	99,69	5,68	46,29	1,51	90,67	31,19	68,56	10,05
6,67	99,79	5,48	49,93	1,81	95,97	47,01	80,04	16,17

A graphical interpretation of experimental results for the waiting time (WT) for user task before its servicing in the virtual environments is shown in Figure 20. The intensity λ of input flow varies over a certain range and it can see in the right side of figure that the waiting time has a strong increasing for the high values of the intensity.

Figure 20. Average assessments for user tasks waiting time

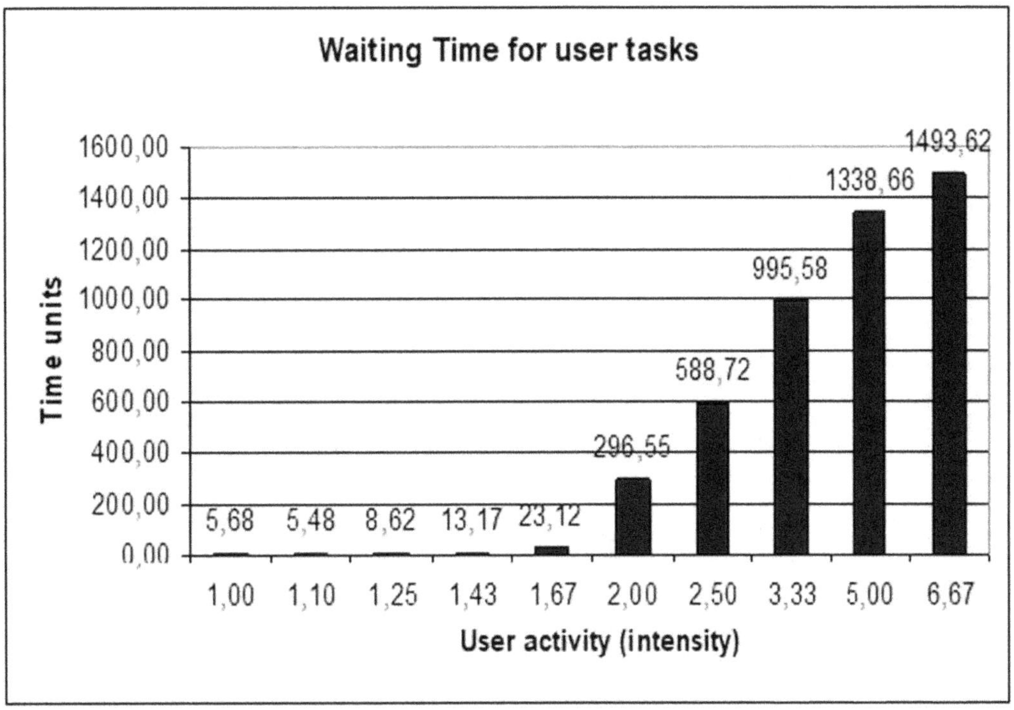

Multiple-user access to multiple resources. Assessments for the parameters WL and WT obtained by simulation are summarized in Table 9. The average values for the factors U_{av} (users) and R_{av} (resources) are calculated on the base of all received experimental results from the simulation. An illustration of this approach is shown in Figure 21. The parameter Net unites all communication tools (D and T) take part in the remote access.

Table 9. Summarized assessments for multi-user access to multiple resources

$\lambda.10^{-2}$	Average assessments for WL			Average assessments for WT		
	U_{av}	R_{av}	Net	U_{av}	R_{av}	Net
1,00	29,19	15,36	20,00	2,87	0,12	0,00
1,10	32,99	16,56	22,06	3,81	0,05	0,16
1,25	36,37	18,54	24,58	4,32	0,18	0,46
1,43	42,95	21,61	27,50	5,95	0,33	0,66
1,67	48,44	24,73	32,22	5,60	0,69	0,70
2,00	59,87	27,80	38,46	18,64	1,69	1,44
2,50	74,60	36,87	52,67	77,29	3,51	3,13
3,33	81,09	43,70	59,30	472,25	6,04	4,06
5,00	99,45	55,47	73,69	854,97	12,17	7,42
6,67	99,66	60,10	79,05	816,25	17,92	7,93

Figure 21. Average assessments for user tasks waiting time

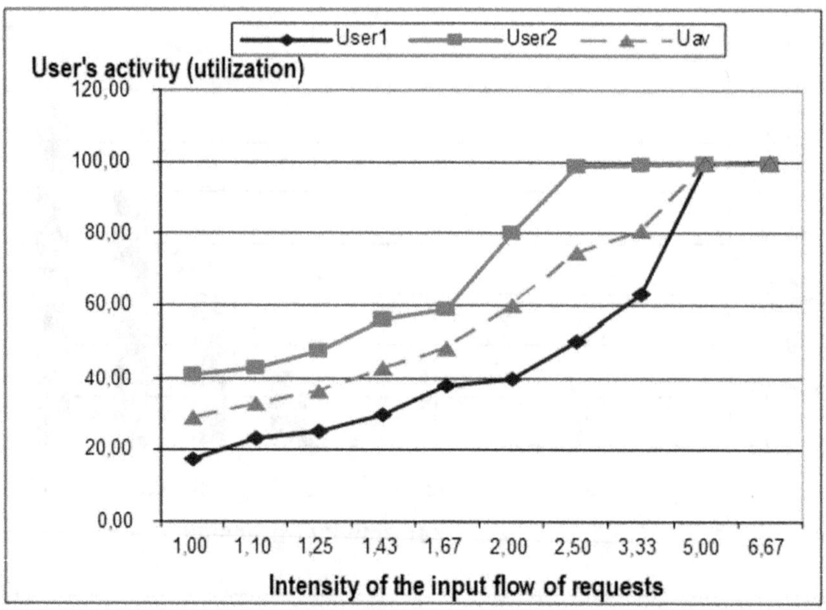

The figures below present graphical interpretation of the functions WL(λ) and WT(λ). Figure 22 shows that no problem with the remote access to the multiple resources in the virtual environment because the maximum average utilization of resources is about 60% in case of 100% average utilization of users.

Figure 23 shows that as the intensity of the input flow of user's requests increases, the waiting time increases, and at higher values of λ it is significantly greater in resources (R_{av}) compared to the communication space (NET) – see the right side of the figure.

Figure 22. Graphical interpretation of the assessments for WL(λ)

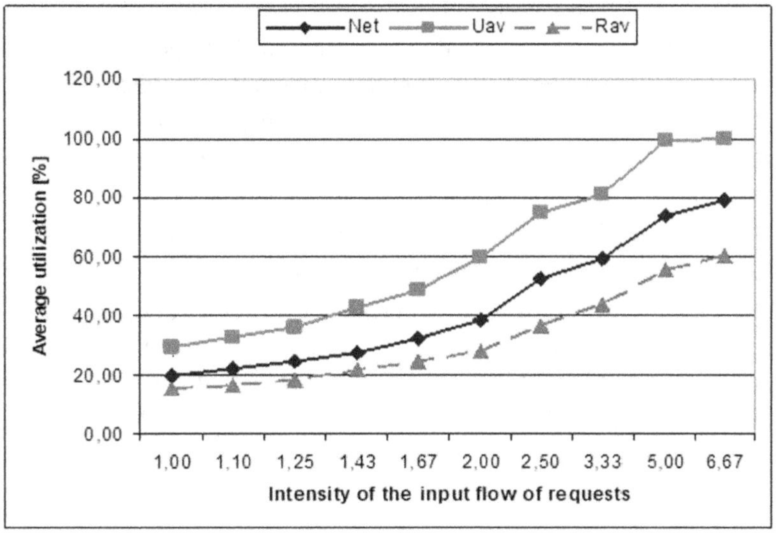

Figure 23. Graphical interpretation of the assessments for WT(λ)

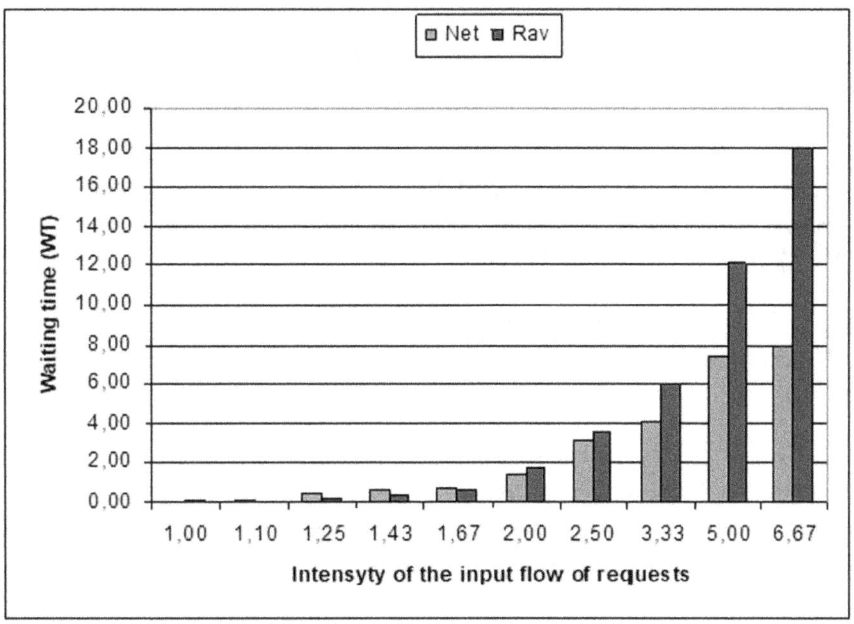

Computer simulation as an approach, apart from simulating information processes, as well as serving a flow of requests, is widely used in other areas as well. One example is the research on the possibilities of physical cloth simulation to facilitate the visualization of realistic clothing on human bodies presented in (Vassilev, 2020). The paper states that the simulation of clothing on a human body boils down to solving two main tasks – using an appropriate cloth model and developing a collision detection (CD) technique on a selected cloth model. To solve the second problem, 3 paradigms are proposed to accelerate computer simulation of fabric using parallel algorithms. The first paradigm proposes realization of the simulation based on multi-thread processor in multiple cores. In this case the Graphics Processing Unit (GPU) accelerates the CD using interference tests in image space. In paradigm 2, all simulation activities (cloth simulation and CD) are implemented in GPU. In paradigm 3, cloth simulation execution is performed on the GPU via NVidia OptiX engine and hard disk usage.

Another approach to applying agent-based simulation modeling for the purposes of ongoing artificial intellectual economy research is presented in (Hakrama & Frasheri, 2018). Agent-based modeling (ABM) is applied to solve complex problems in different areas and is related to the field of artificial intellect. The application of this technology allows a better representation of complex phenomena in various fields, and ABM is the basis of the new approach proposed in the article in the simulation of an artificial economy. A conceptual model by using the capabilities offered by UML (for example use cases for person and for firm) is developed as a preliminary stage of the simulation investigation. The simulation is organized based on sequence diagrams as a part of communication between agents and the implementation is made by using a structure offered by Repast Relogo. This product allows simulation based on a tick mechanism which works as a simulation clock and offers simulation framework for graphical components defining and used as a input data.

REFERENCES

Bartolucci, S., Birchall, P., Bombin, H., Cable, H., Dawson, C., Gimeno-Segovia, M., Johnston, E., Kieling, K., Nickerson, N., Pant, M., Pastawski, F., Rudolph, T., & Sparrow, C. (2023). Fusion-based quantum computation. *Nature Communications*, *14*(1), 912. doi:10.103841467-023-36493-1 PMID:36805650

Ben Ouahi, M., Ait Hou, M., Bliya, A., Hassouni, T., & Al Ibrahmi, E. M. (2021). The effect of using computer simulation on students' performance in teaching and learning physics: Are there any gender and area gaps? *Education Research International*, *2021*, 1–10. doi:10.1155/2021/6646017

Caro, J. J., & Möller, J. (2016). Advantages and disadvantages of discrete-event simulation for health economic analyses. *Expert Review of Pharmacoeconomics & Outcomes Research*, *16*(3), 327–329. doi:10.1586/14737167.2016.1165608 PMID:26967022

Daley, A. J., Bloch, I., Kokail, C., Flannigan, S., Pearson, N., Troyer, M., & Zoller, P. (2022). Practical quantum advantage in quantum simulation. *Nature*, *607*(7920), 667–676. doi:10.103841586-022-04940-6 PMID:35896643

Frichi, Y., Jawab, F., Aboueljinane, L., & Boutahari, S. (2022). Development and comparison of two new multi-period queueing reliability models using discrete-event simulation and a simulation–optimization approach. *Computers & Industrial Engineering*, *168*, 108068. doi:10.1016/j.cie.2022.108068

Gregor, M., Hodoň, R., Grznár, P., & Mozol, Š. (2022). Design of a system for verification of automatic guided vehicle routes using computer emulation. *Applied Sciences (Basel, Switzerland)*, *12*(7), 3397. doi:10.3390/app12073397

Hakrama, I., & Frasheri, N. (2018). Agent-based modelling and simulation of an artificial economy with Repast. *International Journal on Information Technologies and Security*, *10*(2), 47–56.

Huttar, C. M., & BrintzenhofeSzoc, K. (2020). Virtual reality and computer simulation in social work education: A systematic review. *Journal of Social Work Education*, *56*(1), 131–141. doi:10.1080/10437797.2019.1648221

Ivanova, Y. (2022). Applications of simulation modelling method in prevention of jamming attacks. *Proceedings of the 2022 IEEE International Conference on Information Technologies (InfoTech-2022)*, IEEE. 10.1109/InfoTech55606.2022.9897121

Jodayree, M., Abaza, M., & Tan, Q. (2019). A predictive workload balancing algorithm in cloud services. *Procedia Computer Science*, *159*, 902–912. doi:10.1016/j.procs.2019.09.250

Katzfuss, M., Guinness, J., & Lawrence, E. (2022). Scaled Vecchia approximation for fast computer-model emulation. *SIAM/ASA Journal on Uncertainty Quantification*, *10*(2), 537–554. doi:10.1137/20M1352156

López-Santos, O., Salas-Castaño, M. C., & Salazar-Dantonio, D. F. (2022). Continuous simulation of the power flow in AC–DC hybrid microgrids using simplified modelling. *Computation (Basel, Switzerland)*, *10*(4), 52. doi:10.3390/computation10040052

Magalhães, D., Calheiros, R. N., Buyya, R., & Gomes, D. G. (2015). Workload modeling for resource usage analysis and simulation in cloud computing. *Computers & Electrical Engineering*, *47*, 69–81. doi:10.1016/j.compeleceng.2015.08.016

Mansouri, N., Ghafari, R., & Zade, B. M. H. (2020). Cloud computing simulators: A comprehensive review. *Simulation Modelling Practice and Theory*, *104*, 102144. doi:10.1016/j.simpat.2020.102144

Nassar, H. M., & Tekian, A. (2020). Computer simulation and virtual reality in undergraduate operative and restorative dental education: A critical review. *Journal of Dental Education*, *84*(7), 812–829. doi:10.1002/jdd.12138 PMID:32147841

Ram, J. P., Manghani, H., Pillai, D. S., Babu, T. S., Miyatake, M., & Rajasekar, N. (2018). Analysis on solar PV emulators: A review. *Renewable & Sustainable Energy Reviews*, *81*, 149–160. doi:10.1016/j.rser.2017.07.039

Romansky, R., & Noninska, I. (2019b). Cyber space features – security and data protection requirements. *Proceedings of the 2019 IEEE International Conference on Information Technologies (InfoTech-2019)*. IEEE. 10.1109/InfoTech.2019.8860880

Rotunno, G., Lo Zupone, G., Carnimeo, L., & Fanti, M. P. (2023). Discrete event simulation as a decision tool: A cost benefit analysis case study. *Journal of Simulation*, 1–17. doi:10.1080/17477778.2023.2167618

Stahl, I. (2009). *WebGPSS*. Beliber AB. http://www.webgpss.com/

Swarup, S. (2019, April). Adequacy: what makes a simulation good enough? In 2019 Spring Simulation Conference (SpringSim) (pp. 1-12). IEEE. doi:10.23919/SpringSim.2019.8732895

Vassilev, Tz. (2020). Comparison of several paradigms for accelerating physical simulation on a PC. *International Journal on Information Technologies and Security*, *12*(3), 13–24.

Zhang, X., & Furnstahl, R. J. (2022). Fast emulation of quantum three-body scattering. *Physical Review. C*, *105*(6), 064004. doi:10.1103/PhysRevC.105.064004

Chapter 12
Random Number Generators

ABSTRACT

The purpose of the chapter is to introduce basic concepts of random sequence generation that can be used in the simulation modeling of random flows. In the first part, features of random number generators (RNGs) are presented, distinguishing the two directions of realization – true RNGs and pseudo-RNGs. The first type are nondeterministic systems with hardware realization. The second type is realized based on deterministic algorithms with initial values for some parameters. A brief overview of these types was made. The second part of the chapter is devoted to Pseudo-RNGs, which are mainly used in simulation modeling. Their implementation is based on a selected probability distribution law – discrete or continuous. An example for program implementation of for the purposes of random flow modeling is presented and parameters of the most frequently used distributions are summarized. At the end, the three types of distributions for the simulation are discussed in more detail - uniform, normal, and distribution of Poisson.

1. BASIC CONCEPT OF RANDOM NUMBER GENERATORS

Random number generator (RNG) is a tool for generating a sequence of random numbers that can be implemented in a suitable programming environment for processes automatization. RNGs have many uses in different fields in the contemporary digital world, including cryptography, games, scientific research, etc. Specifically in cryptography, they are applied to generate secure keys and codes to protect confidential information in the encryption and decryption processes of messages and data (Crocetti et al., 2023). Due to the applicability of RNGs in various fields of scientific research, their quality in forming reliable and unpredictable sequences is important to ensure the reliability of the entire system. Various methods are applied to evaluate RNG quality, such as statistical tests, hardware evaluations, etc.

RNG is a generalizing concept, but in principle two main types can be defined – Pseudo-Random Number Generator (PRNG) and True Random Number Generator (TRNG) (Johnston, 2018). Generators of the first type are deterministic algorithms for randomly generating random numbers under given initial conditions, which explains the addition of "Pseudo" in the name. The second type of TRNG are nondeterministic systems, but they are not well defined. The implementation also requires the presence

DOI: 10.4018/978-1-6684-8947-5.ch012

of a hardware component because it is known that each computer algorithm is implemented as a sequence of discrete steps (a sequence of instructions), which does not meet the requirement of non-determinism.

Brief Summary of True RNGs

The development of contemporary technologies in the digital society has greatly increased the need for TRNG, especially for encryption purposes. Traditional hardware implementations consume large amounts of power, requiring powerful batteries for the devices, which reduces efficiency. On the other hand, the application of compact and promising on-chip TRNGs are a very good solution for generating secure random sequences used for cryptographic keys in mobile devices and Internet of Things with limited resources. Therefore, in (Equbal et al., 2023) a TRNG with hybrid CMOS-resistive random-access memory using coupled entropy sources is proposed, which improves the stochastic nature of the generated bits and reduces the power consumption in subsequent processing. The associated sources of entropy are of two types, the first being temporal and spatial (cycle-cycle and device-device, respectively). The second is related to the race condition of the reset latch. Examination of the raw generated bits has proven the correctness of the stochastic sequence under varying workload and temperature.

To solve the problem of the amount of power consumed, a TRNG circuit was developed using stable random telegraphic noise implemented by memristors (Pazos et al., 2023). The constructed TRNG uses junction memristors with a small and low-cost commercial microcontroller to produce a stochastic signal with high throughput and low power consumption. The subject of application of TRNGs developed based on memristors in the field of cryptography is also discussed in (Liu et al., 2023). In this case, the idea of developing physical-level primitives such as a two-dimensional electronic fingerprint integrated into a memristor TRNG is advocated. The purpose is to protect the generator from unwanted external influences and falsification. To counter possible falsification, a two-dimensional physical non-cloning function is implemented that provides multiple verification codes, with the authors determining applicability in nanoelectronics.

Another direction of using TRNG related to cloud services is presented in (Li et al., 2023). The reason for offering a new development is related to the restriction imposed by cloud vendors on Field Programmable Gate Array (FPGA). The use of circular oscillator circuits to generate random sequences is prohibited. In this reason, the new design uses clock oscillation as a source of entropy, which is collected with a time-digital convertor and a controllable delay line. A stochastic model was validated, and testing was done with test suites, after which 60 instances were deployed in the cloud.

New technological advances, especially in the field of quantum technologies, have introduced a new understanding of TRNG, forming two sub-areas of development (Guillan-Lorenzo et al., 2023). The first area is with physical construction and TRNG, where the generation is done by measuring parameters of classical systems with chaotic behavior. The second direction is Quantum RNG, which exploits the uncertainty of quantum physics and the inherent randomness in quantum mechanics that makes quantum systems a good source of entropy. Quantum random number generation is one of the modern technologies, offering various alternative methods of generation. In support of this is the review presented in (Herrero-Collantes & Garcia-Escartin, 2017) of various technological solutions proposed over time, starting from the early devices based on radioactive decay and reaching the various possibilities of quantum light devices for quantum entropy collection. Generating correct random number sequences using randomness extraction and amplification can be achieved even with less than reliable hardware devices. This is also confirmed by a statement in (Guillan-Lorenzo et al., 2023) that quantum RNGs provide completely random numbers disturbed only by measurement noise. Additionally, this paper clarifies that after digitization of the received data, binary random number sequences can be formed unrelated to the classical noise in the system. To confirm the correctness of generation, an analysis of the optical generators with three different devices was made in the article. The obtained experimental data were further processed by means of a Toeplitz extractor, which confirmed full fulfillment of the standard requirements for a sequence of random binary numbers.

Quantum RNG forms the random sequence based on a photon hitting a beam splitter and subsequent random detection of a particular output beam, which is inherent in the probabilistic nature of quantum mechanics. One drawback of this process, as pointed out in (Gerry et al., 2022), is the difficulty of producing single photons on demand using pulses of weak laser light. To overcome this, the paper proposes another approach based on moderate coherent light, where the probability of receiving an even or odd number of photons and 0.5. Beam splitters and single-photon detectors are used to count the generated photons. The advantage of the proposed technology is that it does not require the application of classical algorithms for removing deviations or post-processing of the generated bit sequence.

A survey of different types of TRNGs is made in (Yu et al., 2019), where a classification based on three main sources of entropy is proposed. These sources determine the main groups of TRNG – Noise (thermal noise on resistors and capacitors), Phase jitter (phase oscillation of oscillating signals), Chaos and Others. In Noise TRNG, the resistive noise is amplified to the required level by an amplifier, then processed by a comparator to compare the level with the reference – thus forming a digital random signal. In Phase jitter TRNG, the source of phase jitter noise is a ring oscillator in a metal oxide semiconductor (CMOS) circuit. In both methods, disadvantages can be pointed out, which are overcome in chaos due to its characteristics such as absence of periodicity, high degree of unpredictability, wide spectrum, and sensitivity to initial conditions. The article specifically focuses on TRNG based on Chaos, stating that the discovery of spontaneous chaos in semiconductor superlattices at room temperature is a useful step in the field. The reason is that chaos has a good sensitive dependence on initial conditions, randomness, periodicity, and reproduction. Typical structure of TRNG designed by five modules is shown in Figure 1 and includes the following parts (Yu et al., 2019):

- Entropy source for generation of an analog random signal.
- Sampling and quantifying the random signal.
- Signal converter "analog to digital".
- Processing the obtained sequence if it does not satisfy the uniform distribution.
- Random number testing.

Figure 1. Typical structure of TRNG
(Yu et al, 2019)

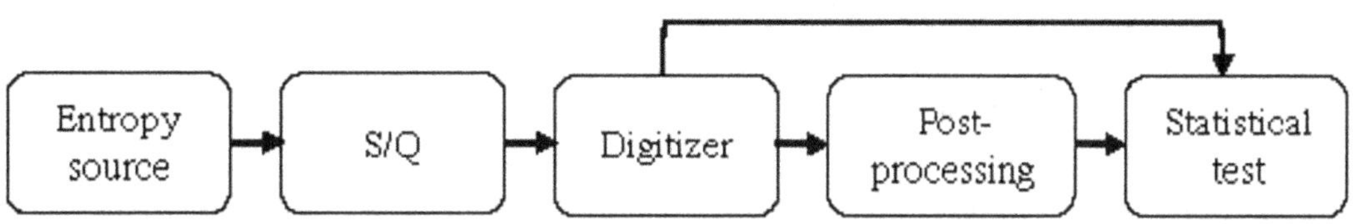

Two main types of time dependent Chaotic TRNGs (discrete and continuous) are defined in (Yu et al., 2019), and a brief description is given below. For each structure, a detailed review of existing development methods has been made, specifying their features and benefits. As a supplement to the overview, explanations of a mathematical model for describing spontaneous chaos in semiconductor lattices are made, a numerical solution is presented and its possible limitations when conducting experiments are discussed.

Chaotic TRNG with discrete time. This type of systems is widely used in non-linear sciences (physics, chemistry, biology, etc.) when it is necessary to generate reliable random numerical sequences. A one-dimensional nonlinear discrete-time dynamic system is defined by the expression $x_{k+1}=\tau(xk_{)}$, were the parameters xk (for k = 0, 1, 2, …) present concrete states, and τ is a mapping the transition $xk_{\rightarrow}xk_{+1}$. The successive passage through the states for k = 0, 1, 2,…, called a trajectory of discrete system in discrete time, starts from x0 and is determined by τ. In different discrete-time chaotic systems, different mapping methods can be applied – logistic, Bernoulli, etc. A study of RNG-related problems based on a discrete-time chaotic map is presented in (Magfirawaty et al., 2022). A new generator is proposed applying the Siponi Chaotic Map, which is a modification of the Logistic Chaotic Map. The base method could generate only positive real numbers, allows easy implementation, and has a better level of security compared to other nonlinear functions. The modification of the chaotic map proposed in the article has a deterministic nature with the possibility of processing positive and negative real values. The designed chaotic RNG is implemented on an FPGA board and the necessary testing is done to prove the functional efficiency.

Chaotic TRNG with continuous time. In this type of generator, the chaotic system is based on observation time series, and the state depends on time. The mathematical model of a chaotic dynamic system with continuous time is defined by the equation $\dot{x} = f\left(x,t\right)$, where $x \in R^n$ is variable of the state and function is f: $R^n \times R^n \rightarrow R^n$. In practice, there are various continuous-time chaotic systems such as Lorentz system, Chua scheme, Jain system, chaotic oscillator, hyperchaotic systems, etc. Main application of these systems, apart from TRNG development, is mainly in the field of neural networks, synchronization, the cheese communication, etc. The possibilities of continuous chaotic system-based pseudo-random number generation are discussed in (Karakaya, 2022), using two different chaotic systems as the source of entropy. The design of the electronic circuits of the chaotic generators is supported by a simulation run in the Orcad-Pspice environment. The goal is for the state variables to obtain values in fractional number format to be fed into a developed post-processor algorithm, with statistical testing of the generated random stream. Another design of a 3D continuous chaotic Lorentz system, "*improved by introducing a saw-tooth and sine functions*" and presented in (Azzaz et al., 2021). The design of chaotic TRNG is performed according to the Euler method using hardware description language (VHDL) and validated on Xilinx Virtex-II-Pro FPGA platform. Fixed-point machine arithmetic is used to represent the 32-bit data. The hardware architecture has a two-stage processing organization using a pipelined and parallel structure implemented in only 2 clock cycles.

Features of Pseudo-RNGs

Pseudo-random number generators (PRNGs) form a random sample (sequence of numerical values) that has a certain repetition step, which distinguishes it from the TRNGs presented above. The reason is that some initial condition is used which defines a limit of the generated values. In principle, PRNGs are

developed both in hardware implementation and as a software solution based on a discrete algorithm. A review of proposed PRNGs with verification of their capabilities and appropriate classification is done in (Bhattacharjee & Das, 2022). The proposed classification forms the following three groups of generators: ✓ Based on linear congruent generators; ✓ Based on linear shift registers with feedback; ✓ Based on cellular automata.

The Hausa theory discussed above in the TRNG overview also applies to creating PRNGs based on chaotic maps. These are mathematical functions that can generate a random numerical pattern based on a given initial value, making it pseudo-random. A review of implementations in this direction is given in (Naik & Singh, 2022), where a summary of the capabilities of chaotic maps for generating pseudo-random numbers is made. The comparative analysis of published results showed borderline estimates for important parameters, such as: ✓ the lowest correlation coefficient of 0.00006 was achieved with a chaotic Iked map; ✓ highest entropy value of 7.999995 bits per byte is when using quantum chaotic map; ✓ lowest execution time of 0.23 seconds is on Zaslavsky chaotic map; ✓ the maximum achieved speed of 15.367 Mbits/sec is with a hyperchaotic card.

In conclusion, the application areas of PRNG with chaotic maps are indicated, and besides the typical encryption, video games and digital marketing, the simulation of stochastic systems is also indicated.

A continuation of the Hausa idea is the development of a hardware PRNG based on chaotic maps presented in (de la Fraga et al., 2017). Bifurcation diagrams of four chaotic maps were used, and a rationale was made for choosing the best parameter values to ensure high entropy and a positive Lyapunov exponent. The resulting binary sequences were analyzed for the possibility of being applied to build PRNGs, both in a software version and as a hardware device. 32 and 64-bit microprocessors and binary floating-point and fixed-point arithmetic for them were used for the implementation of the software variant. The hardware implementation uses a programmable architectural concept based on FPGA circuits, and a connection is made between the circuit implementation and a personal computer through a communication interface. The result of the conducted research confirms the usefulness of chaotic maps for the implementation of PRNG, as the analyzes determined as the best solution a Bernoulli displacement map.

PRNGs have different applications to solve basic problems in the stochastic modeling of random events, for generation of discrete and continuous random values, random sequences of numbers, forming stochastic flows, simulation organization, etc. Due to the fact, that when solving such tasks, algorithmizing of the process and program implementation of the calculations is used, applications in the field of computer simulation and analytical research mainly find software implementations of PRNG. Based on this clarification, the abbreviation of the general form RNG will be used below for convenience.

A simple example of the applicability of RNGs in computer simulation and analytical modeling is as follows. For example, when determining a random time-interval Δt with a log-normal probability distribution with a time mean of 5.42 and a root-mean-square deviation of 1.41, the formula ” $t = EXP[\frac{1}{4}_{\ln " t} + R\tilde{A}_{\ln " t}]$ is used, where $\mu_{\ln \Delta t}$ is the mathematical expectation (in the case of a value of 5, 42), $(\sigma_{\ln \Delta t})^2$ is the variance (mean-square deviation is the square root of it) and R is a normally distributed random number. If this random number is 0.8, then a value of 7.17 is calculated for the random variable Δt.

Usually, pseudo RNG implementation algorithms are based on the uniform distribution of random numbers. The most commonly used is the multiplicative method, which is based on the recurrent dependence $r_i=(A.r_{i-1}-C)\text{mod } M$, where A, C are constants, and M is a sufficiently large positive integer. With a suitable choice of the constants and the initial value r_0, this formula allows the generation of a sequence of integers uniformly distributed in the interval (0, M-1) with a repetition period M. This defines the numbers as pseudo-random. A scaling transformation is used to generate numbers in the range (0, 1). In many areas of research in the field of probability and statistics, the gamma distribution is applied to generate random sequences. The reason is that it provides a good base for the study of digital communication and signal processing. In addition, algorithms for generating pseudo-random numbers, which are becoming faster with the development of computing technology, allow higher quality and efficiency of generation (Luengo, 2022). The article provides a detailed review of the existing algorithms for generating numbers with a gamma distribution, presenting a theoretical and mathematical justification for their validity.

When generating random sequences intended for solving certain more complex tasks, it is not always possible to ensure the necessary precision of the level of randomness. This is especially important in pseudo-random generations, which depend on the set initial value and the sampling algorithm. To overcome such a problem, a nonlinear filter is presented in (Álvarez et al., 2022) to improve the quality of output sequences formed by pseudo-RNGs with respect to statistical randomness. Techniques from symmetric ciphers involving initialization and an evolving internal state register are used. A combination of different types of operations is applied to propagate non-random patterns in the input sequence. The analysis performed shows a high level of efficiency and perfect randomness characteristics, allowing also the transformation of regular non-random sources (e.g., a counter) into pseudo-random sequences of values with moderate power consumption.

Improving the level of randomness of generated pseudo-random sequences is also a goal in (Jeong et al., 2020), where a new type of pseudo-RNG is proposed. Recurrent neural networks were used for the implementation, through which the appearance of a sequence of irrational numbers is imitated. This sequence is used to iteratively generate pseudo-random numbers, with an algorithm designed to ensure the absence of repetition in the output sequence. To determine the stability and level of randomness of the proposed system, experiments were conducted to observe its behavior under different conditions. In conclusion, it is stated that apart from the field of machine learning, the system can also be used in traffic management and the study of stochastic communications in sensor networks. The idea of using neural network and machine learning techniques in creating pseudo-RNGs is also developed in (Pasqualini & Parton, 2020), where a new reinforcement learning approach to features supported by PRNGs is proposed. In this case, the use of machine learning techniques is aimed at training a neural network for policy-based PRNG generation to solve an N-dimensional navigation problem. The N parameter specifies the length of the generation sequence period. It is possible to iteratively improve the supported policy based on the average result of testing performed over a period.

RNGs find application in modeling workload components in the following cases:

- Modeling a random event X occurring with probability p – a random value r uniformly distributed in the interval (0,1) is chosen and compared to p. If $r \leq p$ it is assumed that event X has occurred.
- Modeling of a discrete random variable (RV) X – it is assumed that it can take values x_i with probability p_i ($i = 1, 2, \ldots$, n), such that $x_1 < x_2 < \ldots < x_n$ and $\Sigma p_i = 1$. A random number r with a uniform distribution in the interval (0,1) is generated and an index $k \in [0,n]$, is determined, where $p_1 + p_2 + \ldots + p_{k-1} \leq \boldsymbol{r} \leq p_k + \ldots + p_n$. RV is then assigned the value $X = x_k$.
- Modeling continuous RV – appropriate distribution law with density $f(x)$ and distribution function $F(x)$ is used. To determine one particular value, r is generated with a uniform distribution in the

interval (0,1), by which the particular value x_k is determined for the continuous RV. An example for modeling continuous RV x with uniform distribution is presented in Figure 2.

Figure 2. Continuous random variable modeling

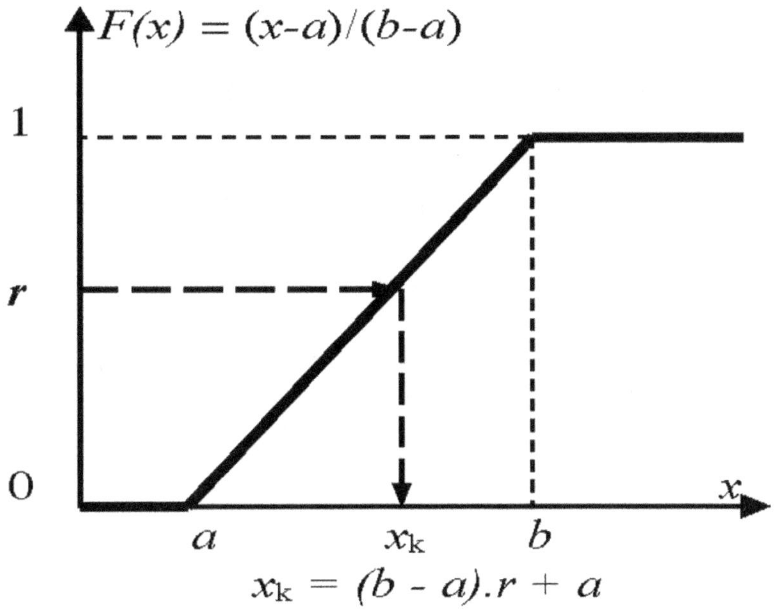

One application of RNG in the generation of pseudo-random numbers was made in the study of the Horadam sequence, for which, despite more than 60 years of research, new properties are still being discovered. This sequence is defined by the follows relation:

$w_n = pw_{n-1}+qw_{n-2}(n\geq 2)$

$w_0=a;\ w_1=b$

where the parameters q, b, p, q are arbitrary complex numbers. It is a general second-order repeatability in a complex plane given two initial values (*a* & *b*) and the repeatability coefficients *p* & *q* (Bagdasar et al., 2023). Paper presents an investigation of the Horadam sequence with a dense orbit in the two-dimensional domain of the complex plane, which is made in the framework of a uniform distribution of the complex argument. This is represented by a ring between two centered circles with radii ||A|−|B||=R1<R2=|A|+|B|, where the parameters A and B are determined by the relation below

$w_n = pw_{n-1}+qw_{n-2}+rw_{n-3}(n\geq 3)$

$w_0=a;\ w_1=b;\ w_2=c$

The case was used to develop a generator of pseudo-random sequence of values with uniform distribution in a fixed interval [-π, +π]. A test check of periodicity and correlation was made, and by applying Monte Carlo simulation the parameter π is estimated. The results of the experiments are compared with similar classical algorithms and for the first time is made investigation of the probability density for the sequence of radii and it is proved that the circle distribution is valid and for Horadam sequences of third order and higher.

2. GENERATION OF PSEUDO-RANDOM NUMBER SEQUENCES

Pseudo-random number generators are involved in the organization of computer simulations to represent stochastic workflows, as well as in systems modeled on the basis of queuing systems and networks. Input workflow initially enters a queue to wait for the release of the server (if it is busy serving a previous request), and the times between successive requests are determined by the set intensity of the flow. These probability parameters depend on the chosen distribution law. A typical example is the service in a car wash (server) of a flow of arriving cars (workflow) that enter initially in a queue. A typical service is FCFS and let us assume that the input flow distribution function is defined by the relationship between the probability of occurrence X_i and the inter-request interval Y_i, represented by the dependence (X_i, Y_i) of Table 1.

Table 1. Distribution function for input workflow

X_i	0	0,1	0,2	0,3	0,4	0,5	0.6	0.7	0.8	0.9	1
Y_i	0	1	3	4	5	7	10	14	20	33	56

Workflow model can be described as a discrete flow of requests based on the function *Y*=*Y*(*X*) from the values in Table 1. A matrix T with two rows is defined – first row T[1;] for the values of *X*, second row T[2;]for the correspondent values of *Y*. Program generation of discrete flow of requests based on the matrix T can be realized by APL2 function ‘TABFUN’ presented in Figure 3.

Figure 3. Program model of discrete flow

```
        ∇TABFUN [□]∇
      ∇ B←A TABFUN T
[1]     →(A<⌊/T[1;]) / OUT
[2]     →(A>)/T[1;]) / OUT
[3]     K←T[1;] ⍳A
[4]     →((ρTP) [2]<K) / ET
[5]     B←T[2;K]
[6]     →0
[7]  ET: I←+/ (A>T[1;])
[8]     C←(A-T[1;I]) × (T[2;I+1]-T[2;I])
[9]     B←T[2;I] + C÷(T[1;I+1]-T[1;I])
[10]    →0
[11]    OUT: 'ARGUMENT'
[12]    A
[13]    'IS TOO SMALL OR LARGE'
[14]    →0
      ∇

        T←2 6ρ1 2 3 4 5 6 12 14 20 28 33 59
        T
 1  2  3  4  5  6
12 14 20 28 33 59

        4.8 TABFUN T
32
```

When specifying an arbitrary value $b=Y(a)$, the following cases are possible.

- The value of b cannot be determined if $a < X_{min}$ and $a > X_{max}$.
- $b=Y_i$, if it exists the value $X_i=a$ – in this case it is made searching for the smallest index "*i*" of the value A in vector B, then B takes value *Y*[*i*].
- If ($X_{min} < a < X_{max}$) the parameter *b* takes a value determined by the method of linear interpolation.

Two initial models are developed which can be used for investigation of the processes of requests (cars) arriving and for transition between sequence of events. The program realized by using APL2 environment are presented in Figure 4. The first function "ARRIVAL" is the model of the input flow, which generates random value 0<A<1 by using system RNG (?N) – see row [1] in the Function 4(a). The second function "NEXT" compares two-time events and transitions at the lower exponent – Figure 4(b).

Figure 4. Models of the investigation organization

```
       ▼ARRIVAL [□]▼
     ▼ NEWTIME←ARRIVAL TIME
[1]    A←(?100)÷100
[2]    DT←A TABFUN T
[3]    NEWTIME←TIME+DT
[4]    Q←Q,NEWTIME
     ▼
```

(a) Model of incoming cars (requests)

```
       ▼NEXT [□]▼
     ▼ EV←TIMERS NEXT STOPTIME
[1]    EV←TIMERSι⌊/TIMERS
     ▼
```

(b) Model for definition of the next event

The program description of the service node as APL2-function "SERVER" is presented in Figure 5. The model parameters are: Q – queue for waiting; QSTAT – statistics which is collected for the queue; SERVSTAT – statistics collected for the server.

Figure 5.

```
       ▼SERVER [□]▼
     ▼ NEWTIME←SS SERVER TIME
[1]    →(Q[1]>TIME)/ET
[2]    NEWTIME←TIME+SS
[3]    QSTAT←QSTAT,Q[1],NEWTIME
[4]    Q←1↓Q
[5]    →0
[6] ET: SERVSTAT←SERVSTAT,Q[1]-TIME
[7]    NEWTIME←Q[1]
     ▼
```

The organization of model investigation of the object, which is described as a single-channel QS can be presented by the APL2-function in Figure 6. The event nature of the processes in serving the input flow in the server is reflected by three components of the CLOCK work vector: CLOCK[1] – current time CLOCK[2] – time for receiving the current request; CLOCK[3] – time during which the service device is free. Additionally, work variables are used: ENDTIME for the general model time; EVENT to determine the next even; SPEED presents the time for service a separate car (request)

Figure 6.

```
        ∇QS [□]∇
       ∇ QS
[1]       'ENTER HOURS FOR SIMULATION'
[2]       ENDTIME←60×□
[3]       'ENTER MINUTES OF SERVICE:'
[4]       SPEED←□
[5]       QSTAT←2 0ρSERVSTAT←Q←0ρ0
[6]       CLOCK←3ρ0
[7]   ET: →(CLOCK[3]≥ENDTIME)/OUT
[8]       EVENT←(1↓CLOCK)NEXT ENDTIME
[9]       →(ARRIVE, DEPART) [EVENT]
[10]  ARRUVE:CLOCK[1]←CLOCK[2]
[11]      CLOCK[2]←ARRIVAL CLOCK[1]
[12]      →ET
[13]  DEPART:CLOCK[1]←CLOCK[3]
[14]      CLOCK[3]←SPEED SERVER CLOCK[1]
[15]      →ET
[16]  OUT:QSTAT STAT SERVSTAT
     ∇
```

The most used probability distributions used in modeling workflow and service processes are summarized in Table 2. In the following parts, the analytical definitions of algorithms for generating pseudo-random number sequences with basic probability distributions applicable to program implementation are presented.

Table 2. Basic probabilistic distributions

Type, density (f) and function (F) of distribution	*Average*	*Argument*	*Parameters*
Uniform * $f(x)=1/(b-a)$ $F(x)=(x-a)/(b-a)$	$(a+b)/2$	$a \leq x \leq b$	a, b
Exponential * $f(x)=a.EXP(-ax)$ $F(x)=1-EXP(-ax)$	$1/a$	$x \geq 0$	$a > 0$
Poisson $f(x)=\left(\frac{a^i}{i!}\right).EXP(-a)$	a	$-\infty < x < \infty$	$a, i > 0$
Normal $f(x)=\left[\frac{1}{\sigma}.(2\pi)^{1/2}\right].EXP\left[\frac{-(x-a)^2}{2\sigma}\right]$	a	$i = 0, 1, 2, \ldots$	$a > 0$
Binomial $f(x)=\binom{n}{i}.p^i.(1-p)^{n-i}$	$n.p$	$i = 0, 1, 2, \ldots, n$	$n.p$

* For values of the argument x outside the specific range $f(x)=F(x)=0$

UNIFORM DISTRIBUTED RANDOM NUMBER SEQUENCE

A uniformly distributed random number u can assume any value in the range [0, 1] with equal probability. The probabilistic characteristics are as follows:

$f(x)=1;\ 0 \leq x \leq 1$ and $F(x)=x;\ 0 \leq x \leq 1$

$$\mu = \int_0^1 x.dx = 0,5 \text{ and } \sigma^2 = \int_0^1 (x-\mu)^2.dx = 0,0833 \rightarrow \sigma = 0,2887$$

Typically, 57.74% of random numbers are in the range $\mu \pm \sigma$. For the generation, multiplicative methods with repeatedly repeated operations are used. Analytical descriptions of three basic algorithms for generating uniformly distributed random numbers are presented below:

ALG_U1: $u_i^* = k.u_{i-1}$, where: $k=8t\pm3$; t is integer; u_0 is an odd number; u_i is the fractional spart of the number u_i^*.
ALG_U2: $u_{i+1} = 1/2|u_1 u_0 (mod\ 2^6)|$.
ALG_U3: $u_{i+1} = (2^a+1)u_1 + c(mod\ 2^{35})$, where $a \geq 2$ and c is odd number.

Criteria for checking the uniform distribution of a stochastic sequence are the following possibilities:

a) frequency criterion – *N* random numbers are taken, for which the count *m* of numbers in the range $[(\mu-\sigma), (\mu+\sigma)]=[0.2113, 0.7878]$ is determined and if *m/N* is about 0.5774 it is accepted, that the generated sample is approximately uniformly distributed.

b) b) internal correlation method – the sample $\{u_i / i=1 \div N\}$ is assumed to be uniformly distributed if the dependence below is realized:

$$¼=\frac{1}{N}\sum_{i=1}^{N} u_i \approx \frac{1}{2} \text{ and } \tilde{A}^2 = \frac{1}{N}\sum_{i=1}^{N} u_i^2 \approx \frac{1}{3}.$$

An example for analytical realization of the "ALG_U1" is shown in Figure 7. The initial parameters of the solution are $N= 10$; $u_0 = 0{,}37843$; $t = 5$; $k = 8t - 3 = 37$ and the sequence of separate discrete steps is determined by $u_i^* = k.u_{i-1}$ for $i = 1, 2, \ldots, N$.

Figure 7. Analytical solution of algorithm "ALG_U1"

$$u_1^* = k.u_0 = 37.0{,}37843 = 14{,}00191 \Rightarrow u_1 = 0{,}00191$$
$$u_2^* = k.u_1 = 37.0{,}00191 = 0{,}07067 \Rightarrow u_2 = 0{,}07067$$
$$u_3^* = k.u_2 = 37.0{,}07067 = 2{,}61479 \Rightarrow u_3 = 0{,}61479 \checkmark$$
$$u_4^* = k.u_3 = 37.0{,}61479 = 22{,}74723 \Rightarrow u_4 = 0{,}74723 \checkmark$$
$$u_5^* = k.u_4 = 37.0{,}74723 = 27{,}64751 \Rightarrow u_5 = 0{,}64751 \checkmark$$
$$u_6^* = k.u_5 = 37.0{,}64751 = 23{,}95787 \Rightarrow u_6 = 0{,}95787$$
$$u_7^* = k.u_6 = 37.0{,}95787 = 35{,}44119 \Rightarrow u_7 = 0{,}44119 \checkmark$$
$$u_8^* = k.u_7 = 37.0{,}44119 = 16{,}32403 \Rightarrow u_8 = 0{,}\ 32403 \checkmark$$
$$u_9^* = k.u_8 = 37.0{,}\ 32403 = 11{,}98911 \Rightarrow u_9 = 0{,}\ 98911$$
$$u_{10}^* = k.u_9 = 37.0{,}\ 98911 = 36{,}59707 \Rightarrow u_{10} = 0{,}\ 59707 \checkmark$$

A check by the frequency criterion shows that 6 numbers (marked with the sign ✓) fall into the range [0.211, 0.787], which determines $m/N = 6/10$, i.e., the estimate is close to the norm 0.5774.

A check by the internal correlation method for the formed sample provides calculated values $\mu=0.486\approx1/2$ и $\sigma^2=0.298\approx1/3$ that satisfy it.

Random Number Sequence with Normal Distribution

The normal distribution is characterized by follows functions:

$$f(x) = \frac{1}{\tilde{A}(2\grave{A})^{1/2}} EXP\left[\frac{(x - \frac{1}{4})^2}{2\tilde{A}^2}\right] \text{and } F(x) = \int_{-\infty}^{x} f(x).dx\,.$$

with parameters $\mu=0$, $\sigma^2=1$, $\mu_3=0$, $\mu_4=3$. With it, the number of random numbers in the range $\mu\pm\sigma$ is no less than 68.3%, 95.5% of random numbers fall in the range ($\mu\pm2\sigma$), and 99.7% numbers fall in the range ($\mu\pm3\sigma$).

Analytical descriptions of three basic algorithms for generating normally distributed random number sequences are presented below.

ALG_N1:

- Generation of random number u_i wit uniform distribution in [0, 1].
- With the normal distribution function $u_i = F(x=r_j)$ successive normally distributed random numbers are determined. For N≥6 the following formula can be applied:

$$r = \frac{\sum_{j=1}^{N} u_j - \frac{N}{2}}{\left(\frac{N}{12}\right)^{1/2}}\,.$$

ALG_N2 (method of Neuman):
If the uniform distributed numbers u_1 and u_2 satisfy the inequality $ln\ u_2 \leq -2b^2(u_1-1/2)^2$,
in this case the number $r=b(2u_1-1)$ is a normal distributed random number ($b=const$).
ALG_N3 (meth od Muller):
If u_1 and u_2 are uniform distributed random numbers for the determination of two normal distributed random numbers r_1 and r_2 the follows formulas can be used:
$r_1 = (-2.ln\ u_1)^{1/2}.cos(2\pi u_2)$;
$r_2 = (-2.ln\ u_1)^{1/2}.sin(2\pi u_2)$.

Criteria for checking the normal distribution of a stochastic sequence $\{r_j\ /\ j=1, \ldots, N\}$:

$$\frac{1}{N}\sum_{j=1}^{N} r_j \cong 0 \text{ and } \frac{1}{N}\sum_{j=1}^{N} r_j^2 \cong 1$$

For an example analytical implementation of ALG_N1, a sequence of 10 generated numbers with a normal distribution was obtained by using program implementation of the algorithm, and the values are summarized in Table 3. A sample of 6 uniformly distributed numbers formed by a program module implementing ALG_U1 at a given starting odd number 0.375 was used to calculate each random normal distributed number r_j.

Table 3. Results obtained by "ALG_N1" program execution

1.103	0.164	– 0.1153	0.4257	– 1.053	0.5419	– 0.5639	1.24	–2.2722	0.447

Inspection of the generated random numbers shows the following results:

$$\frac{1}{N}\sum_{j=1}^{N} r_j = \frac{1}{10}(0.0828) = 0.00828 \cong 0$$

$$\frac{1}{N}\sum_{j=1}^{N} r_j^2 = \frac{1}{10}(10.1073) = 1.01073 \cong 1$$

Random Number Sequence with Poisson Distribution

The parameters of the Poisson distribution and algorithm for generating random numbers are presented below.

$$f(s) = \frac{(np)^s}{s!}e^{-np};\ F(m) = \sum_{i=0}^{m} f(s) = \sum_{s=0}^{m}\frac{(np)^s}{s!}e^{-np};\ ¼ = Ã^2 = ¼_3 = n.p;\ ¼_4 = 3.n.p.(n.p + \frac{1}{3}).$$

ALG_P1:

(1) A sequence $\{u_j / j=1, \ldots, N\}$ of uniform distributed random numbers is formed.
(2) A production of the generated numbers is formed, and the number of factors involved must satisfy the condition:

$$\prod_{j=1}^{N} u_j < e^{-np}.$$

(3) The number (N-1) defines the value s, which belongs to the sequence of random numbers wit Poisson distribution with parameter $\mu = np$. If the inequality is satisfied by the number u_1, then $s=0$ is defined.

<u>Criteria for checking</u> the formed sample with Poisson distribution is:

$$\frac{1}{m}\sum_{i=1}^{m} s_i \cong n.p \quad ; \quad \frac{1}{m}\sum_{i=1}^{m} s_i^2 \cong n.p.(n.p+1)\,.$$

The analytical implementation of algorithm ALG_P1 can be illustrated with the following example. Let us search for a random number belonging to a Poisson random process with μ=2,5. According to the condition of the algorithm, a uniformly distributed stochastic sequence $\{u_1, \ldots, u_N\}$ is formed until the formed product satisfies the inequality $u_1 \times u_2 \times \ldots \times u_N < e^{-np}$ at μ=*np*. The implementation given in Table 4 defines a random number with Poisson distribution *s*=*N*-1=4. Sample evaluation results of the program implementation of "ALG_P1" are summarized in Table 5.

Table 4. An example for realization of "ALG_P1"

i	u_i	$u_1 \times u_2 \times u_3 \times \ldots \times u_N < e^{-2.5} = 0{,}08208$	*Условие*
N=1	0,91646	0,91646	Not fulfilled
N=2	0,89198	0,81746	Not fulfilled
N=3	0,64809	0,52979	Not fulfilled
N=4	0,16376	0,08676	Not fulfilled
N=5	0,91782	0,07963	***Fulfilled***

Table 5. Generalization of the results of program realization

Sample	{3, 4, 3, 3, 2}	{2, 1, 1, 3, 2, 1, 4, 2, 3, 3}
Parameters	m=5; μ=np=2,5 (n=25; p=0,1)	m=10; μ=np=2 (n=20; p=0,1)
Criteria: a) $1/m\sum s_i \approx np$ b) $1/m\sum (s_i)^2 \approx np(np+1)$	15/5 = 3 ⇔ *np* = 2,5 47/5 = 9,4 ⇔ *np*(*np*+1) = 8,75	22/10 = 2,2 ⇔ *np* = 2 58/10 = 5,8 ⇔ *np*(*np*+1) = 6
Conclusion	*Stochastic number sequence approximates Poisson distribution*	*Stochastic number sequence closely reproduces Poisson distribution*

Determining a particular value for parameter μ=*n.p* when modeling a stochastic input flow of requests for an example computer system can be illustrated by the example in Figure 8.

Figure 8. Determination of the Poisson distribution parameters

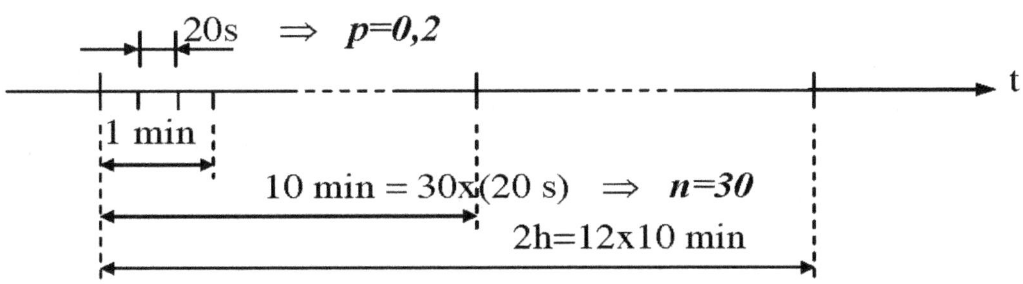

The presented example is related to defining a model of a stochastic input flow of requests arriving at intervals of η = 10min in the rakes of a total time of 2 hours, known to have a Poisson distribution, with the probability of generating a request in an interval τ=20s is p=0,2. The two-time intervals are considered:

τ = 20 s – one request is generated wit probability p=0,2 ⇒ $\mu=np=0,2$;
η = 10 min = 600 s = 30τ ⇒ n = 30 requests with p=0,2 ⇒ $\mu=np=6$.

For the entire period of 2 hours, 12 intervals η of 10 minutes each are determined, and for each of them the condition for forming a Poisson-distributed stochastic sequence is "*less than* e^{-6} = *0.002478*".

REFERENCES

Álvarez, R., Martínez, F., & Zamora, A. (2022). Improving the statistical qualities of pseudo random number generators. *Symmetry*, *14*(2), 269. doi:10.3390ym14020269

Azzaz, M. S., Fellah, R., Tanougast, C., & Kaibou, R. (2021). Design and FPGA implementation of TRNG based on a new multi-wing attractor in Lorenz chaotic system. *The European Physical Journal. Special Topics*, *230*(18), 3469–3480. doi:10.1140/epjs11734-021-00234-6

Bagdasar, O., Chen, M., Drăgan, V., Ivanov, I. G., & Popa, I. L. (2023). On Horadam sequences with dense orbits and Pseudo-Random Number Generators. *Mathematics*, *11*(5), 1244. doi:10.3390/math11051244

Bhattacharjee, K., & Das, S. (2022). A search for good pseudo-random number generators: Survey and empirical studies. *Computer Science Review*, *45*, 100471. doi:10.1016/j.cosrev.2022.100471

Crocetti, L., Nannipieri, P., Di Matteo, S., Fanucci, L., & Saponara, S. (2023). Review of methodologies and metrics for assessing the quality of Random Number Generators. *Electronics (Basel)*, *12*(3), 723. doi:10.3390/electronics12030723

de la Fraga, L. G., Torres-Pérez, E., Tlelo-Cuautle, E., & Mancillas-López, C. (2017). Hardware implementation of pseudo-random number generators based on chaotic maps. *Nonlinear Dynamics*, *90*(3), 1661–1670. doi:10.100711071-017-3755-z

Equbal, M. S., Ketkar, T., & Sahay, S. (2023). Hybrid CMOS-RRAM True Random Number Generator exploiting coupled entropy sources. *IEEE Transactions on Electron Devices*, *70*(3), 1061–1066. doi:10.1109/TED.2023.3241122

Gerry, C. C., Birrittella, R. J., Alsing, P. M., Hossameldin, A., Eaton, M., & Pfister, O. (2022). Proposal for a quantum random number generator using coherent light and a non-classical observable. *Journal of the Optical Society of America. B, Optical Physics*, *39*(4), 1068–1074. doi:10.1364/JOSAB.441210

Guillan-Lorenzo, O., Troncoso-Costas, M., Alvarez-Outarelo, D., Diaz-Otero, F. J., & Garcia-Escartin, J. C. (2023). Optical quantum random number generators: A comparative study. *Optical and Quantum Electronics*, *55*(2), 185. doi:10.100711082-022-04396-y

Herrero-Collantes, M., & Garcia-Escartin, J. C. (2017). Quantum random number generators. *Reviews of Modern Physics*, *89*(1), 015004. doi:10.1103/RevModPhys.89.015004

Jeong, Y. S., Oh, K. J., Cho, C. K., & Choi, H. J. (2020). Pseudo-random number generation using LSTMs. *The Journal of Supercomputing*, *76*(10), 8324–8342. doi:10.100711227-020-03229-7

Johnston, D. (2018). *Random Number Generators—Principles and Practices: A Guide for Engineers and Programmers*. Walter de Gruyter GmbH & Co KG., doi:10.1515/9781501506062

Karakaya, B. (2022, May). Chaotic system-based pseudo random bit generator and post-processor design for image encryption. In *2022 13th National Conference with International Participation (ELECTRONICA)* (pp. 1-4). IEEE. 10.1109/ELECTRONICA55578.2022.9874431

Li, X., Stanwicks, P., Provelengios, G., Tessier, R., & Holcomb, D. (2023). Jitter-based adaptive True Random Number Generation circuits for FPGAs in the cloud. *ACM Transactions on Reconfigurable Technology and Systems*, *16*(1), 1–20. doi:10.1145/3487554

Liu, B., Ma, J., Tai, H. H., Verma, D., Sahoo, M., Chang, Y. F., Liang, H., Feng, S., Li, L.-H., Hou, T.-H., & Lai, C. S. (2023). Memristive True Random Number Generator with intrinsic two-dimensional physical unclonable function. *ACS Applied Electronic Materials*, *5*(2), 714–720. doi:10.1021/acsaelm.2c01533

Luengo, E. A. (2022). Gamma Pseudo Random Number Generators. *ACM Computing Surveys*, *55*(4), 1–33. doi:10.1145/3527157

Magfirawaty, M., Lestari, A. A., Nurwa, A. R. A., Mt, S., & Ramli, K. (2022). A novel discrete-time chaos-function-based random-number generator: Design and variability analysis. *Symmetry*, *14*(10), 2122. doi:10.3390ym14102122

Naik, R. B., & Singh, U. (2022). A review on applications of chaotic maps in pseudo-random number generators and encryption. *Annals of Data Science*, 1-26. doi:10.1007/s40745-021-00364-7

Pasqualini, L., & Parton, M. (2020). Pseudo random number generation: A reinforcement learning approach. *Procedia Computer Science*, *170*, 1122–1127. doi:10.1016/j.procs.2020.03.057

Pazos, S., Zheng, W., Zanotti, T., Aguirre, F., Becker, T., Shen, Y., Zhu, K., Yuan, Y., Wirth, G., Puglisi, F. M., Roldán, J. B., Palumbo, F., & Lanza, M. (2023). Hardware implementation of a true random number generator integrating a hexagonal boron nitride memristor with a commercial microcontroller. *Nanoscale*, *15*(5), 2171–2180. doi:10.1039/D2NR06222D PMID:36628646

Yu, F., Li, L., Tang, Q., Cai, S., Song, Y., & Xu, Q. (2019). A survey on true random number generators based on chaos. *Discrete Dynamics in Nature and Society,* 1-10. doi:10.1155/2019/2545123

Chapter 13
Organization and Application of Computer Simulation

ABSTRACT

The goal of the chapter is to present basic principles for organization of simulation and realization of simulation experiment to investigate defined problem. The first part discusses the basic forms of simulation research, and a brief summary of simulation languages and environments is presented. A selection of the large number of specialized languages with comments on their applicability is presented. The second part is devoted to the planning of simulation experiments and the organization of implementation. This is related to determining the main factors and defining their levels, determining the optimal time for simulation, as well as performing pre-processing of the data obtained from the experiments. The last part presents the organization of selected simulation experiments, indicating that two directions can be defined - for basic simulation and for complementary simulation (combination with other research approaches to obtain more accurate and informative results). The idea of the second group is to supplement the results received based on another research approach.

1. BASIC FORMS OF SIMULATION AND SPECIALIZED LANGUAGES

Main Directions of Simulation Research

The simulation approach allows conducting computer experiments to form results based on pseudo-random samples of operating parameters and request flows. In the initial classification of simulation models, two main opportunities of their implementation were indicated, which will be discussed below. They define both main directions in conducting simulation studies by using simulation models.

Statistical simulation models, during the implementation of which statistical information is accumulated about the monitored variables (observable factors) when servicing the flow of requests, through which statistical estimates are formed. Exemplary estimates obtained from the execution of a simulation program are the following: average occupancy of the devices in the system; distribution of processing

DOI: 10.4018/978-1-6684-8947-5.ch013

time in different states; queue lengths and average waiting times for service; average dwell time of a request in the system; usability (load) of individual devices in the system; total system time etc.

To increase the accuracy of the estimates, it is necessary to accumulate enough registrations, which leads to an increase in the total modeling time. In addition, the speed of the model depends on the efficient organization of the event records. List structures are typically used to maintain lists of future events that are populated dynamically.

The task of simulation experiments is to understand the statistical behavior of the studied object, system, or process and to determine a correct estimate for one or several behavioral parameters that are of particular interest. However, this requires a correct definition of the simulation model and organization of the study, as certain guidelines "*for design, execution, analysis, reporting, and presentation*" are presented in (Morris et al., 2019). The real contribution of this article is a proposal of structural approach for statistical simulation organization. The article states that simulation studies are empirical experiments, which defines the meaning of statistical simulation modeling, which in turn requires very good design justification. It further states that Monte Carlo simulation introduces statistical techniques using pseudo-random sampling, which distinguishes it from pure simulation and relates to statistical modeling, which is discussed in Chapter 14 of this book.

Statistical simulation has also been applied in a multi-core coil development investigation which is presented in (Hoole et al., 2021)1 where the main goal is to determine characteristics of wire winding processes. It is pointed out that the traditional practice of basing such studies on finite element analysis cannot account for variability in conductor placement. In this respect, the article proposes a methodology for statistical simulation to correctly represent both the variability of the wire and the alternating current (AC) losses of a distributed winding. After performing an initial calibration of the methodology, results were obtained that determined the main statistical characteristics of the object under study, limiting the undesirable results of mixing threads between the bundles in the coils.

Other applications of the statistical approach in simulation are given in (Eini et al., 2023), where a comparative analysis of two approaches was made to confirm experimental results. The first approach is based on a process model, while the second is a data-driven statistical simulation using a feed-back neural network. Also, in (Li et al., 2021), where the effectiveness of direct statistical simulation (DSS) for dynamo action is investigated. Two models were analyzed – the first of third order with cubic nonlinearity, and the second with only quadratic nonlinearities. The paper presents the developed complete Python symbol package methodology for deriving the statistical governing equations of low-order dynamical systems.

Functional simulation models reflect the structural features and functional characteristics of individual components in the model, representing the behavior of the object over time. Most often, the functional algorithm of a given system (in particular, a computer system) under the influence of control (hardware or software) is described and studied. In general, developing a sufficiently adequate model is laborious, requiring good skills and knowledge of the site. It applies in cases:

- When analyzing the optimal configuration of a given system for a specific workload (when purchasing or when modifying)
- When developing software systems to make a preliminary assessment of the impact on the hardware environment.
- If necessary for further improvements and changes to a system put into operation in order to predict the expected effect.

Functional simulation is used in various fields of research and some examples aimed at contemporary n technological solutions from recent years are presented below.

As stated in (Huo et al, 2023) "*Recently, system functional simulation mostly uses discrete functional logic models, and it is difficult to simulate the impact of continuous behavioral parameter changes on functional logic*". In this reason, a method for realizing processes of functional simulation based on an integration of the two different approaches – combination of discrete and continuous simulation of a system behavior. The goal is to help in the correct and effective analysis of functional design. The main idea is to unite the opportunities of discrete functional modeling and capabilities of continuous simulation approach. The proposed method is experimentally investigated by simulating an aircraft elevator to confirm the effectiveness and usefulness of the solution.

Quantum technology is seriously entering the computing space, with modern quantum computers in many cases implementing real-time control systems. This requires significant software complexity to support more features and greater real-time capabilities. As pointed out in (Riesebos et al., 2022), testing such systems is a complex process, which is why the paper proposes an interactive simulator to study the performance. Signal simulation is performed at the level of application programming interface (API), and experimental results show a significant increase of an average of 6.9 times the number of simulated cores compared to a hardware implementation. In addition, an average simulation accuracy of 97.9% is presented.

An overview of the capabilities of functional simulation of biological functions in memristors and neuromorphic devices is given in (Chen et al, 2023) with a presentation of their biological basis as a starting point. The next steps are related to representing the biological functions of the synapse, neural and intelligent behaviors. The paper's summary of biological function simulation capabilities is due to the growing application of memristors and neuromorphic devices, with the goal of helping to innovate and improve them. The application of functional simulation in the field of medicine is also discussed in (Péan & Goksel, 2020), where "*a surface-based modeling approach for muscles, which simplifies the modeling process and is efficient for computation*" is presented. It is pointed out that many existing shoulder musculoskeletal simulation models implement line segment approximation of muscles with potential constraints. This is the reason why the paper proposes the use of surface geometries for functional simulation modeling of muscles through membrane elements within a finite element simulation framework. The successful reproduction of the shoulder movement was confirmed when applying the functional simulation.

Brief Summary of Simulation Languages and Environments

Regardless of the type of simulation model, when implementing the computer simulation, it is necessary to plan the simulation experiment before the actual implementation and selection of suitable program language. These two important stages in the simulation research process are related to the language environment in which the model is implemented. It is recommended to use specialized programming languages that have constructs for the implementation of entire procedures for servicing and registering the occurrence of events. An example classification of computer simulation programming languages is as follows.

- *Action-oriented languages* (CSL, GSP, FORSIM, etc.) – models are based on checking various conditions under which the corresponding action is performed. If the conditions are met, the control program includes the action in the queue of executables.
- *Event-oriented languages* (SIMSCRIPT, GASP, SIMCON, etc.) – each event is distributed in a certain sequence. Events are instantaneous and are realized when certain conditions are met.
- *Flow and process-oriented languages* (GPSS, SIMAN, SIDIS, etc.) – they support dynamic objects (transactions) that form streams of requests (messages). Flows move between different logical objects (blocks) of the system model, reflecting the structure of the object and the logic of the functional algorithm.

There are many languages and language environments for organizing and conducting simulation studies, even some publications state that they are over 4000. This makes it impossible to make an adequate, accurate and detailed classification of language structures. Furthermore, the advent of new intelligent and network technologies has led to the development of new simulation approaches, placing specific demands on the applied simulation languages. When discussing the information world, the main application finds discrete-event simulation (DES), as an overview of modern solutions in this area with a focus on production systems is presented in (Turner & Garn, 2022). The role and importance of real-time interactive animations, the development of so-called augmented reality, the place of artificial intelligence in interactive simulation, approaches to simulating processes in complex automated decision-making systems, etc. are discussed. It is indicated that commercial language environments for DES (Anylogic, Arena, Simul8, Witness) provide a convenient script language for individual operations when degenerating systems with visual components and indicate the place of older specialized simulation languages of the 4th generation (GPSS, ECSL, Simula, Modsim). Although Simula additionally introduces object-oriented programming (OOP), the conclusion drawn is that there is no universally accepted standard OOP simulation language. The trend is to provide DES libraries for programming languages (e.g., C++ & SystemC, Java & Ptolemy II, JSimpleSime, DESMO-J, MATLAB & SimEvents, Python & SimPy and Julia & SimJulia)

Based on the above, an example of simple classification of computer simulation languages and environments is presented below, and a summary of key features of selected programming languages is given in Table 1.

1. Languages and packets for universal simulation.
    - SLAM; SLAMSYSTEM – Languages for discrete and continuous simulation supported on the base of the Fortran language and the model is described by diagram as a network of graphical elements.
    - AWESIM – Version of the SLAM language but realized in C program language; it has enhanced capabilities for setting up the model, for animation, and for executing object code written in Visual C++ or Visual Basic.
    - TESS – Packet based on the SLAM language which supports data base with different scenarios.
    - ACSL (Advanced Continuous Simulation Language) – Language for simulation study of the behavior of a given system based on a mathematical model. It provides a computer calculation of the dynamic behavior and provides graphical representations and a tabular display of selected model variables.

- X-language – an integrated intelligent modeling and simulation language that supports the description of system layer structure and physical behavior (Zhang et al., 2021). It allows the simulation of continuous discrete models of events and agents to be maintained.
- Silk – a general-purpose simulation language based on the Java programming language that combines process-oriented modeling structures with common object-oriented language functions. Uses a multithreading facility to coordinate concurrent object streams in process-oriented simulation.

2. Specialized simulation languages.
   - Languages based on the object-oriented approach to programming – INSIGHT; PCModel; SIMPLE_1; MODSIM II; SIM++.
   - NETWORK II.5 – Language for simulation of computer systems and networks.
   - MATLAB – (short for "Matrix Laboratory") is a high-level programming language and interactive numerical modeling environment developed by MathWorks. MATLAB is widely used for algorithmic modeling, data processing, simulation, and graphical representation in scientific and engineering disciplines. Simulink is a dynamic systems modeling, simulation, and analysis platform provided as part of the MathWorks product line. Simulink provides a graphical user environment in which users can build models through "flow charts" without the need to write source code, but through the tools and capabilities of visual programming.
   - CSIM – process-oriented language which is a part of the program language C and includes additional functions for system resource modeling, transmission of messages, collection of data, etc. It allows rapid development of short models and provides an environment for their subsequent implementation in an efficient manner.
   - CSL – simulation language oriented to modeling complex logical problems in which groups of objects with similar properties are formed. The first application of the language was in simulating Monte Carlo processes.
   - SLONN (Simulation Language for modeling Of Neural Networks) – a language with a hierarchical organization with the possibility of spatial and temporal representation of a neuron and synaptic plasticity when simulating small and large neural networks. Provides capabilities to introduce branching when modeling the connection between neurons to degenerate networks of different sizes.
   - MAL (Meta Attack Language) – a language framework for the development of domain-specific languages (DSLs) providing the opportunity to generate attack graphs during simulation modeling of infrastructures. Based on it, several language versions with different areas of application have been developed: ● coreLang (Katsikeas et al., 2020) for simulation modeling of infrastructures and analysis of weak points susceptible to attacks in them; ● powerLang (Hacks et al., 2020), designed for simulations of IT infrastructures in the energy field, for which language components of presenting the features of industrial control systems are included; ● vehicleLang (Katsikeas et al., 2022) for modeling vehicles in IT infrastructures, traffic flows and possible impacts of external attacks.
   - ML3 (Modeling Language for Linked Lives) – language environment for simulation of multi-agent systems with discrete events with application in demography and study of migration processes (Reinhardt et al, 2022). The abstract syntax associate agents with guarded objects, and the linguistic semantics is defined according to generalized semi-Markovian processes.
3. Simulation systems and packages focused on manufacturing systems and processes.

- Simulation packets oriented to manufacturing systems and processes – AutoMod II; ProModel; WITNESS; XCELL+; SIMFACTORY II.5.
- XCELL – Specialized simulation language oriented to manufacturing processes which gives good statistics for individual units of the workshop.
- FACTOR-AIM – A system for graphical simulation of production operations with language blocks that repeat (correspond to) the real elements of the production environment.

Table 1. Comparative characteristics of selected simulation languages

Characteristics	GPSS PC	SIMAN (Cinema)	Simscript (ver. II.5)	SLAM II	SLAM SYSTEM
Oriented to the study of processes (P) or events (E)	P	P, E	P, E	P, E	P, E
Hardware platform: MC – microcomputer; M – mainframe machine; PC – workstation	MC	MC M PC	MC M PC	MC M PC	MC
Input for graphical model	No	Yes (Camera)	No	Yes	Yes
Combining discrete and continuous simulation	Yes	Yes	Yes	Yes	Yes
Number of random number sequences supported	Unlimited	10*	10*	10*	10
Standard probability distributions supported **	U	Be; Er; Ex; Ga; L; N; P; T; U; W	Be; Bi; Er; Ex; Ga; L; N; P; T; U; W	Be; Er; Ex; Ga; L; N; P; T; U; W	Be; Er; Ex; Ga; L; N; P; T; U; W
Command to automatically implementation multiple replications	No	Yes	No	Yes	Yes

* – with possible increasing

** probability distributions (Legend):

Be – Beta; Bi – Binomial; Er – Erlang; Ex – Exponential; Ga – Gama; L – Log-normal; N – Normal; P - Poisson; T – Triangle; U – Uniform; W – of Weibull.

2. SIMULATION EXPERIMENTS PLANING

When researching operational strategies and production processes different evaluation methods can be applied such as abstract mathematical calculations or analysis of empirical data from real behavior. In a certain case, the appropriate approach is simulation, as stated in (Kuznetsov et al., 2020), where a different from traditional ways of evaluating a container stacking strategy is discussed. The reason for choosing the simulation approach in conducting the research is that existing strategies cannot be reliable due to the rapid changes in the terminal environment and the limited amount of data. The research model, as a stand-alone unit, is only a separate tool for the analysis, so it is important to properly organize the simulation experiments. This is usually done with multiple iterations with control of the study parameters. Emphasizing this, the article clearly states that "*the stress of the simulation study should be put not on the model itself, but on the way how to use it, i.e., on the experiments planning*". Only a correctly constructed model and a correctly generated simulation scenario can ensure obtaining adequate and statistically reliable results. No less important is the formulated objective function for the optimal execution of the scenario, as well as the planned experimental time and number of single experiments.

When planning the experiment, the goal is to draw up a plan according to which to carry out the computer simulation and the analysis of the obtained experimental data. The plan should address the key summarized below.

- Specification of initial values for the model parameters.
- Selection of optimal number of factors from the complete set.
- Defining observable and controllable factors and specifying the quantum and direction for changing the controllable factors.
- Selection of simulation duration to ensure stationarity of processes with sufficient efficiency (optimization of machine time).
- Ensuring the unambiguity of the conclusions in case of ambiguity of the results.
- Predicting the possibility of convergence of the results in case of excessive dispersion of the observations.

Some of the main activities that need to be performed when running a simulation are presented in the sub-sections below.

Determination of the Factors

Two groups of factors selected for simulation model are determined – controllable factors (the specific levels are selected when planning the experiment, and their combination determines the type of plan) and observable factors (subject to observation during the computer simulation and values are recorded for them). It is known that in simulation modeling, selected parameters of minor importance to the process can be ignored, thus facilitating the model study. The problem is how to organize the selection of relevant factors so that the reliability of the model experiment and the correctness of the results obtained from it do not suffer.

In the analysis of complex systems and processes, especially in industrial production, the correct determination of the most significant factors is of great importance, which is usually done based on research experience or expert judgments in the relevant field (Akberdin et al, 2018a). Such an approach is possible when there is some information about the researched object, while when designing a new process, it is significantly difficult. One possibility is to use experimental methods to select the factors, but this is quite expensive and takes considerable time. The paper suggests that for the purpose of planning, a mathematical experiment should be used to determine the impact of individual factors on the objective function, eliminating the less significant factors. This problem is further discussed in (Akberdin et al., 2018b) when planning thermodynamic modeling, stating that such an experiment is only numerical. This allows a mathematical design of experiment method to be applied to calculate the impact of a group of factors on the objective function. In this way, physical experiments will be avoided and the dynamics of changes in the observed process can be evaluated by setting different values for the controllable factors in the simulation. This will allow to derive a generalized equation of a given process observable parameter (factor) as a function of all or selected controllable factors.

One of the main purposes of an experimental design, after the essential factors for conducting the simulation have been selected, is to determine primary and secondary factors. From the primary factors, controllable factors should be selected for which a minimum, but sufficient, number of levels should be defined. This number determines the duration of the simulation study and the number of single experiments performed. Therefore, it is important to find a compromise between the time required and the effectiveness of the results. Obviously, the simplest option is to run the simulation model once with one selected level for one controllable factor. Unfortunately, this will provide too low a level of the resulting estimates, which will only have a particular character for one possible situation of a probabilistic process. There are different design options for the simulations, some of the factorial design options are presented below.

- One-factor plan – the simplest plan of changing only one factor, keeping the others constant. This is a traditional approach, as noted in (Beg et al., 2019), but generally has limited relevance in today's digital space and the impact of various influences. The reason is that it requires an investment that will not lead to an adequate determination of causal relationships between parameters of the studied process.
- Full factor plan – all possible combinations of the selected levels for the controllable factors are run, which significantly increases the machine time to run the simulation experiments but provides objective estimates of the observed factors (Oreshina, 2020).
- Incomplete (partial) factor plan – only a selected part of the possible combinations for the factor levels are involved (Tian et al., 2018). An example of the application of this variant of partial plan in conducting analysis is presented in (Napoli & García-Téllez, 2016) in the organization of experiments to study the influence of water resources on agricultural productivity.
- Completely randomized factorial plan – an incomplete design in which the selection of levels is random. This variant of factor plan was applied in (Loquercio et al., 2019) for analyzing the constraints of deploying small drones for competition purposes in a dynamic environment. A state-of-the-art planning and control system with a neural network trained by simulation is used in the functionality study. The simulation experiments conducted allowed for the formation of an output sample of results based on a randomized design.

Determination of Factor Levels

Determining the value levels for each factor is related to determining the limits of its definitional area for the particular experiment. Each combination of possible levels for the factors represents a multi-dimensional point in the so-called factor space. Since individual auxiliary domains usually differ in importance from the target domain, their filtering is done, and an algorithm for joint filtering between domains based on latent factor space is proposed in (Yu et al, 2019). Based on the importance of the correct determination of the factor space, the classification of the target domain is initially formulated, after which a sub-domain decomposition is made to expand the functions of the research problem. As stated in (Cui & Ma, 2015), the application of factor space theory has a beneficial impact on the objective set of attributes and values from different domains. The article proposes a graphical approach for presenting a range of attributes of a researched object based on cluster analysis. The grouping itself is carried out in strict compliance with the principles of division in the case of matching features or split grouping in the case of fuzzy similarity.

In planning, experimental points symmetric to the base level of each controllable factor are selected. To determine the base level the following approaches can use.

- If there is information about only one point and the limits of the factors are not known, this point is chosen as the main level.
- If the limits of the factors are known and it is known that the best values of the studied parameter fall into the factor space, any point can be chosen as the basic level.
- If it is known that the studied parameter has several good values, a random point from the factor space can be chosen as the base level.
- If a sub-area of the factor space is known, in which the studied process takes place optimally, the central point of this sub-area is chosen as the main level.

Figure 1 gives an illustration of the latter approach, assuming that the observable factor under study is dependent on the controllable factors X_1 and X_2. The center point $\langle X_1^0, X_2^0 \rangle$ of the sub-region defines the main levels of change of the two controllable factors.

Figure 1. Subdomain of factor space and levels selection

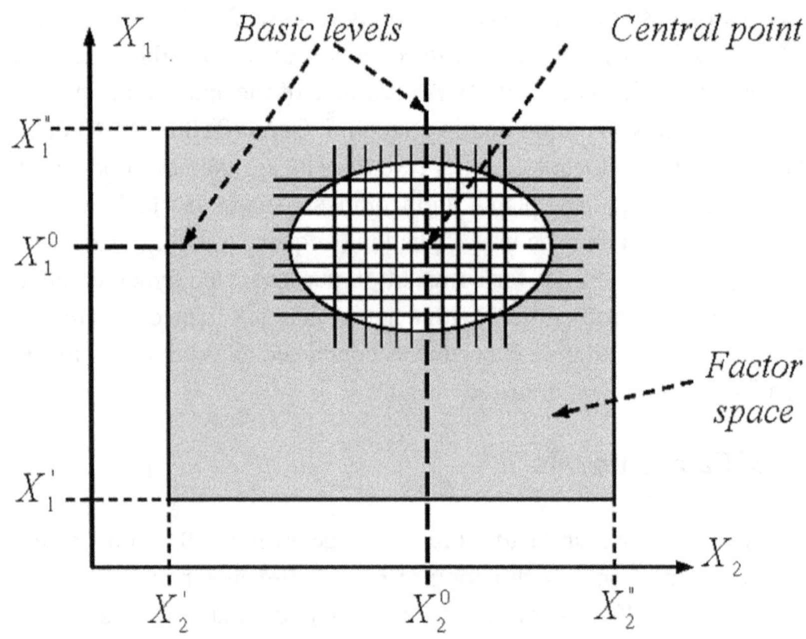

Setting the Simulation Time

The duration of the computer simulation should be determined based on a compromise between the requirement for a larger number of processed events (higher reliability of the results) and the need for the efficiency of the experiment (less machine time). The goal is to determine the duration of the experiment sufficient to ensure the desired accuracy. The specific implementation depends on the stationarity of the investigated processes.

When modeling stationary processes, the preliminary transient mode of non-stationary behavior of the object has an additional influence. For correct estimates of the probability parameters, it is necessary to ensure entry into a stationary regime, where informative observations can be obtained. This necessitates making preliminary studies and estimates for the expected duration of the transient regime and ensuring sufficient time for the model to enter a stationary regime.

This can also be done with an initial control run, monitoring a parameter that is supposed to enter stationary at the latest (for example, a queue in front of the server) – Figure 2. An interval δt is chosen during which readings are taken and the results are compared with the neighboring ones for a stable deviation range around the mean value L_{av}.

In modeling non-stationary processes, the simulation period is usually determined by the duration of the object's existence in the given study area, and the accuracy of the results depends on the number of experiments. Repeated experiments are carried out, the number of which increases as the requirements for the accuracy of the estimates increase.

Figure 2. Stationary process

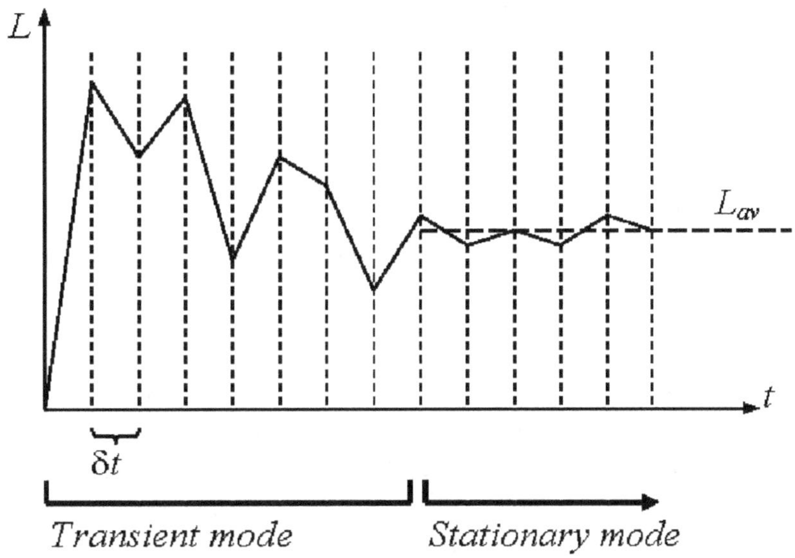

Preliminary Processing of Experimental Data

Conducting a computer simulation according to a defined experimental plan must be realized, in which a sample of experimental results is formed. They allow the relevant conclusions to be drawn, which necessitates the requirement that they be sufficiently informative and comprehensible to the user. It can suggest few actions, presented below, which will ensure the requirement for informative experimental data.

Highly deviating values filtering. Experimental results represent the relationship between observable factors (*Y*) and controllable factors (*X*). It is possible that some of the recorded values differ significantly from the general trend, and the reason has to be analyzed. If it is a random effect, these observations can be filtered (removed), which will increase the informativeness of the sample.

An example illustration is given in Figure 3, where the general trend of the observations defines a linear increasing dependence of $Y(X)$, but a few observations strongly deviate from it. Point C can be assumed to be the result of a random impact, while points A and B must be analyzed to see if they are the result of a transient mode or a transition from another state.

Figure 3. Filtering registrations

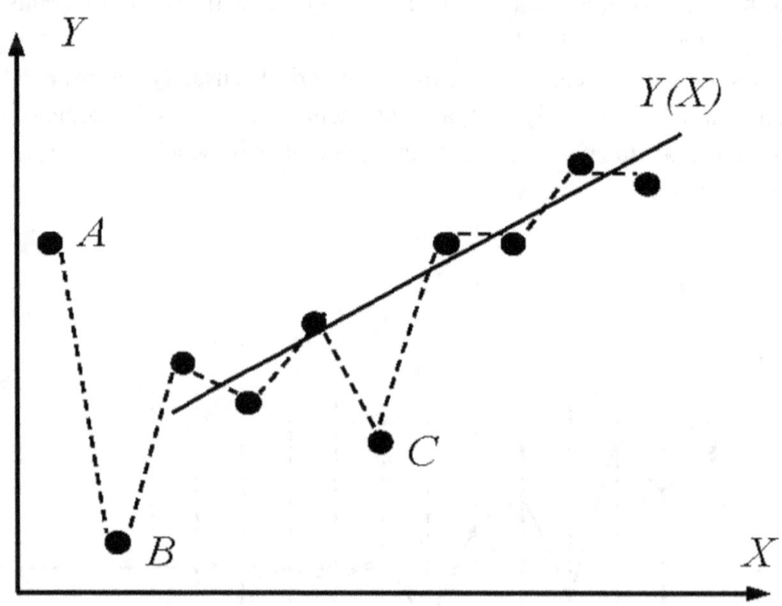

Coordinate system selecting. The question is related to the graphical presentation of the functional dependencies obtained during the modeling and their easier perception. It is known that the linear dependence is the easiest to perceive and more clearly reflects the existing trends. Therefore, it is recommended that the graphical interpretation of the experimental results be carried out in a linear form. Suitable coordinate systems allowing linear interpretation can be found, for example the dependence $Y=k=X^A$ can be represented in a logarithmic coordinate system of the form $Y=\lg k+A.\lg X$. Another possibility is to represent the hyperbolic relation $Y=X/(a+b.X)$ as a straight line by plotting the dependence of $1/Y$ on $1/X$ in a linear coordinate system.

Results analysis. The following is recommended:

- To evaluate the conditions for conducting the simulation experiments.
- Construction of the dependence $Y(X)$, for example by means of a regression model.
- To estimate the significance of factors X on response Y (usually by calculating regression coefficients).
- Determining the adequacy of the regression model and confirming the credibility of the selected trend.

3. REALIZATION OF SIMULATION EXPERIMENTS

The organization of simulation experiments can use different approaches to pre-define the working model through appropriate formalization and representation of the abstract conceptual model. The simulation experiments themselves can be organized as a basic way of conducting the main study of a given object (main simulation) or as an additional simulation study (supplementary simulation) to determine the final assessments and conclusions more accurately. In basic simulation, the experiments are completely conducted according to a set goal and to achieve certain results. In complementary simulation, a combination of approaches is usually done, with the simulation having a supporting role in formulating the conclusions. In the field of computer and electronic technologies, such simulation experiments are often conducted to analyze data received based on another research approach (intelligent software, analytical modeling, monitoring, etc.). These experiments are usually intended to obtain additional estimates of certain behavioral parameters and to extend the conclusions formulated.

Examples of baseline simulation experiments were presented in Chapter 11, where variants of the GPSS language environment were used to create the model, and the initial conceptual model was implemented as QS or QN. A similar approach was also applied in a research project containing a set of interdependent tasks with an arbitrary execution time, presented in (Oleinikova et al., 2021). The goal is to develop a simulation model for estimating the duration of the project, which is a random variable with a certain distribution law, as well as determining the probability of completing the project within a given time. The main task formulated in this way before the simulation experiments determines the use of queuing theory as a basis for organizing the modeling, as the initial conceptual models for implemented as QNs. The AnyLogic software product was used for the implementation of the model, which has a powerful graphical interface and many libraries, allowing to simplify the process of building the model. The specificity of the simulation model is determined by the condition that each new activity starts only after the completion of all related previous tasks. In addition, parallelism of jobs is assumed when performing independent tasks. The requirements and the formulated assumptions allow a description of the main mechanisms for creating a simulation model for evaluating the probabilistic-temporal characteristics of the project with the possibility of determining the critical paths from one node to another in the general structure of the QN. Based on the conducted simulation experiments, estimates for the mathematical expectation of the duration of the project (first main task) were determined and the basic characteristics of this random variable and the probability distribution law were determined,

Other examples of simulation experiments are presented below.

Information Digitalization and Neural Network Simulation

Transformation of information from analog to digital form is often required when performing basic activities related to information technologies, which can be realized through specialized devices. In most cases, repeatability of the digitization process is required, which reflects on the level of data security (Ivanova, 2020). There are universal systems designed for the automatic formation of 3D models that allow the digitization of various objects while maintaining security. An important positive property of 3D digitization during scanning is the level of security provided when storing information about the parameters of the physical object (shape, size, location, etc.). Other applications of 3D digitization are in product testing by comparing the model with a machine-made object, in 3D creation of machine elements, virtualization of prototyping, etc.

Modern digitization technologies oriented to information security are presented in (Ivanova, 2020). The main goal is to ensure the security of the digitization processes themselves, and an adaptive method for digital transformation based on comparative analysis has been proposed. The rationale is that in the digital world there is an increase in attacks targeting the manufacturing sector, mainly industrial control systems that are not sufficiently protected. In addition, digitization can help to strengthen the important sector of cyber security in cyber-space when providing remote access to information resources. This also applies to cloud services, because the cloud is a virtual space shared by many customers (multitenancy), which implies different possibilities for security threats. To counteract this, specialized technologies, new concepts, expert knowledge and especially systems based on Artificial Intelligence (AI) can be used. There are modern 3D scanners with AI, which are real-time intelligent detectors equipped with an imaging system. An artificial neural network allows the use of adaptive learning algorithms to create correct shape and correct depth indication.

The digital transformation method proposed in (Ivanova, 2020) is adaptable because it is applicable for various purposes, including education and training, and because of the possibility of using relatively affordable resources, the total cost of which does not exceed that of a 3D scanner. Such resources can be mobile 3D scanning devices (smartphone, professional camera), specialized tripods, green laser, open-source photogrammetry software products, etc.).

When conducting experiments to confirm the adequacy of a method, specialized photogrammetry software VisualSFM was applied. It was chosen because of its ability to convert different photos of a physical object into its 3D model by calculating missing matches and to simulate different camera poses. To implement the 3D reconstruction, all missing matches are calculated and generated in the Task Viewer environment. The proposed experimental method is an alternative to 3D laser scanning, allowing the replacement of complex and expensive technological solutions with a combination of conventional devices. The proposed approach is applicable for training, for technical analysis through 3D visualization, virtualization of machine elements, etc.

A continuation of the above study was made in (Ivanova, 2021), where an approach for importing the digitized data obtained by scanning in a simulation model of an Artificial Neural Network (ANN) is presented. The realization of the study is by using the capabilities of the SIMBRAIN software, allowing improved self-learning in digital recognition. This product is open source and designed to simulate neural circuits written in Java. The choice was made due to the possibilities offered for conducting research by modeling a self-learning neural network for recognizing digital images within short time-intervals.

Among the capabilities of ANN with backpropagation of error (BPE) is the recognition of different images that have been transformed into a binary code by a scanning software with a conventional type of interface. Such images can be typographic and pictographic symbols, graphically represented emotional expressions, ordered rows of symbols and speech. The product applied in conducting research allows complex objects to be represented as an image with a high level of abstraction and a low level of complexity. It is known that based on digital transformation a given physical object can be represented as a set of points.

The main goal of (Ivanova, 2021) is to propose "*a reliable algorithmic method for optimization of a neural network with BPE using a narrowly tuned simulation model built in a software with capabilities for precise settings*". What is new is the proposed concept of two-stage optimization based on two independent software, and the obtained results will be jointly applied in the simulation process. The two stages support creating the pixel matrix and achieving the maximum efficiency of the ANN, respectively. The neural network is designed to recognize both geometric primitives jointly, represented as 4 orientations of the square and 4 of the cubes. On their basis, the input parameters used in the simulation of two neural networks are defined, respectively before and after image optimization. The cube, unlike the square, can be recognized when partially filled (Figure 4a). The square has 4 visually identical orientations, while the cube has 4 recognizable spatial orientations (Figure 4b). The difference between the two investigated neural networks is determined by the filling of the square represented in two ways. Spatial orientations are sequentially organized from the square to the cube, each successive orientation being a 90-degree clockwise rotation of the image.

Figure 4. Representations in a [8x8] matrix
Source: Ivanova (2021)

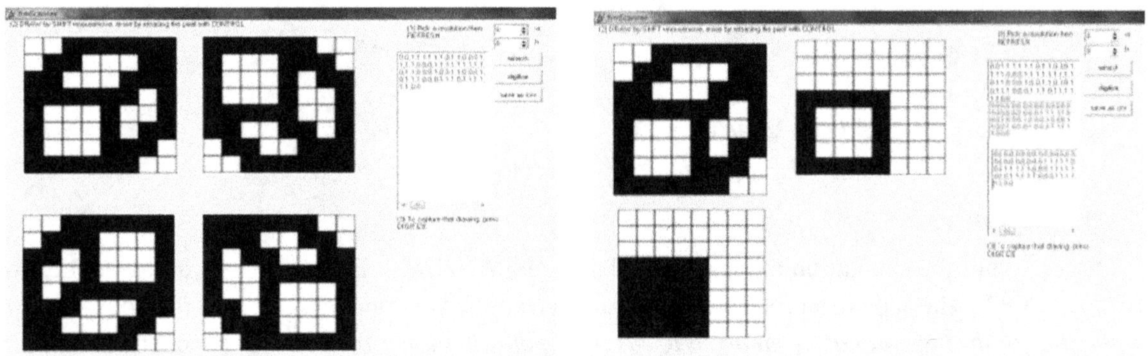

(a) Representation of outlined cube and square *(b) Mine spatial orientations of a cube*

The simulation model in the SIMBRAIN environment is presented in Figure 5, with the binary codes of the participating 64 neurons in the input layer and the two neurons in the output layer defined as Input data and Target data.

The results generated by the simulation are presented, which allow a summary assessment of a total of 4 scenarios and an analysis of the behavior and a comparison between the two variants but simulation models. In the first model (before image optimization), a minimal increase in the parameters *N* (number of processing layers) and n (number of neurons in all layers) reduces the error, but it increases with a 3-fold increase in the second parameter. For the second model, the conclusion is that filling in a figure does not significantly affect the BPE even in the case that the iterations are more compared to the first model. General conclusion is that "*the ANN with two processing layers with 12 neurons per each of them is the most effectively optimized in the current empirical research completed in SIMBRAIN*".

Figure 5. Basic simulation model of the ANN for digital recognition built in SIMBRAIN
Source: Ivanova (2021)

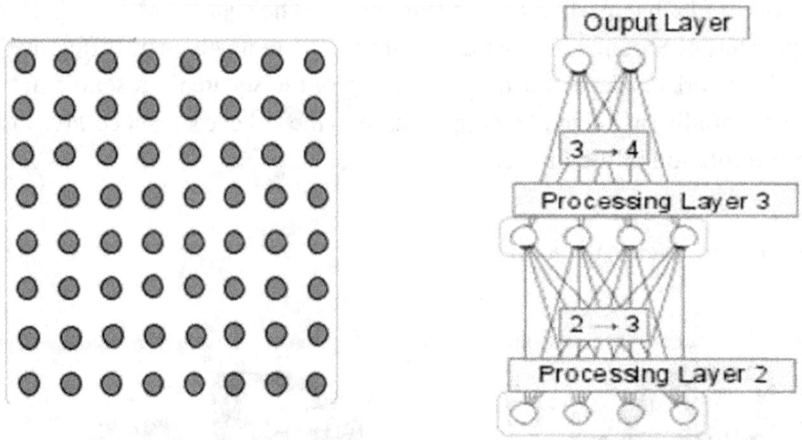

(a) Input Layer *(b) Square Cube*

Discussion of the simulation modeling problem using ANN with BPE continues in the publication (Ivanova, 2022). The approach proposed above has been applied to improve shorthand methods "*that are applicable for the purposes of steganalysis to ensure the information protection*" by conducting empirical research and conducted experimental analysis. It is pointed out that steganography and steganalysis are two separate but interdependent fields. Specifically, the task of steganography is to hide digital data in text, computer graphics, or multimedia, while the task of steganalysis is to detect whether steganography has been applied. Basically, these are the two sides of cyber security, with exposure of protected ("secret") data being a major malicious type of threat. When discussing the formulated problem in the article, a classification of the used shorthand techniques and steganalysis methods is made and experiments are presented for empirical research in the Virtual Steganographic Laboratory software environment, as well as simulation experiments in the SIMBRAIN simulation environment according to the ANN construction sub-step presented above. with BPE to detect and recognize bitmap stego-images.

Analogous to the previous experiment, formation of the main spatial orientations of an original image is applied here in 4 consecutive rotations (Figure 6), which allowed the binary data to be formed, the working input table for the ANN in the SIMBRAIN product is needed. In this case too, the ANN includes 64 input neurons (corresponding to the pixels of the matrix), two hidden layers of 4 neurons each, and 2 output neurons for the fill and outline image, respectively.

Figure 6. Mine spatial orientation in a [8x8] matrix
Source: Ivanova (2022)

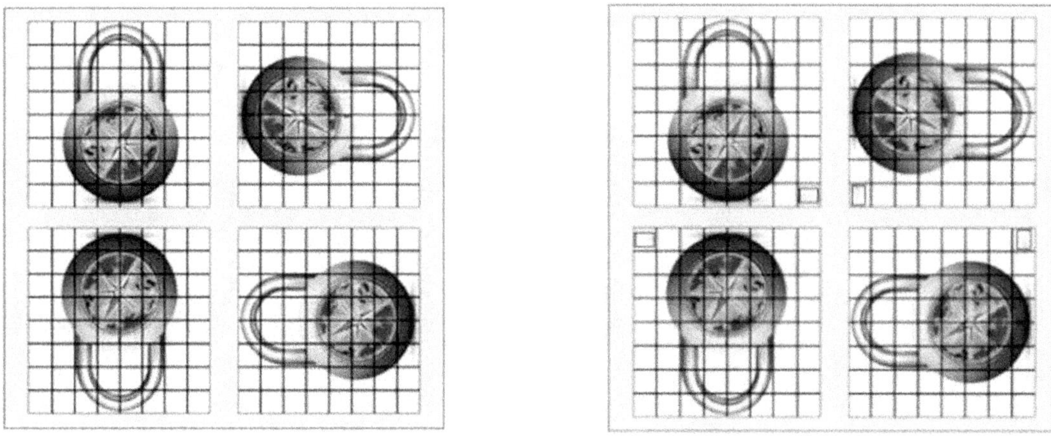

(a) Orientation in the original image (b) Stego-image representation

In the second ANN for "stego-image" recognition, the configuration is the same as for the original image (Figure 6a), with the difference that the least bit of the stego-image is changed from 0 to 1 for its 4 spatial orientations (Figure 6b). In the first of these, the least significant bit is relevant to the 64th pixel (the marked position), and in subsequent rotations its position occupies pixel numbers 57, 8 and 1, respectively.

Modeling and Representation of Power Electronic Converter Parameters

The object of this study is the modeling of Buck DC-DC converter, together with its control system based on PID regulator and Graphical user interface which is discussed in extended version (Gocheva et al., 2020). The proposed software estimates voltage and amperage values on electronic components in stepwise manner by index matrices; it shows a converter scheme and two-dimensional graphs on it on the output voltage in case present in Figure 7 and Figure 8.

Figure 7. Defining the basic parameters of the output voltage necessary to conduct the study

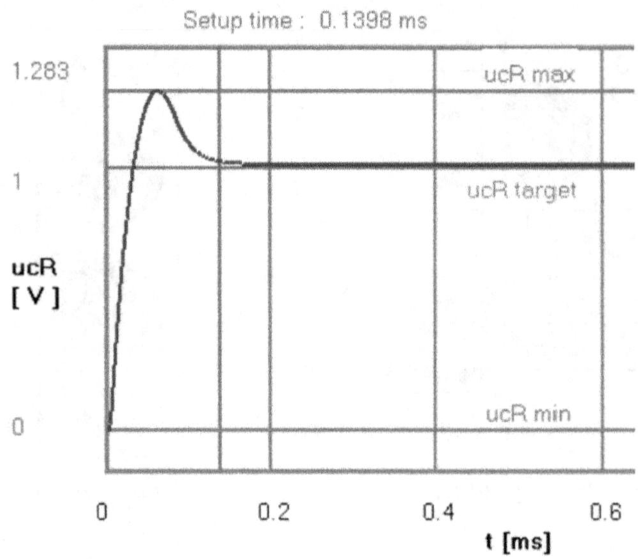

Figure 8. Power diagram of the Buck DC-DC converter and reference curve for the output voltage of the converter

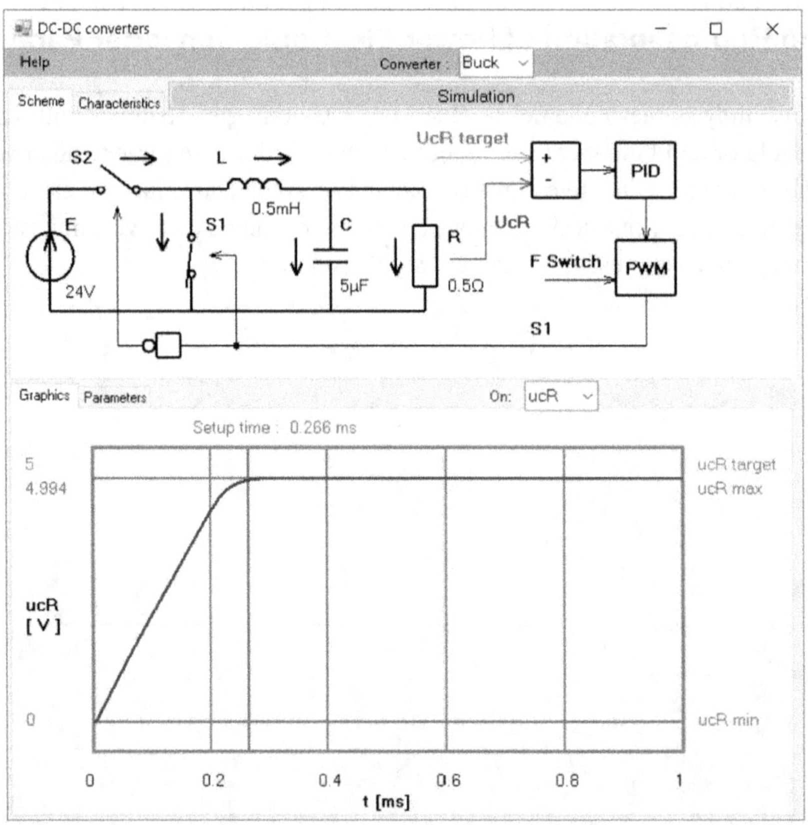

The considered PID controller is used in pulse width modulation (PWM), as the parameter that will be affected by the controller is the conduction time of the transistor. The components are modeled as linear ones, and the following assumptions are made:

- The resistance of the power supply is equal to zero.
- The output impedance is active only.
- The switches are ideal.

These constraints take place in the present paper too. Results of simulations are considered in (Gocheva et al., 2020), but methods for obtaining proper parameters of the converter are not described there.

A step response is modeled below (Balakrishnan et al., 2018). Basic parameters on electric power converters are given on Figure 9, and some ones are marked on Figure 7 and Figure 8 (target value, setup time, minimal and maximal values). All they depend on input characteristics. In the present paper the used duration is equal to 1 ms. An average and a ripple factor (as a percentage) are estimated at the end of this interval.

Figure 9. Control panel of the realized user interface

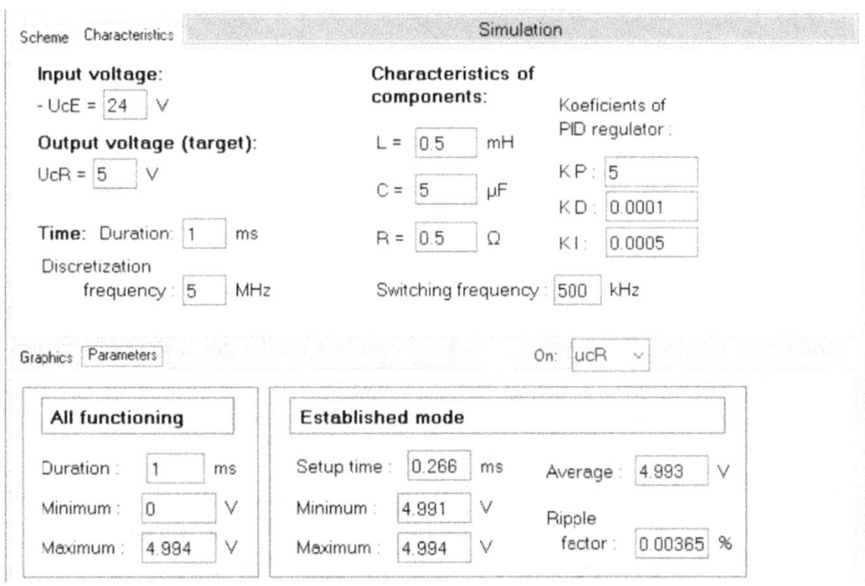

The chosen input voltage is 24 V, and the chosen target output value is 5 V. The output amperage is set in range 1 – 10 A; all surfaces are obtained for the border values of this interval. The switching frequency is set to 1 MHz. It affects all parameters; for example, its reduction to 500 kHz increases the setup time (see Figure 10). The discretization frequency is set to 10 MHz; tests on it show that this frequency, in general, must exceed more than 10 times the switching frequency in order for small errors to be made.

We used well-known values of inductance and capacity and in this sense, the ranges of their change were also selected for the purposes of the study:

L = 0.25 ÷ 0.75 mH & C = 0.5 ÷ 10 μF.

Coefficients of the PID regulator are chosen as below and tests on the first two ones are performed.

✓ Proportional	$K_P = 0.5 \div 10$.
✓ Differential	$K_D = 1.10^{-5} \div 2.10^{-4}$.
✓ Integral	$K_I = 5.10^{-4}$.

Figure 10. Two-dimensional representation of setup time

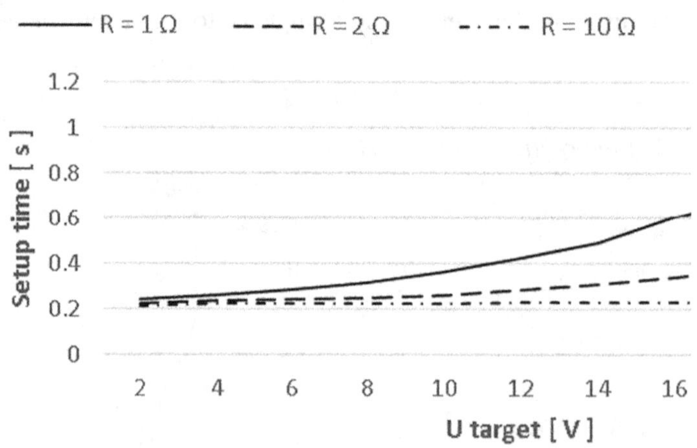

On representation of parameters. Article (Gocheva et al., 2020) contains two-dimensional graphs, which concern dependences of parameters on two characteristics by families of lines. For example, Figure 10 shows a two-dimensional representation of setup time for various target output values and various output impedances. In the present paper an approach based on three-dimensional graphs is presented.

The first simulation concerns dependency on inductance and capacity. In this case the coefficients of the PID regulator are fixed as follows: ✓ $K_P = 5$; ✓ $K_D = 1.10^{-4}$; ✓ $K_I = 5.10^{-4}$.

Since the output current is set in range 1 ÷10 A, the load resistance is set to two border values 0.5 Ω and 5 Ω, for which separate simulations and respective surfaces are performed for each combination of inductance L and capacity C. The results of the numerical experiments are presented in the form of 3D graphs in Figures 11 to 13 as follows: Figure 11 – Output voltage value of the DC-DC converter (maximum and average values); Figure 12 – Study of the dynamics of the DC-DC converter, by determining the duration of the transient process; Figure 13 – Determination of the ripple range of the DC-DC converter, when changing the values of the circuit elements.

Figure 11. Output voltage value of the DC-DC converter

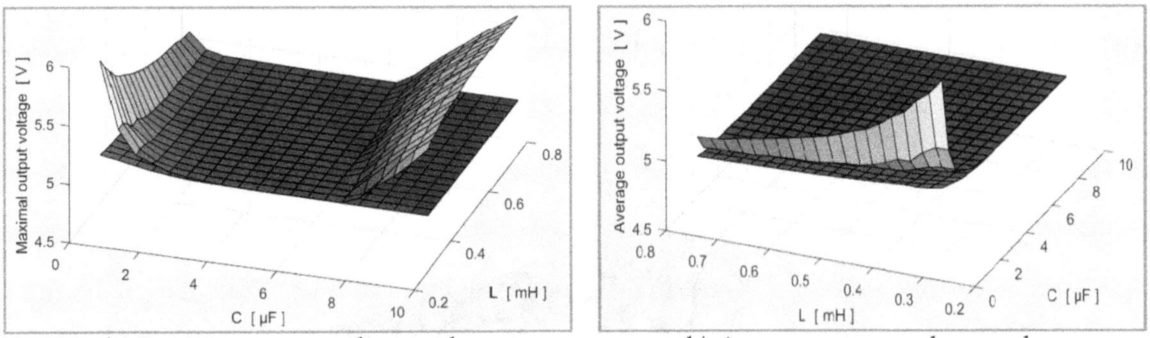

a) Maximum output voltage value *b) Average output voltage value*

Figure 12. Setup time value of the DC-DC converter

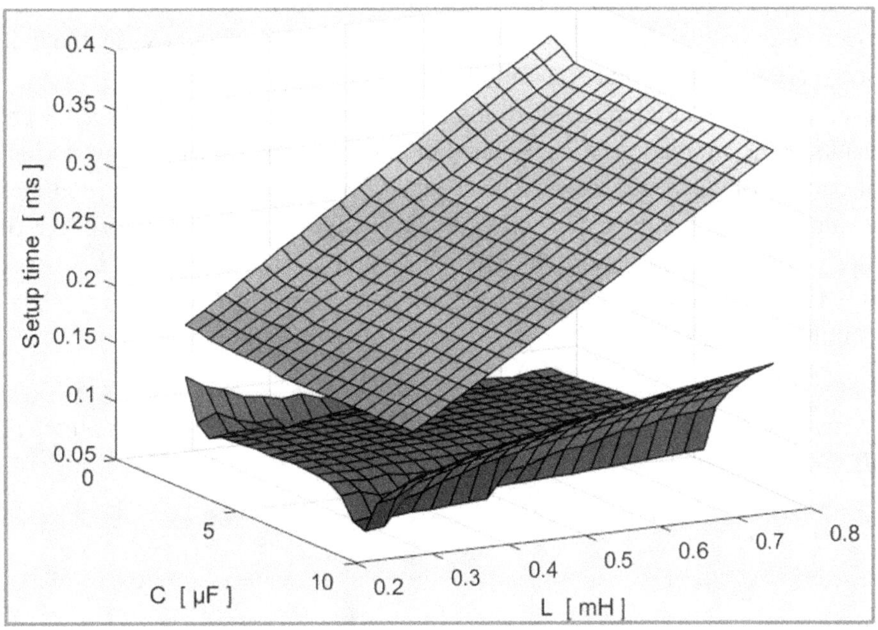

Figure 13. Determination of the ripple range of the DC-DC converter

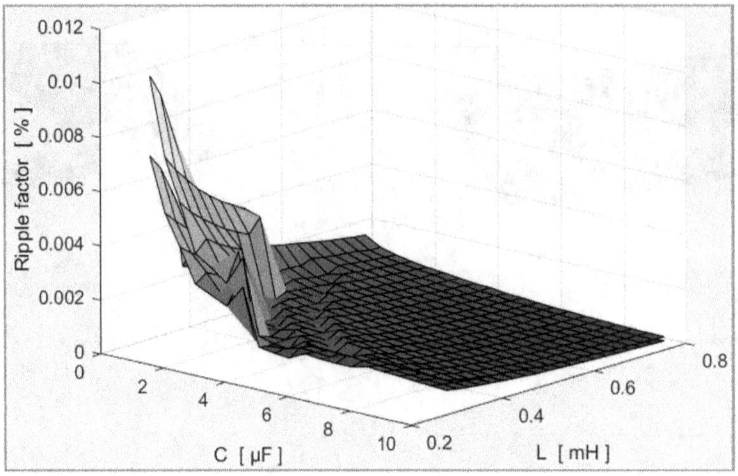

The second investigation, based on simulation, aims to establish the dependence of the output indicators of the device on the coefficients of the PID regulator K_P and K_D. In this case, the inductance L and the capacitance C are fixed with values as follows: ✓ L = 0.5 mH; ✓ C = 5 μF.

In this sense, the already ordered values of the controller integral K_I and the load R are used as constants. The results of the simulations are presented in in Figures 14 to 17 as follows: Figure 14 – Maximum value of output voltage depending on the change of the parameters of the filter elements; Figure 15 – Average value of output voltage depending on the change of the parameters of the filter elements; Figure 16 – Study of the dynamics of the DC-DC converter, by determining the duration of the transient process; Figure 17 – Determination of the ripple range of the DC-DC converter, when changing the values of the circuit elements.

Figure 14. Maximum value of the output voltage of the DC-DC converter when changing the coefficients of the controller K_D and K_P

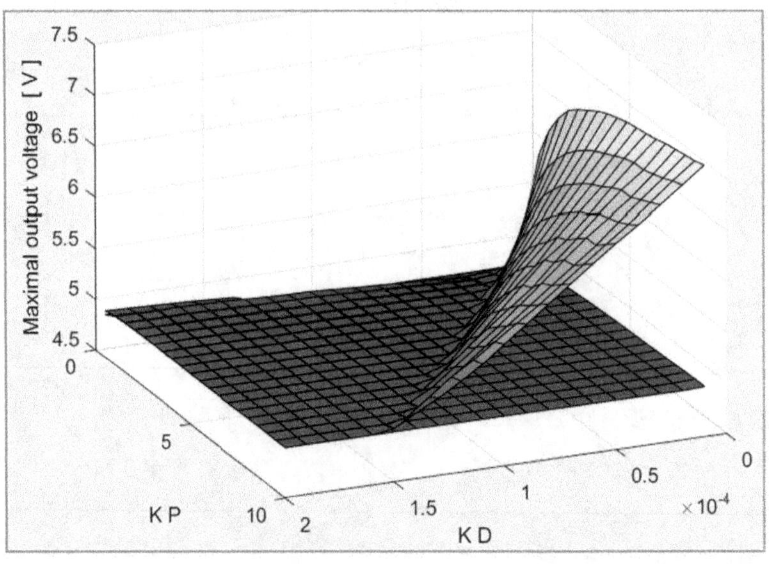

Figure 15. Average output voltage value of the DC-DC converter when changing the coefficients of the controller

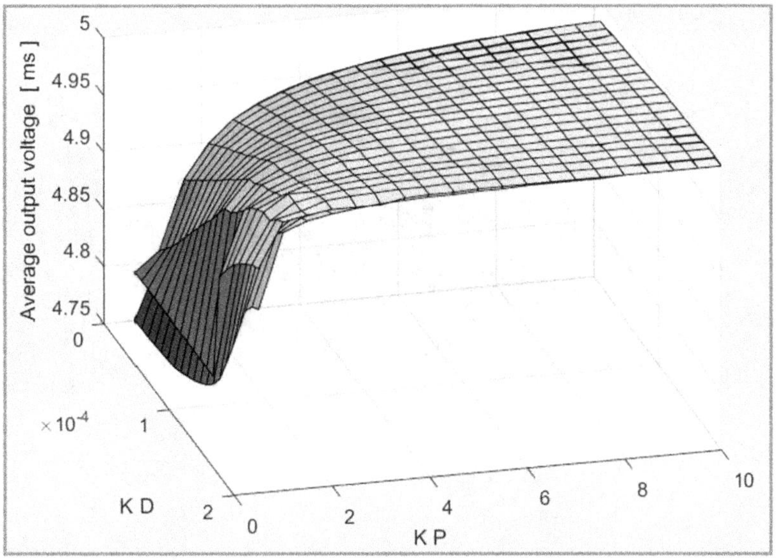

Figure 16. Setup time value of the DC-DC converter when changing the coefficients of the controller

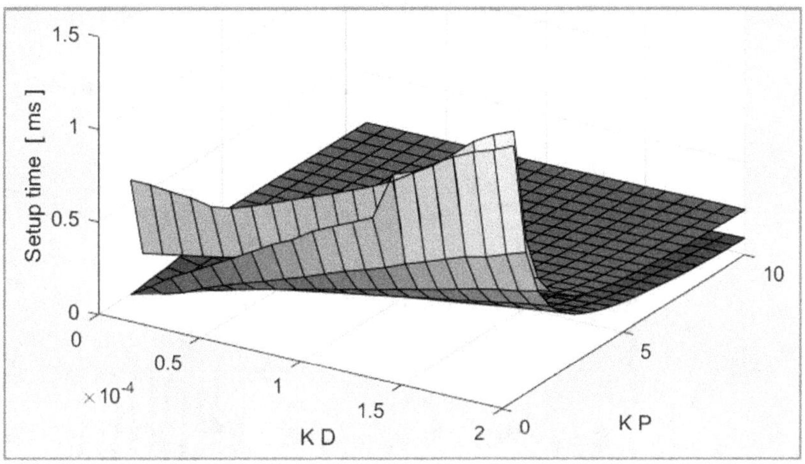

Figure 17. Determination of the ripple range of the DC-DC converter when changing the coefficients of the controller

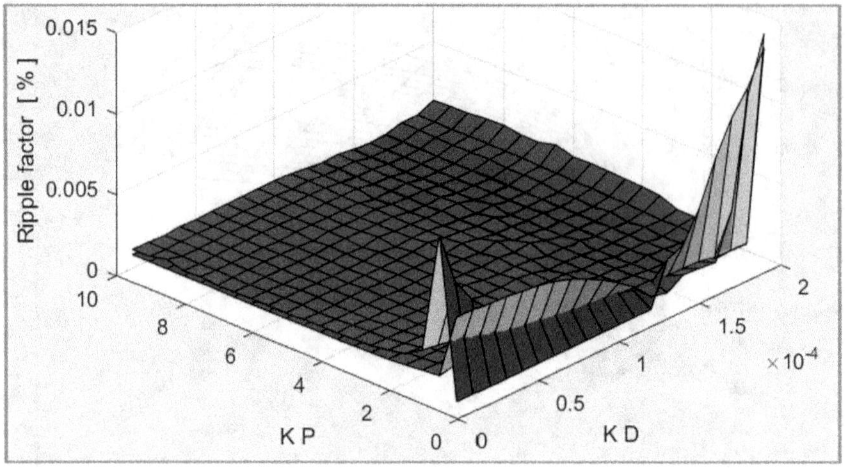

The upper simulations model step responses of a DC-DC buck converter with PID controller. More precisely, they represent graphically parameters of its output voltage. Since the output impedance is chosen to be active only, the output current has the same form. However, the used software estimates parameters of all scheme's components. The change in capacitor current over time is shown in Figure 18.

The displayed graphical results give a clear idea of the influence of on the one hand the elements of the smoothing filter and on the other - the settings of the controller on the operating modes of the converter.

Figure 18. Shape of the current through the output capacitor of the DC-DC converter

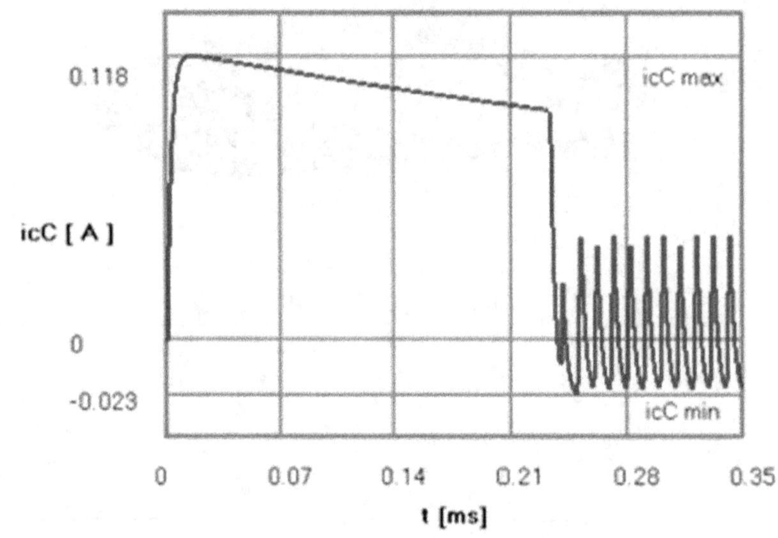

The following conclusions are drawn from the analysis of these results:

- The change of the values of the filter elements (Figures 11 to 13) has the most significant influence on the dynamics of the converter, and when choosing not unreasonably large values leads to inoperable modes, which should be avoided.
- The change of the coefficients of the controller also has a great influence on the operating modes and in practice at very large or very small values of the coefficients inoperable modes are obtained again.
- Optimal operating modes can be realized both by optimization of the circuit elements and by using a suitable controller, but guaranteeing the indicators is realized by joint optimization of both the power circuit and the controller.
- The model-based optimization, realized based on an integrated model of the power part and the controller, is a suitable tool both for the purposes of training and in the development of power electronic devices and systems.
- Indexed matrices and the models created on their basis are a suitable tool for conducting research in the field of power electronics.

On the other hand, the development and use of a graphical user interface greatly facilitates the work with specialized mathematical software. This is very useful for the purposes of training specialists and students, as in recent years there has been a serious outflow of those wishing to work in hardware specialties, and training in mathematics, physics and electrical engineering is becoming weaker. On the other hand, the economy, industry, and society have an increasing need for such specialists, because power electronics is a key factor in achieving the important strategic goals and priorities of both Bulgaria and the European Union. In practice, the concepts and ideas of circular economy, green deal, electric mobility are impossible to implement without the means of power electronics and in particular power electronic devices and systems. Therefore, the intensification of training and the use of attractive forms to create lasting interest in this topic are increasingly necessary and mandatory.

Another significant advantage of the use of modeling of the entire power electronic device is the possibility to apply artificial intelligence techniques for research in the field of power electronics. Based on the model, a database with results would be created relatively quickly, which would be processed to obtain certain optimization procedures with given target functions and constraints.

Conclusion. The proposed approach on visualization can be extended. Other characteristics can be fixed, and new dependencies can be obtained. Nevertheless, the presented graphics can be used properly since they contain essential features (for example, values of inductance and capacity, which are important). Coefficients of PID regulators can be modeled too. Multidimensional analysis can be used on generated data in order forward tasks to be solved. All this activity can be easily automatized since an available software can be modified, including in its graphical user interface (GUI) part.

In the presented example, a controller setting is used based on achieving the best possible dynamics, but with the proposed approach it is possible to implement other settings such as: minimum losses in the elements, maximum efficiency, or a certain level of reliability.

The presented approach is useful for the purposes of power electronics training, where the adjustment of the controller coefficients is a particular difficulty. This is largely a serious barrier to the absorption of the material. The use of an integrated model and a reference curve allow for easy and quick setup of the controller using the graphical user interface. In practice, this approach is applicable to other power electronic devices and is particularly suitable for the implementation of e-learning and distance learning.

Tolerance Analysis of Base Schemes of a DC-DC Converters

Power electronic devices are a very important part of modern electronics. There is virtually no device or process, including the transmission and processing of data and information, where the use of electrical power is not necessary. In this aspect, to a large extent the provision of the required quality of a service, production or information process is determined by the operational capabilities of the power supplies. DC-DC converters are one of the most applied power electronic devices, and in this regard, the requirements for them are quite diverse both in terms of converted power, as well as in terms of weight, size, price and operational requirements. The aim of this study is to compare the most common and commonly used topologies (Buck, Boost and Buck-Boost) in terms of the sensitivity of the operating modes depending on the qualities of the passive elements. The comparison is made based on tolerance analyzes carried out based on developed models. These models are valid for operation in continuous current mode, and during their compilation the conduction losses of the semiconductor switches (transistors) and the filter inductance are considered, and the losses in the filter capacitor are neglected.

For the purposes of the study, three classical circuits of synchronous single-transistor circuits are shown, shown on Figure 19. In order to obtain comparable results, the operation of the converters with the same output voltage - 20V, output current 2A, the same Duty cycle D = 0.5 and operating frequency 300kHz were selected (Hinov & Hranov, 2021). In addition, to ensure the operation of all three converters in continuous conduction mode, the following values of the filter elements L = 330 μH, C = 10 μF are determined according to (Mohan et al, 2003). To describe the real changes in the circuit elements, passive tolerances of ±20% for the inductance and ±10% for the capacitor are given. To achieve the same results in the output, in these topologies it is necessary that the input voltages are as follows: 1. Buck DC-DC converter - Input voltage is equal to 40V, Boost DC-DC converter - Input voltage is equal to 10V and the Buck-Boost DC-DC converter - Input voltage is equal to 20V. Initially, results were obtained for the influence of tolerances in the steady-state mode, and then on the dynamics and transient processes.

Figure 19. Power circuits of the studied DC-DC converters

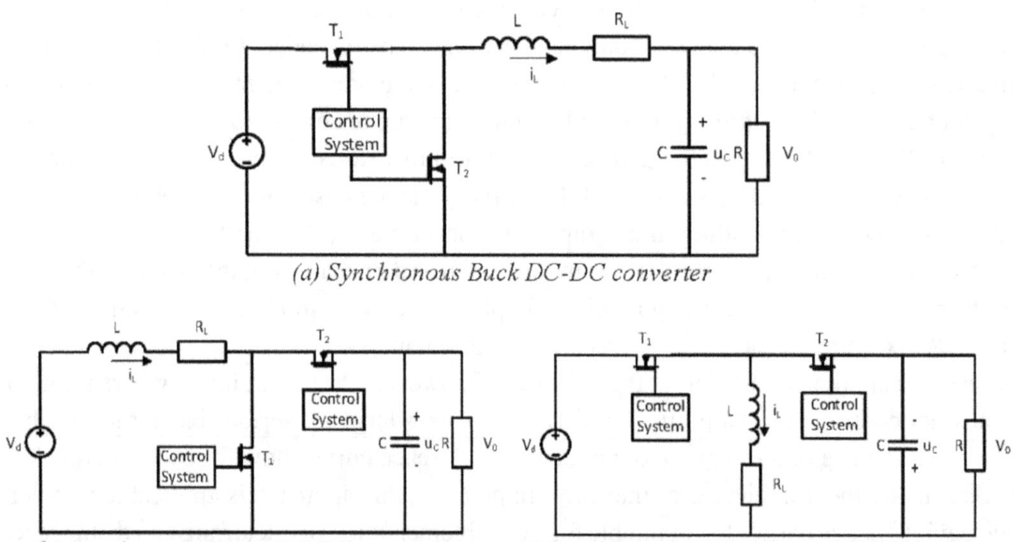

(a) Synchronous Buck DC-DC converter

(b) Synchronous Boost DC-DC converter

(c) Synchronous Buck - Boost DC-DC converter

The models used to describe transient and steady-state modes are as follows:

- for synchronous Buck DC-DC converter:

$$\begin{cases} L\dfrac{di_L}{dt} = F.V_d - i_L(R_{ON} + R_L) - u_C \\ C\dfrac{du_C}{dt} = i_L - \dfrac{u_C}{R} \end{cases},$$

where R_{ON} is the on-state resistance of the transistor, R_L is the active resistance of the inductance, i_L is the current through the inductance, u_C is the capacitor voltage (the output voltage), F is a switching function that determines the two states (turn-on and turn-off of the transistors).

- for synchronous Boost DC-DC converter:

$$\begin{cases} L\dfrac{di_L}{dt} = V_d - i_L(R_{ON} + R_L) - \overline{F}\bullet u_C \\ C\dfrac{du_C}{dt} = \overline{F}\bullet i_L - \dfrac{u_C}{R} \end{cases}.$$

$where\ \overline{F} = 1 - F$

- for synchronous Buck - Boost DC-DC converter:

$$\begin{cases} L\dfrac{di_L}{dt} = F\bullet V_d - i_L(R_{ON} + R_L) - \overline{F}\bullet u_C \\ C\dfrac{du_C}{dt} = \overline{F}\bullet i_L - \dfrac{u_C}{R} \end{cases}.$$

$where\ \overline{F} = 1 - F$

The research is based on the application of MATLAB Simulations using the mathematical models of the DC-DC converters.. MATLAB is used in a variety of fields, such as engineering, science, finance, medicine, and others, where numerical analysis, visualization, and modeling are important components of product or process research and development. In this sense, MATLAB offers many advantages, but it also has some disadvantages that should be considered when choosing this software for specific applications. The main advantages and disadvantages of MATLAB are systematized below.

Advantages of MATLAB:

- Easy to use. MATLAB has an intuitive and easy-to-learn syntax that makes it suitable for both beginners and advanced users.
- Rich library MATLAB has an extensive library of functions and tools for various fields such as mathematics, engineering, statistics, signal, and image processing, etc.

- Interactive environment. MATLAB provides an interactive environment that allows users to run commands and see the results in real time. This makes it easy to experiment and debug.
- Graphical capabilities. MATLAB offers powerful visualization tools, making it suitable for creating high-quality graphs and charts.
- Multitasking. MATLAB allows users to perform multiple tasks simultaneously, which is useful for parallel processing and simulations.
- Integration with other languages and tools. MATLAB can be easily integrated with other programming languages such as C++, Python, and Java, as well as with various tools and libraries.

Disadvantages of MATLAB:

- Licensed Software. MATLAB is paid software, which may limit access to it for students and small organizations.
- Computing resources. Large calculations in MATLAB can require significant computing resources, especially when processing large data.
- Code closure. MATLAB code is proprietary and closed, which limits the freedom of users to change and modify the code as they see fit.
- Execution speed. In some cases, MATLAB can be slower than low-level programming, especially for calculations that require high performance.
- Lack of a rich ecosystem. Compared to some other programming languages and tools, MATLAB has fewer libraries and resources available, which can complicate finding solutions for specific tasks.
- Limitations when working with large data. MATLAB can face challenges when processing large volumes of data due to RAM limitations.

The choice of MATLAB or other software for modeling and simulation depends on the specific requirements of the task to be solved, the availability of hardware means and the skills of the user. MATLAB is a powerful tool for numerical analysis and modeling, but its limitations must be considered when using it.

Tolerance analysis in established mode of operation of DC-DC converters. On Figure 20a (inductor current) and Figure 20b (output voltage) are shown the results of tolerance analysis of the Buck DC-DC converter.

Figure 20. Tolerance analysis of the state variables of a step-down DC-DC converter, when operating in established mode

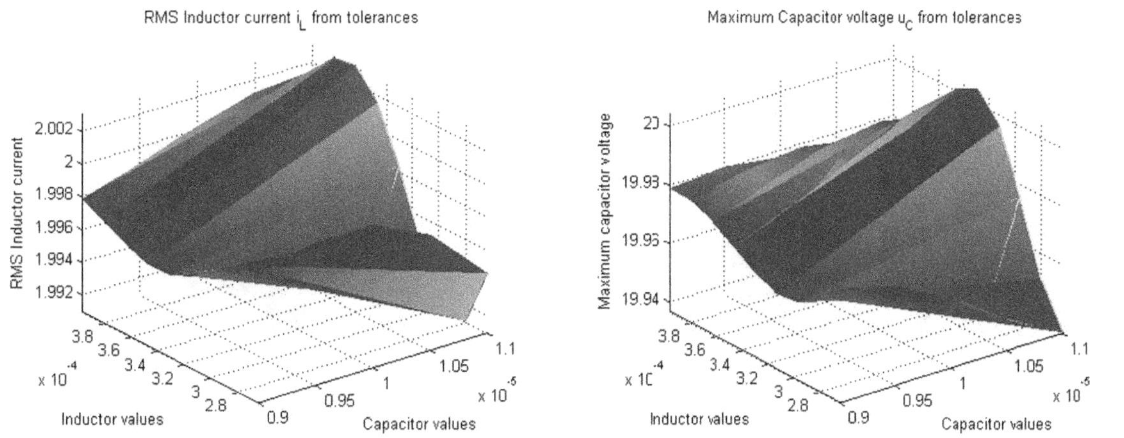

(a) Tolerance analysis RMS Inductor current *(b) Tolerance analysis maximum capacitor voltage*

From the 3D diagrams it is visible that the maximum change in relation to the current of the inductor is 2.003 A at L = 363 μH and C = 11 μF f_0 = 15.83 kHz. The minimum inductor current is 1.991 A at L = 280.5 μH and C = 11 μF with self-resonant frequency of the LC-filter f_0 = 18 kHz. The voltage of the capacitor is maximum 20 V at L = 330 μH and C = 11 μF, f_0 = 16.6 kHz and the minimum is 19.94 V at L = 264 μH and C = 11 μF f_0 = 18.56 kHz respectively.

From the analysis of the presented results for the considered type of DC-DC converter it is observed that the tolerances of the passive circuit elements have a very weak influence on the parameters of the operating regime in the steady-state mode.

On Figure 21 are shown the results of tolerance analysis of the Boost DC-DC converter. From the 3D diagrams it is visible that the maximum change in relation to the current of the inductor is 3.983 A at L = 346.5 μH and C = 11 μF f_0 = 16.2 kHz.

Figure 21. Tolerance analysis of the state variables of a step-up DC-DC converter, when operating in established mode

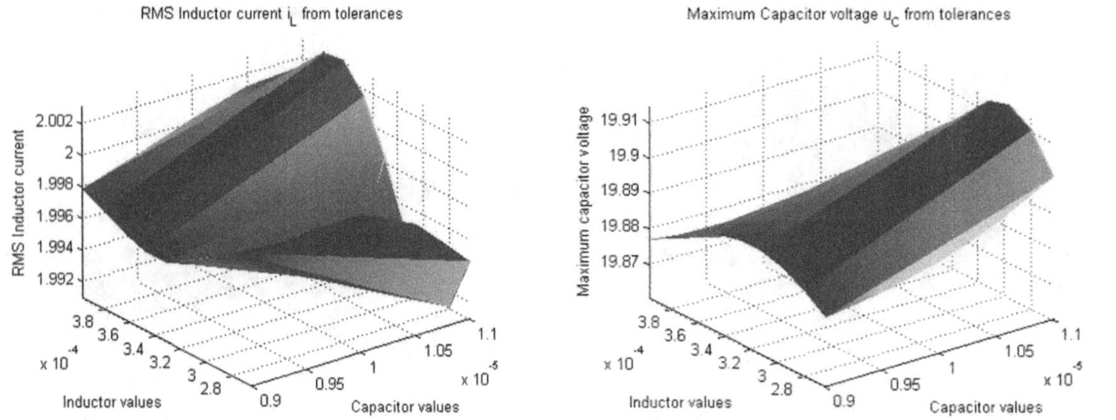

(a) Tolerance analysis RMS Inductor current *(b) Tolerance analysis maximum capacitor voltage*

From the analysis of the presented results for the considered type of DC-DC converter it is observed that the tolerances of the passive circuit elements have a very weak influence on the parameters of the operating regime in the steady-state mode.

On Figure 21a and Figure 21b are shown the results of tolerance analysis of the Boost DC-DC converter. From the 3D diagrams it is visible that the maximum change in relation to the current of the inductor is 3.983 A at L = 346.5 μH and C = 11 μF f0 = 16.2 kHz.

The minimum inductor current is 3.974 A at L = 264 μH and C = 10.5 μF with self-resonant frequency of the LC-filter f_0 = 18.99 kHz. The voltage of the capacitor is maximum 19.91 V at L = 297 μH and C = 11 μF, f_0 = 17.5 kHz and the minimum is 19.86 V at L = 396 μH and C = 11 μF, f_0 = 15.15 kHz respectively.

From the graphical results for the Boost DC-DC converter it is observed that the change of the nominal values of the circuit elements practically does not affect the parameters of the operating regime in the established state. There is a change only in the value of the resonant frequency of the LC-filter, which is quite far from the value of the operating frequency which the semiconductor devices are switched at.

On Figure 22a and Figure 22b are shown the results of tolerance analysis of the Buck-Boost DC-DC converter. From the 3D diagrams it is visible that the maximum change in relation to the current of the inductor is 3.983 A at L = 363 μH and C = 11 μF, f_0 = 15.83 kHz. The minimum inductor current is 3.97 A at L = 264 μH and C = 11 μF with self-resonant frequency of the LC-filter f_0 = 18.56 kHz. The voltage of the capacitor is maximum -19.87 V at L = 264 μH and C = 9 μF, f_0 = 20.52 kHz and the minimum is 19.92 V at L = 313.5 μH and C = 11 μF, f_0 = 17.03 kHz respectively.

The same dependence was observed in the research of this type of power device as in the other two topologies – a very slight change in the parameters of the operating mode both in terms of the output current and in terms of the output voltage. In practice, the observed changes are insignificant and do not lead to significant changes in the operation of the devices. This effect is largely due to the characteristics of the series resonant circuits, which is essentially the output circuit of the converters.

Figure 22. Tolerance analysis of the state variables of a Buck-Boost DC-DC converter, when operating in established mode

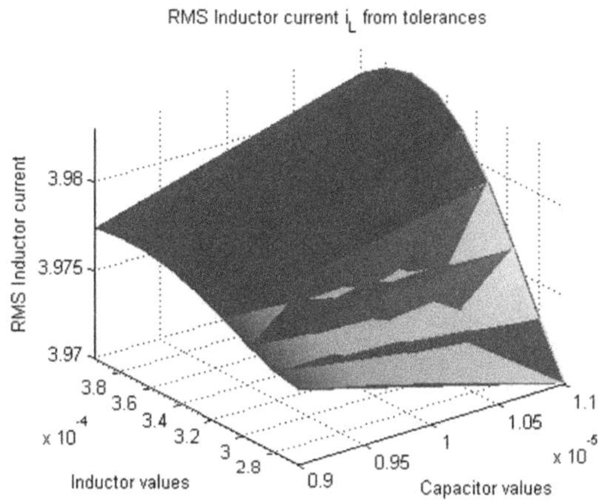

(a) tolerance analysis RMS Inductor current

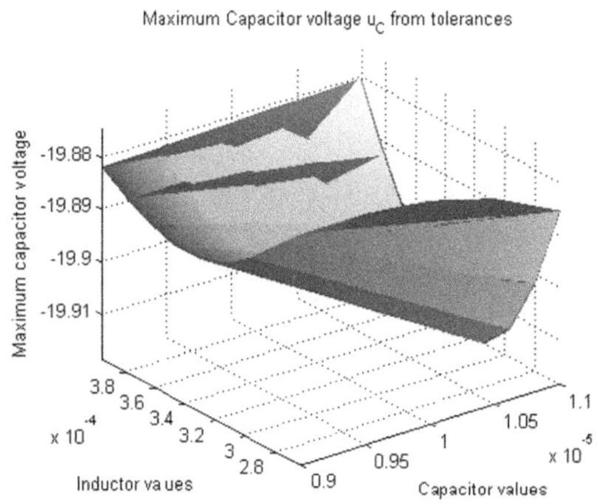

(b) tolerance analysis maximum capacitor voltage

Specific for most of the resonant circuits is that their characteristics are strongly dependent at frequencies close to the resonant (difference up to 20%) and, accordingly, very slightly changing at the frequency far from the resonant (over 3 times difference). As a general conclusion of the research in steady-state modes of the converters, is that with the correct design of the power circuit, the operating mode of the device practically does not depend on the tolerances of the filter elements.

Tolerance analysis in transient mode of operation of DC-DC converters. The study is conducted using the same mathematical models and tolerance values, but the emphasis is on the form of the transient process and its properties, such as the maximum overshoots.

On Figure 23a (inductor current) and Figure 23b (output voltage) are shown the results of tolerance analysis of the Buck DC-DC converter transient process. From the waveforms it is visible that the maximum overshoot in relation to the current of the inductor is 4.63 A at filter elements L = 264 μH and C = 11 μF, f_0 = 18.56 kHz. The minimum overshoot inductor current is 3.58 A at L = 396 μH and C = 9 μF with self-resonant frequency of the LC-filter f_0 = 16.75 kHz. The overshoot transient voltage of the capacitor is at its peak maximum 28.96 V at filter elements L = 264 μH and C = 11 μF, f_0 = 18.56 kHz accordingly. The minimum overshoot output voltage is 26.57 V at L = 396 μH and C = 9 μF with self-resonant frequency of the LC-filter f_0 = 16.75 kHz.

The analysis of the results shows that in the transient process, the differences between the minimum and maximum values of the current through the inductor and the output voltage are significant, respectively above 1A and above 2.4 V, and the influence of tolerances on the output current is more significant.

On Figure 24 (inductor current) and Figure 24 (output voltage) are shown the results of tolerance analysis of the Boost DC-DC converter transient process. From the waveforms it is visible that the maximum overshoot in relation to the current of the inductor is 4.81 A at filter elements L = 264 μH and C = 11 μF, f_0 = 18.56 kHz.

Figure 23. Tolerance analysis of the state variables of a step-down DC-DC converter, when operating in transient mode

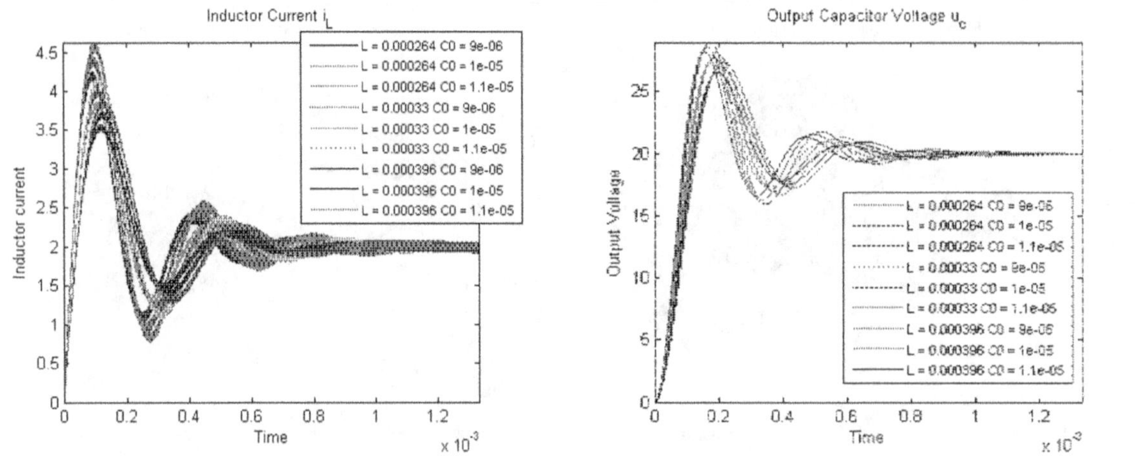

(a) tolerance analysis RMS Inductor current *(b) tolerance analysis maximum capacitor voltage*

Figure 24. Tolerance analysis of the state variables of a step-up DC-DC converter, when operating in transient mode

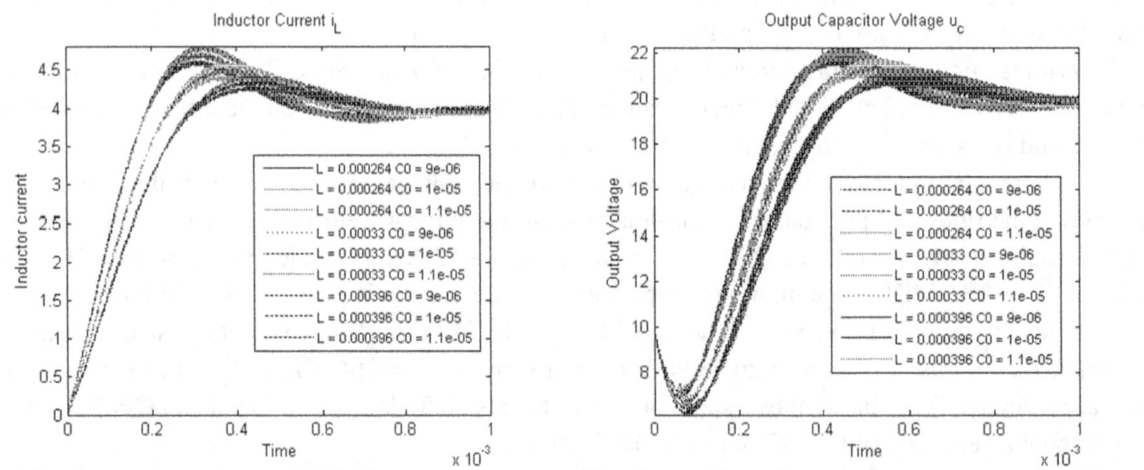

(a) tolerance analysis RMS Inductor current *(b) tolerance analysis maximum capacitor voltage*

The minimum overshoot inductor current is 3.58 A at L = 396 µH and C = 9 µF with self-resonant frequency of the LC-filter f_0 = 16.75 kHz. The overshoot transient voltage of the capacitor is at its peak maximum 22.24 V at filter elements L = 264 µH and C = 11 µF, f_0 = 18.56 kHz and accordingly. The minimum overshoot output voltage is 20.96 V at L = 396 µH and C = 9 µF with self-resonant frequency of the LC-filter f_0 = 16.75 kHz.

Considering the presented graphical dependences of the transients, it is found that in the transient mode the differences between the minimum and maximum values of current through the inductor and the output voltage are significant, respectively over 1.2A and over 1.2 V, and relative to the output voltage are smaller compared to the previous circuit.

On Figure 25a (inductor current) and Figure 25b (output voltage) are shown the results of tolerance analysis of the Buck-Boost DC-DC converter transient process.

Figure 25. Tolerance analysis of the state variables of a Buck - Boost DC-DC converter, when operating in transient mode

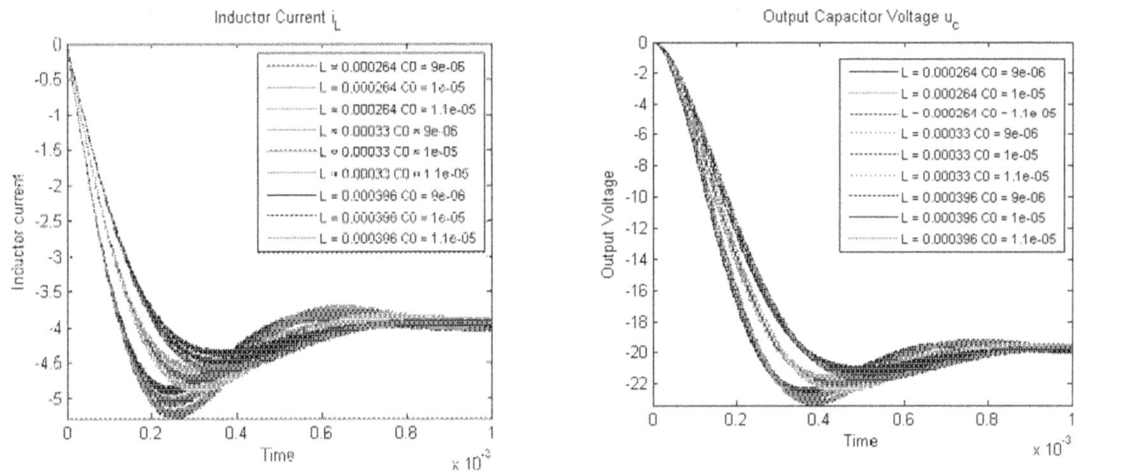

(a) tolerance analysis RMS Inductor current *(b) tolerance analysis maximum capacitor voltage*

From the waveforms it is visible that the maximum overshoot in relation to the current of the inductor is -5.28 A at filter elements L = 264 μH and C = 11 μF f_0 = 18.56 kHz. The minimum overshoot inductor current is – 4.4 A at L = 396 μH and C = 9 μF with self-resonant frequency of the LC-filter f_0 = 16.75 kHz. The overshoot transient voltage of the capacitor is at its peak maximum -23.38 V at filter elements L = 264 μH and C = 11 μF, f_0 = 18.56 kHz accordingly. The minimum overshoot output voltage is -21.27 V at L = 396 μH and C = 9 μF with self-resonant frequency of the LC-filter f_0 = 16.75 kHz.

The comparison of the results shows that in the transient process the differences between the minimum and maximum values of the current through the inductor and the output voltage are below 1A and above 2V, respectively, and the influence of tolerances on the output current is smaller compared to the other two cases.

Discussion. From the analysis of the presented results the following more important conclusions and conclusions can be made:

- The change of the operating modes, due to the tolerances of the filter elements, is insignificant in the established mode of operation and in practice with well executed design of the power circuit, they can be neglected.

- On the other hand, the tolerance analysis shows that the effect in transient processes is significant and in different topologies there are some differences regarding the change of the maximum values, but as a rule the changes are tens of times larger than in the established mode.
- Due to the nature of the operation of these DC-DC converters, serious attention should be paid to the synthesis of the control and the settings of the controller in order to achieve improved characteristics and guarantee the output performance.

Conclusions are made regarding the influence of the changes of the parameters of the passive building elements, in relation to established mode and transient processes. A good approach is presented for the designer to evaluate the values of the circuit elements and parameters of the operating modes already obtained by the calculations, what changes can be expected in them due tolerance and how to make an optimal selection of elements by various complex indicators such as price, quality, reliability, etc.

The achieved results are important from the point of view of the use of different methods for the design of power electronic converters, as the parameters of the elements that change during operation are studied. This is also very useful for improving the practical training in the field of power electronic converters and power electronics in general.

On the other hand, the use of design methods in regard with the lower dependence of the operating modes from the tolerances allows for optimal control synthesis. This is very important according with the increasing requirements for power electronic devices in terms of operating conditions, price, and technical performance.

As a continuation of these studies, we plan to analyze the impact of tolerances on the stability of the controller and hence the synthesis of robust control with minimal hardware requirements.

A Simulation Study of Current and Voltage Resonance in Electrical Circuits

Electrical engineering training for software professionals and engineers is generally aimed at providing the knowledge and skills to enable them to understand and collaborate more effectively with their hardware counterparts. Also, the goal is to give software professionals the opportunity to develop embedded software or participate in projects that require electrical engineering knowledge.

One of the main topics usually included in such training is related to the description and study of resonance processes in electric circuits. On the other hand, most basic electrical engineering courses offered by universities, technical colleges, online platforms, or specialized training centers use Spice based simulators for their training needs. These simulators use a direct analogy between real circuits and models and are very useful in training hardware-oriented engineers. On the other hand, software specialists have developed specific skills related to programming, the use of abstract constructions and representations of real processes, devices and systems. In (Popov, S., & Hinov, N. 2023), a Python-based approach to the study of various electric circuits is proposed, which is very suitable for programmers. One of the most frequently studied phenomena in electrical engineering - resonance processes - has been examined.

Python is a high-level, interpreted programming language with a syntax that is easy to read and understand. It is a powerful software development tool with wide application in various fields. Here are some of the main features and applications of Python (Pine, D. J. 2019):

- Ease of use. Python is known for its easy and readable syntax, which makes programming accessible even to beginners. This fact makes it suitable for training purposes and rapid prototyping.

- Multiple Libraries. Python has a large ecosystem of libraries and modules that allow developers to solve different tasks without starting everything from scratch. Some popular libraries include NumPy (for numerical calculations), Pandas (for data processing), Matplotlib (for visualization), and many others.
- Cross-platform. Python runs on a variety of operating systems, including Windows, macOS, and various Linux distributions. This makes code written in Python easily portable across platforms.
- Interpreted. Python is run through an interpreter that allows for dynamic code execution. This aspect facilitates rapid application development, and the language is suitable for scripting and interactive work.
- Object-oriented and functional. Python supports object-oriented and functional programming style, which allows developers to us e different paradigms depending on the specific requirements of the task.
- Large communities and active feedback. Python has a large community of users and an active feedback forum. This resource helps developers find solutions to problems and share knowledge and experience.
- Used in various fields. Python is used in all kinds of fields, including web development, science, artificial intelligence, machine learning, data processing, scientific computing, and more.
- Python is a very flexible language that can be used for a variety of applications and tasks. This wide range of capabilities makes it one of the most popular programming languages in the world.

Investigation of resonance in processes in a series resonant circuit (voltage resonance).

The schematic of a series resonant circuit is shown in Figure 26. It consists of a sinusoidal voltage source *e(t)*, an inductance *L*, a capacitance *C* and a resistor *R*. The scheme parameters are:

$e(t)=0.1\sin(5773.5t)$, $R=100\Omega$, $L=0.1H$, $C=300nF$.

Figure 26. Series resonant circuit

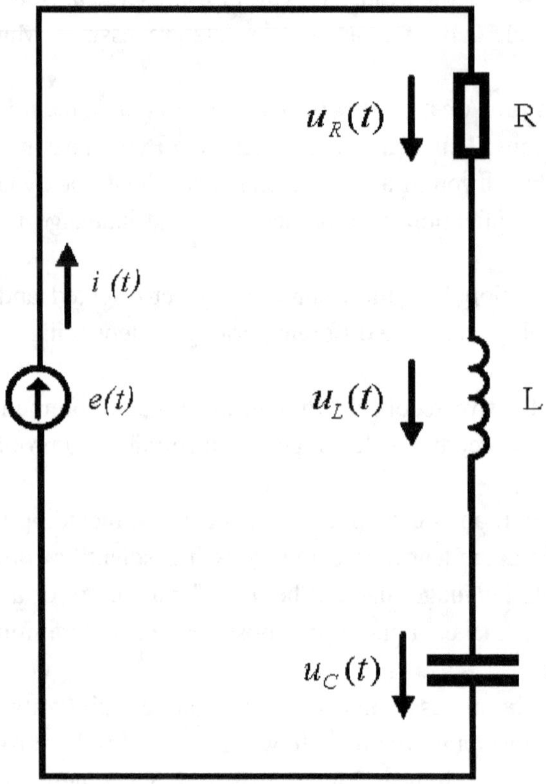

For the modeling of the scheme, a transformation of a ready-made model implemented on LTspise was used. This is convenient in view of the availability of a wide variety of libraries with models of both individual elements and LTspise schemes and systems. In addition, both software are of the "open access" type, which is useful in view of their joint use.

In Figure 27 presents the results of the simulation of a series resonant circuit in Python. Specifically, the dependences of the voltages on the three elements of the series RLC circuit as a function of frequency are given.

From the graphical results presented, it is found that the voltages on all three elements of the series resonant circuit have a maximum value for the same frequency - this is the resonant frequency of the circuit. In this way, the phenomenon of voltage resonance is formed. Subsequently, through the obtained data, various calculations can be made regarding the parameters of the resonant circuit, such as determined by its quality factor and bandwidth.

Figure 27. Frequency dependences of the voltages of the three elements of the series RLC circuit with a voltage resonance

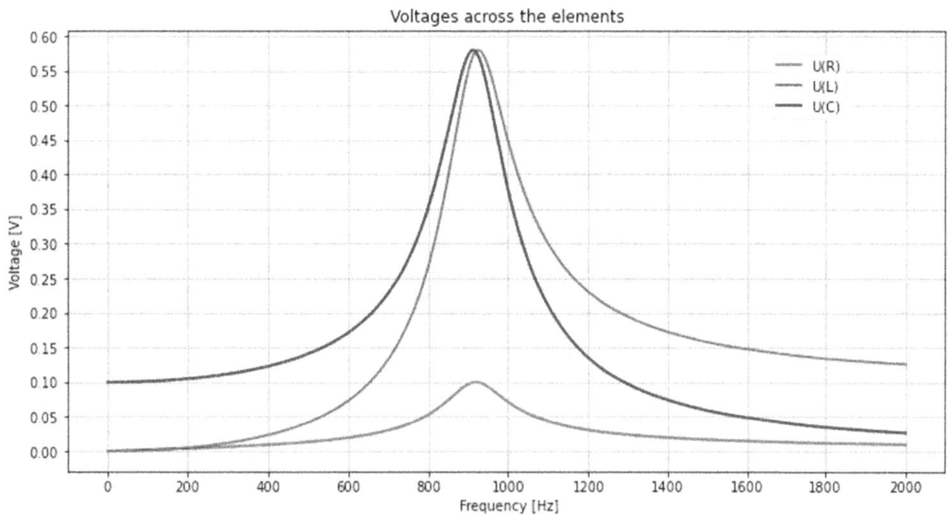

Investigation of resonance in processes in a parallel resonant circuit (current resonance).

The schematic of the parallel resonant circuit is shown in Figure 28. It consists of a sinusoidal voltage source *e(t)*, an inductance, a capacitance, and a resistor.

The scheme and model's parameters are:

$$e(t) = 10.\sqrt{2}.\sin(3162t - 90°), L_1 = 1mH, C_1 = 100\mu F, R_1 = 20© .$$

The implementation of the LTspice model of a parallel resonant circuit in Python is based on the algorithm presented (Popov, S., & Hinov, N. 2023). The transformation of the model was carried out with the help of a developed author's program. In the simulations, parameters are set for conducting an alternating current analysis of the circuit, thereby determining its main characteristics: amplitude and phase.

Figure 28. Parallel resonant circuit

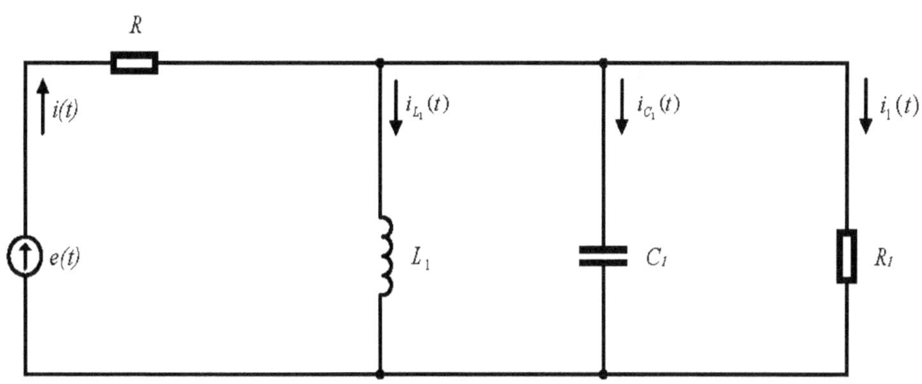

In Figure 29 presents the results of the simulation of the parallel resonant circuit in Python. Specifically, the frequency dependences of the currents through the three elements of the parallel RLC circuit are given.

It can be seen that the currents through all three elements have a maximum value for the same frequency - this is the resonant frequency of the circuit.

Figure 29. Frequency dependences of the currents through the three elements of the parallel RLC circuit

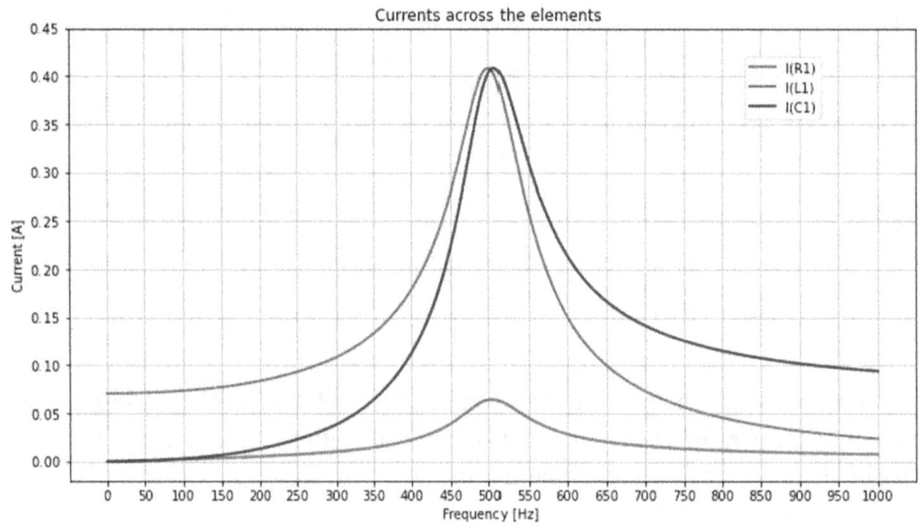

Conclusion. Two examples of current and voltage resonance studies using Python are presented. This programming language was chosen because Python provides many new possibilities compared to traditional software for modeling electrical and electronic circuits. Its use makes it possible to apply achievements in the field of big data processing, optimization, and the application of artificial intelligence techniques to the study and design of electrical and electronic circuits. In this sense, the application of Python does not negate the traditional methods for modeling and automated design of electrical and electronic circuits and systems but offers development and upgrading of the design and prototyping process, integrating modern information and communication technologies and data science into it. It is a good alternative to the classical methods of studying electrical and electronic circuits. With the use of such innovative approaches, these disciplines become more attractive and appealing to young learners who have an affinity for programming.

REFERENCES

Akberdin, A. A., Kim, A. S., & Sultangaziev, R. B. (2018a). Experiment planning in the simulation of industrial processes. *Steel in Translation*, *48*(9), 573–577. doi:10.3103/S0967091218090024

Akberdin, A. A., Kim, A. S., & Sultangaziev, R. B. (2018b). Planning of numerical and physical experiment in simulation of technological processes. *Izvestiya. Ferros Metallurgy*, *61*(9), 737–742. doi:10.17073/0368-0797-2019-9-737-742

Balakrishnan, H., Moreno-Ezuilaz, M., Riba, J. R., Bogarra, S., & Garcia, A. (2018). DC-DC buck converter parameter identification based on a white-box approach. In *2018 IEEE 18th International Power Electronics and Motion Control Conference (PEMC)* (pp. 242-247). IEEE. 10.1109/EPEPEMC.2018.8521981

Beg, S., Swain, S., Rahman, M., Hasnain, M. S., & Imam, S. S. (2019). Application of design of experiments (DoE) in pharmaceutical product and process optimization. In *Pharmaceutical quality by design* (pp. 43–64). Academic Press. doi:10.1016/B978-0-12-815799-2.00003-4

Chen, H., Li, H., Ma, T., Han, S., & Zhao, Q. (2023). Biological function simulation in neuromorphic devices: From synapse and neuron to behavior. *Science and Technology of Advanced Materials*, *24*(1), 2183712. doi:10.1080/14686996.2023.2183712 PMID:36926202

Cui, T. & Ma, Y. (2015). Definition of attribute circle in factor space and its application in object classification. *Computer engineering & science, 37*(11), 2169-2174.

Eini, M. R., Salmani, H., & Piniewski, M. (2023). Comparison of process-based and statistical approaches for simulation and projections of rainfed crop yields. *Agricultural Water Management*, *277*, 108107. doi:10.1016/j.agwat.2022.108107

Gocheva, P. V., Hinov, N. L., & Gochev, V. P. (2020). Index matrices-based modelling of a DC-DC buck converter with PID controller and GUI on it. *2020 International Conference on Information Technologies (InfoTech)*, (pp. 1-4). IEEE.

Hacks, S., Katsikeas, S., Ling, E., Lagerström, R., & Ekstedt, M. (2020). powerLang: A probabilistic attack simulation language for the power domain. *Energy Informatics*, *3*(1), 1–17. doi:10.118642162-020-00134-4

Hinov, N. L., & Hranov, T. H. (2021). Tolerance analysis of common transistor DC-DC converters. In *2021 25th International Conference Electronics* (pp. 1-6). IEEE. 10.1109/IEEECONF52705.2021.9467442

Hoole, J., Mellor, P. H., Simpson, N., & North, D. (2021, May). Statistical simulation of conductor lay and ac losses in multi-strand stator windings. In 2021 IEEE International Electric Machines & Drives Conference (IEMDC) (pp. 1-8). IEEE. doi:10.1109/IEMDC47953.2021.9449582

Huo, Y., Chen, Y., & Su, M. (2023, February). Simulink-integrated representation of functional architectures towards simulation of aircraft systems. In *Proceedings of the International Conference on Aerospace System Science and Engineering 2022* (pp. 170-190). Springer, Singapore. https://doi.org/10.1007/978-981-99-0651-2_14

Ivanova, Y. (2020). Adaptive digitalization methods and digital transformation trends for security. *International Journal on Information Technologies and Security*, *12*(3), 51–62.

Ivanova, Y. (2021). Optimizing simulation models of an artificial neural network for digital recognition. *International Journal on Information Technologies and Security*, *13*(4), 59–70.

Ivanova, Y. (2022). Simulation modelling of artificial neural networks for the purpose of steganalysis. *International Journal on Information Technologies and Security, 14*(2), 99–110.

Katsikeas, S., Hacks, S., Johnson, P., Ekstedt, M., Lagerström, R., Jacobsson, J., Wällstedt, M., & Eliasson, P. (2020, November). An attack simulation language for the IT domain. In *Graphical Models for Security: 7th International Workshop, GraMSec 2020, Boston, MA, USA, June 22, 2020, Revised Selected Papers* (pp. 67-86). *Lecture Notes in Computer Science*. Cham Springer International Publishing. 10.1007/978-3-030-62230-5_4

Katsikeas, S., Johnsson, P., Hacks, S., & Lagerström, R. (2022). VehicleLang: A probabilistic modeling and simulation language for modern vehicle IT infrastructures. *Computers & Security, 117*, 102705. doi:10.1016/j.cose.2022.102705

Kuznetsov, A. L., Kirichenko, A. V., Semenov, A. D., & Oja, H. (2020). Planning simulation experiments in the tasks of studying the operational strategies of container terminals. *TransNav, the International Journal on Marine Navigation and Safety of Sea Transportation, 14*(4), 845-849.

Li, K., Marston, J. B., & Tobias, S. M. (2021). Direct statistical simulation of low order dynamosystems. *Proceedings of the Royal Society A (Mathematical, Physica; and Engineering Sciences), 477*(2254), 20210427. 10.1098/rspa.2021.0427

Loquercio, A., Kaufmann, E., Ranftl, R., Dosovitskiy, A., Koltun, V., & Scaramuzza, D. (2019). Deep drone racing: From simulation to reality with domain randomization. *IEEE Transactions on Robotics, 36*(1), 1–14. doi:10.1109/TRO.2019.2942989

Mohan, N., Undeland, T. M., & Robbins, W. P. (2003). *Power electronics: converters, applications, and design*. John wiley & sons.

Morris, T. P., White, I. R., & Crowther, M. J. (2019). Using simulation studies to evaluate statistical methods. *Statistics in Medicine, 38*(11), 2074–2102. doi:10.1002im.8086 PMID:30652356

Napoli, C., & García-Téllez, B. (2016). *Energy for water in agriculture: A partial factor productivity analysis. King Abdullah Petroleum Studies and Research Center*, (KAPSARC).

Oleinikova, S.A., Selishchev, I.A., & Kravets, O.Ja., Rahman, P.A., & Aksenov, I.A. (2021). Simulation model for calculating the probabilistic and temporal characteristics of the project and the risks of its untimely completion. *International Journal on Information Technologies and Security, 13*(2), 55–62.

Oreshina, O. A. (2020, April). Full factor plan application to polymer composites hardness investigation. [IOP Publishing.]. *Journal of Physics: Conference Series, 1515*(4), 042031. doi:10.1088/1742-6596/1515/4/042031

Péan, F., & Goksel, O. (2020). Surface-based modeling of muscles: Functional simulation of the shoulder. *Medical Engineering & Physics, 82*, 1–12. doi:10.1016/j.medengphy.2020.04.010 PMID:32709260

Pine, D. J. (2019). *Introduction to Python for science and engineering*. CRC press. doi:10.1201/9780429506413

Popov, S., & Hinov, N. (2023). Software implementation of PSPISE-based models in the Python programming language. *International Journal on Information Technologies and Security, 15*(3), 65–74. doi:10.59035/LYWQ6789

Reinhardt, O., Warnke, T., & Uhrmacher, A. M. (2022). A language for agent-based discrete-event modeling and simulation of linked lives. [TOMACS]. *ACM Transactions on Modeling and Computer Simulation*, *32*(1), 1–26. doi:10.1145/3486634

Riesebos, L., & Brown, K. R. (2022, September). Functional simulation of real-time quantum control software. In *2022 IEEE International Conference on Quantum Computing and Engineering (QCE)* (pp. 535-544). IEEE. 10.1109/QCE53715.2022.00076

Tian, P., Shukla, A., Nie, L., Zhan, G., & Liu, S. (2018). Characteristics' relation model of asphalt pavement performance based on factor analysis. *International Journal of Pavement Research and Technology*, *11*(1), 1–12. doi:10.1016/j.ijprt.2017.07.007

Turner, C. J., & Garn, W. (2022). Next generation DES simulation: A research agenda for human centric manufacturing systems. *Journal of Industrial Information Integration*, *28*, 100354. doi:10.1016/j.jii.2022.100354

Yu, X., Jiang, F., Du, J., & Gong, D. (2019). A cross-domain collaborative filtering algorithm with expanding user and item features via the latent factor space of auxiliary domains. *Pattern Recognition*, *94*, 96–109. doi:10.1016/j.patcog.2019.05.030

Zhang, L., Ye, F., Laili, Y., Xie, K., Gu, P., Wang, X., Zhao, C., Zhang, X., & Chen, M. (2021, July). X language: an integrated intelligent modeling and simulation language for complex products. In *2021 Annual Modeling and Simulation Conference (ANNSIM)* (pp. 1-11). IEEE. 10.23919/ANNSIM52504.2021.9552057

Chapter 14
Organization and Application of Statistical Modeling

ABSTRACT

The object of this chapter is principles of organization of statistical modeling and form of its applications. In this reason, the first section discusses the essence of the statical approach and presents the main form for its application. A summary of published research is made to clarify the features of the statistical models and their role in scientific research. The second part is devoted to mathematical statistics, with a brief presentation of basic concepts such as simple and statistical order, statistical estimates, numerical characteristics, and basic methods used in statistical approach. Particular attention is paid to the method of least squares applied in determining the statistical estimates. In the third part, the characteristic surface is presented and the two main groups of statistical models—univariate and multivariate—are discussed. Realization of statistical experiments is discussed in the last part, starting with presentation of a program application for statistical analysis organization and continuing with several concrete examples of regression models.

1. ESSENCE AND APPLICATION OF STATISTICAL MODELING

Statistical modeling is a method of presenting empirical data from conducted experiments through mathematical dependencies, describing the relationships between the studied parameters sufficiently accurately (Kroese & Chan, 2014). This modeling is applied in various fields of scientific research when it is necessary to represent the causal relationships in the studied object. It uses empirical values to study the behavior of an object, which is why it is often called empirical modeling or the empirical approach. For example, classical modeling of spatial extremes uses asymptotic models for block maxima or peaks above high thresholds, and (Huser & Wadsworth, 2022) provides an overview of modern statistical modeling capabilities in the study of spatial extremes. This article discusses different "conditional spatial extremes model, which have recently been getting increasing attention within the statistics of extremes community". The empirical approach is often applied when conducting test experiments, as done in (Pelivani et al., 2022). The task is to formulate a correct general framework for software test

DOI: 10.4018/978-1-6684-8947-5.ch014

optimization, supporting different types of technical extensions. This is an important software engineering task that will allow to establish the quality of the software in terms of detected errors for various characteristics such as reliability, efficiency, usability, and maintainability. Empirical research is mainly focused on execution time, using two important indicators - number of steps and number of objects. The appearance of windows in the work process, the identification of objects, database testing, use of artificial intelligence, etc. have been analyzed.

Another area of application of statistical modeling is presented in (Dvir & Ben-Zvi, 2023), where the influence of statistical uncertainty in the contemporary digital age is investigated. The basis for this is the claim that probabilistic language and tools in the formal accounting of statistical uncertainty tend to make it difficult for researchers, including in pedagogical activities. For this reason, the article has a goal to demonstrate the potential of statistical modeling in the research of several processes in the real word.

Statistical modeling is based on a mathematical apparatus oriented towards statistical processing of accumulated data, and the main features of the theory and practice of statistical modeling are presented in (Freedman, 2009). Statistical models defined in this way reflect the relationships between the studied parameters of the objects, and the initial data for their development are obtained from experiments conducted in a real environment or in another type of model. Basically, any experimental investigation is a set of purposeful actions, through which the essence of the state and functioning of the object of modeling and research is revealed. It consists of two parts – experiments and observations. An experiment is a part of carried out investigation that is realized under a certain set of conditions. As a result of realizing the experiment, a random event occurs. Each trial is represented by a point in the factor space. Observation is the repeated repetition of the experiment at one point and the collection of data about it. Each experimental investigation contains many trials, and each trial includes one or more observations. Statistical modeling typically involves the following two steps.

(1) Planning and conducting experiments to collect empirical data.
(2) Applying an appropriate method for mathematical analysis of empirical data and obtaining statistical estimates.

Statistical modeling is based on the Monte Carlo method consisting of the following:

- A sample of *N* random numbers (RNs) $<x^{(i)},y^{(i)}>$, uniformly distributed in the interval [0,1] is formed.
- The area of impact in the *XY* plane is determined.
- A statistical solution to the problem is sought based on the formed sample.

EXAMPLE: Calculation of the integral $y = \int_0^1 x.dx$

- The sample $<x^{(i)},y^{(i)}> / i=1,2,\dots,N$ of uniformly distributed in the interval [0,1] of equal probability random numbers is formed by using of RNG.
- From them, the area of impact in the *XY* coordinate system is determined and is allocated *M* numbers that fall under the line $y=f(x)=x$ connecting the points <0,0> and <1,1>.
- The ratio *M/N* determines the approximate solution of the problem, i.e., $y \approx M/N$.

The Monte Carlo method allows finding approximate solutions for various dependencies by forming statistical samples (Shreider, 2014; Landau & Binder, 2021). Statistical modeling develops this method by calculating statistical estimates from the formed sample based on a selected statistical method for analyzing empirical data. The main steps are presented below.

- Planning a statistical experiment, in which the levels (values) for the individual variables and the number of trials conducted to accumulate a representative sample are determined.
- Formation of sample, which is done by conducting *N* repetitions of the experiment under identical conditions and taking observations (recording individual values) for each of the observed variables. The number of observations *N* in a specific sample depends on the number of observed variables and can be determined in two ways: ✓ classical (preliminary determined number *N*); ✓ dynamic determination (*N* is determined in the process of forming the sample itself by analyzing the data already obtained).
- Selection of a mathematical method for statistical processing of the formed sample and obtaining statistical estimates for the studied parameters while minimizing the error.

The statistical approach is applied in cases where it is necessary to present the cause-and-effect relationships in the research object. The reason for this may be the diversity or complexity of the object, as well as the unclear nature of the processes, etc. This is a widely used approach, and an example of its application in the study of Distributed Denial-of-Service (DDoS) attacks is presented in (Nooribakhsh & Mollamotalebi, 2020). It is known that DDoS is a common type of attack against the services provided in the network space, where it is quite difficult to detect or prevent in time. This necessitates increasing the interest of researchers to search for effective countermeasures. The article identifies the negative impact on the process of countering attacks at low traffic levels, loss of duration, and high sampling volumes. To overcome these shortcomings, statistical methods are considered, which observe the receiving traffic in different periods of time and analyze its distribution. The result of the study is that the statistical approach is useful in detecting DDoS attacks, especially if the relevant statistical system operates on the victim's side.

The statistical approach is applied in (Kleven, 2021) to evaluate the policy and its welfare effect of the policy changes. The idea of representing the changes by computable elasticities is applied, which will allow to obtain an estimate without estimating the structural primitives of fully specified models. One of the main conclusions of the publication is that "*it is possible to develop transparent sufficient statistics formulas under very general conditions, but the estimation requirements increase greatly*" and the conclusion is made that "*feasible empirical implementations are in fact structural approaches*".

In recent years, the statistical approach has had significant application in the fields of biology, health care and medicine. The accumulation of real empirical data from patient observations allows a statistical model to be applied to analyze a given condition and draw conclusions about the next course of action. For example, gene set analysis allows obtaining a clear insight into the biology of diseases and as stated in (Das et al., 2020) it reduces the complexity of statistical analysis with increased significance of the results obtained. To facilitate the applicability of the statistical approach, the paper reviews the possibilities in this direction, proposing a statistical framework and sequential steps for gene set analysis methods used for microarray and genome-wide association data analysis. In addition, a classification of possible approaches and used analysis tools is made by type of genomic study, null hypothesis, sampling model, and features of test statistics. Biological and statistical challenges are identified to help develop further statistical approaches in genomic analysis, and key factors in potential analysis are identified. Application of testing processes and in particular integration testing with test containers in determining best practices is discussed in (Shukla, 2023d).

Another health issue is related to dementia prevention, which should receive research priority because evidence has been obtained of the relationships between individual health behaviors, cognitive function, and dementia risk. In this reason, the main objective in (Dingle et al., 2023) was to identify and characterize the statistical approaches applied in the study of the multitude of risk factors, assessing their relationship with cognitive outcomes in adults. To solve the set goal, a thorough review of electronic databases was made and numerous publications on the subject were reviewed, looking for the relationship between the co-occurrence of risk factors, their possible grouping, as well as combining the two approaches. In conclusion, it is concluded that the dominant statistical approach in conducting research and the co-occurrence approach for aggregating health-related risk factors and examining associations with cognitive outcomes in adults, and a recommendation is made to move to applying more advanced statistical methods for studying clustering-based approaches.

The research presented in (Zhao et al., 2021) is dedicated to analyzing dietary patterns and their relationship to health. It is emphasized again that despite the existence of many statistical methods, the classical methods related to diet quality assessment, principal component analysis, factor analysis and reduced rank regression analysis are mainly applied. To expand the applicability of statistical methods, the paper provides a complete review of existing statistical methods, with a focus on finite mixture model, tree transformation, data mining, etc. Their advantages and disadvantages are presented, indicating and available application software. As a result of the conducted research, it was concluded that all statistical methods have their unique characteristics and suitability for a specific field, but emerging methods have a higher importance for modern research. For their specific application, it is necessary to make a proper assessment of their effectiveness, validity, and predictive ability, which can be done by conducting experiments.

In a different direction is the research conducted in (Noprisson, 2020), which discusses the topic of business failure. The topic is important because it can be used as a basis for the formation of the policy of a given company or government structures. Results of the prediction of business failure risks can be used in making decisions about taking preventive measures. To suggest possible research options, methods, and results for predicting business failure in different sectors and regions are reviewed. The effect of the study is detailing a few statistical methods for predicting business failures, some of which are hybrid forecasting, Altman Z-score model, data envelopment analysis, logistic regression, neural networks (NN), support vector machine, etc. A comparative analysis of the achieved degree of accuracy for the studied methods was made.

2. MATHEMATICAL STATISTICS AND MODEL INVESTIGATION

Statistical modeling is directly related to the application of methods and techniques from mathematical statistics to calculate estimates for random variables and random processes. Below are basic concepts of mathematical statistics directly related to statistical modeling.

Sample and Statistical Order

A general set *W* is the complete set of elements *w* that possess a certain characteristic, on which the study is carried out. A sample is the part of this set that is subject to specific study, since it is impossible in a given experiment to obtain all elements of *W*. Events are defined as a subset of the sample space. Figure 1 presents an example where the events E should be taken as "*the set for x where x is not less than 40 and not greater than 60*".

Figure 1. General set W and specific event E

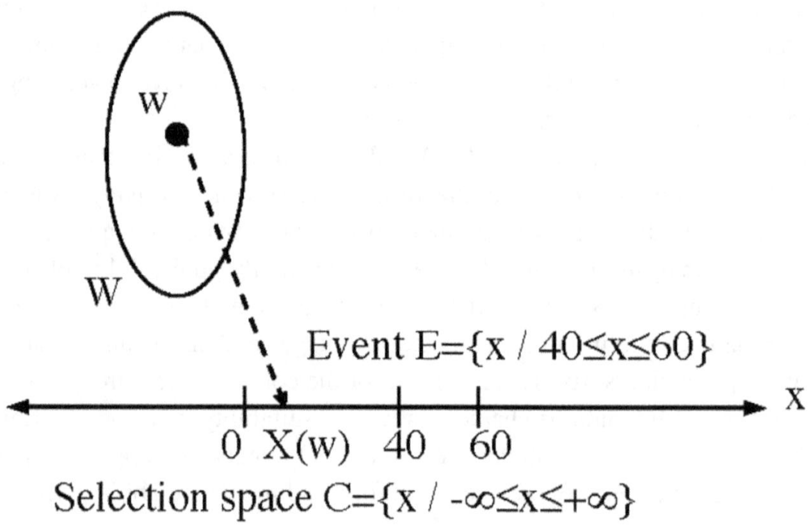

The sample values, arranged in ascending order, form a statistical order. Such an arrangement makes it possible to directly determine the minimum and maximum value of the sample, as well as in what interval to look for the values of the random variable. The number of observations for a given sample can be determined in two ways.

- Predefined value for *N* (classical method but does not always ensure research efficiency).
- Dynamic determination of *N* in the process of forming the sample itself by analyzing the already obtained data (allows smaller values for *N* but complicates the preliminary planning of the experiment).

The formation of a sample is an essential issue in conducting statistical research and empirical modeling, which requires the necessary correctness in conducting observations. According to (Brus, 2021), there are at first glance quite persistent, but misconceptions about sampling design and the formation of statistical inferences that are based on classical sampling theory. According to the author, these misconceptions are the result from a misunderstanding of basic statistical concepts such as independence, expectation, and deviation and variance of estimators or predictors. When designing a statistical model, these concepts have their own specific and different meaning because they consider different sources of randomness. For example, the population mean should not be confused with the sample mean, which is also true of the population variance and the sample variance. This confusion can lead to erroneous formulas for the variance and for estimating the population mean. To confirm the statement in the article, a study of the fundamental differences of these two approaches is done by applying simulation. In addition, an overview of the possibilities of using knowledge about the spatial structure of a research variable in designing a sample and a statistical model is presented.

Statistical Estimates

Each sample has a range of value changes determined by the two extreme possibilities for the investigated parameter X – minimum value (x_{min}) and maximum value (x_{max}). This allows to form a distribution of the actually registered observations falling within the defined value intervals for the spice ($X_{MIN} \div X_{MAX}$), which is represented by a histogram. The histogram is a model of the sample density and is calculated by the formula (the parameters are presented in Table 1):

$$h_i = \frac{m_i}{N.'_i} = \frac{p_i}{'_i} = f_i(X), \text{ for } i=1,2,\ldots,k$$

Table 1. Basic parameters of a histogram

Parameter	*Comment*
m_i	number of observations falling into the interval "*i*"
k	number of separate intervals
$N = \sum_{i=1}^{k} m_i$	total number of observations in the sample
p_i	frequency for the *i*-th interval
δ_i	size of the *i*-th interval determined by ($x_i - x_{i-1}$)

The graphical representation of a histogram is bars (rectangles) for each defined interval, which correspond to the number of observations falling within it without overlapping (Figure 2). Typically, interval parameters are represented on the X-axis and frequency values on the Y-axis. Each bar (rectangle) corresponds to the relative frequency for the interval it represents.

Figure 2. Histogram

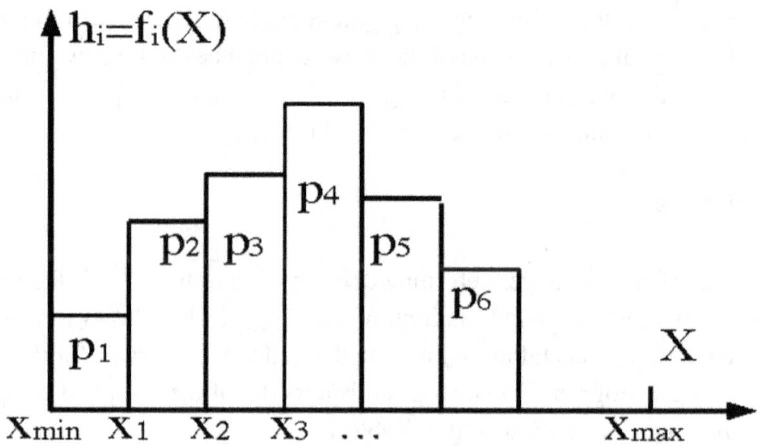

If the size δ_i is small enough (i.e., $\delta_i \to 0$), the histogram represents the density distribution of the population with some accuracy.

Statistical inferences about the general population can be made based on the sample by calculating estimates and testing hypotheses.

Statistical estimation is the calculation of a given function $g(x)$ for a variable based on its individual sample values. The estimates $g_1(x)$, $g_2(x)$, ..., obtained based on different samples C_1, C_2, ..., form the set of estimates, which has the same distribution as W, i.e., as well as the random variable (RV) X (Figure 3). One desirable property of the estimator is its unbiasedness (the mathematical expectation of the estimator $g(X)$ coincides with the value of the estimated parameter G, i.e., $E[g(X)] = G$. In some experiments, it is possible to determine several unbiased estimates. Different methods are used to calculate the estimates, for example, method of moments, maximum likelihood method, Pearson criterion, Kolmogorov criterion, etc., but to the extent that there is no best estimation procedure, the most common method is the small squares method.

Figure 3. Assessments forming

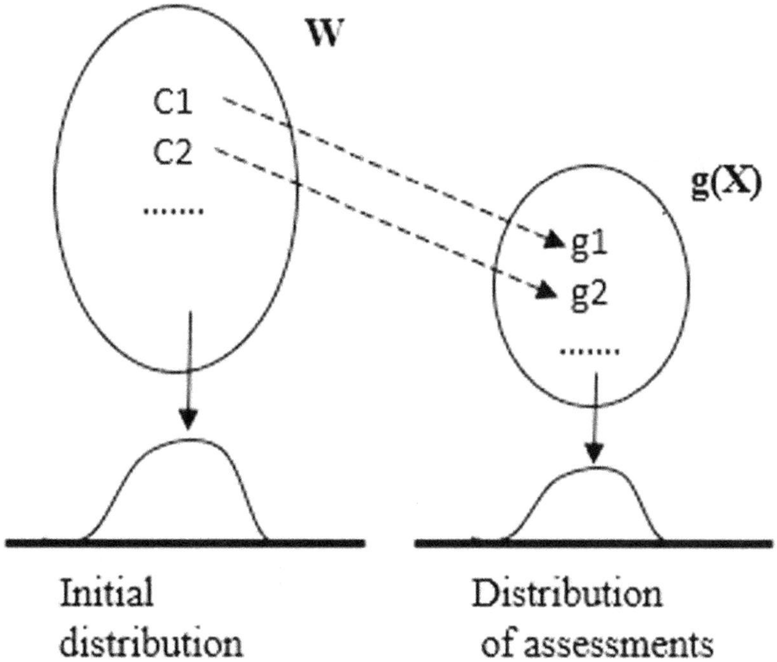

Hypothesis testing is a procedure for accepting or rejecting a statement about the values of one or more parameters for a given distribution or about its shape. Therefore, the statistical hypothesis is also a statement about the general population.

Numerical Characteristics

For each discrete RV *X* based on a formed sample of the values xi that it has assumed in *N* consecutive registrations, numerical characteristics can be calculated. Their description using the apparatus of mathematical statistics is presented in Table 2.

Statistical modeling is related to the application of methods and means of mathematical statistics to calculate numerical estimates for the characteristics of random variables and random processes. For their determination, several basic methods are applied, briefly presented below.

Method of stochastic approximation. The method is based on a general iterative procedure to determine a desired characteristic based on a previous value (from a previous iteration), an additional estimate for the process, and a coefficient depending on the iteration number (Kushner & Clark, 2012). A certain part of the developed stochastic approximation methods confirms convergence of the iterations, but not necessarily of the optimal base levels. This necessitates the development of more optimized methods to increase the advantages of the approach, including when forming samples of random values for the monitored variables. For example, in (Chau & Fu, 2014) a general overview of the applied stochastic approximation methods with orientation towards simulations is made.

Table 2. Basic numerical characteristics

Mathematical expectation	$E[X] = x_{av} = \frac{1}{N}\sum_{i=1}^{k}(m_i.x_i)$
Variance (the standard deviation is $\sigma_X = +\sqrt{V[X]}$)	$V[X] = \sigma^2 = \frac{1}{N}\sum_{i=1}^{k}[m_i(x_i - x_{av})^2]$
j-th starting (initial) moment	$\mu_j = \frac{1}{N}\sum_{i=1}^{k}(m_i.x_i^j)$
j-th central moment	$\sigma_j = \frac{1}{N}\sum_{i=1}^{k}[m_i(x_i - x_{av})^j] = \sigma^j$
Coefficients of asymmetry and excess	($A_S = \frac{\mu_3}{\sigma^3}$) and ($E_X = \frac{\mu_4}{\sigma^4} - 3$)

At its core, stochastic approximation is an iterative search algorithm, viewed as the stochastic analogue of a sequence in deterministic optimization. The subject of discussion are the classic methods of Robbins–Monro and Kiefer–Wolfowitz, as well as very used variants such as Kesten's rule, iterate averaging, varying bounds, simultaneous perturbation stochastic approximation, robust stochastic approximation, accelerated stochastic approximation (AC-SA) for convex and strongly convex functions, and Secant-Tangents AveRaged stochastic approximation (STAR-SA). Concrete studies of the capabilities of selected algorithms have been made through numerical examples. An application of the Robbins–Monro algorithm to develop a recursive value distribution estimator is presented in (Jmaei et al., 2017). The proposed recursive estimator, accepted as a competitor of Vitale's distribution estimator, has been investigated for its capabilities and properties, and with optimal parameters, the new proposal shows better results regarding the mean integrated squared error.

Another application of stochastic approximation is presented in (Ghadimi et al., 2020) with the purpose to investigate bounded nested approximation with application of a composition of two smooth functions with unknown exact values. A single time-scale stochastic approximation algorithm is proposed to find an approximate stationary point of the problem. Two auxiliary averaging filters are included in the algorithm to estimate the gradient of the composite objective function, and a special Lyapunov function is used to estimate the complexity. The proposed method is applicable to both unconstrained and constrained problems without the need for changes and batch samples. The performed analysis has confirmed that a simplified variant of the parameter-free algorithm is applicable in solving constrained optimization problems, and the complexity estimate is the same as for an unconstrained variant.

Method of parametric functions. Usually, this method is reduced to the above by introducing an additional parameter *q* depending on the number of iterations. A few studies related to this method can be briefly presented. Parametric functions were used in the development of an image enhancement method presented in (Bianco et al., 2029). The parameters are derived from a down sampled version of the raw input image, and the resulting transformation is applied to the input of the full-resolution algorithm. The separation of the parameters and the color transformation allow the speed and accuracy of the method to be achieved. Variants of approximations of the method can be generated by changing the parametric functions used as color transformations (polynomial, fractional, cosine and radial).

Another investigation based on parametric modeling and optimization of a permanent-magnet brushless DC (BLDC) motor by using artificial neural network is presented in (Sadrossadat & Rahmani, 2020). To achieve higher speed, an objective function is set to minimize the volume and cost parameters in developing the magnetic head. A gradient-based method with nonlinear magnetic constraints was applied to determine optimal geometric parameters. Experimental results have confirmed the effectiveness of the BLDC model, determining it to be more accurate than other existing analytical models and faster than simulation models. Parametric optimization of involved parameters is applied in (Mehmood & Ochs, 2020) in computing the derivative of the solution to a given problem. For a class of highly convex functions, this can be achieved by automatically differentiating iterative minimization algorithms. In the case of algorithm convergence to pointwise, it is proved that "*the derivative sequence also converges pointwise to the derivative of the minimizer with respect to the parameters*". In addition, the levels of convergence are given for both studied levels, proving that its average speed leads to an acceleration of the calculation of the derivatives. The theoretical results are confirmed by conducting experiments with L2-Regularized Logistic Regression.

Method of least squares (MLS). Briefly, the method consists of obtaining estimates for the parameters *A*, *B*, *C*, … by minimizing the sum of squares:

$$S = \sum_{i=1}^{N} [x_Q^{(i)} - A.x_A^{(i)} - B.x_B^{(i)} - \ldots - Z.x_Z^{(i)}]^2 = min$$

This method is perhaps the most common in statistical research, the main reasons for this being that different estimates can be formed within this framework, and the use of a quadratic makes it mathematically applicable to divisive problems. The first reason will be discussed below, and for the second reason, it is related to the Pythagorean theorem, which allows, when the error is independent of the size of the calculation, to use the squares of the error and the size estimate. An additional reason for the prevalence of MLS is that the mathematical tools it uses are well researched and readily available.

A brief historical overview of the development of MLS is presented in (Abdi, 2007), noting that it is one of the oldest techniques used in modern statistics. The role of Greek mathematics and Galileo, who is considered the first modern predecessor, is specified, and the modern approach was presented at the beginning of the 19th century by the mathematicians Legendre and Gauss. Galton (1886) later used MLS in his research that established the foundations of correlation and coined the name regression analysis (Galton, 1886). For the development of the method in different contexts, factor analysis for Pearson and experimental design for Fisher can be indicated.

Modern research uses MLS to determine the numerical values of various statistical parameters, to characterize their statistical properties, and to relate a given function to a set of numerical data. The theory has been extended with additional variants, such as the Ordinary Least Squares (OLS) version, a more sophisticated version Weighted Least Squares (WLS), s well as the more modern versions Alternate Least Squares (ALS) and Partial Least Squares (PLS). An analysis of a class of randomized algorithms for solving overdetermination least squares problems is done in (Lacotte & Pilanci, 2020), where "*the gradients are pre-conditioned by an approximation of the Hessian, based on a subspace embedding of the data matrix*". The considered class of algorithms covers several randomized versions of MLS for fast problem solutions, mainly targeting classical Gaussian projections and subsampled randomized

Hadamard transforms (SRHT). What is new in the paper is the derivation of the limiting spectral density of SRHT embeddings, using the result to derive a family of normalized orthogonal polynomials of the SRHT density. In addition, the optimal first-order preconditioned method is determined along with its convergence rate. The analysis of Gaussian embeddings using the classical random matrix theory is similarly organized.

The research presented in (Cohen & Migliorati, 2017) is related to the reconstruction of an unknown bounded function u defined in a domain $X \subset R^d$ with n points (x^i), for $i=1,\ldots,n$ and measuring the error at the norm $L^2(X,d\rho)$ for a given probability measure ($d\rho$). The study was conducted in a set space with $m<n$ independent random samples, and to ensure the correctness of the least squares approximations, the value of m should not be very close to n. In a theoretical aspect, the contribution is establishing the mathematical expectation and determination of weighted least squares probabilities in general approximation spaces. From a numerical point of view, the proposed method for generating independent and identically distributed samples is useful in solving multivariate fitting problems. To illustrate the contributions of the method, concrete examples of approximation spaces of polynomial type with domain X motivated by parametric and stochastic approximations are implemented.

Brief Presentation of MLS

To illustrate MLS, let us hypothesize that there is a linear relationship $Y = Y(X)$ between the random processes Y and X, which is approximated by the estimation equation $y = a + b.x$ (Figure 4). A sample of N values $<X^{(i)},Y^{(i)}>$ is formed, with each observation $Y(i)$ deviating from the approximating value $y(i)$ by $y^{(i)}$ c $d^{(i)}=Y^{(i)}-y^{(i)}$. According to the MLS, according to the hypothesis, an estimate $y(x)$ is sought for the dependence $Y(X)$ for which:

$$S = \sum_{i=1}^{N} \left[d^{(i)} \right]^2 = min$$

Figure 4. Illustration of the method of least squares

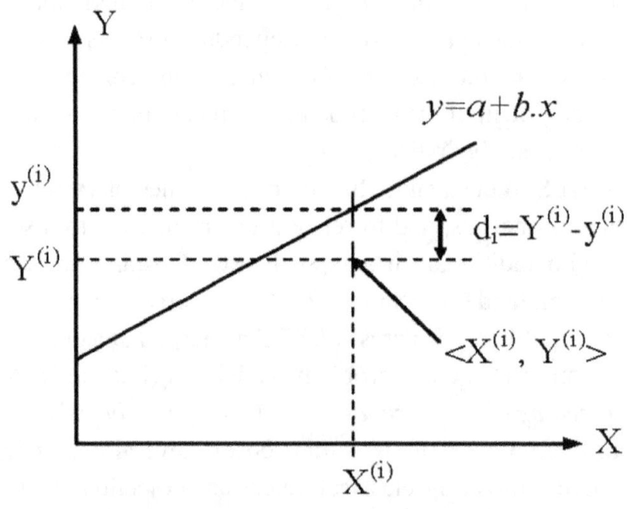

The following sum-of-squares minimization is applied to estimate the vector β:

$$S^2 = \sum_{i=1}^{n}(y_i - \beta_0 - \beta_1 x_{i1} - \beta_2 x_{i2} - \ldots - \beta_p x_{ip})^2 = (y - X\beta)'' \left(y - X\beta\right)$$

It is known from the theory that in this case minimization of S^2 can be done if the following conditions are fulfilled:

$$\begin{cases} \dfrac{\partial S}{\partial \beta_0} = 0 \\ \dfrac{\partial S}{\partial \beta_1} = 0 \\ \ldots\ldots \\ \dfrac{\partial S}{\partial \beta_p} = 0 \end{cases}$$

Also, according to statistical theory, the transformations presented below can be made. To begin with, applying differential calculus and after conversion to normal form, the system of equations presented below is formed, and its matrix entry will have the form $X^T y = X^T X\beta$.

$$\begin{cases} \sum_{i=1}^{n} y_i = n{}^{2}{}_{0} + {}^{2}{}_{1}\sum_{i=1}^{n} x_{i1} + {}^{2}{}_{2}\sum_{i=1}^{n} x_{i2} + \ldots + {}^{2}{}_{p}\sum_{i=1}^{n} x_{ip} \\ \sum_{i=1}^{n} x_{i1} y_i = {}^{2}{}_{0}\sum_{i=1}^{n} x_{i1} + {}^{2}{}_{1}\sum_{i=1}^{n} x_{i1}^2 + {}^{2}{}_{2}\sum_{i=1}^{n} x_{i1}x_{i2} + \ldots + {}^{2}{}_{p}\sum_{i=1}^{n} x_{i1}x_{ip} \\ \sum_{i=1}^{n} x_{i2} y_i = {}^{2}{}_{0}\sum_{i=1}^{n} x_{i2} + {}^{2}{}_{1}\sum_{i=1}^{n} x_{i1}x_{i2} + {}^{2}{}_{2}\sum_{i=1}^{n} x_{i2}^2 + \ldots + {}^{2}{}_{p}\sum_{i=1}^{n} x_{i2}x_{ip} \\ \vdots \\ \sum_{i=1}^{n} x_{ip} y_i = {}^{2}{}_{0}\sum_{i=1}^{n} x_{ip} + {}^{2}{}_{1}\sum_{i=1}^{n} x_{i1}x_{ip} + {}^{2}{}_{2}\sum_{i=1}^{n} x_{i2}x_{ip} + \ldots + {}^{2}{}_{p}\sum_{i=1}^{n} x_{ip}^2 \end{cases}$$

Each j-th element $\sum_{i=1}^{n} x_{ij} y_i$ of $X^T y$ denotes the product of the vectors x_j and y, and the vector $X^T y$ denotes a vector of $(p+1)$ products of the matrix X and the vector y. The diagonal elements of $X^T.X$ are the sum of the squares of the elements in each column of the matrix X; the elements of the inverse diagonal are the cross products of the different columns of the matrix X. The solution of the system of normal equations has the form $\hat{\beta} = (X^T X)^{-1} X^T y$

For this solution to exist, the matrix $(X^T X)^{-1}$ must exist (i.e., this matrix must have rank $(p+1)$. A necessary and sufficient condition is that the columns of X are linearly independent (i.e., that any column of the matrix X cannot be expressed as a linear combination of the remaining columns).

MLS gives unshifted estimates $\hat{\beta}_0, \hat{\beta}_1, \ldots, \hat{\beta}_p$ of the vector $\hat{\beta}$: $E\left(\hat{\beta}_j\right) = \beta_j$, j=1,2,…,p respectively in matrix form $E\left(\hat{\beta}\right) = \beta$. Moreover, the estimates obtained by the method have the smallest variance. The variances of the elements of $\hat{\beta}$ and the covariances between all possible pairs of the elements of $\hat{\beta}$ can be summarized in the so-called covariance matrix, the elements of which have the general representation: $Cov\left(\hat{\beta}_i, \hat{\beta}_j\right) = E[\left(\hat{\beta}_i - E\left[\hat{\beta}_i\right]\right)\left(\hat{\beta}_j - E\left[\hat{\beta}_j\right]\right)^T]$. The complete matrix is symmetric and contains all possible pairs:

$$E[\left(\hat{\beta} - \beta\right)\left(\hat{\beta} - \beta\right)^T] = \begin{bmatrix} V\left(\hat{\beta}_1\right) & Cov\left(\beta_0, \beta_1\right) & \cdots & Cov\left(\beta_0, \beta_p\right) \\ Cov\left(\beta_0, \beta_1\right) & V\left(\hat{\beta}_2\right) & \cdots & Cov\left(\beta_1, \beta_p\right) \\ \vdots & \vdots & \ddots & \vdots \\ Cov\left(\beta_0, \beta_p\right) & Cov\left(\beta_1, \beta_p\right) & \cdots & V\left(\hat{\beta}_p\right) \end{bmatrix}$$

The covariation matrix is obtained by $\sigma_\varepsilon^2 (X^T X)^{-1} = \sigma_\varepsilon^2 C^2$, where $C^2=(X^TX)^{-1}$ is denoted. The elements of the matrix $C^2=(X^TX)^{-1}$ are c_{ij}. The indexes i,j=1,2,…,p are such that $Cov\left(\hat{\beta}_i, \hat{\beta}_J\right) = \sigma_\varepsilon^2 c_{ij}^2$. An unbiased estimate s^2 of σ_ε^2, is given by the expression:

$$s^2 = \frac{(y - X\hat{\beta})^T \left(y - X\hat{\beta}\right)}{(n - p - 1)}$$

The numerator in this representation is the minimum value of the expression $(y–X\beta)$T$(y–X\beta)$, which was stated above.

3. CARACTERISTIC SURFACE AND BASIC STATISTICAL MODELS

Characteristic Surface

The characteristic surface is a geometric representation of the dependence of a given quantity on several other factors, determined as a result of a sample of random observations. This dependence is described by an equation of the characteristic surface, which connects the parameters of the system and the input effects { x_1, …, x_n } with some criterion $C=f(x_1, \ldots, x_n)$ characterizing the system – Figure 5. The characteristic surface can be represented in the form of a system of equations; set of graphs; table; stratifications for variable estimation.

Figure 5. Characteristic surface

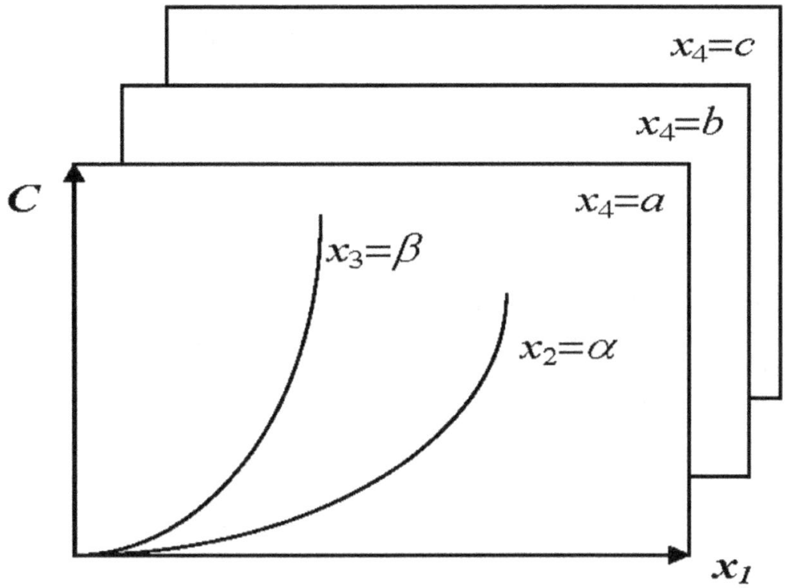

A key point in statistical modeling is the choice of a method for determining the equation of the characteristic surface and its graphical representation. Most often, this is related to the selection of a curve as close as possible to the empirically obtained points (registrations). Along with the functional approximation, another type of statistical models is also used to represent a characteristic surface, some of which are presented in the following parts. Interestingly, the statistical approach is applicable to solving tasks from various fields where a decision must be made based on a sample of data collected by various methods. An example of this is the following investigation.

Statistical modeling was used in solving the task of object monitoring and collection of sparse information stored at various remote objects in (Atlasov et al., 2020). Events that generate this type of information have a small probability p of happening, which increases its value determined by the formula $H = -\ln(p)$. This importance of information requires increasing the speed of its collection, processing, and protection, which is the task of a built system of multiple servers. In this respect, the article presents a solution to the problem for minimization of the general time for searching and processing in the system for which a simulation modeling is applied. The idea is to initially search servers that control a group of other servers, which will reduce the number of objects visited. To do this, a random variable corresponding to the number of search requests is generated by specifying several random numbers. Their product defines a value k, which determines the number of groups at the first division of the total number of servers n. Consecutively formed groups are divided into subgroups forming a hierarchical structure of servers distributed on separate sequential levels. The conducted statistical modeling and calculated estimates have shown that with a probability of $p<0.5$ and n servers, the number of realized requests for access to the desired informant will be about $2np^{1/2}$. If an additional level is created, this will reduce the number of requests to $3np^{2/3}$, and it has been shown that when choosing 3 levels, the mathematical expectation of requests can reach a value of $4np^{3/4}$.

Univariate (Descriptive) Statistical Models

These are statistical models formed by calculating one-dimensional statistical estimates based on the empirical data from the accumulated sample $\{x^{(1)}, x^{(2)}, \ldots, x^{(N)}\}$ for a separate factor (variable) X. Examples of such type of statistical models are presented below (Figure 6):

a) Basic statistics (range of variation, range, mean and limit values, variance, standard deviation, coefficient of variation.
b) Stratification – a statistical model for the object's behavior in a given range by distributing the values from the sample into three groups according to set lower (L) and upper (H) limits.
c) Histogram – distribution of the sample values by groups depending on the entry into them at a given step δ_h to define subintervals of the range $[x_{min}, x_{max}]$.

Figure 6. Univariate statistical assessments

$x^{(1)}$	$x^{(2)}$	$x^{(3)}$	$x^{(4)}$	$x^{(5)}$	$x^{(6)}$	$x^{(7)}$	$x^{(8)}$	$x^{(9)}$	$x^{(10)}$
2	7	6	4	7	5	5	6	4	5

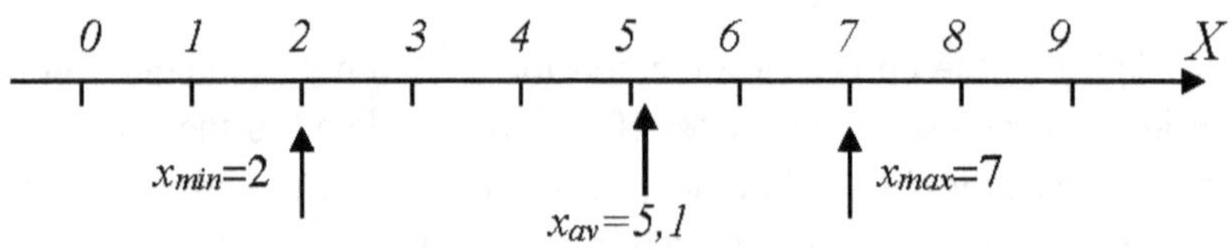

N=10
x_{min}=2; x_{max}=7;
x_{av}=5,1
$\Delta = x_{max} - x_{min}$=5
σ^2 = 2,322
σ= 1,5238
$C_v = \sigma / x_{av}$=0,2988

a) Basic statistical assessments

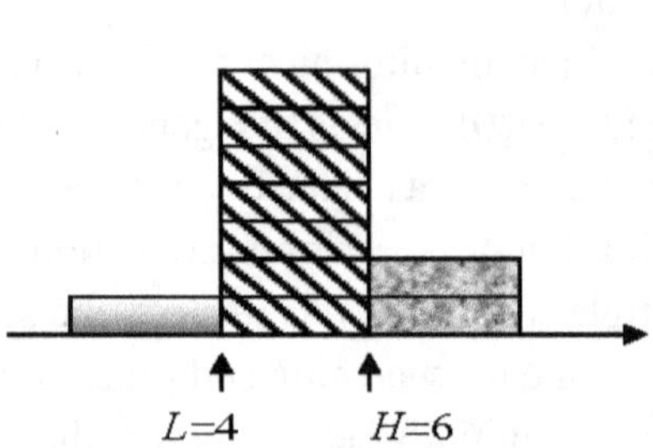

b) Stratification

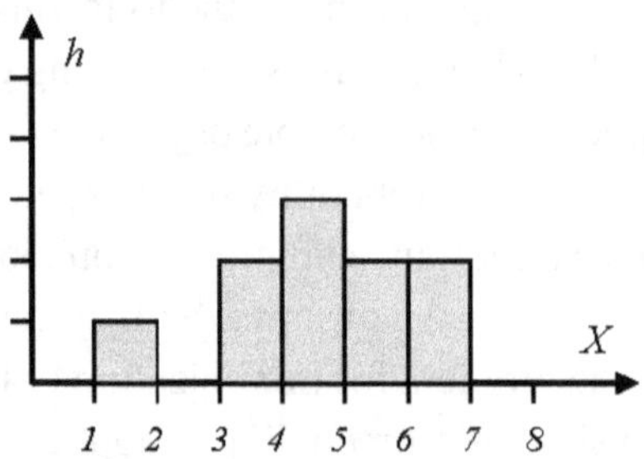

c) Histogram

Multivariate Statistical Models

This group of statistical models represents the interrelationship of two or more factors in a multivariate experiment. They consist in the acceptance of a given hypothesis and subsequent statistical analysis of the sample to determine estimates allowing the proof or rejection of the hypothesis. The main analyzes applied in the construction of statistical models are correlation, regression, and dispersion. There is some relationship between the first two analyses, with correlational providing an estimate of the interdependence between pairs of variables, and regression defining causal relationships between two or more than two variables as a linear or non-linear functional relationship. A discussion of their specific features in analyzing statistical data and results from various experiments is presented in Chapter 15.

Correlation analysis in statistical modeling. The main purpose is to investigate the relationship between two variables X and Y, and the modeling is based on the calculation of correlation and covariance coefficients for each pair of model parameters. Its main task is to establish the significance (strength) of the relationship between the values taken by each of the studied RVs, for which statistical estimates of correlation coefficients and correlation dependence are calculated. Depending on the number of simultaneously acting factors, it can be single (dependency for only a pair of factors) or multiple (research for a group of more than two factors), in form it can be linear or non-linear, and as a value characteristic it is a straight dependence (positive value) or inverse relationship (negative value). The correlation coefficient is a number that quantifies the existing relationship between two random variables X and Y, taking values in the interval $[-1, +1]$. The sign of the coefficient reflects the direction of correlation – negative or positive, and the specified value defines the degree of correlation (a value of 0 defines the absence of correlation dependence). The absolute value of the correlation coefficient determines the correlation ratio, which is in the interval $[0,1]$. Similarly, with a zero value of the correlation ratio, there is no correlation dependence between the two investigated RVs.

Regression statistical models. A regression model is introduced, assuming the existence of a certain dependence $Y=Y(X_1, \ldots, X_k)$ between one variable Y chosen as dependent and one or several independent variables. The dependencies defining the regression models can be different linear or non-linear functions (quadratic, cubic parabola, logarithmic, exponential), depending on the number of independent variables, they are defined as simple ($k=1$) and multiple ($k>1$).

The choice of regression model type is based on the scatterplot – the set of multivariate points $<x^{(i)}, y^{(i)}>$ for the variables under study. For example, in the illustration of Figure 7, the presented scatter diagram gives reason to choose a hypothesis that the variable Y depends linearly on the variable X.

Figure 7. Relationship between the sample and linear regression model

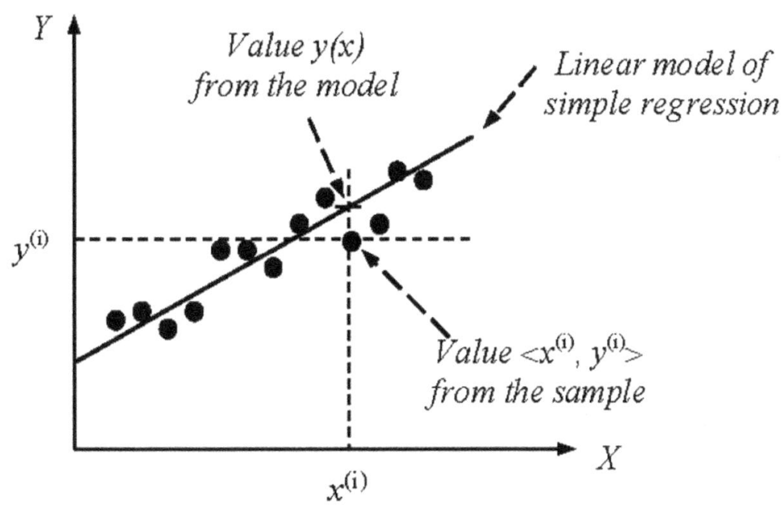

Linear Statistical Model of Simple Regression (Figure 7)

A straight-line equation $y=b_0 + b_1x$ is chosen to represent the statistical model, and the values of the coefficients are calculated to ensure the minimization of the error of approximation. The following expressions for the coefficients were determined by the method of least squares:

displacement (relative to the center of the coordinate system):	→	$b_0=y_{av} - b_1 x_{av}$
coefficient of regression:	→	$b_1 = \dfrac{\sum_{i=1}^{N}\left[x^{(i)} - x_{av}\right]\cdot\left[y^{(i)} - y_{av}\right]}{\sum_{i=1}^{N}\left[x^{(i)} - x_{av}\right]^2}$

Linear Statistical Multiple Regression Model

A hypothesis of linear dependence of Y, but on several independent variables X_1, ..., X_k, is assumed, which determines a regression model of the following type $y=b_0 + b_1x_1 + b_2x_2 + \ldots + b_kx_k$. The following generalized expressions are used to calculate the coefficients: $b_j = \beta_j.\dfrac{\sigma_Y}{\sigma_{X_j}}$ and $b_0 = y_{av} - \sum_{j=1}^{k} b_j.x_{j-av}$.

The calculation of β_j is done through the directs and inverse correlation matrices for variables.

In a more general aspect, one can assume that n observations are made on the variables $\left(x_{1i}, x_{2i}, \ldots, x_{ki}, y_i\right)$, $i=1,2,\ldots,n$, where y is the dependent variable out of k independent variable. This allows a complete linear regression model to be defined: $y_i = b_0 + b_1x_{1i} + b_2x_{2i} + \ldots + b_kx_{ki} + \varepsilon_i$, $i=1,2,\ldots,n$ where εi, $i=1,2,\ldots$,n are the random error values that are uncorrelated and uniformly distributed with expectation E(εi)=0 and variance $V\left(\varepsilon_i\right) = \sigma_\varepsilon^2 > 0$. The error distribution does not depend on the joint distribution of X1, X2, ...,Xk and therefore the conditional expectation is E[Y|X1, X2, ...,Xk]= β0+β1X1+...+βkXk and the conditional variance is $V[Y \mid X_1, X_2, \ldots, X_p] = \sigma^2_{Y \cdot X_1, X_2, \ldots, X_p} = \sigma_\varepsilon^2$. The unknown parameters β0,β1, ...,β_k are constants. Clearly, the results depend on the observed values of X1,X2,... Xk. Then a system of n equations is defined for the accumulated n observations of the variables involved in the investigation:

$$y_1 = \beta_0 + \beta_1x_{11} + \beta_2x_{21} + \ldots + \beta_kx_{k1} + \varepsilon_1$$
$$y_2 = \beta_0 + \beta_1x_{12} + \beta_2x_{22} + \ldots + \beta_kx_{k2} + \varepsilon_2$$

....

$$y_n = \beta_0 + \beta_1x_{1n} + \beta_2x_{2n} + \ldots + \beta_kx_{kn} + \varepsilon_n$$

The matrix form of the above system of equations is y=Xβ+ε, where:

$$y = \begin{bmatrix} y_1 \\ y_2 \\ \vdots \\ y_n \end{bmatrix}_{(n\times1)} ; X = \begin{bmatrix} 1 & x_{11} & x_{12} & \cdots & x_{1p} \\ 1 & x_{21} & x_{22} & \cdots & x_{2p} \\ \vdots & & & & \\ 1 & x_{n1} & x_{n2} & \cdots & x_{np} \end{bmatrix}_{n\times(p+1)} ; \beta = \begin{bmatrix} \beta_0 \\ \beta_1 \\ \beta_2 \\ \vdots \\ \beta_p \end{bmatrix}_{(p+1)\times1} ; \varepsilon = \begin{bmatrix} \varepsilon_1 \\ \varepsilon_2 \\ \vdots \\ \varepsilon_n \end{bmatrix}_{(n\times1)}$$

To make inferences about the multivariate linear regression using the results of the MLS, it is necessary to make the additional assumption that the elements εi ($_i$=1÷n) *o*f the error vector ε *h*a*v*e a normal distribution, which holds due to the above mathematical expectation 0 and corresponding variance σ_ε^2. Inferences about each individual regression coefficient can be made from the T-statistic (Student's T-test)

$$T = \frac{\left(\hat{b}_j - b_j\right)}{\sqrt{s^2 c_{jj}^2}}$$

which has a *t* distribution with (*n*-*k*-1) degrees of freedom. This statistic is used to test the null hypothesis H_0: $b_j = b^*$ by calculating на $T = \frac{\left(\hat{b}_j - b^*\right)}{\sqrt{s^2 c_{jj}^2}}$.

Regression statistical modeling for predicting energy consumption in smart homes and applied in (Haboubi & Salem, 2022), developing two predictive analysis models based on linear regression and support vector regression. Smart homes generate a continuous flow of process data based on time development over months, days, and hours. It was specified that the most suitable for conducting the research is the regression model, because it allows the analysis of the relationship between dependent RV and independent RV. The developed two regression models were used for machine learning with initial data cleaning and processing and subsequent visualization to reveal hidden information about the behavior of smart home appliances. Thus, the two regression models have been applied to predict energy consumption, and the comparison of the two techniques has confirmed the reliability of the smart home platform. The linear regression model *Y=a+bX* examines the dependence of the energy consumption *Y* of the system on temperature *X*, for which it is necessary to calculate the two coefficients of the regression line. The second model, based on the Support Vector Regression (SVR) is related to the Support vector Machine (SVM), which is a popular model for supervised learning. In this case the SVR is selected because it allows calculation of the regression parameters. This is an effective algorithm for predicting a set of data with real values. Linear regression is applied to minimize the cost function, and vector regression aims to determine the error within an optimal threshold.

Non-Linear Simple Regression Statistical Model

Given a suitable scatterplot, hypotheses for non-linear dependencies can be constructed by applying a selected statistical model from those shown in Table 3.

Table 3. Statistical model for non-linear dependency

Statistical model	Simple regression	Multiple regression
Exponential	$Y = b_0.e^{b_1 X}$	$Y = b_0.e^{\Sigma b_j.X_j}$
Logarithmic	$Y = b_0.X^{b_1}$	$Y = b_0 \prod_{j=1}^{u} X_j^{b_j}$
Inverse	$Y = [b_0 + (b_1.X_1)]^{-1}$	$Y = [b_0 + \sum(b_j.X_j)]^{-1}$
Polynomial	$Y = b_0 + b_1 X + b_2 X^2 + \ldots + b_k X^k$	

Variance statistical models. These models refer to estimating the variance of the population by calculating statistical estimates based on the sample. Typically, a comparison of the means of individual strata of a formed sample of statistical observations is applied (one-factor model) or the differential effects of two factors are investigated (two-factor model). Most used for models based on univariate analysis of variance, where each variable is represented as a sum of mean and error. In the two-factor variance analysis, two types of relationships presented below are investigated.

- Intersection (the two factors are represented by all possible combinations for the individual levels).
- Grouping (each individual level of one factor participates in combinations with only one level of the other factor).

The main tasks of dispersion analysis are to establish a quantitative relationship between the output value and the input factors when modeling multifactor objects and to form a qualitative assessment (presence or absence) of the influence of one or a group of factors on the output value.

One-Factor Dispersion Statistical Model

The sample of the statistical variable Y is stratified into k groups $\{y_{1j}\}, \{y_{2j}\}, \ldots, \{y_{kj}\}$, each with power L_i (i=1,…,k), such that $\Sigma L_i = N$. Each group has the same variance, with group means being $y_{i\text{-}av}$. Estimates of the differential effects $\alpha_i = y_{i\text{-}av} - y_{av}$ are determined for each group (the deviation of the mean for each group from the overall mean for the sample). An interpretation of this model for k=4 is shown in Figure 8.

One-factor covariance statistical model

It is based on the previous model $\alpha_i = y_{i\text{-}av} - y_{av}$, introducing the condition $\Sigma L_i.\alpha_i = 0$ and hypothesis of linear dependence $Y(X)$ in the groups.

Confirmation of the hypothesis is sought by calculating estimates of the overall mean, the within-group regression coefficient, the adjusted differential effects, and the adjusted mean for each subgroup (Figure 9).

Figure 8. One-factor variance model

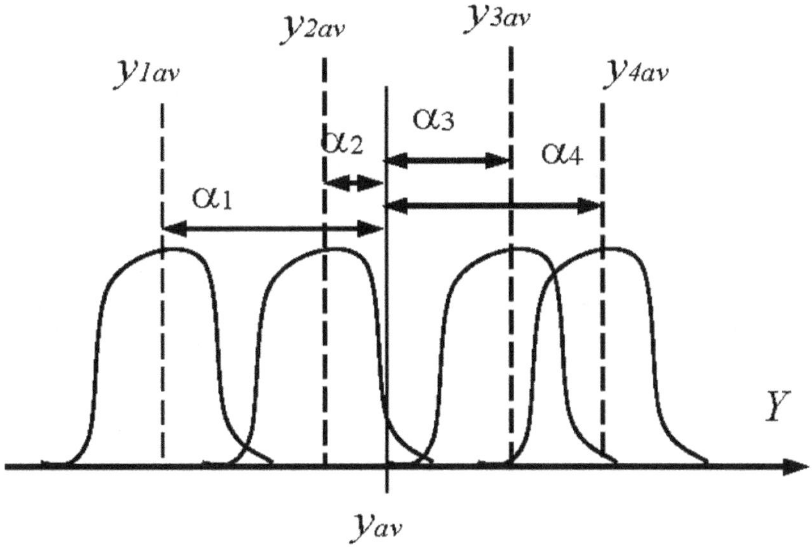

Figure 9. One-factor covariance model

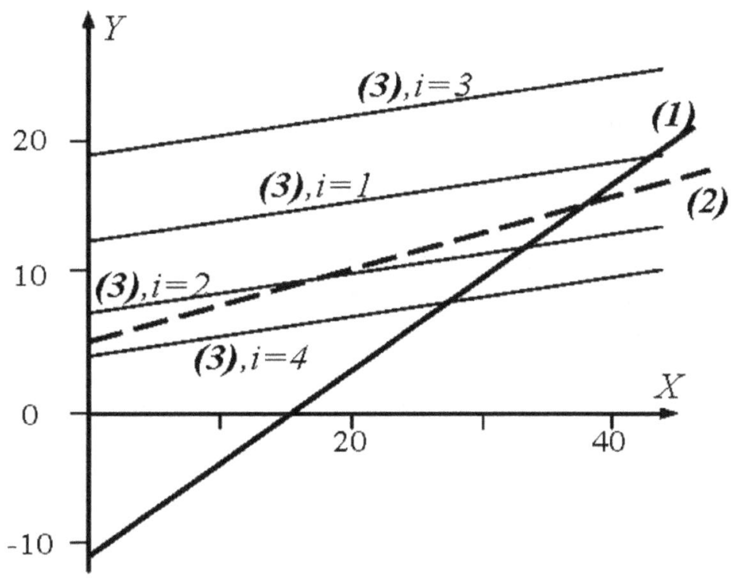

4. REALIZATION OF STATISTICAL EXPERIMENTS

Program environment for statistical analysis based on the regression model.

Analyzer is a Windows-based application developed in the Microsoft Visual Studio .Net environment and implemented in the C# language. The main purpose is to perform statistical modeling of accumulated empirical data regarding various applications. Hypotheses for the presence of interdependence between parameters are investigated by building regression statiscal models. The analytical organization of statistical modeling uses the principles of the regression method to study a possible relationship between two or more random variables. To prove or reject a given hypothesis, the Analyzer calculates statistical estimates based on an initial sample, which are summarized in Table 4.

Table 4. Statistical assessments

Main assessment	*Analytical expression*
Sample size (power).	$N = \left\| x_i^{(1)}, x_i^{(2)}, \ldots, x_i^{(N)} \right\|$
Range of variation and sweep	$[x_{i\_min}, x_{i\_max}]$; $(x_{i\_max} - x_{i\_min})$
Mathematical expectation for x_i (mean)	$EX = \frac{1}{N} \sum_{q=1}^{N} x_i^{(q)} = x_{i\_av} = \overline{x}_i$
Variance for x_i ($\overline{x}_i$ is mean)	$DX = \frac{1}{N} \sum_{q=1}^{N} \left[x_i^{(q)} - \overline{x}_i \right]^2 = \sigma_i^2$
Mean square (standard) deviation	$\sigma = +\sqrt{DX}$
Deviation factor	$C_V = \frac{\sigma}{x_{av}}$.
Covariance (joint variance) between x_i and x_j	$\sigma_{ij} = \frac{1}{N} \sum_{q=1}^{N} \left[x_i^{(q)} - \overline{x}_i \right] . \left[x_j^{(q)} - \overline{x}_j \right]$

The importance of covariance analysis in teaching mathematical disciplines is discussed in (Bagossi et al., 2022). It is emphasized that covariational reasoning enables the formation of conceptual knowledge and supports innovative methods for evaluating processes and events. In addition to the assessments presented above and in continuation of the covariance processes, there are correlation and regression statistical estimates (Table 5), which can be calculated based on empirical data. An example for calculation of correlation assessments is the experimental research described in (Mechta et al., 2022). To solve a defined machine testing problem, two predictive models were developed and experimentally evaluated for accuracy and predictive performance. The obtained results showed a strong correlation between the two types of errors investigated in the testing. Another example of applying regression statistical analyzes and determining regression estimates is presented in (Bevilacqua et al., 2023). The aim was to investigate the performance parameters of three weak acids, applying two statistical approaches - multifactorial ANOVA and multiple regression. Multiple regression analysis was applied in the study of parameters of the studied objects.

Table 5. Correlation and regression statistical assessments

Main assessment	*Analytical expression*
Correlation coefficient	$r_{ij} = \frac{\sigma_{ij}}{\sigma_i . \sigma_j}$
A simple linear regression model (b_0 – displacement; b_1 – regression coefficient)	$x_j = b_0 + b_1 . x_i$; $b_1 = \frac{\sigma_{ij}}{\sigma_i^2}$; $b_0 = \bar{x}_j - b_1 . \bar{x}_i$
A multiple linear regression model	$x_j = b_0 + b_1 . x_i + b_2 . x_p + \ldots + b_k . x_k$; k<n

The operation of the application is organized as a sequence of forms (Figure 9). At startup, the main form MainForm is displayed, in which the studied factors (dependent and independent) are defined, and after selecting the Next button, the user is taken to the form for entering the sample for statistical analysis InForm. The InForm (data entry form) displays the dependent and independent variables defined in the previous step (MainForm). By selecting a button, a mode for entering empirical data (Import Data or Import File) is established, and by using the Back and Next buttons, the user can choose to go back or perform the next step.

Figure 10. Initial forms of the application

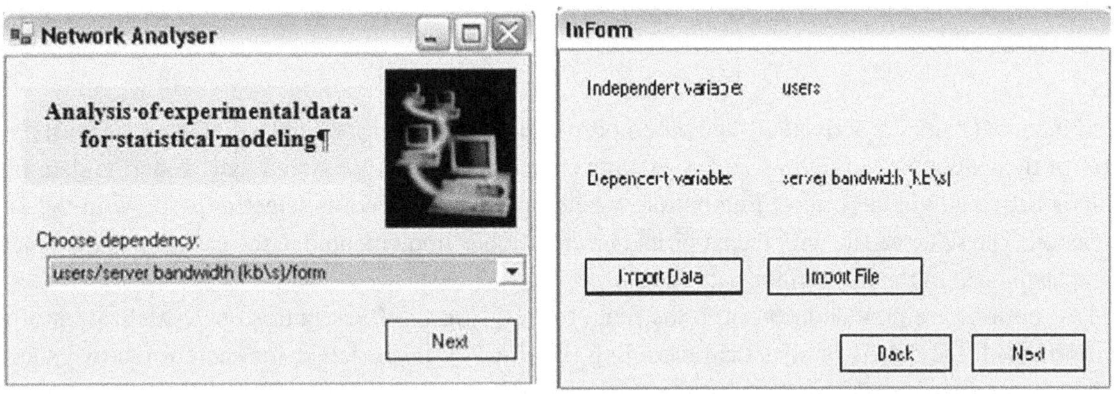

(a) Selection of parameters *(b) Input of empirical data*

If there is no empirical data entered for the selected variables in the studied dependency, an error message appears, while in the presence of data in the InForm form, a DataGrid is dynamically created in the form of a table, which is filled with the data for the variables (Figure 10)

Figure 11. Empirical data entry form

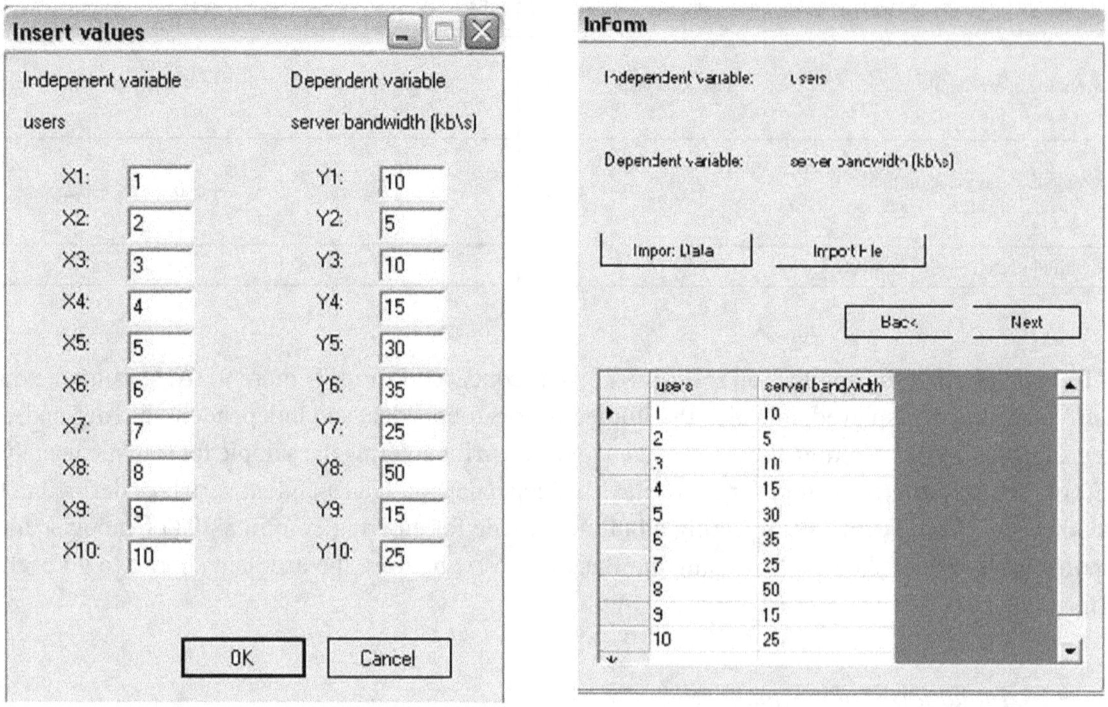

(a) Manual data entry

(b) Import a data file

Manual data entry is activated when the Import Data button is pressed and a dialog opens for manual entry of the values for the two variables, making checks for correctly entered data. Entering data from a file is activated via the Import File button, which opens a window for selecting a file with the *.xls extension. The selected file with the empirical information is implemented in the program and tabulated in the Data Grid in the InForm format.

Two options are provided: return to the main form (using the Back button) or visualization of the OutForm with the results of statistical modeling (Figure 11). In the latter, for each of the two studied variables, estimates for the mathematical expectation and variance are derived, as well as the statistical estimates for the studied statistical model of simple linear regression, including the covariance and correlation coefficients between the two variables and the estimates for the two parameters of the regression model - the shift (b_0) and regression (b_1) coefficients. Graphical interpretation involves plotting the regression line in a traditional coordinate system with scaling depending on the values of the analyzed sample and the estimated estimates.

Figure 12. OutForm with results of statistical model

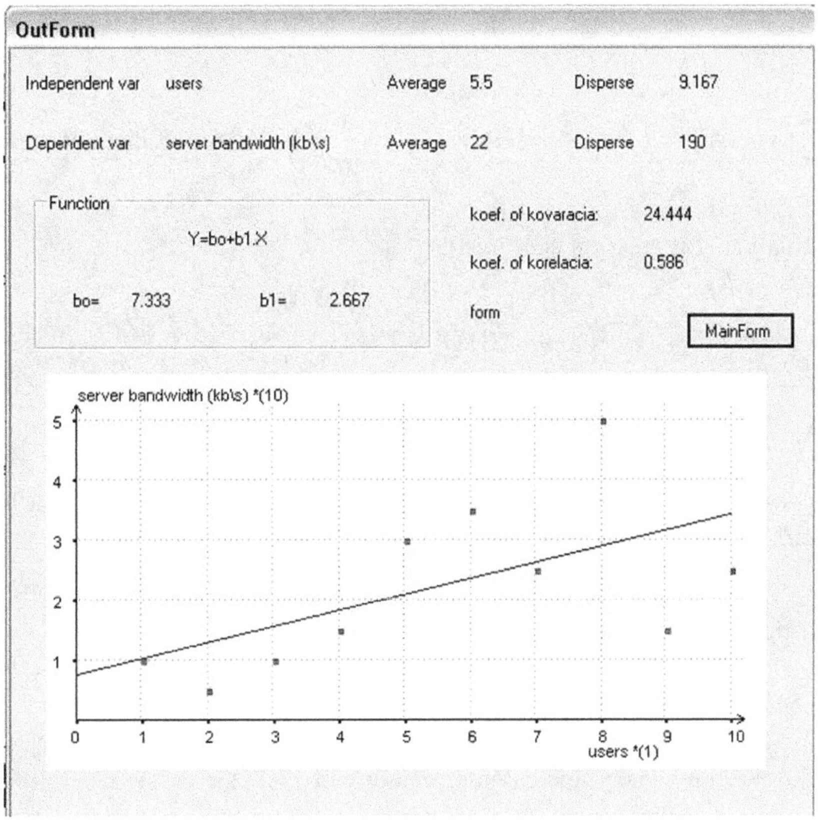

An example for statistical model of simple linear regression

Step (1) Sample formation and statistical model definition.

N	*1*	*2*	*3*	*4*	*5*	*6*	*7*	*8*	*9*	*10*
X_i	2	3	5	4	2	4	3	5	2	3
Y_i	5	7	12	9	7	8	7	9	4	5

Diagram of <x, y>➔

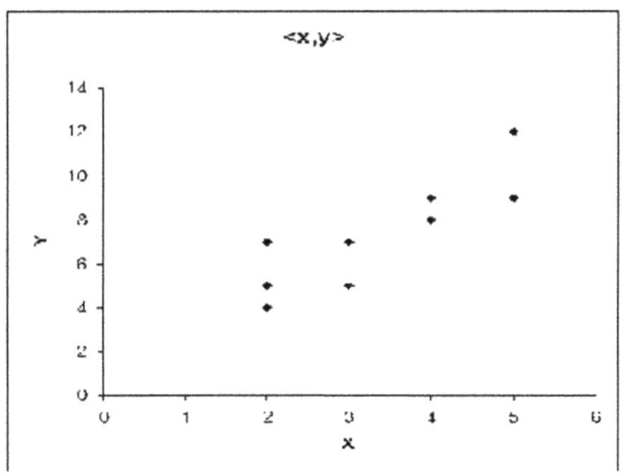

Linear model of simple regression definition:
$Y(X) \Rightarrow y = b_0 + b_1.x$

Step (2) Determining descriptive assessments.

N = 10; $\overline{x} = 33/10 = 3.3$; $\overline{y} = 73/10 = 7.3$;

$\Delta_X = X_{MAX} - X_{MIN} = 3$; $\Delta_Y = Y_{MAX} - Y_{MIN} = 8$

$$\tilde{A}_X^2 = \frac{1}{N-1}\sum(x_i - \overline{x})^2 = 1{,}34 \Rightarrow \sigma = 1.16 \Rightarrow C_V = \frac{\tilde{A}}{\overline{x}} = 0{,}352 \text{ (coefficient of variation)}$$

Step (3) Calculation of regression coefficients.

$$b_1 = \frac{\sum_{i=1}^{N}(X_i - \overline{x}).(Y_i - \overline{y})}{\sum_{i=1}^{N}(X_i - \overline{x})^2} = \frac{\sum_{i=1}^{N}(X_i - 3{,}3).(Y_i - 7{,}3)}{\sum_{i=1}^{N}(X_i - 3{,}3)^2} = \frac{21{,}1}{12{,}1} = 1{,}744$$

$$b_0 = \overline{y} - b_1.\overline{x} = 7.3 - (1.744, 3.3) = 1.545$$

Step (4) Statistical regression model $Y = Y(X)$ defining.

$y = b_0 + b_1 x = 1.545 + 1.744x$

Step (5) Model execution and graphical interpretation of $Y = Y(X)$.

N	1	3	5	7	9
X_i	2	5	2	3	2
Y_i	5	12	7	7	4
y_i	5,033	10.265	5,033	6,765	5,033
Y_i-y_i	-0,033	+1,735	+1,967	+0,235	-1,03

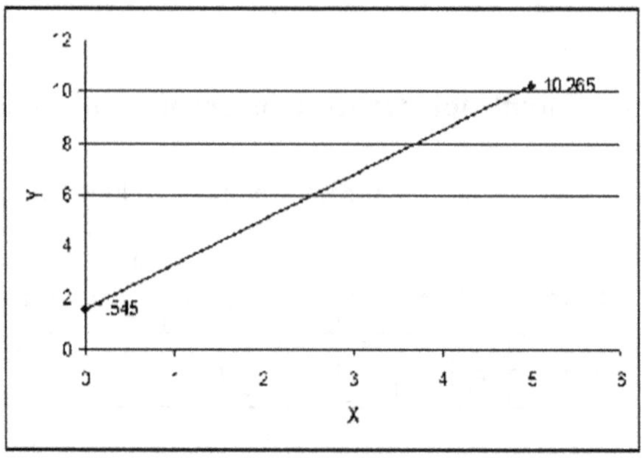

An example for statistical model of simple exponential regression

Step (1) Sample formation and statistical model definition.

N	*1*	*2*	*3*	*4*	*5*	*6*	*7*	*8*	*9*	*10*
X_i	1	2	3	4	5	6	7	8	9	10
Y_i	3	6	10	21	32	56	82	154	257	482

Diagram of <x, y>➔

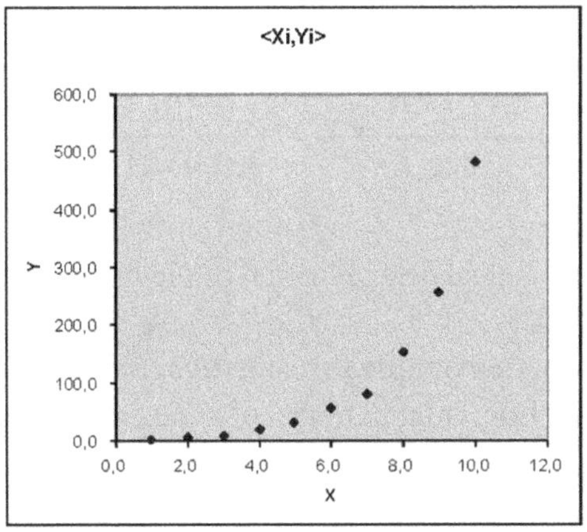

Exponential model of simple non-linear regression:

$$y = a.e^{b.x} = a.EXP(b.x)$$

Step (2) Determining the coefficients

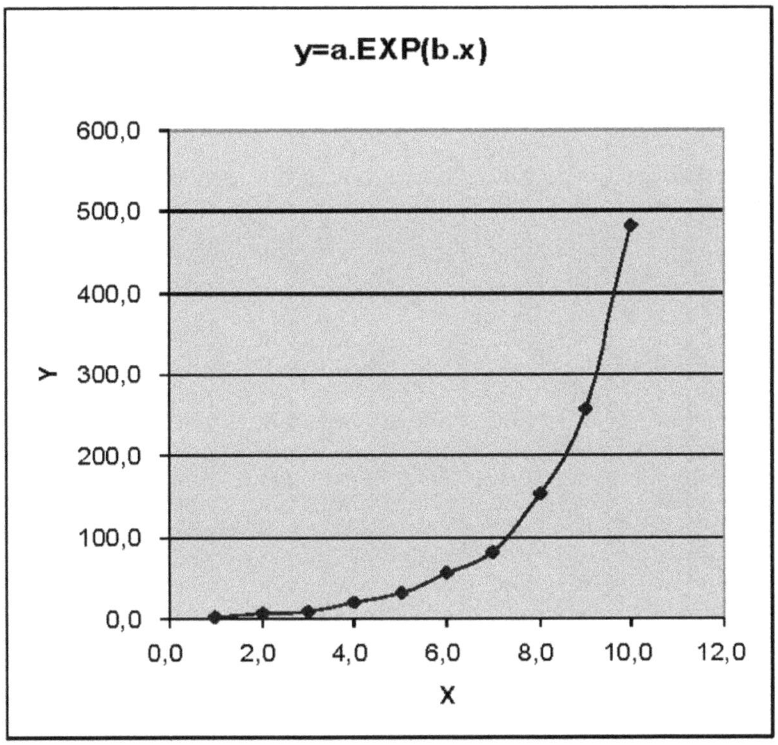

Step (3) Graphical model of exponential regression

Step (4) Example empirical determination of statistical model

- the diagram from step (1) can be made the conclusion $X=0 \Rightarrow Y\approx 2 \Rightarrow a=2$;
- this determines the model equation $y=2.e^{b.x}$;
- selection of point from the sample, for example <$x=3$, $y=10$> and the equation is defined as $10=2.e^{3b}$;

- solving the last equation:

$$b = \frac{\log 10 - \log 2}{3.\log EXP} = \frac{1 - 0,30103}{3.0,4343} = \frac{0,69897}{1,3029} = 0,5365 \approx 0,54$$

- final determination of the model $y=2.e^{0.54x}$

An example for statistical model of simple logarithmic regression
Logarithmic model of simple non-linear regression:

$$\boldsymbol{y = b_0 . x^{b_1}}$$

$$b_1 = \frac{\sum_{i=1}^{N} \log X_i \sum_{i=1}^{N} \log Y_i - N \sum_{i=1}^{N} \log X_i . \log Y_i}{\left(\sum_{i=1}^{N} \log X_i \right)^2 - N \sum_{i=1}^{N} (\log X_i)^2}$$

$$\log b_0 = \frac{1}{N} \left\{ \sum_{i=1}^{N} \log Y_i - b_1 . \sum_{i=1}^{N} \log X_i \right\}$$

Sample forming:

N	*1*	*2*	*3*	*4*	*5*	*6*
X_i	1	2	3	4	5	6
Y_i	3	12	27	48	75	108

Diagram of <x,y>:

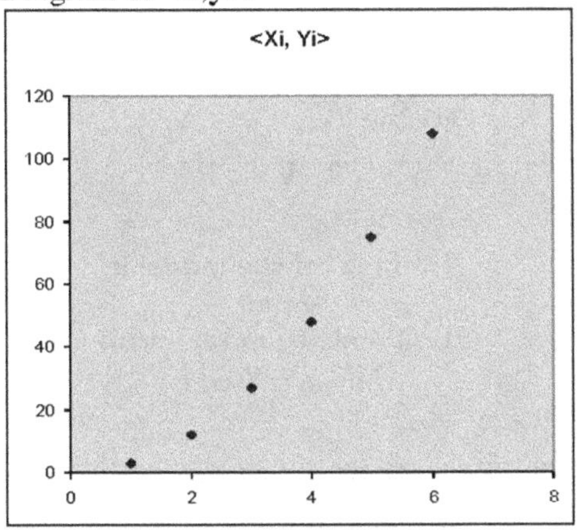

Calculation of the coefficients:

$$b_1 = \frac{2{,}8573.8{,}5774 - 6.4{,}9017}{(2{,}8573)^2 - 6.1{,}7744} = 1{,}97469 \approx 2$$

$$log\, b_0 = 1/6(8{,}5774 - 2.2{,}8577) = 0{,}476984$$

$$log\, b_0 \approx 0{,}477 \Rightarrow b_0 = 3$$

Calculation of intermediate values:

$$\sum_{i=1}^{N} \log X_i = 2{,}8573 \qquad \sum_{i=1}^{N} \log Y_i = 8{,}5774$$

$$\sum_{i=1}^{N} (\log X_i)^2 = 1{,}7744 \qquad \sum_{i=1}^{N} \log X_i , \log Y_i = 4{,}9017$$

Logarithmic model: $y=3.x^2$

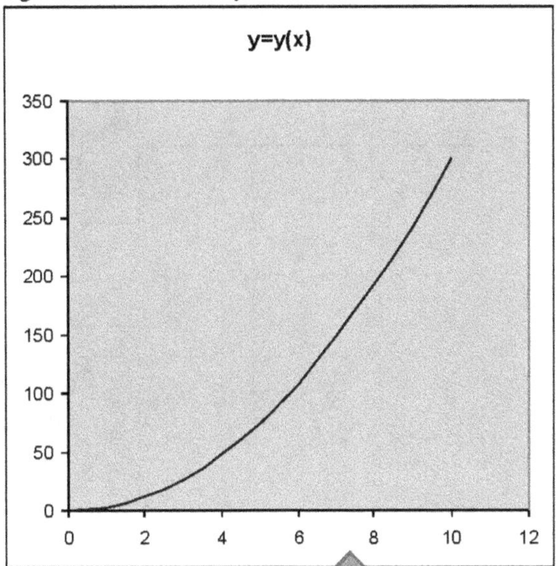

Calculated model values for the variables to construct the nonlinear dependency:

x	1	2	3	4	5	6	7	8	9	10	11	12
y	3	12	27	48	75	108	147	192	243	300	363	432

5. DATA DRIVEN MODELING

The development and application of models are common practices in most fields of science, technology and society. On the other hand, in management theory and systems engineering, the model is viewed from the perspective of the relationship between the variables that describe the system through certain mathematical relationships. Modeling is very widely used for various research purposes, such as dynamic system model is used for: simulation, design, optimization, prediction, control, error and fault detection.

Four main approaches to modeling dynamic systems are known: analytical approach, numerical approach, data-driven approach and hybrid approach (Habib et al., 2021). In analytical modeling, the system is described with mathematical expressions (algebraic, differential or partial differential equations). These mathematical equations are derived based on various physical laws and regularities such as laws of conservation of mass, energy, momentum, etc. (Estrada-Flores et al., 2006). Numerical modeling approaches such as Finite Element Method (FEM) or Finite Difference Method (FDM) are applied when the resulting mathematical equations cannot be solved analytically or the solution requires a lot of computational resources and time, and also in cases where the determination of an exact solution is not necessary.

In the data-driven approach, the relationships between the input and output variables of the system are determined directly from the experimental data obtained from the study of the very system or process we are modeling. Figure 13 shows schematically the basic idea of data-driven modeling.

Figure 13. Block diagram representing the main idea of data-driven modeling

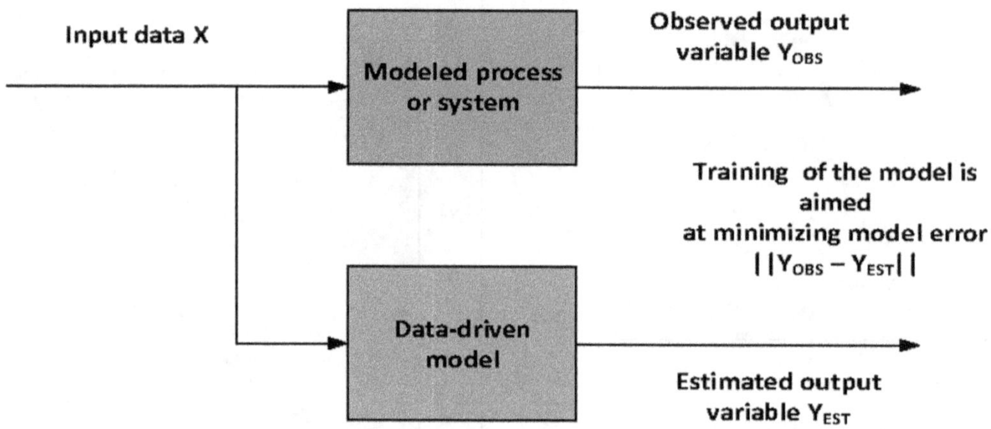

Data-driven models (DDM) are classified as parametric and non-parametric. In the parametric method of model synthesis, its structure is initially selected, and after the system parameters are estimated and determined, the system model itself is completed. Non-parametric methods are without a specific model structure and rely mainly on the obtained response of the modeled system (Kashiwagi, H. 2009). In this regard, parametric methods require large computing resources, but they provide qualitative models with a complex system structure. On the other hand, nonparametric methods are simple and their model coefficients usually reflect the main characteristics of the system. In this sense, data-driven modeling gives the best results in cases where the behavior of the studied system is unknown or very complex to describe with mathematical expressions. Thus, this modeling is also called the black box model.

A hybrid modeling approach requires prior partial knowledge of the system or process under study, combining elements of analytical modeling with data-driven modeling. In addition, the hybrid approach contains both the use of physical principles to model individual parts of the system or process that are well known, and experimental data that are used to model the remaining parts of the system. In this regard, the hybrid model is also called the gray box model.

Significant progress in the methods and means of data collection, processing and storage and artificial intelligence techniques have led to an abundance of data of various natures. In this sense, data-driven modeling develops, builds on, and complements the approach by which processes, devices, systems, and phenomena are modeled (Montáns et al., 2019). On the other hand, most investigated systems or processes (e.g. decentralized electric power generation, electric vehicles, electronic technology systems) are hybrid and include subsystems with different time constants. They are becoming increasingly complex, so that it is practically difficult to implement or impossible to model them by applying classical analytical models. In accordance with control theory and systems engineering, data-driven modeling includes the following main stages: system identification process (obtaining input-output data); choice of model class; estimation of model parameters and verification of the created model.

On the other hand, dynamic systems have different inputs and outputs, but they also have standard model structures and algorithms for evaluating linear and nonlinear systems of varying complexity. The main purpose of this section is to introduce the data-driven modeling paradigm, concepts and applied techniques from the point of view not so much of theory as of its application, advantages, disadvantages and limitations.

Data-driven modeling approaches. Mathematical models are used to capture the underlying, dynamic and real-time behavior of real-world systems. This means that mathematical models help to understand the input-output relationship of physical systems. The model-based design workflow supports the fast and cost-effective design of dynamic systems because tests are performed in a simulation environment before deployment. In addition, by applying model-based optimization, a final product or article is obtained optimally according to one or several pre-selected criteria.

Engineering systems and processes are modeled by applying different approaches. On the other hand, modeling approaches are classified into four main categories: analytical, numerical, data-driven, and hybrid models (Nelles, 2020). Figure 14 illustrates the mentioned modeling approaches. Furthermore, the four modeling approaches are compared in Table 6 based on the level of physical insight, model accuracy, assumptions, ease of development, and development cost.

Table 6. Comparison of the main properties of the main modeling methods considered

Property	Analytical Model	Numerical Model	Data-driven Model	Hybrid Model
Physical compliance	A good physical understanding and analogy for the behavior of the system or process	Physical compatibility and understanding to a great extent	The physical analogy is not essential	The physical understanding of the system is partial
Accuracy of the model	Low, even for complex systems	When modeling complex systems and processes, it is relatively high.	When modeling complex systems and processes, it is relatively high.	When modeling complex systems and processes, it is relatively high.
Assumptions used in modeling	Lots of assumptions	Lots of assumptions	Almost no assumptions are made	Relatively few assumptions are used
Complexity of creating the model	Low complexity, which leads to simplicity in development and implementation	Medium level of complexity	High level of complexity	Medium level of complexity
Development Resources	Very small usable resources, basically using known physical laws	Due to the application of numerical calculation methods, medium-level resources are required	Considerable resources are required due to the collection of experimental data	Considerable resources are required due to the collection of experimental data

Figure 14. Approaches to modeling systems and processes
Source: Habib et al. (2021)

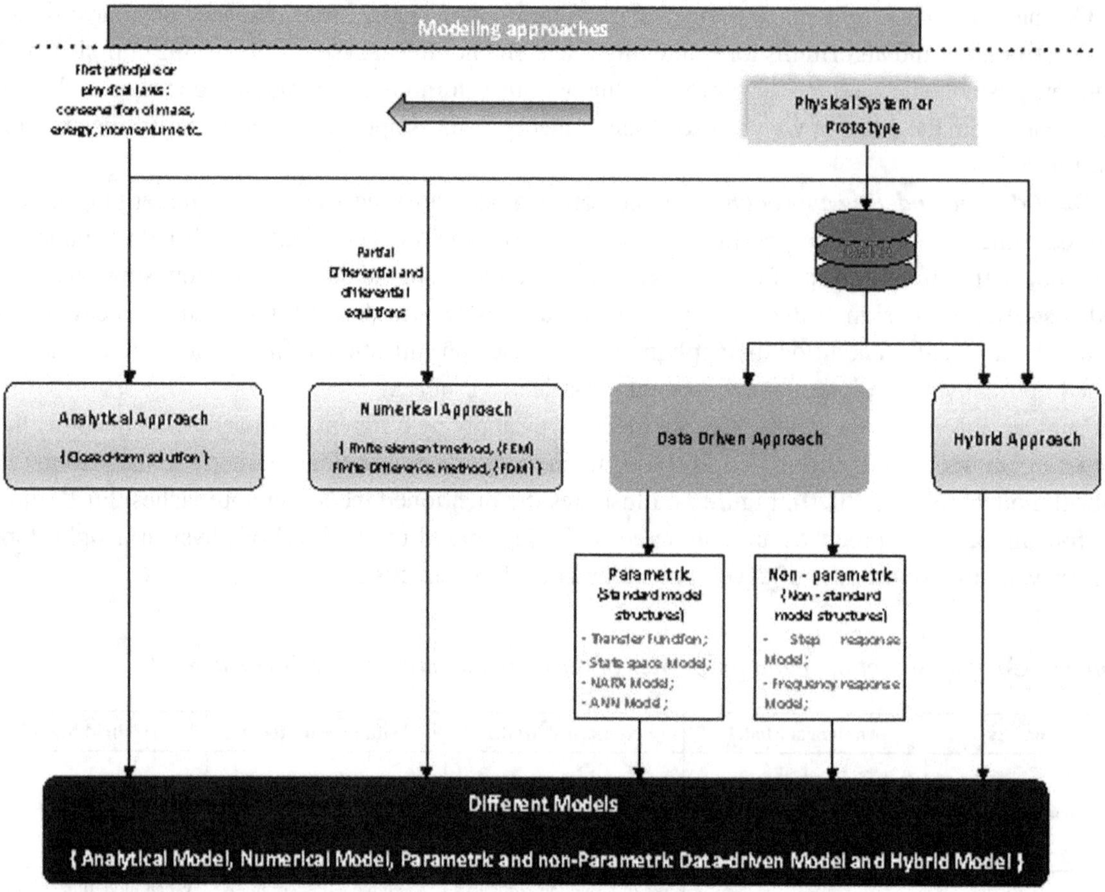

Data-driven parametric model synthesis. The data-driven parametric modeling approach is illustrated by the diagram shown in Figure 15. The considered process takes place in the following sequence: it starts with the collection of experimental data, which should materialize the relationship between a given input and the obtained reaction of the outputs; continues with the selection of a specific model structure; then an estimation of the parameters of the thus structured model is carried out and ends with testing and verification of the model. These steps of creating a data-driven parametric model will be detailed below.

A. Applying inputs to the inputs and experimentally extracting outputs from the outputs

From the descriptions of the design properties of data-driven models, it is clear that the quality of the developed model depends very much on both the measured output data and the configuration and synthesis of the input signal(s). In this sense, to ensure that the main properties or behavior of the modeled system are faithfully reflected in the collected data, experiment planning is performed. When planning the experiment, it is extremely important to correctly define the operating range of the system (based on the characteristics and properties of its inputs) in order to avoid deviations that may arise from the experimental data.

On the other hand, the collection of data used for system identification is carried out through a closed-loop or open-loop experiment. The most common implementation method is open system identification (without feedback). In cases of an unstable system (from the point of view of physical safety), the identification process must be implemented in closed-loop conditions. Another important aspect of this stage of modeling is the processing of the collected data (filtering, normalization or resampling) in order to remove anomalies, deviations and errors from the measurements.

Figure 15. Illustration of the individual stages of the process of creating a data-driven parametric model
Source: Habib et al. (2021)

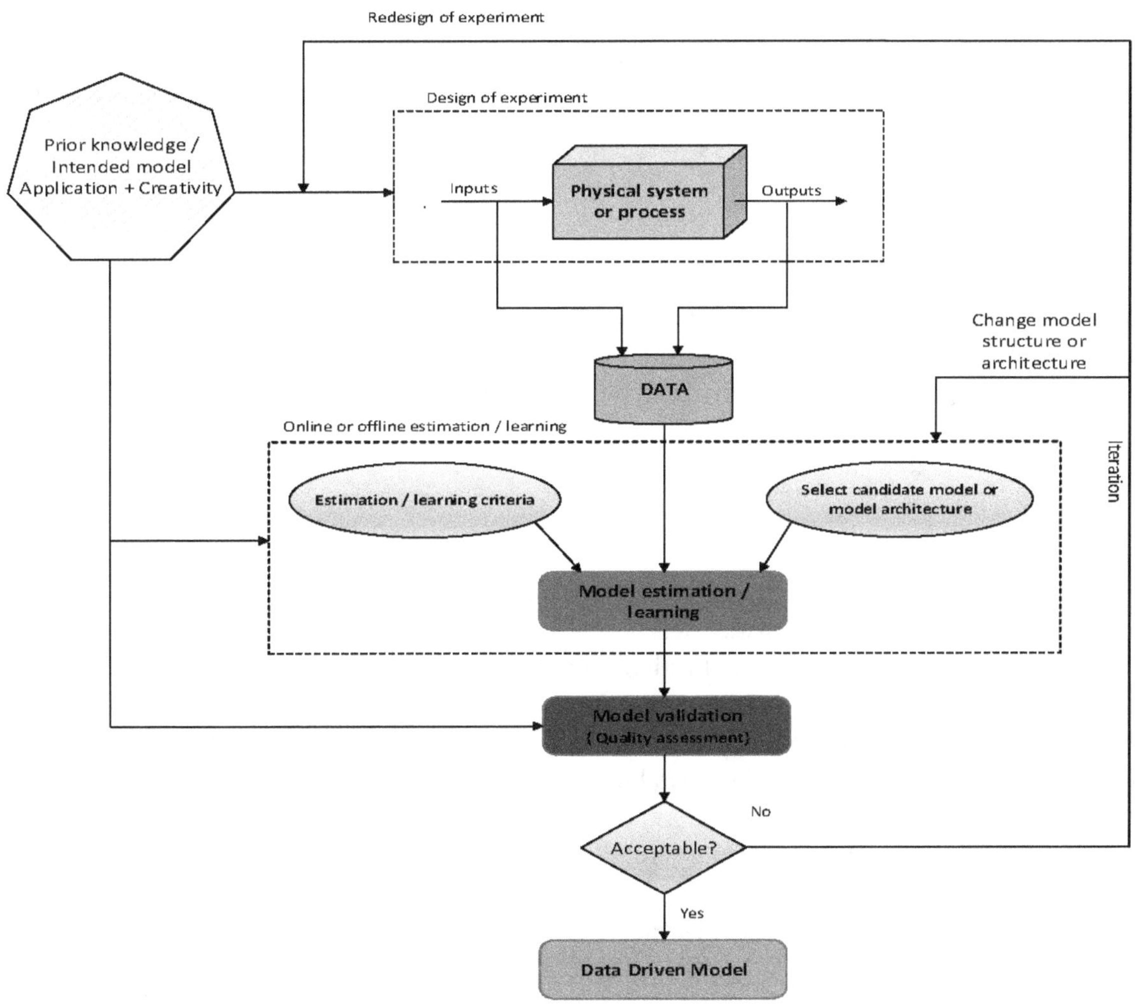

B. Model structure selection

Standard model structures are usually applied, which are used for both linear and non-linear models. The structures used in linear models are as follows: transfer function (TF), linear solvent strength (LSS), output error (OE), autoregressive-exogenous (ARX), autoregressive moving averaging with exogenous inputs (ARMAX), Box-Jenkins (BJ). In contrast to them, nonlinear model structures are used: nonlinear state-space (NLSS), nonlinear autoregressive exogenous (NARX), nonlinear autoregressive moving averaging with exogenous inputs (NARMAX) and Hammerstein-Wiener (HW). (Patcharaprakiti, et al., 2011). Determining the parameters of the model structure should also be considered as setting the model itself in order to achieve a maximum fit with the data. On the other hand, the choice of a linear or non-linear model structure for the identification of the system under study can become a problem due to the significant number of different structures available (the ones already mentioned and also their modifications). Usually, the behavior and character of the studied system and user preferences (as far as they have previous experience of working with similar structures) are used in the implementation of the structure selection process (Schoukens & Ljung, 2019). Good results for the identification of a linear system are obtained by applying a linear state space structure or a transfer function of a certain order. Moreover, these two structures give good results and are suitable for application in classical and modern control synthesis methods.

C. Estimation of model parameters

This involves using the acquired experimental data and a selected model structure to numerically estimate the parameter values of the selected model. Most of the available assessment methods are based on fit criteria between the measured output(s) of the physical system $y(t)$ and output(s) of a selected model that is predicted mathematically. This can be mathematically expressed as,

$$\hat{\theta}_n = \arg\min V(\theta_n) = \sum_{i=1}^{t} \left\| y(i) - \hat{y}(i|\theta_n) \right\|^2$$

where $V(\theta_n, t)$ is the objective function; θ is a vector of unknown parameter; t is the number of data points and n is the estimated number of parameters.

The evaluation algorithms for identification of the modeled system, regardless of the model structure used (linear or non-linear) are as follows: least squares, recursive least squares algorithm, Subspace algorithms for the identification of combined deterministic-stochastic systems (N4SID) and Neural Networks (NN). Algorithms involving neural networks are classified into radial-based NN (RBNN) or multilayer perceptron (MLP) neural network. Conditions for fulfilling the parameter estimation criteria include the achievement of certain pre-set errors; number of iterations, number of epochs, etc.

D. Model validation and quality assessment

The assessment of the quality of the system model is based on establishing the effectiveness of the developed model. This is realized when reproducing the received data through the created model. The problem that can arise at this stage of system identification is that unsatisfactory results are obtained from the use of the developed model. In most cases, this is due to the use of insufficiently informative data and/or an inappropriate model structure is chosen. To resolve this issue, a different model structure is chosen and/or additional experiments are performed to gather adequate data.

Data-driven non-parametric modeling. In nonparametric data-driven modeling, there is no specific model structure. This is also the main difference compared to the parametric approach to data-driven modeling. Also, in data-driven nonparametric models, the number of their parameters increases with the size of the data used for identification. On the other hand, data-driven parametric models have a finite number of parameters that are not affected by the size of the data (Nelles, 2020). In addition, the parameters/coefficients used in non-parametric models reflect the physical characteristics of the modeled system (Tangirala, 2018). In this sense, the basic information for a model that describes the step effect of inputs, such as gain, time constant, and time delay, is determined by transient response analysis (this is essentially a non-parametric method for data-driven modeling). Other examples of nonparametric data-driven modeling methods include impulse response analysis, correlation analysis, Fourier analysis, and spectral analysis.

Criteria to follow in data-driven modeling. The main criteria considered when developing data-driven models are:

- What is the problem being solved: Process or system identification.
- Purpose and functions of the developed model: the main applications are related to: conducting simulations, forecasting, optimization, control and monitoring, etc.
- Complexity and dimensionality of the system: The number of inputs and outputs has a significant Influence on the choice of a rational structure of the model. For example, a single-input-single-output (SISO) system model structure is easiest to choose and implement in modeling.
- Quality and quantity of available data: the two main factors to consider in designing the experiment and choosing the evaluation method are the size of the data set available and its quality.
- Offline and Online Training: Offline training method is applicable for all model structures. The method for online identification of the investigated system gives good results in identifying systems that change over time. Online learning is key in control synthesis and is used in adaptive and model predictive control. In these forms of management, gathering information about the system in real time is very important (Schoukens, et al., 2018).

Data-driven modeling, evaluation/training methods. Essentially, model parameter estimation in data-driven modeling is an optimization task because once a particular model structure is chosen, its model parameters can be determined by a given algorithm. With the transition from linear to nonlinear and hybrid modeling systems, the degree of complexity of both the models themselves and the methods for their evaluation increases. In Figure 16 shows the main categories of data-driven model estimation methods. Classical and statistical estimation methods such as the method of least squares, recursive least squares (RLS), and spectral analysis are used to identify the parameters of a linear regression model. Some non-linear models in the time domain and frequency response models are also evaluated with their help. Subspace methods such as canonical variable analysis (CVA), multivariate state space output error (MOESP), and numerical algorithms for state subspace identification (N4SID) are based on geometric interpretations and linear algebra. They give excellent results when evaluating linear Multi Input Multi Output (MIMO) and single-input multiple-output (SIMO) systems.

Figure 16. Data-Driven modeling methods
Source: Habib et al. (2021)

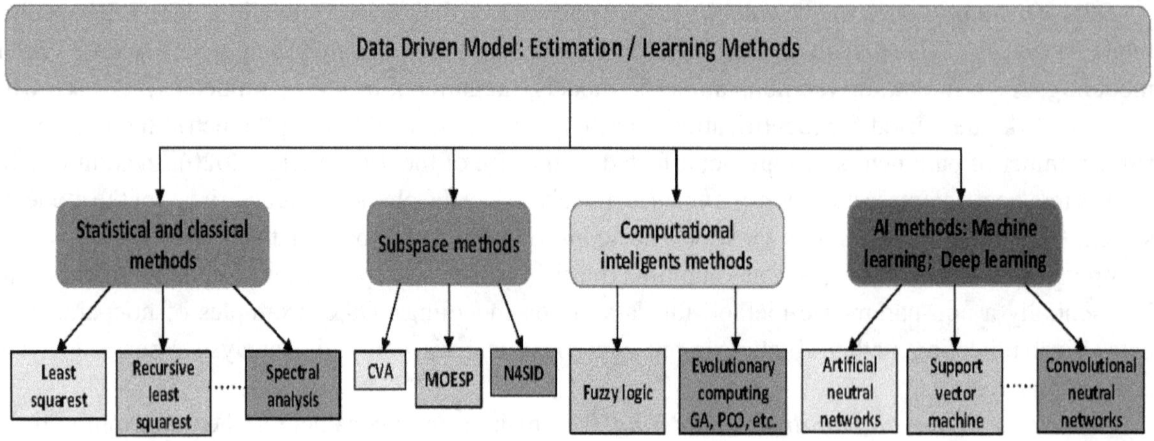

Model evaluation methods belonging to the computational intelligence (CI) group are as follows: fuzzy logic system, genetic algorithm (GA), and particle swarm optimization (PSO). The fuzzy logic system has been successfully applied to model nonlinear functions of arbitrary complexity based on interpretations (IF-Then rules), while GA is an optimal search algorithm to determine model parameters. To the evaluation group, there are also methods based on artificial intelligence, such as: machine learning (ML) and deep learning (DL), such as artificial neural network (ANN), support vector machine (SVM) and convolutional neural network (CNN). These methods are successfully applied in the presence of a large set of data and are therefore used to study complex relationships and interactions in complex systems composed of different subsystems. In essence, the success of applying a NARX-type model structure in data-driven modeling of various nonlinear systems is mainly due to the use of an ANN estimator in the model architecture. In this sense, the current trend in creating data-driven models is the use of hybrid estimation methods such as neuro-fuzzy, PSO-SVM, GA-ANN, etc. Table 7 compares data-driven modeling evaluation methods based on the amount of data, the level of computation, and the complexity of the application.

Table 7. A comparison of data-based modeling evaluation/training methods

Property	Classical and Statistical	Subspace	Evolutionary Computing	AI: Machine Learning and Deep Learning
Amount of data	Small	Small or large	Large	Very large
Level of computation	Low	Medium	Medium	High
Complexity of Application	A linear system with a lower degree of complexity	Linear deterministic and stochastic system	Linear and non-linear functions of medium complexity	Systems of high complexity

Source: Habib et al. (2021)

Current status, challenges and perspectives of data driven modeling. The development of computational mathematics, data science and the capabilities of computer systems has accelerated the application of data-driven modeling and computational intelligence to solving various problems related to the sustainable development of society: such as optimal resource management, intelligent electricity generation, environmental transport and agriculture etc. Data-driven models are very effective in solving a problem or modeling a particular system or process if:

- there is a significant amount of reliable data describing this research object or process;
- no significant changes are observed in the studied object or process during the period covered by the collection and processing of the data for the model.

Data-driven models are a good alternative to other possible modeling methods if it is difficult to build knowledge-based simulation models or the available models are not adequate enough. This is most often observed due to the difficulty of understanding and finding a physical analogy of the main processes in the system. On the other hand, it is useful to use alternative models both to validate the simulation results of data-driven models. When applying data-driven models, it should be taken into account that they do not take into account and have no connection with the physics of the modeled process, but are simply an abstract description that, thanks to the analysis of the accumulated data, captures the relevant relationships between the input and output variables. However, data-driven models can in some cases be more accurate than process description-based models because they are based on the use of objective information (ie, the data). In this sense, models based on the physical analogy and notion of projections can be either incomplete (when applying many assumptions) or very complex and labor-intensive to implement.

In order to be able to use the advantages of different types of models, modern modeling is developed as a combination of them, most often combining data-driven models with other types of models. In this way, hybrid models are obtained that optimally combine physically based and data-driven models. In this regard, it should be emphasized that data-driven modeling does not negate the remaining physics-based models, but develops and expands the field of application of modeling in general. One of the challenges for modeling specialists in this regard is to ensure the proper inclusion of data-driven models in the process of modeling the systems under study, and thus to fully realize their potential to support concrete decision-making.

Perspectives on data-driven modeling include:

- Its application in smart grids for decentralized electricity generation and other decentralized processes.
- In Industry 4.0 and 5.0 when developing digital twins, with the aim of their quick and easy adaptation to changes that occur in production systems and processes.
- Modeling of energy storage cells where their state and charge level depends on many factors and conditions.

REFERENCES

Abdi, H. (2007). The method of least squares. Encyclopedia of measurement and statistics, 1, 530-532

Atlasov, I.V., Bolnokin, V.E., & Kravets, O.Ja., Mutin, D.I. & Nurutdinov, G.N. (2020). Statistical models for minimizing the number of search queries. *International Journal on Information Technologies and Security*, *12*(3), 3–12.

Bagossi, S., Ferretti, F., & Arzarello, F. (2022). Assessing covariation as a form of conceptual understanding through comparative judgement. *Educational Studies in Mathematics*, *111*(3), 469–492. doi:10.100710649-022-10178-w

Bevilacqua, A., Speranza, B., Petruzzi, L., Sinigaglia, M., & Corbo, M. R. (2023). Using regression and Multifactorial Analysis of Variance to assess the effect of ascorbic, citric, and malic acids on spores and activated spores of Alicyclobacillus acidoterrestris. *Food Microbiology*, *110*, 104158. doi:10.1016/j.fm.2022.104158 PMID:36462814

Bianco, S., Cusano, C., Piccoli, F., & Schettini, R. (2019). Learning parametric functions for color image enhancement. In *Computational Color Imaging: 7th International Workshop, CCIW 2019, Chiba, Japan, March 27-29, 2019* [Springer International Publishing.]. *Proceedings*, *7*, 209–220.

Brus, D. J. (2021). Statistical approaches for spatial sample survey: Persistent misconceptions and new developments. *European Journal of Soil Science*, *72*(2), 686–703. doi:10.1111/ejss.12988

Chau, M., & Fu, M. C. (2014). An overview of stochastic approximation. Handbook of simulation optimization, 149-178.

Cohen, A., & Migliorati, G. (2017). Optimal weighted least-squares methods. *The SMAI journal of computational mathematics, 3*, 181-203.

Das, S., McClain, C. J., & Rai, S. N. (2020). Fifteen years of gene set analysis for high-throughput genomic data: A review of statistical approaches and future challenges. *Entropy (Basel, Switzerland)*, *22*(4), 427. doi:10.3390/e22040427 PMID:33286201

Dingle, S. E., Bujtor, M. S., Milte, C. M., Bowe, S. J., Daly, R. M., & Torres, S. J. (2023). Statistical approaches for the analysis of combined health-related factors in association with adult cognitive outcomes: A scoping review. *Journal of Alzheimer's Disease*, February, 1-25. doi:10.3233/JAD-221034

Dvir, M., & Ben-Zvi, D. (2023). Fostering students' informal quantitative estimations of uncertainty through statistical modeling. *Instructional Science*, *51*(3), 1–28. doi:10.100711251-023-09622-y

Estrada-Flores, S., Merts, I., De Ketelaere, B., & Lammertyn, J. (2006). Development and validation of "grey-box" models for refrigeration applications: A review of key concepts. *International Journal of Refrigeration*, *29*(6), 931–946. doi:10.1016/j.ijrefrig.2006.03.018

Freedman, D. A. (2009). *Statistical models: theory and practice*. Cambridge University Press. doi:10.1017/CBO9780511815867

Galton, F. (1886). Regression towards mediocrity in hereditary stature. *Journal of the Anthropological Institute of Great Britain and Ireland*, *15*, 246–263. doi:10.2307/2841583

Ghadimi, S., Ruszczynski, A., & Wang, M. (2020). A single timescale stochastic approximation method for nested stochastic optimization. *SIAM Journal on Optimization*, *30*(1), 960–979. doi:10.1137/18M1230542

Habib, M. K., Ayankoso, S. A., & Nagata, F. (2021, August). Data-Driven Modeling: Concept, Techniques, Challenges and a Case Study. In *2021 IEEE International Conference on Mechatronics and Automation (ICMA)* (pp. 1000-1007). IEEE. 10.1109/ICMA52036.2021.9512658

Haboubi, S., & Salem, O. B. (2022). Predictive analytics for energy consumption in smart homes with fog and cloud computing using support vector regression. *International Journal on Information Technologies and Security*, *14*(1), 49–60.

Huser, R., & Wadsworth, J. L. (2022). Advances in statistical modeling of spatial extremes. *Wiley Interdisciplinary Reviews: Computational Statistics*, *14*(1), e1537. doi:10.1002/wics.1537

Jmaei, A., Slaoui, Y., & Dellagi, W. (2017). Recursive distribution estimator defined by stochastic approximation method using Bernstein polynomials. *Journal of Nonparametric Statistics*, *29*(4), 792–805. doi:10.1080/10485252.2017.1369538

Kashiwagi, H. (2009). Nonparametric System Identification. *CONTROL SYSTEMS, ROBOTICS AND AUTOMATION*, *6*, 524.

Kleven, H. J. (2021). Sufficient statistics revisited. *Annual Review of Economics*, *13*(1), 515–538. doi:10.1146/annurev-economics-060220-023547

Kroese, D. P., & Chan, J. C. (2014). *Statistical modeling and computation*. Springer. doi:10.1007/978-1-4614-8775-3

Kushner, H. J., & Clark, D. S. (2012). *Stochastic approximation methods for constrained and unconstrained systems* (Vol. 26). Springer Science & Business Media.

Lacotte, J., & Pilanci, M. (2020, November). Optimal randomized first-order methods for least-squares problems. In *International Conference on Machine Learning* (pp. 5587-5597).

Landau, D., & Binder, K. (2021). *A guide to Monte Carlo simulations in statistical physics*. Cambridge University Press. doi:10.1017/9781108780346

Mechta, A., Slamani, M., Zaoui, M., Mayer, R., & Chatelain, J. F. (2022). Correlation assessment and modeling of intra-axis errors of prismatic axes for CNC machine tools. *International Journal of Advanced Manufacturing Technology*, *120*(7-8), 5093–5115. doi:10.100700170-022-09074-7

Mehmood, S., & Ochs, P. (2020, June). Automatic differentiation of some first-order methods in parametric optimization. In *International Conference on Artificial Intelligence and Statistics* (pp. 1584-1594). Proceedings of Machine Learning Research (PMLR).

Montáns, F. J., Chinesta, F., Gómez-Bombarelli, R., & Kutz, J. N. (2019). Data-driven modeling and learning in science and engineering. *Comptes Rendus. Mécanique*, *347*(11), 845–855. doi:10.1016/j.crme.2019.11.009

Nelles, O. (2020). *Nonlinear system identification: from classical approaches to neural networks, fuzzy models, and gaussian processes*. Springer Nature. doi:10.1007/978-3-030-47439-3

Nooribakhsh, M., & Mollamotalebi, M. (2020). A review on statistical approaches for anomaly detection in DDoS attacks. *Information Security Journal: A Global Perspective, 29*(3), 118-133.

Noprisson, H. (2020). The business failure prediction using statistical approach. *International Journal of Scientific Research in Science, Engineering and Technology*, *7*(5), 161–168.

Patcharaprakiti, N., Kirtikara, K., Tunlasakun, K., Thongpron, J., Chenvidhya, D., Sangswang, A., & Muenpinij, B. (2011). Modeling of photovoltaic grid connected inverters based on nonlinear system identification for power quality analysis. *Electrical Generation and Distribution Systems and Power Quality Disturbances*, 53-82.

Pelivani, E., Besimi, A., & Cico, B. (2022). An empirical study of user interface testing tools. *International Journal on Information Technologies and Security*, *14*(1), 37–48.

Sadrossadat, S. A., & Rahmani, O. (2020). ANN-based method for parametric modelling and optimizing efficiency, output power and material cost of BLDC motor. *IET Electric Power Applications*, *14*(6), 951–960. doi:10.1049/iet-epa.2019.0686

Schoukens, J., Godfrey, K., & Schoukens, M. (2018). Nonparametric data-driven modeling of linear systems: Estimating the frequency response and impulse response function. *IEEE Control Systems*, *38*(4), 49–88. doi:10.1109/MCS.2018.2830080

Schoukens, J., & Ljung, L. (2019). Nonlinear system identification: A user-oriented road map. *IEEE Control Systems*, *39*(6), 28–99. doi:10.1109/MCS.2019.2938121

Shreider, Y. A. (Ed.). (2014). *The Monte Carlo method: the method of statistical trials* (Vol. 87). Elsevier.

Shukla, S. (2023d). Streamlining integration testing with test containers: Addressing limitations and best practices for implementation. [IJLEMR]. *International Journal of Latest Engineering and Management Research*, *8*(3), 19–26. doi:10.56581/IJLEMR.8.3.19-26

Tangirala, A. K. (2018). *Principles of system identification: theory and practice*. Crc Press. doi:10.1201/9781315222509

Zhao, J., Li, Z., Gao, Q., Zhao, H., Chen, S., Huang, L., Wang, W., & Wang, T. (2021). A review of statistical methods for dietary pattern analysis. *Nutrition journal, 20*(1), 1-18. doi:10.1186/s12937-021-00692-7

Chapter 15
Experimental Data Processing

ABSTRACT

Conducting an experiment is related to providing certain experimental results, mostly numerical data. The chapter discusses the possibilities for adequate processing of such data, as well as the appropriate interpretation of results and assessments from the analysis conducted. In its main part, the processing of experimental data is based on conducted statistical analysis and graphical interpretation of the estimates. This defines the three main parts included in the chapter. In the first part, an approach is proposed for primary processing of experimental data (determination of definitional areas, individual criteria for evaluation, logical completeness of the evaluation and the need for preliminary filtering of the sample). The second part is dedicated to the basic processing of experimental data, where statistical approaches for this are discussed (basic statistics, correlation and regression analysis, logistic regression, variance analysis). In the last part, the basic methods are tabular and graphical interpretation of estimates from the analysis of experimental data.

1. INTRODUCTION

When conducting experiments using different approaches (modeling, monitoring, benchmark, etc.), information (registrations) is provided for the set of observable variables forming a vector of the studied observable parameters $\vec{x} = <X_1, X_2, \ldots, X_n>$. Each parameter is considered as a random variable, and the recorded values – as its specific manifestation during the experiment. Conclusions are drawn based on the analysis of the accumulated information, which requires that the modeling data be highly informative. This is also related to the effectiveness of the experiments, which is determined by a properly defined plan. Guidelines for planning experiments are presented in the editorial (Curtis et al., 2022), the purpose of which is to specify the current requirements and good practices in this direction. It is a follow-up to two previous publications on the topic, with the article referenced here addressing new issues and emphasizing three aspects of design: randomization, blinded analysis, and balance of group sizes.

The statistical interpretation of experimental results is an important part of conducting research in any field, it is stated in (Lúcio & Sari, 2017). Of great importance is the correct determination of the main factors that may be subject to variability when conducting the experiments. To avoid this drawback

DOI: 10.4018/978-1-6684-8947-5.ch015

and to reduce the experimental error, it is necessary to pay important attention to the planning of the experiment. The application of statistical tests for the analysis of experimental data should be based on their good knowledge and statistical validity for the specific field of study. The article suggests possible alternatives for reducing factorial variability due to uncontrolled effects in a given experiment with a recommendation for appropriate statistical analyses. The aim is to provide information for the proper planning and conduct of experiments and in the analysis and interpretation of experimental results.

An example of conducting statistical analysis of experimental data is presented in (Romansky, 2023). In this case, the main goal is to analyze the delay factors in accessing remote information resources in active learning processes. It is known that in the latter network communications and access to distributed resources in the network space are of primary importance. It is assumed that clients access different types and content of resources in the digital space, maintained and offered by different servers – general purpose and specialized. To investigate parameters (variable factors) of these communications between client and server, monitoring was conducted to collect empirical data used to form initial samples of values (registrations). These experimental data are subjected to statistical analysis to determine estimates of significant service parameters. Different servers supporting purely text objects and resources containing text and images were analyzed, and statistical estimates were determined for total time for dynamic resources and influence of transfer parameters at different types of resources. A summary of the obtained results is made in graphic form of formed functional dependences of the studied time parameters on the volume of available resources.

Modeling also provides data that can be further processed based on developed plan for statistical experimentation. Regarding the final phase for analyzing the model results, the plan should include the following main steps:

- Determining the meaningful meaning of the vector of model parameters.
- Planning of primary processing of model information.
- Planning the main processing of the model information.
- Analysis and interpretation of the obtained estimates.

While the first step is the object of decision already at the initial stage of organization of experiments (including modeling), the remaining three are related to the two main parts of the "Analysis and interpretation of model results" phase presented below (see Figure 10 in Chapter 3).

2. PRIMARY PROCESSING OF EXPERIMENTAL DATA

Each conducted experiment for the study of selected parameters provides different registrations for the observable factors from the vector $\vec{x} = \langle X_1, X_2, \ldots, X_n \rangle$, which form a two-dimensional sample $\{\xi_{ij}\}$ for the variables $X_i \in \vec{x}$ at the moments N_j of successive registrations. This initial (raw) sample must be subjected to primary processing to increase its informative nature. It is carried out in successive steps presented below.

Specifying the Definitional Areas

For each variable X_i (i=1,…,n) there exists a hit region for its values recorded at fixed time intervals t_j within the period T of the experiment, i.e.

for $\forall X_i$ $(i = 1,\dots, n)$ $\exists R_i$,
so that $\forall \xi_{ij} \in R_i$ $(i = 1 \div n; j = 1 \div N)$, where $\xi_{ij} = X_i(t_j)$; $t_j \in T$.

Then, if T is the interval of the experiment, and t_j $(j = 1 \div N)$ are the individual moments of registration, for the number of virtual clocks can be used determination $N=T/\Delta t$, where Δt = ú$t_j - t_{j-1}$ú; t_j, $t_{j-1} \in T$. The interval Δt expresses the frequency of registration when conducting the specific experiment.

Based on the nature of the processes in a computer system, which is a heterogeneous system, it can be assumed that the spaces R_i are the subspaces of the set of rational numbers, i.e., $R_i \subset R$ $(i=1,\dots, n)$, or can be reduced to such by a simple relation. For each variable X_i, the recorded information $X_i(t_j)=\xi_{ij}$ for the successive moments in time (j=1,…, N) falls into the area R_i, and minimum ($\xi_{i\,min}$) and maximum (($\xi_{i\,max}$) can be determined value, i.e.

$\exists$ ($\xi_{i\,min}$, $\xi_{i\,max}$) $\in R_i$, such that $\xi_{i\,min} \leq X_i \leq \xi_{i\,max}$,
or for $\forall \xi_{ij}$ $(j=1 \div N) \Rightarrow \xi_{ij} \in [\xi_{i\,min}, \xi_{i\,max}] \Rightarrow R_i \equiv [\xi_{i\,min}, \xi_{i\,max}] \subset$ R.

This can be written as the following condition for defining the individual definitional domains that form the overall definitional domain for the functionality of the system $R = \bigcup_{i=1}^{n} R_i$:

IF $\xi_{i\,min} \leq x_i \leq \xi_{i\,max}$ THEN $R_i = [\xi_{i\,min}, \xi_{i\,max}]$, for $\forall \xi_{ij}$ $(j=1 \div N) \in [\xi_{i\,min}, \xi_{i\,max}]$.

The actions envisaged in this step are represented in the structural model of Figure 1. The formation of the data structures 'DATA' and 'DEF' can be done dynamically (in real-time), so that at any moment the two arrays will reflect the actual current situation regarding registrations and their definition areas. This allows dynamic determination of the experiment duration N.

Figure 1. Structural model of the process for registration of values and determination of definition areas

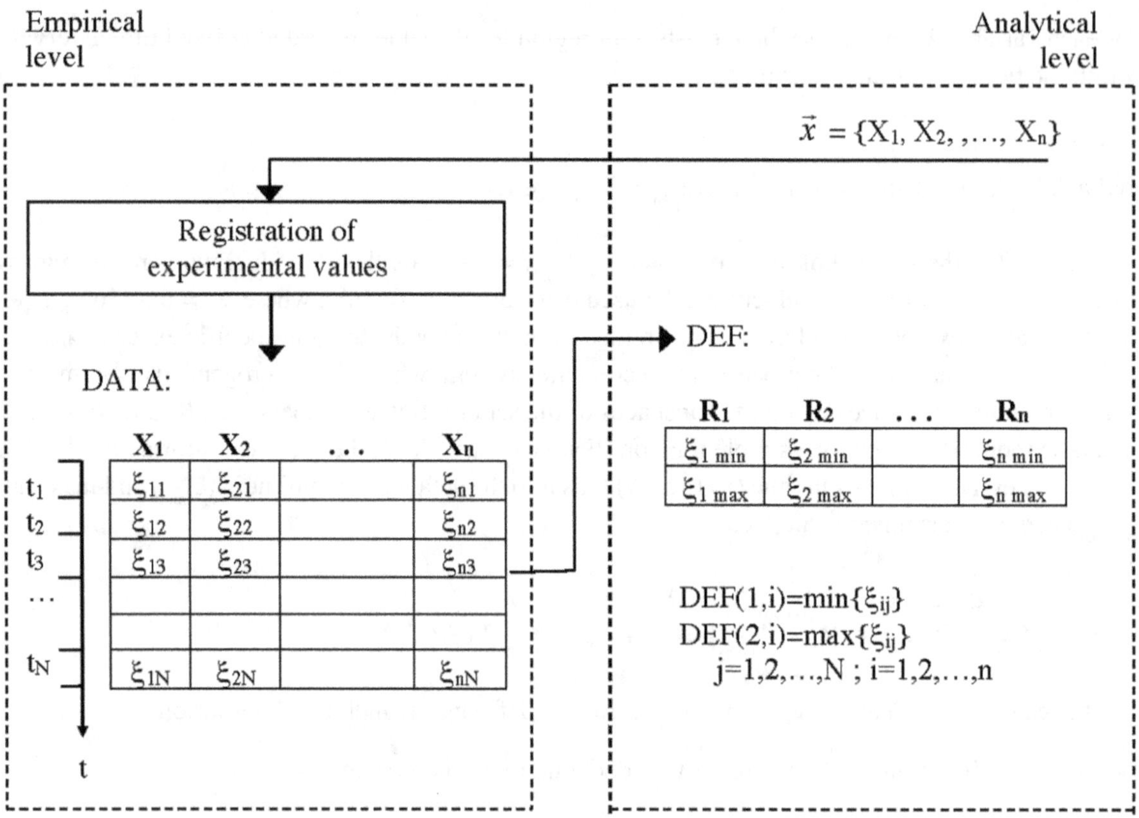

Determination of Individual Assessment Criteria

These criteria make it possible to determine the individual evaluations for the observed parameters and are the basis for forming a global evaluation criterion. Regardless of the relatively subjective nature when defining the criteria, each of them can be considered as an image of the corresponding definition area R_i in a certain co-area (evaluation area) I, considering: ✓ the type and nature of the variable; ✓ the purpose and purpose of the evaluation; ✓ the role and significance of the variable on the overall performance of the computer system.

For this purpose, it is necessary to define the vector of variables, define the definition areas for each variable and specify the weighting coefficients of significance. The unambiguous definition of the individual criteria through the images g_i: $R_i \rightarrow I$ $(i=1 \div n)$ allows determining an individual evaluation $E_i = g_i(X_i)$ for each variable falling into a common evaluation area I. From the point of view of the global evaluation, it is appropriate to the closed interval $I = [0, 1]$ is chosen. This will also allow a possible percentage representation of the evaluation coefficients $E_i \in I$, which will affect the degree of satisfaction of the set requirements for the i-th variable when forming the overall evaluation E. The two main situations presented can be indicated.

(1) It is assumed that during the registration an array 'DATA' (Figure 1) is formed with values $\xi_{ij} \in \{0,1\}$ for $\forall X_i \in \vec{x}$ and for $\forall t_j \in T$. This corresponds to the registration of occurrence or not of a given event

in a computer environment named by a variable from the vector of variables. Then $R_1 \equiv R_2 \equiv \ldots \equiv R_n \equiv \{0,1\}$ will be satisfied for the definition areas, and the determination of the individual criteria will be based on the functional transformation g_i: $\{0,1\} \rightarrow [0,1]$, for $I = 1, 2, \ldots, n$. The algorithmic description of the procedure is given in Figure 2.

Figure 2. Algorithmic presentation of the procedure for situation (1)

```
FOR i=1 TO n
   BEGIN
           count=0;
           FOR j=1 to N
                   IF ξij=1 THEN count=count+1;
           END_j
           Ei:=count/N
   END
END_i
```

(2) It is assumed that 'DATA' contains quantitative information about the number of occurrences of each event, named with a variable X_i, appearing as a random variable (RV) and assuming different values in the successive intervals $\Delta t_j = t_j - t_{j-1} \in T = (t_N - t_0)$. In this situation, the specification of the individual criteria and determination of the corresponding evaluations can be done based on the dispersion of the RV X_i in the area R_i and the corresponding standard deviation. For this purpose, the expectation, variance, and standard deviation must be calculated for X_i. Then, if $I=[0,1]$, the relation $E_i = \frac{\sigma_i}{\overline{x}_i} \in I$ can be used to evaluate the variable X_i.

To ensure real-time estimation formation along with ongoing recording of experimental values, the procedure of Figure 3 is recommended. Through it, at any moment, information can be obtained about the formed statistical estimates of the monitored variables based on the conducted j time registrations of the experiment at an empirical level. The advantages of this approach are as follows:

a) the specification of the definitional areas is combined with the specification of the criteria, and it is also possible to specify the individual assessments (full or partial);
b) there is parallelism between the empirical level and the abstract-analytical level, and the update of the statistical estimates calculated at the second level takes place simultaneously with the registration of data for the current tact of the experiment.

Logical Completeness of Assessment and Filtering of Primary Information

The goal is to determine a global evaluation criterion G, defined as a functional transformation $G{:}(R_1 x R_2 x \ldots x R_n) \rightarrow I \in [0, 1]$ and allowing the determination of a general performance evaluation $E = G(\vec{x}) = F(E_1, E_2, \ldots, E_n) = F[g_1(x_1), \ldots, g_n(x_n)]$. To determine G, it is necessary to analyze the purpose of the evaluation and to define coefficients of significance (weight coefficients) τ_i of the variables at an abstract-evaluative level. This defines a formation of the global estimate as a function $E{=}\Omega(\tau_1, E_1, \ldots, \tau_n, E_n)$. One way to determine the overall score is as follows:

$$E = \frac{1}{n}\sum_{i=1}^{n} \ddot{A}_i.E_i$$

Figure 3. Procedure for situation (2)

```
FOR i=1 TO n
 BEGIN
     ξi min=100; ξi max=0; A=0; B=0;
     FOR j=1 to N
         READ(DATA[Δtj]);
         A=A+ξij;
         B=B+ξij* ξij;
         C=A * A;
         IF ξij>ξi max THEN ξi max= ξij;
         IF ξij<ξi min THEN ξi min= ξij;
         IF j>1 THEN
             BEGIN
                 σi²=(B-C/j)/(j-1);
                 x̄i=A/j;
                 σi=+(σi²)^1/2
             END
     END_j
     Ri=[ξi min , ξi max];
     Ei=σi / x̄i
 END
END_i
```

A major issue for the experimenter to decide is determining the range of weighting coefficients for the individual variables. Alternatives (but not limited to) are:

- range [0, 1], as a single value means full participation of the estimate E_i for the variable X_i;
- range [0.5, 1.5] where three comparative significance levels (high, medium, and low) are entered.

Filtering is "purging" the initial sample of non-informative registrations based on certain criteria. It is performed by defining a filtering vector ***S*** (Figure 4), equal to the analyzed sample. Each element ***S***[*j*]=s_j is a Boolean variable reflecting the rule:

$$s_j = \begin{cases} 1, \textit{participation of a column 'j'} \\ 0, \textit{non-participation of a column 'j'} \end{cases}$$

To determine the specific value of each element of the filtering vector, a check is made whether the values of the variables at the *j*-th registration meet the specified conditions. Since a certain dependence between the parameters is assumed, a negative answer for a given value removes (filters) the entire record (column).

To determine the elements of the filtering vector, a logical analysis of the data is applied based on a defined matrix of conditions ***C*** (Figure 5). Its dimension is [3, L], where $L \leq n$, and individual rows have the following meaning: ***C***[1,*] – the number of checked variables; ***C***[2,*] – relation code for logical verification; ***C***[3,*] – logical check values.

Figure 4. Filtering a sample

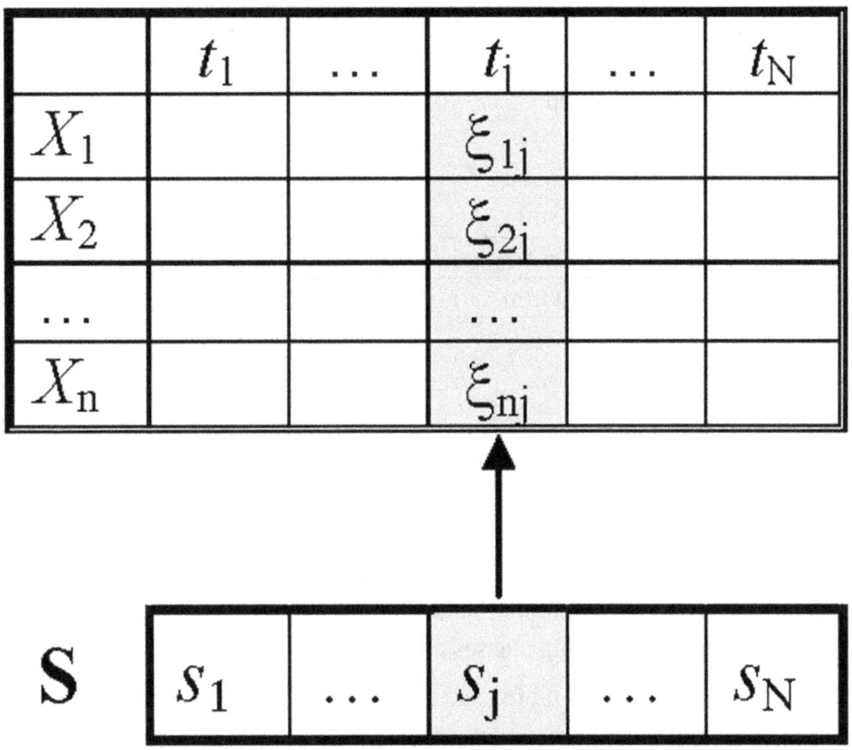

Figure 5. Matrix of conditions

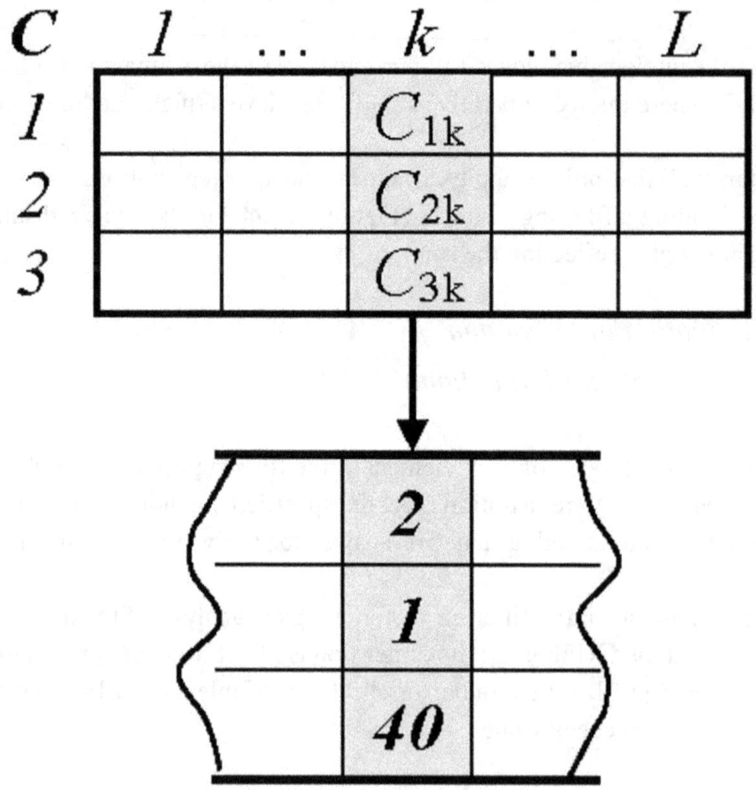

An example of defining the code of the logical conditions applied in a check is presented in Table 1. Based on this determination, the condition presented for column "*k*" in Figure 5 should be interpreted as follows: "For parameter X_2 is required $\forall \xi_{2j} < 40$, for $j=1 \div N$".

Table 1. Codes of logical conditions

Code	1	2	3	4	5	6
Condition	L (<)	LE (≤)	E (=)	NE (≠)	GE (≥)	G (>)

3. BASIC PROCESSING OF EXPERIMENTAL DATA

It has already been noted that the most widely used methods for basic analysis of the results of model research are based on statistical data processing. Some of them are presented below.

Frequency distribution. For the examined (and filtered) sample $\Sigma = \{\xi_{ij}\}$ of *N* registered values for the vector of variables, *k* non-intersecting classes of registrations are determined. The power m_i of each class A_i ($i=1 \div k$) determines the relative hit frequencies $p_i = m_i/N$ for which $\sum_{i=1}^{k} p_i = 1$. The percentage frequency is defined by $f_i = 100.p_i$ [%]. An example illustration is given in Figure 6. The broken line defines the so-called unnormalized empirical distribution function $F(X)$ obtained by accumulating the frequencies starting from 0.

Basic statistics. The main statistical estimates calculated over a formed sample $\Sigma_i=\{\xi_{iq}\}$ of registrations for a given variable $X_i \in \vec{x}$ are presented in Table 2. To the basic characteristics presented in this table it is possible to add the numerical characteristic presented in Chapter 14 (Table 2 and Table 6).

Figure 6. Frequency distributions

A_i	m_i	p_i	f_i
[0,1)	10	0,1	10%
[1,2)	30	0,3	30%
[2,3)	25	0,25	25%
[3,4)	15	0,15	15%
[4,5)	10	0,1	10%
[5,6)	8	0,08	8%
[6,7)	0	0	0%
[7,8)	2	0,02	2%
Sum	*N=100*	*Σp=1*	*100%*

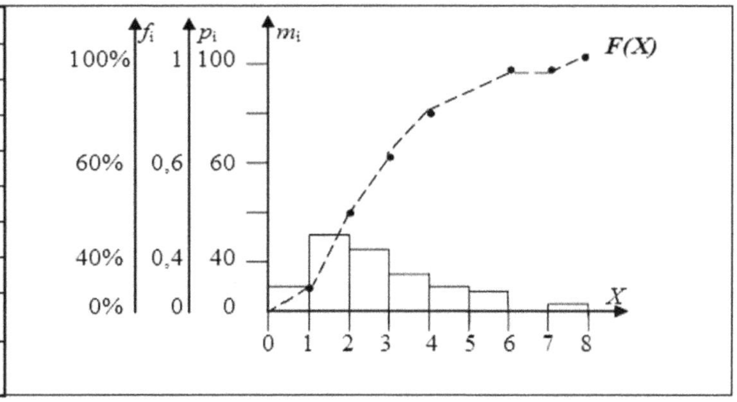

Table 2. Basic statistical assessments

Sample size	$N = ú\Sigma_iú$
Sample span (range of values)	$[\xi_{MIN}, \xi_{MAX}]$
Arithmetic mean value of the sample	$X_{i-AV} = \frac{1}{N}\sum_{q=1}^{N} ¾_{iq}$
Mean deviation	$'X_i = \frac{1}{N}\sum_{q=1}^{N} \left\| ¾_{iq} - X_{i-AV} \right\|$
Mean square deviation	$\tilde{A}_i = \left\{ \frac{1}{N}\sum_{q=1}^{N} \left[¾_{iq} - X_{i-AV} \right]^2 \right\}^{1/2}$
Coefficient of variation	$C_V = \frac{\sigma_i}{X_{i-AV}}$

Correlation analysis. Correlation analysis examines the relationship between two variables $X_i, X_j \in \vec{x}$ by calculating the correlation and covariance coefficients for each set pair of model parameters. As stated in (Xu et al., 2015), "*it reveals the correlations between the factors and the system status through statistical properties of data*" under the conditions of applying a random matrix as a mathematical framework. The application of this analysis defines a purely statistical approach without the need for prior knowledge of the behavior of the object under study, allowing for universal application and robustness to the recorded data. The calculated correlation matrix is of size $n \times n$, being symmetric about the main diagonal. For each pair of parameters $<X_i, X_j>$, a correlation coefficient is calculated:

$$r_{ij} = \frac{\tilde{A}_{ij}}{\tilde{A}_i . \tilde{A}_j}; \text{ като } -1 \le r_{ij} = r_{ji} \le +1 \text{ и } \forall r_{ii} = 1, \text{ за } i,j = 1 \div n.$$

To form the correlation matrix, it is necessary to calculate:

- the arithmetic mean values $X_{i\text{-}AV}$; $i=1,2,\ldots,n$;
- the standard (mean-square) deviations $\sigma i_{;}$ i=1,2,..,n;
- the variances for the variables X_i, and the $1/(N\text{-}1)$ correction is made for unbiased estimation:

$$\tilde{A}_i^2 = \frac{1}{N-1}\sum_{q=1}^{N}\left[\text{¾}_{1q} - X_{i-AV}\right]^2 = \frac{1}{N-1}\left[\sum_{q=1}^{N}\left(\text{¾}_{1q}\right)^2 - N.\left(X_{i-AV}\right)^2\right];$$

- the joint variances (covariances) between two variables $<X_i, X_j>$:

$$\tilde{A}_{ij} = \frac{1}{N-1}\sum_{q=1}^{N}\left[\text{¾}_{1q} - X_{i-AV}\right].\left[\text{¾}_{jq} - X_{j-AV}\right] = \frac{1}{N-1}\left[\sum_{q=1}^{N}\left(\text{¾}_{1q}.\text{¾}_{jq}\right) - N.\left(X_{i-AV}.X_{j-AV}\right)\right].$$

Correlation analysis finds wide application in various fields, one example being the study of the correlation of human thermal sensation depending on multiple physiological factors presented in (Li et al., 2023). An initial collection of the necessary data was carried out for six physiological factors related to temperature, blood pressure, heart rate, etc. The thus formed matrix of recorded values for multiple objects allows us to determine estimates for the studied factors and to make the corresponding decisions in a generalized sense.

Regression analysis. Regression analysis examines the causal relationship between set quantitative facts as interacting variables (Arkes, 2023). The relationship is determined based on a preliminary definition of a dependent variable Y, which examines the existence of direct dependence on one (X – simple regression) or several ($X_1, \ldots, X_k$ – multiple regression) variables defined as independent, i.e. generalization function $Y = Y(X_1, \ldots, X_k)$. is investigated. One possibility is to investigate the presence of linear dependence (Montgomery et al., 2021), which was discussed in Chapter 14, where the possibility of applying a non-linear dependence model was also presented. It was also stated there that one of the commonly used methods for determining error-minimizing estimators is the method of least squares (MLS). When approximating the dependence $Y(X)$ by a regression line, the error is distributed $N(0,\sigma^2)$.

When using a regression method for processing experimental data, the study can be conducted based on the estimates presented below.

Mean square of the regression error (variance of the estimate) $MS_R = SS_R/V_R$ and standard error of the estimate $\sqrt{MS_R}$, where $SS_R = \sum_q \left[y_q - b_0 - b_1.x_q \right]^2$ is the residual sum of the squares of the error (regression deviation), and is the number of residual degrees of freedom.

Residual sum of squares depending on the regression $SS_D = b_1^2 \sum_{q=1}^{N} \left(x_q - x_{AV} \right)^2$ SS_D=b_1^2

F-ratio for hypothesis testing: $F = \dfrac{MS_D}{MS_R} = \dfrac{SS_D.V_R}{SS_R.V_D}$, where V_D=1.

Regression analyzes can have their own specific varieties used in conducting scientific research. One example of a specific application of regression as a method is the one presented in (Sopov & Semenkina, 2021) Symbolic regression using genetic programming (SRGP), presented as "*one of the most popular machine learning approaches for building human-readable interpretable models*". However, the drawback is also pointed out that SRGP has problems when solving large-volume tasks, which lead to "*rapid bloating of trees*" and require additional specialized techniques to overcome the problem. To solve such a problem, a new approach for decomposing a large-scale symbolic regression problem into several subparts with a smaller number of variables using cooperative coevolution is proposed in the paper.

One possible situation is when, when conducting a multiple regression analysis, in addition to correlating multiple independent variables with the dependent variable, there is also a relationship between the independent variables themselves. This determines the presence of multicollinearity (strong correlation of independent variables), which can make some of the significant variables (factors) into those of lower significance and even statistically insignificant. A discussion of various techniques for detecting multicollinearity is offered in (Shrestha, 2020). Three main techniques are considered, with the calculation of correlation coefficients, the variance inflation factor, and the eigenvalue method, respectively. In addition, it states that advanced methods such as "*principal components regression, weighted regression, and ridge regression method*" can be used to determine the presence of multicollinearity.

Regression analysis is the most widely used statistical method in various practical experiments to study the influence between certain factors of behavior. Nevertheless, in (Chukhrova & Johannssen, 2019) it is stated that problems are possible, for example with too small a sample of values, the assumptions of a correct distribution are not fulfilled, with an unclear relationship between the dependent and independent variables being studied, as well as with insufficient certain essence of a studied event. A possible solution when such problems arise is sought based on a modification of the classic statistical regression analysis, and the application of the theory of fuzzy sets is discussed in the article. For example, fuzzy linear regression analysis looks for regression models that correctly fit all analyzed data based on a specific criterion. The paper also points out that other extensions to application-based fuzzy least squares or machine learning, as well as probabilistic approaches, cluster fuzzy regression methods, logistic regression (presented below), and others. One such extension involves using an adaptive neuro-fuzzy inference system for the analysis and prediction of non-parametric fuzzy regression functions presented in (Naderkhani et al., 2021). Two new hybrid algorithms are proposed using a fuzzy variant of MLS and linear programming to optimize the secondary weights. A comparison of three nonparametric fuzzy regression methods with crisp inputs and asymmetric trapezoidal fuzzy outputs using the statistical techniques of local linear smoothing, K-nearest neighbor smoothing, and kernel smoothing with trapezoidal fuzzy data is also made.

The application of fuzzy set theory is the subject of research in the following two publications. In (Demidova & Stepanov, 2021a) the problem of choosing the length of the time series in the construction of a fuzzy regression model is discussed by conducting a preliminary analysis of the time series to allow identification of all attractors represented in them. This will allow the truncated time series created, containing only a fraction of the trailing attractors, to be applied in the construction of fuzzy regression models. The second paper discusses the application of sparse set theory in developing a multifactor predictive model (Demidova & Stepanov, 2021b). The proposal is to present each of the analyzed factors with the same source of origin, but different development trajectories, in the form of a fuzzy time series. This will allow the creation of groups of fuzzy logical dependencies for each trajectory to be used in developing the multivariate regression prediction model based on operations of union and intersection. The optimal values of the parameters of the prediction model were determined using a genetic algorithm, which ensured a minimum average percentage of the prediction error.

A continuation of the idea of fuzzy regression analysis is done in (Chachi & Taheri, 2022), pointing out the importance of the MLS least squares method for estimating the parameters of this analysis. To obtain stable estimates of the investigated parameters, the article proposes an extended M-estimation approach for fuzzy regression analysis, which ensures the correctness of the results in the case of outliers or the presence of changes in the analyzed data set. A weighted algorithm is proposed that allows reducing the impact of deviations on the model and allows easy implementation in a theoretical and computational sense. The Monte-Carlo method was applied to analyze the sensitivity of the evaluated parameters.

Logistic regression. It is a method of studying the behavior of one or more predictors of a dichotomous (two-outcome) dependent variable that can have two specified values, as (Connely, 2020) states that declared independent variables can be "*nominal, ordinal (ranked), interval, or ratio level (or continuous) data*". These two values of the two-outcome model are usually 0 and 1, which are defined for a chosen dependent variable *Y*. This implies that the subject under study (the dependent variable) can qualify in one of two exclusionary categories with probabilities of falling into each of them *p* and (1-*p*) respectively. The thus defined dummy random variable *Y* is represented by the values 0 and 1, the two categories based on the expression $f(Y|p) = p^{Y}(1-p)^{(1-Y)}$. The model assumes that the probability p depends on the following linear function:

$$d(x_1, x_2, ..., x_c) = {}^{2}{}_{0} + \sum_{i=1}^{c} {}^{2}{}_{i} x_i ,$$

where x_i, $i=1,2,...,c$ are the factors (the independent variables) and β_i are the regression coefficients to be estimated from the experimental data. For each individual observation (total n in number), the conditional probability of y_j is expressed by $p(d_j)$ according to the expression:

$$f(y_j \mid p(d_j)) = [p(d_j)]^{y_j} [1 - p(d_j)]^{(1-y_j)}$$

Then for *n* observations the joint conditional probability will be:

$$f(y_1, y_2, \ldots, y_n \mid p(d_1), p(d_2), \ldots, p(d_n)) = [p(d_1)]^{y_1}[1-p(d_1)]^{(1-y_1)}$$
$$[p(d_2)]^{y_2}[1-p(d_2)]^{(1-y_2)} \ldots [p(d_n)]^{y_n}[1-p(d_n)]^{(1-y_n)} = \quad .$$
$$= \prod_{j=1}^{n}[p(d_j)]^{y_j}[1-p(d_j)]^{(1-y_j)}$$

To be able to relate the *y* value of the response *Y* to the *d* value requires a more specific assumption about the function *p(d)*. There are usually three alternatives for this feature.

(1) (1) Linear probability model defined by the following conditions (Figure 7):

$p(d)=0$, if $d<d_0$; $p(d)=a_0+a_1 d$, if $d_0 \leq d \leq d_1$; $p(d)=1$, if $d_1<d$;

(2) (2) Probit model: $p(d)=F(d)$; $-\infty \leq d \leq \infty$, where *F(d)* is distribution function at the standard normal distribution with density *f(d)*. The function *p(d)* is called probit-transformation and tends to the two limits – lower and upper (Figure 8), and the graph gives a more realistic view, because when applying MLS for *p(d)* the obtained results may be outside the interval (0.1). The definition of this model is as follows:

$$p(d) = \int_{-\infty}^{d} \frac{1}{\sqrt{2\pi}} e^{\frac{-\omega^2}{2}} d\omega$$

(3) (3) Logistic model: $p(d)=G(d)$; $-\infty \leq d \leq \infty$, where *G(d)* is the distribution function at the logistic distribution with density *g(d)*, and can be defined by $G(d) = \frac{e^d}{1+e^d}, 0 < G(d) < 1$, which permit to determine the expression:

$$p(d) = \frac{e^d}{1+e^d} .$$

Figure 7. Linear model

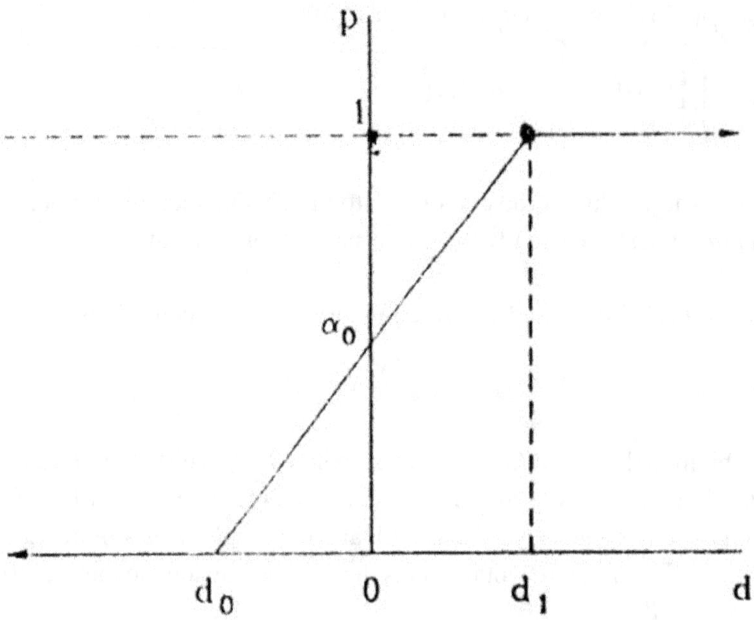

Figure 8. Distribution function under normal and logistic distribution

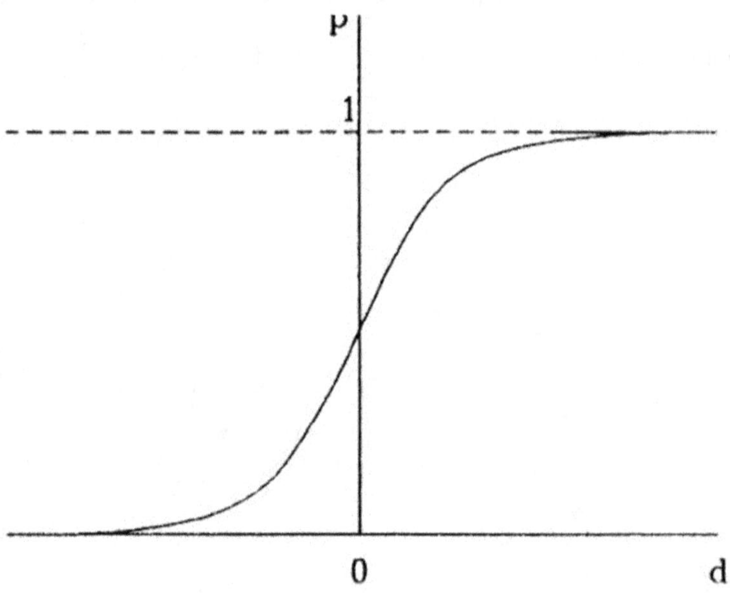

To estimate the two-outcome logistic regression model, the maximum likelihood estimation method is applied, solving the system of equations:

$$\sum_{i=1}^{n} p_i x_i = \sum_{i=1}^{n} y_i x_i \ .$$

In this case, we consider n random observations over a population and a dummy variable Y (value 0 or 1) that indicates to which group each of the observations belongs. Thus, one observes (y_i=0) or (y_i=1) for each i=1,2,...n. Observations over all c factors are denoted by $\left(x_{i1}, x_{i2}, \ldots x_{ic}\right)$. Thus the likelihood for the i-th observation is $p^{y_i}{}_i \left[1 - p_i\right]^{(1-y_i)} = \left[\, p\left(d_i\right)\right]^{y_i} [1 - p\left(d_i\right)]^{(1-y_i)}$.

As with classical regression analysis, logistic regression can use a multivariate variant where the dependent variable (unlike the two-output model) can take on more than two values, i.e., to have more than two categories (multinomial logit model).

One application of logistic regression is in the analysis of binary outcomes in psychology and education (Huang, 2022). The paper states that in certain cases the results are difficult to understand and to overcome this problem an alternative approach is proposed for the analyzes of experimental data in a linear probability model, a log-binomial model, and a modified Poisson regression model. A simulation based on the Monte Carlo method was applied to estimate the bias of the estimates and standard errors.

Variance analysis. It is used to determine the relative influence of various facts among the elements of external practices. It is usually used to compare the means of individual strata of a sample (one-factor analysis) or to examine the differential effects of two factors (two-factor analysis). In the analysis of statistical data, it is also applied in hypothesis testing to assess the influence of a given factor or group of factors on a certain behavior and how statistically significant it is. The influence of the factors is determined based on the evaluation of their variances. Variance analysis can only determine whether there is a relationship between the studied parameters. To apply variance analysis, the conditions presented below must be met.

- Preliminary analysis should be done to establish which parameters are interrelated, which are factors, and which are outcomes..
- The distribution of units in the population should be normal or close to normal.
- The samples must have equal variances.
- The data used must be from independent random samples.

4. INTERPRETATION OF ASSESSMENTS

Experiments conducted in the investigation of objects (processes and systems) in the fields of electronics and computer processing usually provide numerical estimates, which is typical for both modeling and monitoring. The various specialized software environments offer good possibilities for the interpretation and visualization of the experimental results and for the interpretation of the formed estimates. However, in certain cases, further interpretation of the estimates is necessary to reflect the specificity of the study. The main purpose of the interpretation is to present the estimates in a convenient and easy-to-understand form (numerical or graphical). It is a well-known maxim that tabular data is easier to perceive than verbal interpretation, and graphical representation is easier to perceive than tabular. This is also confirmed in (Mishra et al., 2023) with the statement "*graphs have been a ubiquitous way of representing heterogeneous data*", by analyzing the various possibilities of interpreting numerical estimates and extracting data from graphs. As technology advances, graphical data summarization achieves greater expansion and diversity, helping to properly understand the hidden details of data, including applying "deep learning". A conclusion is also presented that "*extraction of relevant data from massive and complex graph structure, enables the data to be used by many application areas*". The article discusses various techniques for graphical data visualization and provides an overview of the features, application areas, and benefits of graphical visualization, presenting existing tools for graphical interpretation.

A simple example for interpretation of hypothetical assessment from a conducted experiment is presented in Figure 9. Traditional interpretations are presented in tabular and graphical form, which can be easily obtained from general-purpose software. A typical example of graphical interpretation is line graphs and bar charts, and in both cases it is possible to present them in 2-D and 3-D versions. An example for 3-D interpretation of the assessment calculated based on monitoring experiments for investigation of the factor "waiting time" for case of different active users is presented in Figure 10 (Romansky, 2022c) – see Chapter 1 (Figures 8, 9 & 10).

A pie chart (Figure 9d) usually represents a given relative characteristic by the area of a circle, with individual components represented by sectors (that is why the name "sector diagram" is used somewhere). Each sector depicts a fraction or percentage relating to a given variable within the system-wide characteristic. This diagram allows to obtain a visual interpretation of the weight of the individual components, or the significance of the variables based on the relative scores obtained from the analyses.

The Kiviat diagram (Figure 9e) is a multibeam plot. In it, the rays are uniformly distributed radii of a circle. The radii are the coordinate axes along which the values of the individual factors are displayed. The arrangement of the values along the axes is arbitrary. It is recommended to alternate adjacent axes of "good" and "bad" characteristics. An example for illustration of the assessments for two independent factors is presented in Figure 11.

Figure 9. Examples for interpretation of experimental assessment

A	B	C	D	E	F
60	20	80	25	55	20

(a) numerical interpretation is a table

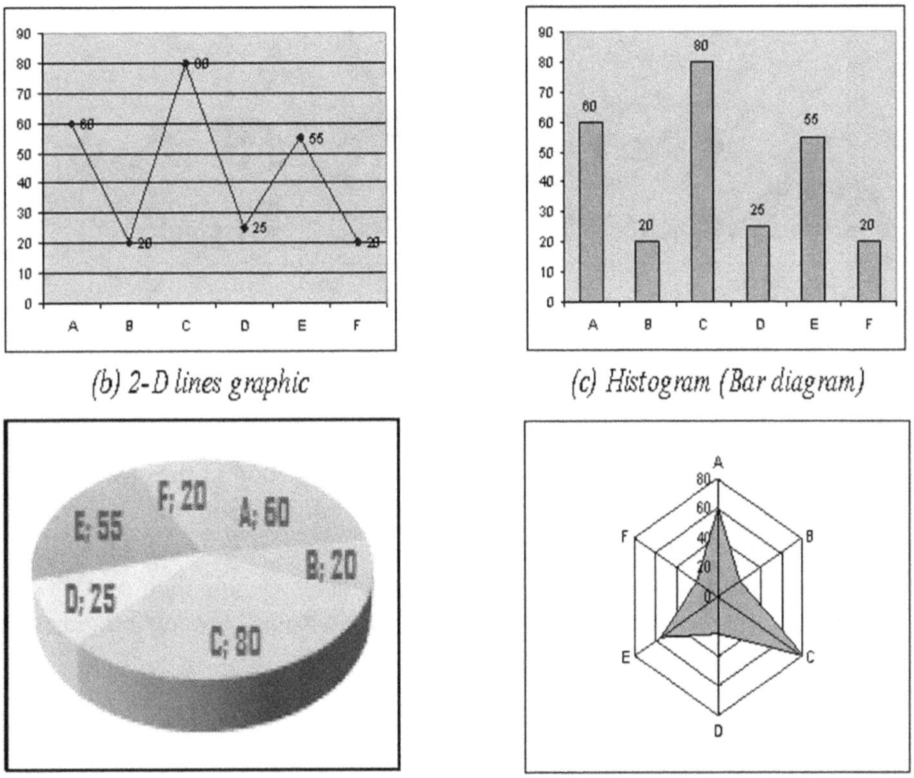

(b) 2-D lines graphic *(c) Histogram (Bar diagram)*

(d) Pie chart (sector diagram) *(e) Kiviat diagram (Radar diagram)*

Figure 10. Distribution of "waiting time"/"users" in information service

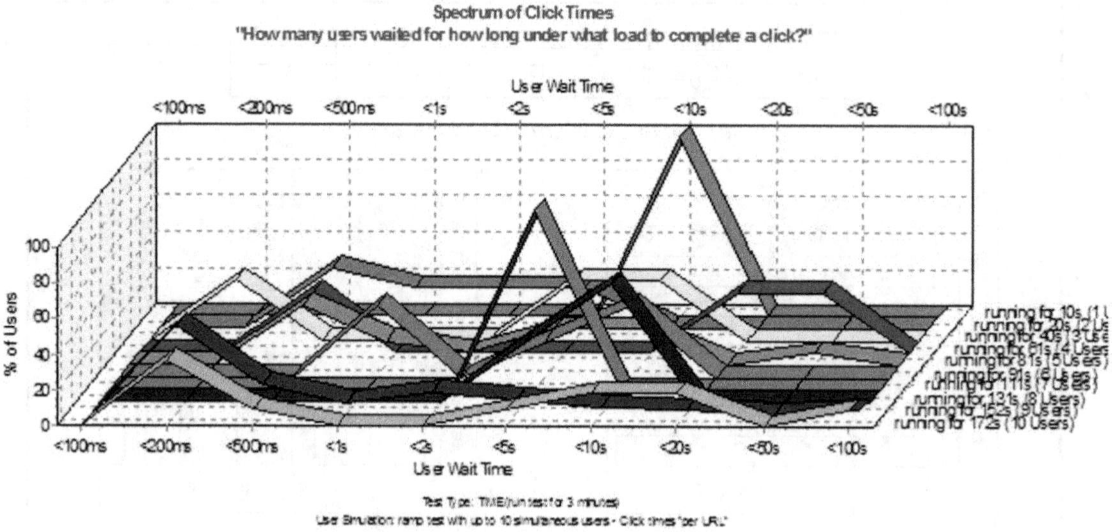

The Gantt chart (Figure 12) is applied to the aggregate presentation of the characteristics related to the use of the individual components of the system. For the construction, the exact moments of time (registrations of the model timer) are needed for the realization of the events. This diagram provides graphical information about the adequacy of the systems components and their overlap over time. The Gantt chart is an effective means of visualizing the time relationships between individual components.

Figure 11. Kiviat diagram for two factors

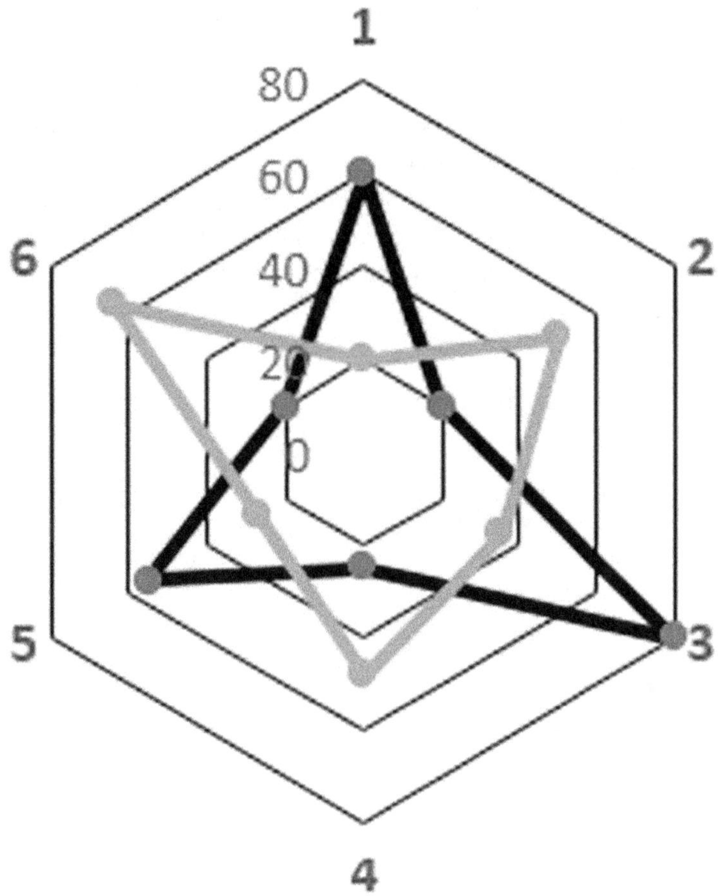

Figure 12. Gantt diagram

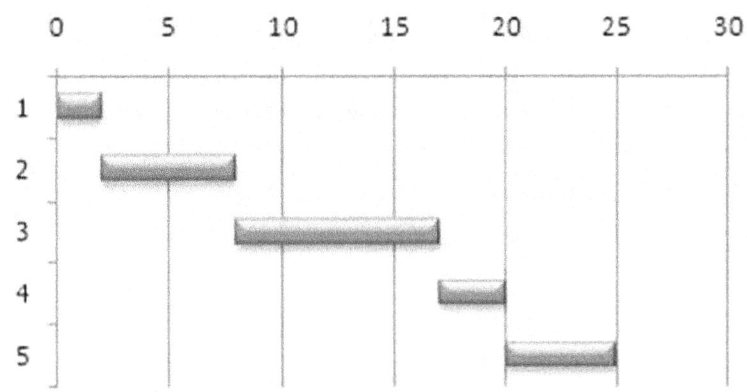

An example for graphical interpretation of experimental data obtained based on carried out simulation modeling is presented below. This is result of the induction machine protection analysis discussed in (Paskalovski & Digalovski, 2022). The purpose of this article is "*to analysis the parameters of an induction motor when there are faults and to project the adequate protection system for them in order to have a safe operation mode that is, not to damage the induction machine*". In the purpose of discussion in this section only part of the research will be shortly presented mainly with used graphical illustration of the simulation results. One of the discussed in the article models of protection system is shown in Figure 13. An overcurrent relay is used in the model, and the objective is to simulate an overcurrent fault and an overcurrent protection system, and the current values are the object of observation. The model in the figure is developed in Matlab Simulink environment, and the overcurrent fault is simulated by the block three-phase fault.

Figure 13. Model overcurrent fault and protection system

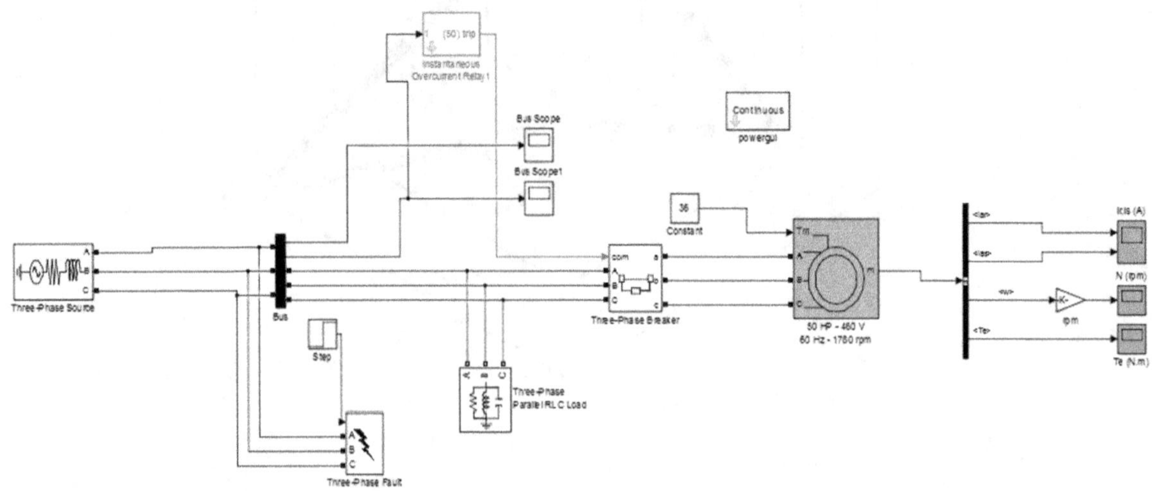

When simulating a failure in the current working state of the machine, a strong increase in the current values is noticed (left side of the Figure 14). This activates the voltage follower relay which sends a signal to the trip switch and the machine remains operational (no damage occurs). The current relay also monitors the amplitude of the current in the system and sends a trip signal to the circuit breaker if it detects an amplitude whose value is twice the nominal value of the amplitude of the system current. Interpretation of the simulation results of the activated protection system are shown in Figure 15 and Figure 16.

Figure 14. Stator and rotor current after activating the protection system

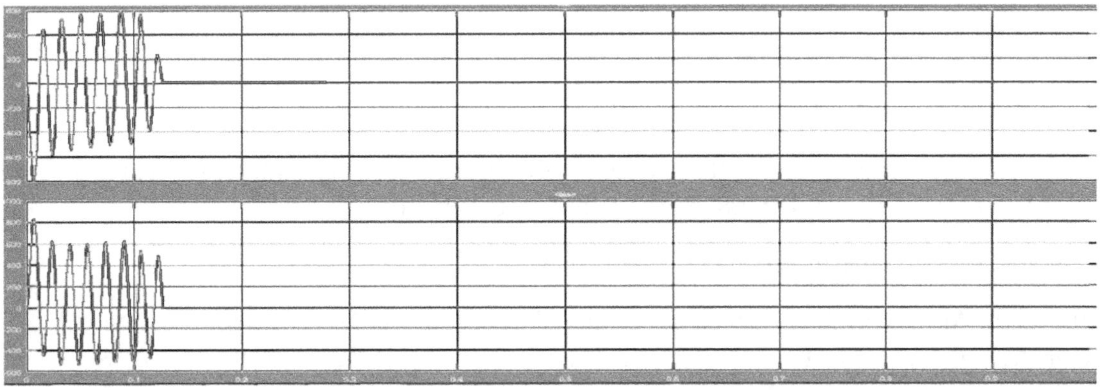

Figure 15. Rotor speed after activating the protection system

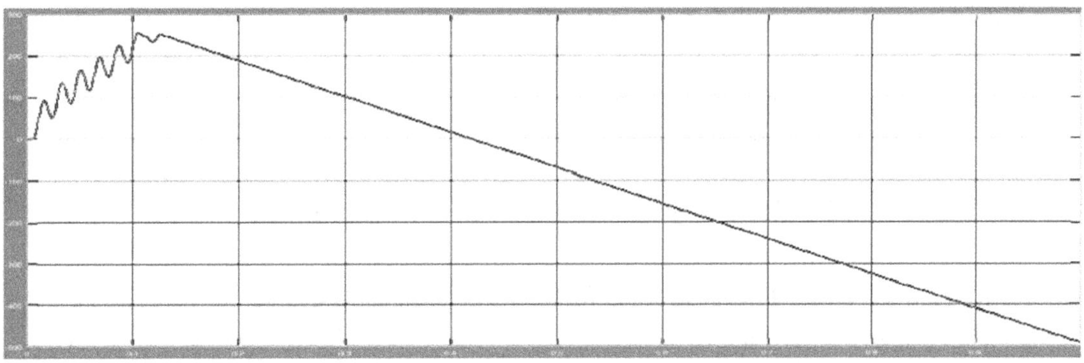

Figure 16. Torque after activating the protection system

REFERENCES

Arkes, J. (2023). *Regression analysis: a practical introduction*. Taylor & Francis.

Chachi, J., Taheri, S. M., & D'Urso, P. (2022). Fuzzy regression analysis based on M-estimates. *Expert Systems with Applications*, *187*, 115891. doi:10.1016/j.eswa.2021.115891

Chukhrova, N., & Johannssen, A. (2019). Fuzzy regression analysis: Systematic review and bibliography. *Applied Soft Computing*, *84*, 105708. doi:10.1016/j.asoc.2019.105708

Connelly, L. (2020). Logistic regression. *Medsurg Nursing*, *29*(5), 353–354.

Curtis, M. J., Alexander, S. P., Cirino, G., George, C. H., Kendall, D. A., Insel, P. A., Izzo, A. A., Ji, Y., Panettieri, R. A., Patel, H. H., Sobey, C. G., Stanford, S. C., Stanley, P., Stefanska, B., Stephens, G. J., Teixeira, M. M., Vergnolle, N., & Ahluwalia, A. (2022). Planning experiments: Updated guidance on experimental design and analysis and their reporting III. *British Journal of Pharmacology*, *179*(15), 3907–3913. doi:10.1111/bph.15868 PMID:35673806

Demidova, L. A., & Stepanov, M. A. (2021a). The aspects of choosing the time series length when constructing a fuzzy regression model. *2021 IEEE International Conference on Information Technologies (InfoTech-2021)*. IEEE. 10.1109/InfoTech52438.2021.9548398

Demidova, L. A., & Stepanov, M. A. (2021b). Development of multifactor forecasting model based on fuzzy time series. *2021 IEEE International Conference on Information Technologies (InfoTech-2021)*. IEEE. 10.1109/InfoTech52438.2021.9548389

Huang, F. L. (2022). Alternatives to logistic regression models in experimental studies. *Journal of Experimental Education*, *90*(1), 213–228. doi:10.1080/00220973.2019.1699769

Li, K., Yu, R., Liu, Y., Wang, J., & Xue, W. (2023). Correlation analysis and modeling of human thermal sensation with multiple physiological markers: An experimental study. *Energy and Building*, *278*, 112643. doi:10.1016/j.enbuild.2022.112643

Lúcio, A. D. C., & Sari, B. G. (2017). Planning and implementing experiments and analyzing experimental data in vegetable crops: Problems and solutions. *Horticultura Brasileira*, *35*(3), 316–327. doi:10.15900102-053620170302

Mishra, P., Kumar, S., & Chaube, M. K. (2023). Graph interpretation, summarization and visualization techniques: A review and open research issues. *Multimedia Tools and Applications*, *82*(6), 8729–8771. doi:10.100711042-021-11582-9

Montgomery, D. C., Peck, E. A., & Vining, G. G. (2021). *Introduction to linear regression analysis*. John Wiley & Sons.

Naderkhani, R., Behzad, M. H., Razzaghnia, T., & Farnoosh, R. (2021). Fuzzy regression analysis based on fuzzy neural networks using trapezoidal data. *International Journal of Fuzzy Systems*, *23*(5), 1267–1280. doi:10.100740815-020-01033-2

Paskalovski, S., & Digalovski, M. (2022). Simulation models for induction machine protection analysis. *International Journal on Information Technologies and Security*, *14*(2), 63–74.

Romansky, R. (2022c). Statistical analysis of empirical network traffic data from program monitoring. *International Journal on Information Technologies and Security*, *14*(3), 15–24.

Romansky, R. (2023). Empirical evaluation of the transfer of information resources in active learning. *International Journal on Information Technologies and Security*, *15*(1), 39–48. doi:10.59035/WZFU1905

Shrestha, N. (2020). Detecting multicollinearity in regression analysis. *American Journal of Applied Mathematics and Statistics*, *8*(2), 39–42. doi:10.12691/ajams-8-2-1

Sopov, E., & Semenkina, M. (2021). Genetic programming using cooperative coevolution and problem decomposition for solving large-scale symbolic Regression problems. *2021 IEEE International Conference on Information Technologies (InfoTech-2021)*, (pp. 124-127). IEEE. 10.1109/InfoTech52438.2021.9548435

Xu, X., He, X., Ai, Q., & Qiu, R. C. (2015). A correlation analysis method for power systems based on random matrix theory. *IEEE Transactions on Smart Grid*, *8*(4), 1811–1820. doi:10.1109/TSG.2015.2508506

Appendix

Table 1. Abbreviations in the text

AI Artificial Intelligence	MC Markov Chain
AM Adjacency Matrix	MIMD Multiple Instruction stream, Multiple Data stream
ANN Artificial Neural Network	MIPS Million Instructions per Second
AQT Analytic Queuing Theory	ML Machine Learning
CI Computational Intelligence	MLS Method of Least Squares
CNN Convolutional Neural Network	MSA Matrix Scheme of Algorithm
CPS Cyber-Physical System	MTU Model Time Unit
CPU Central Processor Unit	NN Neural Network
CS Computer System	OGSA Ordered Graph Scheme of Algorithm
CVA Canonical Variable Analysis	OS Operating System
DAM Deterministic Analytical Model	OSPF Open Shortest Path First
DC Direct Current	PAM Probabilistic Analytical Model
DDM Data Driven Model	PID Proportional-Integral-Derivative
DES Discrete Event Simulation	PN Petri Net
DFD Data flow diagram	PRNG Pseudo-Random Number Generator
DL Deep Learning	PSO Particle Swarm Optimization
DLE Distributed Learning Environment	PT Probability Theory
DRMS Digital Rights Management System	QN Queuing Network
DTN Dialog Transition Network	QS Queuing System
EM External Memory	QT Queuing Theory
ERD Entity-Relationship Diagrams	RLS Recursive Least Squares
FCFS First Come First Served	RM Reachability Matrix
FDA Finite Deterministic Automata	RNG Random Number Generator
FIFO First In First Out	RP Random Process
FLOPS Floating Point Operations per Second	RV Random Variable
FNA Finite Nondeterministic Automata	SAM Stochastic Analytical Model
GA Genetic Algorithm	SSQ Single-Server Queue
GN Generalized Net	STN State Transition Network
GPSS General Purpose Simulation System	SU Service Unit
GSA Graph Scheme of Algorithm	SVM Support Vector Machine
GUI Graphical User Interface	TRNG True Random Number Generator
ICT Information and Communication Technology	UML Unified Modeling Language
IDEF Icam DEFinition	UMT Unit of Machine Time
I/O Input/Output	WF Workflow
IoT Internet of Tings	WFA Wave Front Array
IT Information Technology	WFM Workflow Model
LSA Logical Scheme of Algorithm	WL Workload

Compilation of References

Abdi, H. (2007). The method of least squares. Encyclopedia of measurement and statistics, 1, 530-532

Afolalu, S. A., Ikumapayi, O. M., Abdulkareem, A., Emetere, M. E., & Adejumo, O. (2021). A short review on queuing theory as a deterministic tool in sustainable telecommunication system. *Materials Today: Proceedings*, *44*, 2884–2888. doi:10.1016/j.matpr.2021.01.092

Ahmed, A. E., & Ibrahim, M. E. (2018). Colored Petri net models for clustered and tree-based data aggregation in wireless sensor networks. *International Journal on Information Technologies & Security*, *10*(3), 25–36.

Akberdin, A. A., Kim, A. S., & Sultangaziev, R. B. (2018a). Experiment planning in the simulation of industrial processes. *Steel in Translation*, *48*(9), 573–577. doi:10.3103/S0967091218090024

Akberdin, A. A., Kim, A. S., & Sultangaziev, R. B. (2018b). Planning of numerical and physical experiment in simulation of technological processes. *Izvestiya. Ferros Metallurgy*, *61*(9), 737–742. doi:10.17073/0368-0797-2019-9-737-742

Akram, A., Giannakou, A., Akella, V., Lowe-Power, J., & Peisert, S. (2021). Performance analysis of scientific computing workloads on general purpose TEEs. *2021 IEEE International Parallel and Distributed Processing Symposium (IPDPS)*, (pp. 1066-1076). IEEE. 10.1109/IPDPS49936.2021.00115

Akther, A., Ushakov, Y., Balanov, A. G., & Savel'ev, S. E. (2021). Deterministic modeling of the diffusive memristor. *Chaos (Woodbury, N.Y.)*, *31*(7), 073111. doi:10.1063/5.0056239 PMID:34340321

Allen, J., Galescu, L., Teng, C. M., & Perera, I. (2020). Conversational agents for complex collaborative tasks. *AI Magazine*, *41*(4), 54–78. doi:10.1609/aimag.v41i4.7384

Alshareef, H., Stucki, S., & Schneider, G. (2021, December). Refining privacy-aware Data Flow Diagrams. In *Software Engineering and Formal Methods: 19th International Conference, SEFM 2021, Virtual Event, December 6–10, 2021, Proceedings* (pp. 121-140). Cham: Springer International Publishing. 10.1007/978-3-030-92124-8_8

Alshareef, H., Tuma, K., Stucki, S., Schneider, G., & Scandariato, R. (2022, August). Precise analysis of purpose limitation in Data Flow Diagrams. In *Proceedings of the 17th International Conference on Availability, Reliability and Security* (pp. 1-11). 10.1145/3538969.3539010

Alturki, M. A., Ban Kirigin, T., Kanovich, M., Nigam, V., Scedrov, A., & Talcott, C. (2022). On the formalization and computational complexity of resilience problems for Cyber-Physical Systems. *International Colloquium on Theoretical Aspects of Computing* (pp. 96-113). Springer, Cham.

Álvarez, R., Martínez, F., & Zamora, A. (2022). Improving the statistical qualities of pseudo random number generators. *Symmetry*, *14*(2), 269. doi:10.3390ym14020269

Andries, V. D., Goras, L., Buzo, A., & Pelz, G. (2017, July). Automatic tuning for a DC-DC buck converter with adaptive controller. In *2017 International Symposium on Signals, Circuits and Systems (ISSCS)* (pp. 1-5). IEEE. 10.1109/ISSCS.2017.8034938

Antonelli, F., Cortellessa, V., Gribaudo, M., Pinciroli, R., Trivedi, K. S., & Trubiani, C. (2020). Analytical modeling of performance indices under epistemic uncertainty applied to cloud computing systems. *Future Generation Computer Systems*, *102*, 746–761. doi:10.1016/j.future.2019.09.006

Arifuzzaman, M., & Arslan, E. (2021). Online optimization of file transfers in high-speed networks. *Proceedings of the International Conference for High Performance Computing, Networking, Storage and Analysis*. IEEE. 10.1145/3458817.3476208

Arkes, J. (2023). *Regression analysis: a practical introduction*. Taylor & Francis.

Atanassov K., (2001). Generalized nets as tools for modelling in the area of the artificial intelligence. *Advanced Studies in Contemporary Mathematics*, Vol. 3, 2001, No. 1, 21-42.

Atanassov, K. (2019). On two-way generalized nets. In *Advanced Computing in Industrial Mathematics: 12th Annual Meeting of the Bulgarian Section of SIAM,* (pp. 51-62). Springer International Publishing.

Atangana, A., & Araz, S. I. (2021). *Deterministic-Stochastic modeling: A new direction in modeling real world problems with crossover effect*. Hal Science. https://hal.science/hal-03201318

Atlasov, I. V., Bolnokin, V. E., Kravets, O. J., Mutin, D. I., & Nurutdinov, G. N. (2020). Statistical models for minimizing the number of serch queries. *International Journal on Information Technologies & Security*, *12*(3), 3–12.

Atlasov, I.V., Bolnokin, V.E., & Kravets, O.Ja., Mutin, D.I. & Nurutdinov, G.N. (2020). Statistical models for minimizing the number of search queries. *International Journal on Information Technologies and Security*, *12*(3), 3–12.

Atoum, I. (2019, January). A scalable operational framework for requirements validation using semantic and functional models. In *Proceedings of the 2nd International Conference on Software Engineering and Information Management* (pp. 1-6). 10.1145/3305160.3305166

Azzaz, M. S., Fellah, R., Tanougast, C., & Kaibou, R. (2021). Design and FPGA implementation of TRNG based on a new multi-wing attractor in Lorenz chaotic system. *The European Physical Journal. Special Topics*, *230*(18), 3469–3480. doi:10.1140/epjs11734-021-00234-6

Baccelli, F., Foss, S., & Gaujal, B. (1996). Free-choice Petri nets-an algebraic approach. *IEEE Transactions on Automatic Control*, *41*(12), 1751–1778. doi:10.1109/9.545714

Baek, H., Jun, B. H., Yoon, S. M., & Noh, M. (2019). Study on identification procedure for unidentified underwater targets using small ROV based on IDEF method. *Journal of Ocean Engineering and Technology*, *33*(3), 289–299. doi:10.26748/KSOE.2019.022

Bagdasar, O., Chen, M., Drăgan, V., Ivanov, I. G., & Popa, I. L. (2023). On Horadam sequences with dense orbits and Pseudo-Random Number Generators. *Mathematics*, *11*(5), 1244. doi:10.3390/math11051244

Bagossi, S., Ferretti, F., & Arzarello, F. (2022). Assessing covariation as a form of conceptual understanding through comparative judgement. *Educational Studies in Mathematics*, *111*(3), 469–492. doi:10.100710649-022-10178-w

Bahga, A., & Madisetti, V. K. (2011). Synthetic workload generation for cloud computing applications. *Journal of Software Engineering and Applications*, *4*(07), 396–410. doi:10.4236/jsea.2011.47046

Baker, E., Barbillon, P., Fadikar, A., Gramacy, R. B., Herbei, R., Higdon, D., Huang, J., Johnson, L. R., Ma, P., Mondal, A., Pires, B., Sacks, J., & Sokolov, V. (2022). Analyzing stochastic computer models: A review with opportunities. *Statistical Science*, *37*(1), 64–89. doi:10.1214/21-STS822

Balakrishnan, H., Moreno-Ezuilaz, M., Riba, J. R., Bogarra, S., & Garcia, A. (2018). DC-DC buck converter parameter identification based on a white-box approach. In *2018 IEEE 18th International Power Electronics and Motion Control Conference (PEMC)* (pp. 242-247). IEEE. 10.1109/EPEPEMC.2018.8521981

Bartolucci, S., Birchall, P., Bombin, H., Cable, H., Dawson, C., Gimeno-Segovia, M., Johnston, E., Kieling, K., Nickerson, N., Pant, M., Pastawski, F., Rudolph, T., & Sparrow, C. (2023). Fusion-based quantum computation. *Nature Communications*, *14*(1), 912. doi:10.103841467-023-36493-1 PMID:36805650

Barwasser, A., Lentes, J., Riedel, O., Zimmermann, N., Dangelmaier, M., & Zhang, J. (2022). Method for the development of Software-Defined Manufacturing equipment. *International Journal of Production Research*, 1–18.

Batool, K., & Niazi, M. A. (2017). Modeling the internet of things: A hybrid modeling approach using complex networks and agent-based models. *Complex Adaptive Systems Modeling*, *5*(4), 4. Advance online publication. doi:10.118640294-017-0043-1

Beg, S., Swain, S., Rahman, M., Hasnain, M. S., & Imam, S. S. (2019). Application of design of experiments (DoE) in pharmaceutical product and process optimization. In *Pharmaceutical quality by design* (pp. 43–64). Academic Press. doi:10.1016/B978-0-12-815799-2.00003-4

Ben Ouahi, M., Ait Hou, M., Bliya, A., Hassouni, T., & Al Ibrahmi, E. M. (2021). The effect of using computer simulation on students' performance in teaching and learning physics: Are there any gender and area gaps? *Education Research International*, *2021*, 1–10. doi:10.1155/2021/6646017

Bevilacqua, A., Speranza, B., Petruzzi, L., Sinigaglia, M., & Corbo, M. R. (2023). Using regression and Multifactorial Analysis of Variance to assess the effect of ascorbic, citric, and malic acids on spores and activated spores of Alicyclobacillus acidoterrestris. *Food Microbiology*, *110*, 104158. doi:10.1016/j.fm.2022.104158 PMID:36462814

Bhati, D., Chakraborty, S., & Lateef, S. G. (2020). A discrete probability model suitable for both symmetric and asymmetric count data. *Filomat*, *34*(8), 2559–2572. doi:10.2298/FIL2008559B

Bhattacharjee, K., & Das, S. (2022). A search for good pseudo-random number generators: Survey and empirical studies. *Computer Science Review*, *45*, 100471. doi:10.1016/j.cosrev.2022.100471

Bianco, S., Cusano, C., Piccoli, F., & Schettini, R. (2019). Learning parametric functions for color image enhancement. In *Computational Color Imaging: 7th International Workshop, CCIW 2019, Chiba, Japan, March 27-29, 2019* [Springer International Publishing.]. *Proceedings*, *7*, 209–220.

Bourgey, F. (2020). *Stochastic approximations for financial risk computations* [Doctoral dissertation, Institut polytechnique de Paris].

Bożek, A., Rak, T., & Rzonca, D. (2022). Timed Colored Petri Net-based event generators for Web systems simulation. *Applied Sciences (Basel, Switzerland)*, *12*(23), 12385. doi:10.3390/app122312385

Brezovan, M., Precup, R. E., Selişteanu, D., & Stănescu, L. (2022). Colored Petri nets-based control and experimental validation on three-tank system level control. *International Journal of General Systems*, 1–47.

Brus, D. J. (2021). Statistical approaches for spatial sample survey: Persistent misconceptions and new developments. *European Journal of Soil Science*, *72*(2), 686–703. doi:10.1111/ejss.12988

Burke, A., Leemans, S. J., & Wynn, M. T. (2021). Stochastic process discovery by weight estimation. In *Process Mining Workshops: ICPM 2020 International Workshops, Padua, Italy, October 5–8, 2020, Revised Selected Papers 2* (pp. 260-272). Springer International Publishing 10.1007/978-3-030-72693-5_20

Butler, T., Jakeman, J., & Wildey, T. (2018). Convergence of probability densities using approximate models for forward and inverse problems in uncertainty quantification. *SIAM Journal on Scientific Computing*, *40*(5), A3523–A3548. doi:10.1137/18M1181675

Bychkov, I., Kazakov, A., Lempert, A., & Zharkov, M. (2021). Modeling of railway stations based on queuing networks. *Applied Sciences (Basel, Switzerland)*, *11*(5), 2425. doi:10.3390/app11052425

Calzarossa, M. C., Massari, L., & Tessera, D. (2016). Workload characterization: A survey revisited. *ACM Computing Surveys*, *48*(3), 1–43. doi:10.1145/2856127

Cardelli, L., Kwiatkowska, M., & Laurenti, L. (2016). Stochastic analysis of chemical reaction networks using linear noise approximation. *Bio Systems*, *149*, 26–33. doi:10.1016/j.biosystems.2016.09.004 PMID:27816736

Carlucci, S., Causone, F., Biandrate, S., Ferrando, M., Moazami, A., & Erba, S. (2021). On the impact of stochastic modeling of occupant behavior on the energy use of office buildings. *Energy and Building*, *246*, 111049. doi:10.1016/j.enbuild.2021.111049

Caro, J. J., & Möller, J. (2016). Advantages and disadvantages of discrete-event simulation for health economic analyses. *Expert Review of Pharmacoeconomics & Outcomes Research*, *16*(3), 327–329. doi:10.1586/14737167.2016.1165608 PMID:26967022

Carstens, B. C., Smith, M. L., Duckett, D. J., Fonseca, E. M., & Thomé, M. T. C. (2022). Assessing model adequacy leads to more robust phylogeographic inference. *Trends in Ecology & Evolution*, *37*(5), 402–410. doi:10.1016/j.tree.2021.12.007 PMID:35027224

Cassandras, C. G., Lafortune, S., Cassandras, C. G., & Lafortune, S. (2021). Petri nets. *Introduction to discrete event systems*, 259-302.

Chachi, J., Taheri, S. M., & D'Urso, P. (2022). Fuzzy regression analysis based on M-estimates. *Expert Systems with Applications*, *187*, 115891. doi:10.1016/j.eswa.2021.115891

Chapple, M. (2021). *Entity-Relationship Diagram Definition- Use ER diagrams to illustrate relationships between database entities.* LifeWire. https://www.lifewire.com/entity-relationship-diagram-1019253

Chau, M., & Fu, M. C. (2014). An overview of stochastic approximation. Handbook of simulation optimization, 149-178.

Cheng, Q., Liu, Z., Guo, J., Wu, X., Pendyala, R., Belezamo, B., & Zhou, X. S. (2022). Estimating key traffic state parameters through parsimonious spatial queue models. *Transportation Research Part C, Emerging Technologies*, *137*, 103596. doi:10.1016/j.trc.2022.103596

Chen, H., Li, H., Ma, T., Han, S., & Zhao, Q. (2023). Biological function simulation in neuromorphic devices: From synapse and neuron to behavior. *Science and Technology of Advanced Materials*, *24*(1), 2183712. doi:10.1080/14686996.2023.2183712 PMID:36926202

Chen, J. C., & Wu, Y. J. (2020). Discrete-time Markov chain for prediction of air quality index. *Journal of Ambient Intelligence and Humanized Computing*, 1–10. doi:10.100712652-020-02036-5

Chen, J., & Shi, Y. (2021). Stochastic model predictive control framework for resilient cyber-physical systems: Review and perspectives. *Philosophical Transactions - Royal Society. Mathematical, Physical, and Engineering Sciences*, *379*(2207), 20200371. doi:10.1098/rsta.2020.0371 PMID:34398650

Cheryshov, A.B., Choporov, O.N., Preobrazhenskiy, A.P., & Kravets, O.Ja. (2020). The development of optimization model and algorithm for support of resources management in organizational system. *International Journal on Information Technologies and Security*, *12*(2), 25–36.

Chkalova, M., & Pavlidis, V. (2020). Modeling of grain raw stuff grinding process. *Engineering for Rural Development*, *19*, 341–348. doi:10.22616/ERDev.2020.19.TF080

Chukhrova, N., & Johannssen, A. (2019). Fuzzy regression analysis: Systematic review and bibliography. *Applied Soft Computing*, *84*, 105708. doi:10.1016/j.asoc.2019.105708

Ciccozzi, F., Malavolta, I., & Selic, B. (2019). Execution of UML models: A systematic review of research and practice. *Software & Systems Modeling*, *18*(3), 2313–2360. doi:10.100710270-018-0675-4

Cohen, A., & Migliorati, G. (2017). Optimal weighted least-squares methods. *The SMAI journal of computational mathematics, 3*, 181-203.

Collier, Z. A., Gaskins, A., & Lambert, J. H. (2022). Business process modeling for semiconductor production risk analysis using [*IEEE Engineering Management Review.*]. *Ide (São Paulo)*, F0.

Connelly, L. (2020). Logistic regression. *Medsurg Nursing*, *29*(5), 353–354.

Crocetti, L., Nannipieri, P., Di Matteo, S., Fanucci, L., & Saponara, S. (2023). Review of methodologies and metrics for assessing the quality of Random Number Generators. *Electronics (Basel)*, *12*(3), 723. doi:10.3390/electronics12030723

Cruz, F. R., Quinino, R. D. C., & Ho, L. L. (2017). Bayesian estimation of traffic intensity based on queue length in a multi-server M/M/s queue. *Communications in Statistics. Simulation and Computation*, *46*(9), 7319–7331. doi:10.1080/03610918.2016.1236953

Cui, T. & Ma, Y. (2015). Definition of attribute circle in factor space and its application in object classification. *Computer engineering & science, 37*(11), 2169-2174.

Curtis, M. J., Alexander, S. P., Cirino, G., George, C. H., Kendall, D. A., Insel, P. A., Izzo, A. A., Ji, Y., Panettieri, R. A., Patel, H. H., Sobey, C. G., Stanford, S. C., Stanley, P., Stefanska, B., Stephens, G. J., Teixeira, M. M., Vergnolle, N., & Ahluwalia, A. (2022). Planning experiments: Updated guidance on experimental design and analysis and their reporting III. *British Journal of Pharmacology*, *179*(15), 3907–3913. doi:10.1111/bph.15868 PMID:35673806

D'Apice, C., Dudin, A., Dudin, S., & Manzo, R. (2022). Priority queueing system with many types of requests and restricted processor sharing. *Journal of Ambient Intelligence and Humanized Computing*, •••, 1–12. doi:10.100712652-022-04233-w

Daley, A. J., Bloch, I., Kokail, C., Flannigan, S., Pearson, N., Troyer, M., & Zoller, P. (2022). Practical quantum advantage in quantum simulation. *Nature*, *607*(7920), 667–676. doi:10.103841586-022-04940-6 PMID:35896643

Daradkeh, T., Agarwal, A., Zaman, M., & Manzano, R. (2021, June). Analytical modeling and prediction of cloud workload. In *2021 IEEE International Conference on Communications Workshops (ICC Workshops)* (pp. 1-6). IEEE. 10.1109/ICCWorkshops50388.2021.9473619

Darmayunata, Y., Devega, M., & Yuhelmi, Y. (2023). Development of web-based single channel multi steps online queuing system with model view controller. *Sinkron: jurnal dan penelitian teknik informatika, 8*(1), 390-397.

Das, S., McClain, C. J., & Rai, S. N. (2020). Fifteen years of gene set analysis for high-throughput genomic data: A review of statistical approaches and future challenges. *Entropy (Basel, Switzerland)*, *22*(4), 427. doi:10.3390/e22040427 PMID:33286201 de la Fraga, L. G., Torres-Pérez, E., Tlelo-Cuautle, E., & Mancillas-López, C. (2017). Hardware implementation of pseudo-random number generators based on chaotic maps. *Nonlinear Dynamics*, *90*(3), 1661–1670. doi:10.100711071-017-3755-z

Demidova, L. A., & Stepanov, M. A. (2021a). The aspects of choosing the time series length when constructing a fuzzy regression model. *2021 IEEE International Conference on Information Technologies (InfoTech-2021)*. IEEE. 10.1109/InfoTech52438.2021.9548398

Demidova, L. A., & Stepanov, M. A. (2021b). Development of multifactor forecasting model based on fuzzy time series. *2021 IEEE International Conference on Information Technologies (InfoTech-2021)*. IEEE. 10.1109/InfoTech52438.2021.9548389

Desel, J., & Esparza, J. (1995). Free choice Petri nets (40). Cambridge university press.

Dhingra, V., Kumawat, G. L., Roy, D., & de Koster, R. (2018). Solving semi-open queuing networks with time-varying arrivals: An application in container terminal landside operations. *European Journal of Operational Research*, *267*(3), 855–876. doi:10.1016/j.ejor.2017.12.020

Digalovski, M., & Rafajlovski, G. (2020). Distribution transformer mathematical model for power losses minimization. *International Journal on Information Technologies and Security*, *12*(2), 57–68.

Dijkstra, E. W. (1972). Notes on structured programing. In O. J. Dahl, E. W. Dijkstra, & C. A. R. Hoare (Eds.), *Structured programming* (pp. 1–82). Academic Press Ltd. https://dl.acm.org/doi/pdf/10.5555/1243380

Dimishkovska, N., & Iliev, A. (2019). Markov chain model for small hydropower plant reliability and operation evaluation. In *33rd International Conference on Information Technologies (InfoTech-2019)* (pp. 98-109).

Dingle, S. E., Bujtor, M. S., Milte, C. M., Bowe, S. J., Daly, R. M., & Torres, S. J. (2023). Statistical approaches for the analysis of combined health-related factors in association with adult cognitive outcomes: A scoping review. *Journal of Alzheimer's Disease*, February, 1-25. doi:10.3233/JAD-221034

Distinct Corporation. (2022). *Distinct Network Monitor*. District Corporation. https://www.distinct.com/monitor/monitor.htm (Visited on 5 Jan 2023)

do Amaral, J. V. S., Montevechi, J. A. B., de Carvalho Miranda, R., & de Sousa, W. T. Junior. (2022). Metamodel-based simulation optimization: A systematic literature review. *Simulation Modelling Practice and Theory*, *114*, 102403. doi:10.1016/j.simpat.2021.102403

Du Bois, K., Schaeps, T., Polfliet, S., Ryckbosch, F., & Eeckhout, L. (2011). SWEEP: Evaluating computer system energy efficiency using synthetic workloads. In *HiPEAC'11 Proceedings of the 6th International Conference on High Performance and Embedded Architectures and Compilers*, (pp. 159-166). IEEE.)10.1145/1944862.1944886

Dubois, D., & Prade, H. (2006). *Possibility theory and its applications: a retrospective and prospective view*. Springer Vienna.

Dudin, A. N., Klimenok, V. I., & Vishnevsky, V. M. (2020). *The theory of queuing systems with correlated flows*. Springer. doi:10.1007/978-3-030-32072-0

Durrett, R. (2019). *Probability: theory and examples* (Vol. 49). Cambridge university press. doi:10.1017/9781108591034

Dvir, M., & Ben-Zvi, D. (2023). Fostering students' informal quantitative estimations of uncertainty through statistical modeling. *Instructional Science*, *51*(3), 1–28. doi:10.100711251-023-09622-y

Efatmaneshnik, M., Shoval, S., & Joiner, K. (2019). System test architecture evaluation: A probabilistic modeling approach. *IEEE Systems Journal*, *13*(4), 3651–3662. doi:10.1109/JSYST.2019.2899697

Eini, M. R., Salmani, H., & Piniewski, M. (2023). Comparison of process-based and statistical approaches for simulation and projections of rainfed crop yields. *Agricultural Water Management*, *277*, 108107. doi:10.1016/j.agwat.2022.108107

Elliott, J., Kelly, D., Best, N., Wilde, M., Glotter, M., & Foster, I. (2013, July). The parallel system for integrating impact models and sectors (pSIMS). In *Proceedings of the Conference on Extreme Science and Engineering Discovery Environment: Gateway to Discovery* (pp. 1-8). 10.1145/2484762.2484814

Elliriki, M., Reddy, C. S., Anand, K., & Saritha, S. (2022). Multi server queuing system with crashes and alternative repair strategies. *Communications in Statistics. Theory and Methods*, *51*(23), 8173–8185. doi:10.1080/03610926.2021.1889603

Englander, I., & Wong, W. (2021). *The architecture of computer hardware, systems software, and networking: An information technology approach.* John Wiley & Sons.

Epishkin, V. E. (2023). Mathematical model of technological equipment while analyzing the capacity of car server stations. *Transportation Research Procedia*, *68*, 622–629. doi:10.1016/j.trpro.2023.02.085

Equbal, M. S., Ketkar, T., & Sahay, S. (2023). Hybrid CMOS-RRAM True Random Number Generator exploiting coupled entropy sources. *IEEE Transactions on Electron Devices*, *70*(3), 1061–1066. doi:10.1109/TED.2023.3241122

Estrada-Flores, S., Merts, I., De Ketelaere, B., & Lammertyn, J. (2006). Development and validation of "grey-box" models for refrigeration applications: A review of key concepts. *International Journal of Refrigeration*, *29*(6), 931–946. doi:10.1016/j.ijrefrig.2006.03.018

Evgenev, G. B. (2021). Intelligent system of computer aided processes planning. In Z. Hu, B. Wang, S. Petoukhov, & M. He (Eds.), *Advances in Artificial Systems for Power Engineering. AIPE 2020. Advances in Intelligent Systems and Computing* (Vol. 1403, pp. 26–39). Springer Int. Publ. doi:10.1007/978-3-030-80531-9_3

Faix, M., Laurent, R., Bessière, P., Mazer, E., & Droulez, J. (2016). Design of stochastic machines dedicated to approximate Bayesian inferences. *IEEE Transactions on Emerging Topics in Computing*, *7*(1), 60–66. doi:10.1109/TETC.2016.2609926

Fathalla, S., Auer, S., & Lange, C. (2020, March). Towards the semantic formalization of science. In *Proceedings of the 35th Annual ACM Symposium on Applied Computing* (pp. 2057-2059). ACM. 10.1145/3341105.3374132

Feitelson, D. G. (2015). *Workload modeling for computer systems performance evaluation.* Cambridge University Press. doi:10.1017/CBO9781139939690

Fonseca, E. M., Duckett, D. J., Almeida, F. G., Smith, M. L., Thomé, M. T. C., & Carstens, B. C. (2022). Assessing model adequacy for Bayesian Skyline plots using posterior predictive simulation. *PLoS One*, *17*(7), e0269438. doi:10.1371/journal.pone.0269438 PMID:35877611

Freedman, D. A. (2009). *Statistical models: theory and practice.* Cambridge University Press. doi:10.1017/CBO9780511815867

Frichi, Y., Jawab, F., Aboueljinane, L., & Boutahari, S. (2022). Development and comparison of two new multi-period queueing reliability models using discrete-event simulation and a simulation–optimization approach. *Computers & Industrial Engineering*, *168*, 108068. doi:10.1016/j.cie.2022.108068

Fu, M., Wang, D., Wang, J., & Li, M. (2018, October). Modeling method of operational task combined with IDEF and UML. In *2018 IEEE 3rd Advanced Information Technology, Electronic and Automation Control Conference (IAEAC)* (pp. 1443-1447). IEEE. 10.1109/IAEAC.2018.8577660

Gabrielli, A., & Wüthrich, V., M. (. (2018). An individual claims history simulation machine. *Risks*, *6*(2), 29. doi:10.3390/risks6020029

Galton, F. (1886). Regression towards mediocrity in hereditary stature. *Journal of the Anthropological Institute of Great Britain and Ireland, 15*, 246–263. doi:10.2307/2841583

Garzon, J., Ameigeiras, P., Ramos-Munoz, J. J., Navarro-Ortiz, J., Andres-Maldonado, P., & Lopez-Soler, J. M. (2019). Performance modeling of softwarized network services based on queuing theory with experimental validation. *IEEE Transactions on Mobile Computing*, *20*(4), 1558–1573.

Gerry, C. C., Birrittella, R. J., Alsing, P. M., Hossameldin, A., Eaton, M., & Pfister, O. (2022). Proposal for a quantum random number generator using coherent light and a non-classical observable. *Journal of the Optical Society of America. B, Optical Physics*, *39*(4), 1068–1074. doi:10.1364/JOSAB.441210

Ghadimi, S., Ruszczynski, A., & Wang, M. (2020). A single timescale stochastic approximation method for nested stochastic optimization. *SIAM Journal on Optimization*, *30*(1), 960–979. doi:10.1137/18M1230542

Girardin, V., & Limnios, N. (2022). *Applied probability: from random experiments to random sequences and statistics*. Springer Nature. doi:10.1007/978-3-030-97963-8

Giua, A., & Silva, M. (2018). Petri nets and automatic control: A historical perspective. *Annual Reviews in Control*, *45*, 223–239. doi:10.1016/j.arcontrol.2018.04.006

Gnedenko, B. V. (2018). *Theory of probability*. Routledge. doi:10.1201/9780203718964

Gocheva, P. V., & Gochev, V. P. (2018a), Application of generalized networks for modeling mass service systems. In *National Forum Electronics 2018*.

Gocheva, P. V., Hinov, N. L., & Gochev, V. P. (2018b). Modeling of electronic circuits with Generalized Nets. In *2018 IX National Conference with International Participation ELECTRONICA*, (pp. 1-4). IEEE. 10.1109/ELECTRONICA.2018.8439168

Gocheva, P. V., Hinov, N. L., & Gochev, V. P. (2020). Index matrices-based modelling of a DC-DC buck converter with PID controller and GUI on it. *2020 International Conference on Information Technologies (InfoTech)*, (pp. 1-4). IEEE.

Gochev, V. P., Gocheva, P. V., & Hinov, N. L. (2021, March). NET implementation of electronic circuit design. In. AIP Conference Proceedings, (p. 070016). AIP Publishing LLC.

Gochev, V. P., & Hinov, N. L. (2022). Generalized nets representing C based programming constructs. In *2022 IEEE International Conference on Information Technologies (InfoTech-2022)*, (pp.13-16). IEEE. 10.1109/InfoTech55606.2022.9897111

Gonçalves, G. D., Drago, I., Vieira, A. B., da Silva, A. P. C., Almeida, J. M., & Mellia, M. (2016). Workload models and performance evaluation of cloud storage services. *Computer Networks*, *109*, 183–199. doi:10.1016/j.comnet.2016.03.024

Gortsev, A. M., & Nezhelskaya, L. A. (2022). Analytical investigation of a single-channel QS with incoming asynchronous event flow. *Automation and Remote Control*, *83*(8), 1200–1212. doi:10.1134/S0005117922080045

Gotovos, A., Burkholz, R., Quackenbush, J., & Jegelka, S. (2021). Scaling up continuous-time Markov chains helps resolve underspecification. *Advances in Neural Information Processing Systems*, *34*, 14580–14592.

Gouel, M., Vermeulen, K., Mouchet, M., Rohrer, J. P., Fourmaux, O., & Friedman, T. (2022). Zeph & Iris map the internet: A resilient reinforcement learning approach to distributed IP route tracing. *Computer Communication Review*, *52*(1), 2–9. doi:10.1145/3523230.3523232

Gregor, M., Hodoň, R., Grznár, P., & Mozol, Š. (2022). Design of a system for verification of automatic guided vehicle routes using computer emulation. *Applied Sciences (Basel, Switzerland)*, *12*(7), 3397. doi:10.3390/app12073397

Guillan-Lorenzo, O., Troncoso-Costas, M., Alvarez-Outarelo, D., Diaz-Otero, F. J., & Garcia-Escartin, J. C. (2023). Optical quantum random number generators: A comparative study. *Optical and Quantum Electronics*, *55*(2), 185. doi:10.100711082-022-04396-y

Habib, H., Khanam, S. H., Azhar, A., & Khalid, Z. (2017). *Distributed Memory Architecture (Non Shared MIMD Architecture)*. Slideshare. (https://www.slideshare.net/HBukhary/distributed-memory-architecture-non-shared-mimd-architecture)

Habib, M. K., Ayankoso, S. A., & Nagata, F. (2021, August). Data-Driven Modeling: Concept, Techniques, Challenges and a Case Study. In *2021 IEEE International Conference on Mechatronics and Automation (ICMA)* (pp. 1000-1007). IEEE. 10.1109/ICMA52036.2021.9512658

Haboubi, S., & Salem, O. B. (2022). Predictive analytics for energy consumption in smart homes with fog and cloud computing using support vector regression. *International Journal on Information Technologies and Security*, *14*(1), 49–60.

Hacks, S., Katsikeas, S., Ling, E., Lagerström, R., & Ekstedt, M. (2020). powerLang: A probabilistic attack simulation language for the power domain. *Energy Informatics*, *3*(1), 1–17. doi:10.118642162-020-00134-4

Hakrama, I., & Frasheri, N. (2018). Agent-based modelling and simulation of an artificial economy with Repast. *International Journal on Information Technologies and Security*, *10*(2), 47–56.

Hanson, J. R., & Walker, S. I. (2021). Formalizing falsification for theories of consciousness across computational hierarchies. Neuroscience of Consciousness.) doi:10.1093/nc/niab014

Hassani, M. M., & Berangi, R. (2018). An analytical model to calculate blocking probability of secondary user in cognitive radio sensor networks. *International Journal on Information Technologies & Security*, *10*(2), 3–12.

Hazır, Ö. (2015). A review of analytical models, approaches and decision support tools in project monitoring and control. *International Journal of Project Management*, *33*(4), 808–815. doi:10.1016/j.ijproman.2014.09.005

Hennessy, J., & Patterson, D. (2017). *Computer Architecture: A Quantitative Approach* (6th ed.). Morgan Kaufman / Elseviar Inc.

Herrero-Collantes, M., & Garcia-Escartin, J. C. (2017). Quantum random number generators. *Reviews of Modern Physics*, *89*(1), 015004. doi:10.1103/RevModPhys.89.015004

He, Z., Zhang, R., Ran, N., & Gu, C. (2022). Path planning of multi-type robot systems with time windows based on timed colored Petri nets. *Applied Sciences (Basel, Switzerland)*, *12*(14), 6878. doi:10.3390/app12146878

Hinov, N. L., & Hranov, T. H. (2021). Tolerance analysis of common transistor DC-DC converters. In *2021 25th International Conference Electronics* (pp. 1-6). IEEE. 10.1109/IEEECONF52705.2021.9467442

Hinov, N. L., Gocheva, P. V., & Gochev, V. P. (2022a). Fuzzy reasoning on buck DC-DC power converter parameters. *International Journal on Information Technologies & Security*, *14*(4), 33–44.

Hinov, N., Gocheva, P., & Gochev, V. (2022b). Index matrices-based software implementation of power electronic circuit design. *Electronics (Basel)*, *11*(5), 675. doi:10.3390/electronics11050675

Hoole, J., Mellor, P. H., Simpson, N., & North, D. (2021, May). Statistical simulation of conductor lay and ac losses in multi-strand stator windings. In 2021 IEEE International Electric Machines & Drives Conference (IEMDC) (pp. 1-8). IEEE. doi:10.1109/IEMDC47953.2021.9449582

Huang, B., Zhou, M., Lu, X. S., & Abusorrah, A. (2023). Scheduling of resource allocation systems with timed Petri nets: A Survey. *ACM Computing Surveys*, *55*(11), 1–27. doi:10.1145/3570326

Huang, F. L. (2022). Alternatives to logistic regression models in experimental studies. *Journal of Experimental Education, 90*(1), 213–228. doi:10.1080/00220973.2019.1699769

Huang, J. (2022, May). Fundamental theorems in the probability theory. In *2nd International Conference on Applied Mathematics, Modelling, and Intelligent Computing (CAMMIC 2022)* (Vol. 12259, pp. 126-132). SPIE.

Huo, Y., Chen, Y., & Su, M. (2023, February). Simulink-integrated representation of functional architectures towards simulation of aircraft systems. In *Proceedings of the International Conference on Aerospace System Science and Engineering 2022* (pp. 170-190). Springer, Singapore. https://doi.org/10.1007/978-981-99-0651-2_14

Huser, R., & Wadsworth, J. L. (2022). Advances in statistical modeling of spatial extremes. *Wiley Interdisciplinary Reviews: Computational Statistics, 14*(1), e1537. doi:10.1002/wics.1537

Huttar, C. M., & BrintzenhofeSzoc, K. (2020). Virtual reality and computer simulation in social work education: A systematic review. *Journal of Social Work Education, 56*(1), 131–141. doi:10.1080/10437797.2019.1648221

Hwangbo, S., Heo, S., & Yoo, C. (2022). Development of deterministic-stochastic model to integrate variable renewable energy-driven electricity and large-scale utility networks: Towards decarbonization petrochemical industry. *Energy, 238*, 122006. doi:10.1016/j.energy.2021.122006

Ivanova, Y. (2020). Adaptive digitalization methods and digital transformation trends for security. *International Journal on Information Technologies and Security, 12*(3), 51–62.

Ivanova, Y. (2021). Optimizing simulation models of an artificial neural network for digital recognition. *International Journal on Information Technologies and Security, 13*(4), 59–70.

Ivanova, Y. (2022). Applications of simulation modelling method in prevention of jamming attacks. *Proceedings of the 2022 IEEE International Conference on Information Technologies (InfoTech-2022)*, IEEE. 10.1109/InfoTech55606.2022.9897121

Ivanova, Y. (2022). Simulation modelling of artificial neural networks for the purpose of steganalysis. *International Journal on Information Technologies and Security, 14*(2), 99–110.

Jena, D., & Ramana, V. V. (2015). Modeling of photovoltaic system for uniform and non-uniform irradiance: A critical review. *Renewable & Sustainable Energy Reviews, 52*, 400–417. doi:10.1016/j.rser.2015.07.079

Jeong, Y. S., Oh, K. J., Cho, C. K., & Choi, H. J. (2020). Pseudo-random number generation using LSTMs. *The Journal of Supercomputing, 76*(10), 8324–8342. doi:10.100711227-020-03229-7

Jia, J., & Heragu, S. S. (2009). Solving semi-open queuing networks. *Operations Research, 57*(2), 391–401. doi:10.1287/opre.1080.0627

Jiang, Q., Fu, X., Yan, S., Li, R., Du, W., Cao, Z., Qian, F., & Grima, R. (2021). Neural network aided approximation and parameter inference of non-Markovian models of gene expression. *Nature Communications, 12*(1), 2618. doi:10.103841467-021-22919-1 PMID:33976195

Jin, C. (2020). Global classical solutions and convergence to a mathematical model for cancer cells invasion and metastatic spread. *Journal of Differential Equations, 269*(4), 3987–4021. doi:10.1016/j.jde.2020.03.018

Jitmit, C., & Vatanawood, W. (2021, April). Simulating Artificial Neural Network using hierarchical Colored Petri Nets. In *2021 6th International Conference on Machine Learning Technologies* (pp. 127-131).

Jmaei, A., Slaoui, Y., & Dellagi, W. (2017). Recursive distribution estimator defined by stochastic approximation method using Bernstein polynomials. *Journal of Nonparametric Statistics, 29*(4), 792–805. doi:10.1080/10485252.2017.1369538

Jodayree, M., Abaza, M., & Tan, Q. (2019). A predictive workload balancing algorithm in cloud services. *Procedia Computer Science*, *159*, 902–912. doi:10.1016/j.procs.2019.09.250

Johnson, D. A., Melo, V., & Lambert, J. H. (2022, October). Risk identification with entity attributes diagrams in business process modeling. In *2022 IEEE International Symposium on Systems Engineering (ISSE)* (pp. 1-8). IEEE.

Johnston, D. (2018). *Random Number Generators—Principles and Practices: A Guide for Engineers and Programmers*. Walter de Gruyter GmbH & Co KG., doi:10.1515/9781501506062

Kamsky, A. (2019). Adapting TPC-C benchmark to measure performance of multi-document transactions in MongoDB. *Proceedings of the VLDB Endowment International Conference on Very Large Data Bases*, *12*(12), 2254–2262. doi:10.14778/3352063.3352140

Kang, S., Ko, Y., & Seo, J. (2013). A dialogue management system using a corpus-based framework and a dynamic dialogue transition model. *AI Communications*, *26*(2), 145–159. doi:10.3233/AIC-130552

Karakaya, B. (2022, May). Chaotic system-based pseudo random bit generator and post-processor design for image encryption. In *2022 13th National Conference with International Participation (ELECTRONICA)* (pp. 1-4). IEEE. 10.1109/ELECTRONICA55578.2022.9874431

Karaseva, T., & Semenkin, E. (2021). On the automatic identification of differential equations using a hybrid evolutionary approach. In *2021 IEEE International Conference on Information Technologies (InfoTech-2021)*, (pp. 128-133). IEEE. 10.1109/InfoTech52438.2021.9548643

Karian, Z. A., & Dudewicz, E. J. (2020). Modern statistical, systems, and GPSS simulation (2nd ed.), CRC press, Taylor & Francise group (94 p.). doi:10.1201/9781003067993

Kashiwagi, H. (2009). Nonparametric System Identification. *CONTROL SYSTEMS, ROBOTICS AND AUTOMATION*, *6*, 524.

Katsikeas, S., Hacks, S., Johnson, P., Ekstedt, M., Lagerström, R., Jacobsson, J., Wällstedt, M., & Eliasson, P. (2020, November). An attack simulation language for the IT domain. In *Graphical Models for Security: 7th International Workshop, GraMSec 2020, Boston, MA, USA, June 22, 2020, Revised Selected Papers* (pp. 67-86). *Lecture Notes in Computer Science*. Cham Springer International Publishing. 10.1007/978-3-030-62230-5_4

Katsikeas, S., Johnsson, P., Hacks, S., & Lagerström, R. (2022). VehicleLang: A probabilistic modeling and simulation language for modern vehicle IT infrastructures. *Computers & Security*, *117*, 102705. doi:10.1016/j.cose.2022.102705

Katzfuss, M., Guinness, J., & Lawrence, E. (2022). Scaled Vecchia approximation for fast computer-model emulation. *SIAM/ASA Journal on Uncertainty Quantification*, *10*(2), 537–554. doi:10.1137/20M1352156

Kermanshachi, S., Safapour, E., Anderson, S., Goodrum, P., Taylor, T., & Sadatsafavi, H. (2019, January). Development of multi-level scoping process framework for transportation infrastructure projects using IDEF modeling technique. In *Proceedings of Transportation Research Board 98th Annual Conference*. IEEE.

Khubaev, G. N., Scherbakov, S. M., & Shirobokova, S. N. (2015). Conversion of IDEF3 models into UML-diagrams for the simulation in the SIM system-UML. *European science review*, (11-12), 20-25.

Kleven, H. J. (2021). Sufficient statistics revisited. *Annual Review of Economics*, *13*(1), 515–538. doi:10.1146/annurev-economics-060220-023547

Klimenok, V., Dudin, A., & Vishnevsky, V. (2020). Priority multi-server queueing system with heterogeneous customers. *Mathematics*, *8*(9), 1501. doi:10.3390/math8091501

Kounev, S., Lange, K. D., & von Kistowski, J. (2020). *Systems benchmarking: for scientists and engineers* (Vol. 1). Springer International Publishing. doi:10.1007/978-3-030-41705-5

Krak, I., Barmak, O., & Manziuk, E. (2022). Using visual analytics to develop human and machine-centric models: A review of approaches and proposed information technology. *Computational Intelligence*, *38*(3), 921–946. doi:10.1111/coin.12289

Krak, T., De Bock, J., & Siebes, A. (2017). Imprecise continuous-time Markov chains. *International Journal of Approximate Reasoning*, *88*, 452–528. doi:10.1016/j.ijar.2017.06.012

Kravets, O. J., Atlasov, I. V., Aksenov, I. A., Molchan, A. S., Frantsisko, O. Y., & Rahman, P. A. (2021). Increasing efficiency of routing in transient modes of computer network operation. *International Journal on Information Technologies & Security*, *13*(2), 3–14.

Kravets, O.Ja., Aksenov, I.A. & Rahman, P.A. (2022). Algorithmization of analytical methods for finding motion vectors when processing image series. *International Journal on Information Technologies and Security*, *14*(2), 39–50.

Krishnamurthy, D., Rolia, J. A., & Majumdar, S. (2006). A synthetic workload generation technique for stress testing session-based systems. *IEEE Transactions on Software Engineering*, *32*(11), 868–882. doi:10.1109/TSE.2006.106

Kroese, D. P., & Chan, J. C. (2014). *Statistical modeling and computation*. Springer. doi:10.1007/978-1-4614-8775-3

Kuchárik, M., & Balogh, Z. (2019). Modeling of uncertainty with petri nets. In *Intelligent Information and Database Systems: 11th Asian Conference*, (pp. 499–509). Springer.

Kumar, N. N., Patil, S. R., Srikanth, S., & Ravichandran, S. (2021, September). Feedback controller design for a DC-DC buck converter. In *2021 Fourth International Conference on Electrical, Computer and Communication Technologies (ICECCT)* (pp. 1-5). IEEE.

Kumawat, G. L., & Roy, D. (2021). A new solution approach for multi-stage semi-open queuing networks: An application in shuttle-based compact storage systems. *Computers & Operations Research*, *125*, 105086. doi:10.1016/j.cor.2020.105086

Kung, S. Y. (2020). Wavefront array processors. In *Systolic Signal Processing Systems* (pp. 97–160). CRC Press.

Kuntz, J., Thomas, P., Stan, G. B., & Barahona, M. (2021). Stationary distributions of continuous-time Markov chains: A review of theory and truncation-based approximations. *SIAM Review*, *63*(1), 3–64. doi:10.1137/19M1289625

Kürkçü, Ö. K., Aslan, E., & Sezer, M. (2017). A numerical method for solving some model problems arising in science and convergence analysis based on residual function. *Applied Numerical Mathematics*, *121*, 134–148. doi:10.1016/j.apnum.2017.06.015

Kurniawan, N. B., Bandung, Y., & Yustianto, P. (2019). Services computing systems engineering framework: A proposition and evaluation through soa principles and analysis model. *IEEE Systems Journal*, *14*(3), 3105–3116. doi:10.1109/JSYST.2019.2939433

Kushner, H. J., & Clark, D. S. (2012). *Stochastic approximation methods for constrained and unconstrained systems* (Vol. 26). Springer Science & Business Media.

Kuznetsov, A. L., Kirichenko, A. V., Semenov, A. D., & Oja, H. (2020). Planning simulation experiments in the tasks of studying the operational strategies of container terminals. *TransNav, the International Journal on Marine Navigation and Safety of Sea Transportation*, *14*(4), 845-849.

Lacotte, J., & Pilanci, M. (2020, November). Optimal randomized first-order methods for least-squares problems. In *International Conference on Machine Learning* (pp. 5587-5597).

Lanchier, N. (2017). *Stochastic modeling*. Springer. doi:10.1007/978-3-319-50038-6

Landau, D., & Binder, K. (2021). *A guide to Monte Carlo simulations in statistical physics*. Cambridge University Press. doi:10.1017/9781108780346

Laomettachit, T., Kraikivski, P., & Tyson, J. J. (2022). A continuous-time stochastic Boolean model provides a quantitative description of the budding yeast cell cycle. *Scientific Reports, 12*(1), 20302. doi:10.103841598-022-24302-6 PMID:36434030

Law, A. M. (2019, December). How to build valid and credible simulation models. In 2019 Winter Simulation Conference (WSC), USA (pp. 1402-1414). IEEE. doi:10.1109/WSC40007.2019.9004789

Lee, S., Dudin, A., Dudina, O., & Kim, C. (2022). Analysis of a priority queueing system with the enhanced fairness of servers scheduling. *Journal of Ambient Intelligence and Humanized Computing*, 1–13. doi:10.100712652-022-03903-z

Lemiale, V., Huston, C., Mead, S., Alexander, D. L., Cleary, P. W., Adhikary, D., & Delaney, G. W. (2022). Combining statistical design with deterministic modelling to assess the effect of site-specific factors on the extent of landslides. *Rock Mechanics and Rock Engineering, 55*(1), 259–273. doi:10.100700603-021-02674-x

Li, K., Marston, J. B., & Tobias, S. M. (2021). Direct statistical simulation of low order dynamosystems. *Proceedings of the Royal Society A (Mathematical, Physica; and Engineering Sciences), 477*(2254), 20210427. 10.1098/rspa.2021.0427

Liang, B., Gregory, M. A., & Li, S. (2022). Multi-access Edge Computing fundamentals, services, enablers, and challenges: A complete survey. *Journal of Network and Computer Applications, 199*, 103308. doi:10.1016/j.jnca.2021.103308

Li, K., Yu, R., Liu, Y., Wang, J., & Xue, W. (2023). Correlation analysis and modeling of human thermal sensation with multiple physiological markers: An experimental study. *Energy and Building, 278*, 112643. doi:10.1016/j.enbuild.2022.112643

Likhttsinder, B. (2017, October). Conditional average value of queues in queuing systems with bath request flows. In *2017 4th International Scientific-Practical Conference Problems of Infocommunications. Science and Technology (PIC S&T)*, IEEE. 10.1109/INFOCOMMST.2017.8246347

Likhttsinder, B. Y., & Bakai, Y. O. (2022, March). *Development of an interval method for queue analysis in queueing systems. In 2022 Systems of Signals Generating and Processing in the Field of on Board Communications*. IEEE.

Likhttsinder, B. Y., Blatov, I. A., & Kitaeva, E. V. (2022). On estimates of the mean queue length for single-channel queuing systems in terms of statistical unconditional second-order moments of the modified arrival flow. *Automation and Remote Control, 83*(1), 92–105. doi:10.1134/S0005117922010076

Lipsett, R., Schaefer, C. F., & Ussery, C. (2012). *VHDL: Hardware description and design*. Springer Science & Business Media.

Liu, B., Ma, J., Tai, H. H., Verma, D., Sahoo, M., Chang, Y. F., Liang, H., Feng, S., Li, L.-H., Hou, T.-H., & Lai, C. S. (2023). Memristive True Random Number Generator with intrinsic two-dimensional physical unclonable function. *ACS Applied Electronic Materials, 5*(2), 714–720. doi:10.1021/acsaelm.2c01533

Liu, H. C., You, J. X., Li, Z., & Tian, G. (2017). Fuzzy Petri nets for knowledge representation and reasoning: A literature review. *Engineering Applications of Artificial Intelligence, 60*, 45–56. doi:10.1016/j.engappai.2017.01.012

Liu, X., Zhao, M., Wei, Z., & Lu, M. (2022). The energy management and economic optimization scheduling of microgrid based on Colored Petri net and Quantum-PSO algorithm. *Sustainable Energy Technologies and Assessments, 53*, 102670. doi:10.1016/j.seta.2022.102670

Liu, Y., Zhang, S., Mu, X., Ding, Z., Schober, R., Al-Dhahir, N., Hossain, E., & Shen, X. (2022). Evolution of NOMA toward next generation multiple access (NGMA) for 6G. *IEEE Journal on Selected Areas in Communications*, *40*(4), 1037–1071. doi:10.1109/JSAC.2022.3145234

Li, X., Stanwicks, P., Provelengios, G., Tessier, R., & Holcomb, D. (2023). Jitter-based adaptive True Random Number Generation circuits for FPGAs in the cloud. *ACM Transactions on Reconfigurable Technology and Systems*, *16*(1), 1–20. doi:10.1145/3487554

López-Santos, O., Salas-Castaño, M. C., & Salazar-Dantonio, D. F. (2022). Continuous simulation of the power flow in AC–DC hybrid microgrids using simplified modelling. *Computation (Basel, Switzerland)*, *10*(4), 52. doi:10.3390/computation10040052

Loquercio, A., Kaufmann, E., Ranftl, R., Dosovitskiy, A., Koltun, V., & Scaramuzza, D. (2019). Deep drone racing: From simulation to reality with domain randomization. *IEEE Transactions on Robotics*, *36*(1), 1–14. doi:10.1109/TRO.2019.2942989

Lúcio, A. D. C., & Sari, B. G. (2017). Planning and implementing experiments and analyzing experimental data in vegetable crops: Problems and solutions. *Horticultura Brasileira*, *35*(3), 316–327. doi:10.15900102-053620170302

Luengo, E. A. (2022). Gamma Pseudo Random Number Generators. *ACM Computing Surveys*, *55*(4), 1–33. doi:10.1145/3527157

Lvovich, K. I., Preobrazhenskii, A. P., Choporov, O. N., Aksenov, I. A., & Ivaschenko, A. V. (2021). Modelling of optimization process of personnel adaptation to digital management in organizational systems. *International Journal on Information Technologies and Security*, *13*(4), 71–82.

Madan, K. C. (2011). A non-preemptive priority queueing system with a single server serving two queues M/G/1 and M/D/1 with optional server vacations based on exhaustive service of the priority units. *Applied Mathematics*, *2*(06), 791–799. doi:10.4236/am.2011.26106

Magalhães, D., Calheiros, R. N., Buyya, R., & Gomes, D. G. (2015). Workload modeling for resource usage analysis and simulation in cloud computing. *Computers & Electrical Engineering*, *47*, 69–81. doi:10.1016/j.compeleceng.2015.08.016

Magfirawaty, M., Lestari, A. A., Nurwa, A. R. A., Mt, S., & Ramli, K. (2022). A novel discrete-time chaos-function-based random-number generator: Design and variability analysis. *Symmetry*, *14*(10), 2122. doi:10.3390ym14102122

Mansouri, N., Ghafari, R., & Zade, B. M. H. (2020). Cloud computing simulators: A comprehensive review. *Simulation Modelling Practice and Theory*, *104*, 102144. doi:10.1016/j.simpat.2020.102144

Martyshkin, A. I., Pashchenko, D. V., Trokoz, D. A., Sinev, M. P., & Svistunov, B. L. (2020). Using queuing theory to describe adaptive mathematical models of computing systems with resource virtualization and its verification using a virtual server with a configuration similar to the configuration of a given model. *Bulletin of Electrical Engineering and Informatics*, *9*(3), 1106–1120. doi:10.11591/eei.v9i3.1714

Marzolla, M. (2022). A Software package for queueing networks and Markov chains analysis. *arXiv preprint arXiv:2209.04220.*

Massini, A. (2017). *Advanced parallel architectures*. Uni Romal. (https://twiki.di.uniroma1.it/pub/AAP/WebHome/2017-lesson14-Multiproc-InterconnectionNetworks.pdf)

Matyushenko, S., & Ermolayeva, A. (2021). On stationary characteristics of a multiserver exponential queuing system with reordering of requests. In *2021 13th International Congress on Ultra Modern Telecommunications and Control Systems and Workshops (ICUMT)*, Brno, Czech Republic, pp. 98-103. 709.10.1109/ICUMT54235.2021.9631709

McIntire, M. G., Keshavarzi, E., Tumer, I. Y., & Hoyle, C. (2016). Functional models with inherent behavior: Towards a framework for safety analysis early in the design of complex systems. In *ASME International Mechanical Engineering Congress and Exposition,* Volume 11*: Systems, Design, and Complexity*. IEEE. 10.1115/IMECE2016-67040

Mechta, A., Slamani, M., Zaoui, M., Mayer, R., & Chatelain, J. F. (2022). Correlation assessment and modeling of intra-axis errors of prismatic axes for CNC machine tools. *International Journal of Advanced Manufacturing Technology, 120*(7-8), 5093–5115. doi:10.100700170-022-09074-7

Mehmood, S., & Ochs, P. (2020, June). Automatic differentiation of some first-order methods in parametric optimization. In *International Conference on Artificial Intelligence and Statistics* (pp. 1584-1594). Proceedings of Machine Learning Research (PMLR).

Mehta, R. (2022). Discrete-time simulation for performance modelling of FIFO single-server queuing system. *International Journal of Systems. Control and Communications, 13*(2), 112–132.

Mishra, P., Kumar, S., & Chaube, M. K. (2023). Graph interpretation, summarization and visualization techniques: A review and open research issues. *Multimedia Tools and Applications, 82*(6), 8729–8771. doi:10.100711042-021-11582-9

Mkhinini, M. M., Labbani-Narsis, O., & Nicolle, C. (2020). Combining UML and ontology: An exploratory survey. *Computer Science Review, 35*, 100223. doi:10.1016/j.cosrev.2019.100223

Mohan, N., Undeland, T. M., & Robbins, W. P. (2003). *Power electronics: converters, applications, and design.* John wiley & sons.

Moiseeva, S. P., Bushkova, T. V., Pankratova, E. V., Farkhadov, M. P., & Imomov, A. A. (2022). Asymptotic analysis of resource heterogeneous QS under equivalently increasing service time. *Automation and Remote Control, 83*(8), 1213–1227. doi:10.1134/S0005117922080057

Montáns, F. J., Chinesta, F., Gómez-Bombarelli, R., & Kutz, J. N. (2019). Data-driven modeling and learning in science and engineering. *Comptes Rendus. Mécanique, 347*(11), 845–855. doi:10.1016/j.crme.2019.11.009

Montgomery, D. C., Peck, E. A., & Vining, G. G. (2021). *Introduction to linear regression analysis*. John Wiley & Sons.

Mora, M., Adelakun, O., Galvan-Cruz, S., & Wang, F. (2022). Impacts of IDEF0-based models on the usefulness, learning, and value metrics of Scrum and XP project management guides. *Engineering Management Journal, 34*(4), 574–590. doi:10.1080/10429247.2021.1958631

Morris, T. P., White, I. R., & Crowther, M. J. (2019). Using simulation studies to evaluate statistical methods. *Statistics in Medicine, 38*(11), 2074–2102. doi:10.1002im.8086 PMID:30652356

Muminov, B. B., & Bekmurodov, U. B. (2020, October). IDEF models and innovative system for search data in stochastic information environment. *2020 IEEE 14th International Conference on Application of Information and Communication Technologies (AICT),* (pp. 1-6). IEEE.

Muminov, B. B., & Bekmurodov, U. B. (2020, October). IDEF models and innovative system for search data in stochastic information environment. In *2020 IEEE 14th International Conference on Application of Information and Communication Technologies (AICT)* (pp. 1-6). IEEE. 10.1109/AICT50176.2020.9368581

Munir, S., Jami, S. I., & Wasi, S. (2020). Knowledge graph based semantic modelling for profiling in Industry 4.0. *International Journal on Information Technologies and Security, 12*(1), 37–50.

Naderkhani, R., Behzad, M. H., Razzaghnia, T., & Farnoosh, R. (2021). Fuzzy regression analysis based on fuzzy neural networks using trapezoidal data. *International Journal of Fuzzy Systems, 23*(5), 1267–1280. doi:10.100740815-020-01033-2

Naik, R. B., & Singh, U. (2022). A review on applications of chaotic maps in pseudo-random number generators and encryption. *Annals of Data Science*, 1-26. doi:10.1007/s40745-021-00364-7

Nai-Zhi, G., Ming-Ming, Z., & Bo, L. (2022). A data-driven analytical model for wind turbine wakes using machine learning method. *Energy Conversion and Management*, *252*, 115130. doi:10.1016/j.enconman.2021.115130

Najim, A. H., Mansour, H. S., & Abbas, A. H. (2022). Characteristic analysis of queue theory in Wi-Fi applications using OPNET 14.5 modeler. *Eastern-European Journal of Enterprise Technologies, 2*(9), 116, 35-46. doi:10.15587/1729-4061.2022.255520

Napoli, C., & García-Téllez, B. (2016). *Energy for water in agriculture: A partial factor productivity analysis. King Abdullah Petroleum Studies and Research Center*, (KAPSARC).

Narang, P., & Mittal, P. (2022). Performance analysis of DevOps based hybrid models integrated with different automation tool chains for quality software development. *International Journal on Information Technologies and Security*, *14*(4), 25–32.

Nassar, H. M., & Tekian, A. (2020). Computer simulation and virtual reality in undergraduate operative and restorative dental education: A critical review. *Journal of Dental Education*, *84*(7), 812–829. doi:10.1002/jdd.12138 PMID:32147841

Nelles, O. (2020). *Nonlinear system identification: from classical approaches to neural networks, fuzzy models, and gaussian processes*. Springer Nature. doi:10.1007/978-3-030-47439-3

Nikolov, N. (2004). The SIM2000 simulation package for generalized nets: Architecture and Language. Issues in Intuitionistic Fuzzy Sets and Generalized Nets. WSISiZ, Warsaw.

Nooribakhsh, M., & Mollamotalebi, M. (2020). A review on statistical approaches for anomaly detection in DDoS attacks. *Information Security Journal: A Global Perspective, 29*(3), 118-133.

Noprisson, H. (2020). The business failure prediction using statistical approach. *International Journal of Scientific Research in Science, Engineering and Technology*, *7*(5), 161–168.

Nora, S. & Minc, A. (1978). *L'informatisation de la société* (*Vol. 11*). Paris: La documentation française.

Oleinikova, S.A., Selishchev, I.A., & Kravets, O.Ja., Rahman, P.A., & Aksenov, I.A. (2021). Simulation model for calculating the probabilistic and temporal characteristics of the project and the risks of its untimely completion. *International Journal on Information Technologies and Security*, *13*(2), 55–62.

Om Parkash, M. (2021). Two new parametric entropic models for discrete probability distributions. [TURCOMAT]. *Turkish Journal of Computer and Mathematics Education*, *12*(6), 2949–2954.

Oreshina, O. A. (2020, April). Full factor plan application to polymer composites hardness investigation. [IOP Publishing.]. *Journal of Physics: Conference Series*, *1515*(4), 042031. doi:10.1088/1742-6596/1515/4/042031

Ozkaya, M., & Erata, F. (2020). A survey on the practical use of UML for different software architecture viewpoints. *Information and Software Technology*, *121*, 106275. doi:10.1016/j.infsof.2020.106275

Panda, R., Song, S., Dean, J., & John, L. K. (2018, February). Wait of a decade: Did SPEC CPU 2017 broaden the performance horizon? In *2018 IEEE International Symposium on High Performance Computer Architecture (HPCA)* (pp. 271-282). IEEE. 10.1109/HPCA.2018.00032

Pandit, R. A., Waghmare, S. A., & Bhagat, P. M. (2022). History of Probability Theory. *Journal of Social Sciences and Humanities*, *4*(5), 140–142. doi:10.53469/jssh.2022.4(05).29

Parker, W. S. (2020). Model evaluation: An adequacy-for-purpose view. *Philosophy of Science*, *87*(3), 457–477. doi:10.1086/708691

Park, S. W., Lee, K., & Kwon, J. (2022). Neural markov controlled SDE: Stochastic optimization for continuous-time data. In *International Conference on Learning Representations (ICLR 2022)*. IEEE.

Parrot, R., Briday, M., & Roux, O. H. (2023). Design and verification of pipelined circuits with Timed Petri Nets. *Discrete Event Dynamic Systems*, *33*(1), 1–24. doi:10.100710626-022-00371-7

Paskalovski, S., & Digalovski, M. (2022). Simulation models for induction machine protection analysis. *International Journal on Information Technologies and Security*, *14*(2), 63–74.

Pasqualini, L., & Parton, M. (2020). Pseudo random number generation: A reinforcement learning approach. *Procedia Computer Science*, *170*, 1122–1127. doi:10.1016/j.procs.2020.03.057

Patcharaprakiti, N., Kirtikara, K., Tunlasakun, K., Thongpron, J., Chenvidhya, D., Sangswang, A., & Muenpinij, B. (2011). Modeling of photovoltaic grid connected inverters based on nonlinear system identification for power quality analysis. *Electrical Generation and Distribution Systems and Power Quality Disturbances*, 53-82.

Pazos, S., Zheng, W., Zanotti, T., Aguirre, F., Becker, T., Shen, Y., Zhu, K., Yuan, Y., Wirth, G., Puglisi, F. M., Roldán, J. B., Palumbo, F., & Lanza, M. (2023). Hardware implementation of a true random number generator integrating a hexagonal boron nitride memristor with a commercial microcontroller. *Nanoscale*, *15*(5), 2171–2180. doi:10.1039/D2NR06222D PMID:36628646

Péan, F., & Goksel, O. (2020). Surface-based modeling of muscles: Functional simulation of the shoulder. *Medical Engineering & Physics*, *82*, 1–12. doi:10.1016/j.medengphy.2020.04.010 PMID:32709260

Pelivani, E., Besimi, A., & Cico, B. (2022). An empirical study of user interface testing tools. *International Journal on Information Technologies and Security*, *14*(1), 37–48.

Pereira, P., Araujo, J., Melo, C., Santos, V., & Maciel, P. (2021). Analytical models for availability evaluation of edge and fog computing nodes. *The Journal of Supercomputing*, *77*(9), 9905–9933. doi:10.100711227-021-03672-0

Pine, D. J. (2019). *Introduction to Python for science and engineering*. CRC press. doi:10.1201/9780429506413

Popov, G. (2017). GPSS language as tool for reliability simulations. *2017 15th International Conference on Electrical Machines, Drives and Power Systems (ELMA)*, Sofia, Bulgaria, 2017, pp. 461-463, 10.1109/ELMA.2017.7955486

Popov, S., & Hinov, N. (2023). Software implementation of PSPISE-based models in the Python programming language. *International Journal on Information Technologies and Security*, *15*(3), 65–74. doi:10.59035/LYWQ6789

Porat, M. U. (1977). The Information Economy: Sources and Methods for Measuring the Primary Information Sector (Detailed Industry Reports). *The Office*, *77*(12).

Porshnev, S., & Korelin, I. (2017). Non-stationary single-channel queuing system features research in context of number of served queries. *ITM Web of Conferences (2017 Seminar on Systems Analysis) 10*, 03006. DOI: 10.1051/itmconf/20171003006

Privault, N. (2018). *Discrete-time Markov chains. Understanding Markov Chains: Examples and Applications*. Springer. doi:10.1007/978-981-13-0659-4

Pushpalatha, S. S., & Math, S. (2022). Human activity recognition using hybrid model. *International Journal on Information Technologies and Security*, *14*(4), 55–66.

Raghupathi, W., Raghupathi, V., & Ren, J. (2022). Reproducibility in computing research: An empirical study. *IEEE Access : Practical Innovations, Open Solutions*, *10*, 29207–29223. doi:10.1109/ACCESS.2022.3158675

Ram, J. P., Manghani, H., Pillai, D. S., Babu, T. S., Miyatake, M., & Rajasekar, N. (2018). Analysis on solar PV emulators: A review. *Renewable & Sustainable Energy Reviews*, *81*, 149–160. doi:10.1016/j.rser.2017.07.039

Ramos, T. M. F., Almeida, A. A., & Ayala-Rincón, M. (2022). Formalization of the computational theory of a Turing complete functional language model. *Journal of Automated Reasoning*, *66*(4), 1031–1063. doi:10.100710817-021-09615-x

Rao, A. R., & Reimherr, M. (2023). Non-linear functional modeling using neural networks. *Journal of Computational and Graphical Statistics*, 1–20. doi:10.1080/10618600.2023.2165498

Rathi, S., Nagpal, R., Mehrotra, D., & Srivastava, G. (2022). A metric focused performance assessment of fog computing environments: A critical review. *Computers & Electrical Engineering*, *103*(October), 108350. doi:10.1016/j.compeleceng.2022.108350

Razzaq, M. A., Mahar, J. A., Ahmad, M., Saher, N., Mehmood, A., & Choi, G. S. (2021). Hybrid auto-scaled service-cloud-based predictive workload modeling and analysis for smart campus system. *IEEE Access : Practical Innovations, Open Solutions*, *9*, 42081–42089. doi:10.1109/ACCESS.2021.3065597

Reinhardt, O., Warnke, T., & Uhrmacher, A. M. (2022). A language for agent-based discrete-event modeling and simulation of linked lives. [TOMACS]. *ACM Transactions on Modeling and Computer Simulation*, *32*(1), 1–26. doi:10.1145/3486634

Riesebos, L., & Brown, K. R. (2022, September). Functional simulation of real-time quantum control software. In *2022 IEEE International Conference on Quantum Computing and Engineering (QCE)* (pp. 535-544). IEEE. 10.1109/QCE53715.2022.00076

Ritha, W., & Yasodai, P. (2021). Exploration of fuzzy preemptive-resume priority queuing system using robust ranking method. *International Journal of Aquatic Science*, *12*(2), 3224–3230.

Romanov, O., Nikolaev, S., & Orliuk, Y. (2022). Radio monitoring complex model as multi-phase queuing system. *Radioelectronics and Communications Systems*, *65*(3), 155–164. doi:10.3103/S0735272722030050

Romansky, R. (2020b). Formalization and discrete modelling of communication in the digital age by using graph theory. In M. Pal, S. Samanta, & A. Pal (eds) Handbook of Research on Advanced Applications of Graph Theory in Modern Society. IGI Global. doi:10.4018/978-1-5225-9380-5.ch013

Romansky, R. (2020c). Formalization and discrete modelling of communication in the digital age by using graph theory. In M. Pal, S. Samanta, & A. Pal (eds) Handbook of Research on Advanced Applications of Graph Theory in Modern Society. IGI Global. doi:10.4018/978-1-5225-9380-5.ch013

Romansky, R. (2023). Mathematical model investigation of a technological structure for personal data protection. Axioms, 12(2), 102. doi:10.3390/axioms12020102

Romansky, R., & Noninska, I. (2021b). Deterministic modelling of a management system with protected access to virtual resources. AIP Conference Proceedings (American Institute of Physics), 2333, art.090003, 090003-1-090003-10 () doi:10.1063/5.0041760

Romansky, R. (2017). *Information servicing in distributed learning environments. Formalization and model investigation.* LAP LAMBERT Academic Publishing.

Romansky, R. (2020a). An approach for mathematical modelling and investigation of computer processes at a macro level. *Mathematics*, *8*(10), 1838. doi:10.3390/math8101838

Romansky, R. (2021a). Informatization of the society in the digital age. *Biomedical Journal of Scientific & Technical Research*, *33*(3), 25902–25910. doi:10.26717/BJSTR.2021.33.005418

Romansky, R. (2021b). Mathematical modelling and study of stochastic parameters of computer data processing. *Mathematics*, *9*(18), 2240. doi:10.3390/math9182240

Romansky, R. (2021c). Program environment for investigation of micro-level computer processing. *International Journal on Information Technologies and Security*, *13*(1), 83–92.

Romansky, R. (2021c). Program environment for investigation of micro-level computer processing. *International Journal on IT and Security*, *13*(1), 83–92.

Romansky, R. (2022a). Stochastic approach to investigate protected access to information resources in combined e-learning environment. *Mathematics*, *10*(16), 2909. doi:10.3390/math10162909

Romansky, R. (2022b). Evaluation of experimental data from monitoring and simulation of network communication parameters. *International Journal on Information Technologies and Security*, *14*(2), 75–86.

Romansky, R. (2022c). Statistical analysis of empirical network traffic data from program monitoring. *International Journal on Information Technologies and Security*, *14*(3), 15–24.

Romansky, R. (2022d). Investigation of network communications by using statistical processing of monitored data. *2022 IEEE International Conference on Information Technologies (InfoTech-2022)*, (pp. 37-40). IEEE. 10.1109/InfoTech55606.2022.9897115

Romansky, R. (2022e). Formalization and investigation of parallel processes dispatching. *2022 IEEE International Conference on Information Technologies (InfoTech-2022)*, (pp. 94-97). IEEE. 10.1109/InfoTech55606.2022.9897104

Romansky, R. (2022f). An approach for program investigation of computer processes presented by Markov models. *International Journal on Information Technologies and Security*, *14*(4), 45–54.

Romansky, R. (2023). Empirical evaluation of the transfer of information resources in active learning. *International Journal on Information Technologies and Security*, *15*(1), 39–48. doi:10.59035/WZFU1905

Romansky, R., & Noninska, I. (2016). Discrete formalization and investigation of secure access to corporative resources. *International Journal of Engineering Research and Management*, *13*(5), 97–101.

Romansky, R., & Noninska, I. (2019). Technological organization of the access management to information resources in a combined e-learning environment. *International Journal on Information Technologies and Security*, *11*(4), 51–62.

Romansky, R., & Noninska, I. (2019b). Cyber space features – security and data protection requirements. *Proceedings of the 2019 IEEE International Conference on Information Technologies (InfoTech-2019)*. IEEE. 10.1109/InfoTech.2019.8860880

Romansky, R., & Noninska, I. (2021a). Investigation of communication parameters in multicomputer architecture with ring topology. *Proceedings of the 2021 IEEE International Conference on Information Technologies (InfoTech-2021)*, 16-17 Sept 2021, Bulgaria, pp. 119-123. 10.1109/InfoTech52438.2021.9548514

Romansky, R., & Noninska, I. (2022). Deterministic model investigation of processes in a heterogeneous e-learning environment. *International Journal of Human Capital and Information Technology Professionals*, *13*(1), 1–16. doi:10.4018/IJHCITP.293228

Roth, C. H. Jr, & John, L. K. (2016). *Digital systems design using VHDL*. Cengage Learning.

Rotunno, G., Lo Zupone, G., Carnimeo, L., & Fanti, M. P. (2023). Discrete event simulation as a decision tool: A cost benefit analysis case study. *Journal of Simulation*, 1–17. doi:10.1080/17477778.2023.2167618

Roy, D. (2016). Semi-open queuing networks: A review of stochastic models, solution methods and new research areas. *International Journal of Production Research*, *54*(6), 1735–1752. doi:10.1080/00207543.2015.1056316

Rumpe, B. (2016). *Modeling with UML*. Springer. doi:10.1007/978-3-319-33933-7

Ryan, O., & Hamaker, E. L. (2022). Time to intervene: A continuous-time approach to network analysis and centrality. *Psychometrika*, *87*(1), 214–252. doi:10.100711336-021-09767-0 PMID:34165691

Saatkamp, K., Breitenbücher, U., Kopp, O., & Leymann, F. (2020). Method, formalization, and algorithms to split topology models for distributed cloud application deployments. *Computing*, *102*(2), 343–363. doi:10.100700607-019-00721-8

Sadrossadat, S. A., & Rahmani, O. (2020). ANN-based method for parametric modelling and optimizing efficiency, output power and material cost of BLDC motor. *IET Electric Power Applications*, *14*(6), 951–960. doi:10.1049/iet-epa.2019.0686

Safyannikov, N., Chepasov, A., & Bondarenko, P. (2021). *Functional organization of elements of stream converters with actualization of states*. 10th Mediterranean Conference on Embedded Computing (MECO), Budva, Montenegro. ()10.1109/MECO52532.2021.9460167

Saritha, S., Mamatha, E., Reddy, C. S., & Rajadurai, P. (2022). A model for overflow queuing network with two-station heterogeneous system. *International Journal of Process Management and Benchmarking*, *12*(2), 147–158. doi:10.1504/IJPMB.2022.121592

Schneider, A. (2016). *Modelling of data uncertainties on hybrid computers* (No. GRS-392). Gesellschaft fuer Anlagen-und Reaktorsicherheit (GRS) gGmbH.

Schoukens, J., Godfrey, K., & Schoukens, M. (2018). Nonparametric data-driven modeling of linear systems: Estimating the frequency response and impulse response function. *IEEE Control Systems*, *38*(4), 49–88. doi:10.1109/MCS.2018.2830080

Schoukens, J., & Ljung, L. (2019). Nonlinear system identification: A user-oriented road map. *IEEE Control Systems*, *39*(6), 28–99. doi:10.1109/MCS.2019.2938121

Seifermann, S., Heinrich, R., Werle, D., & Reussner, R. (2022). Detecting violations of access control and information flow policies in data flow diagrams. *Journal of Systems and Software*, *184*, 111138. doi:10.1016/j.jss.2021.111138

Shirsat, A., Elamvazhuthi, K., & Berman, S. (2020, November). Multi-robot target search using probabilistic consensus on discrete Markov chains. In *2020 IEEE International Symposium on Safety, Security, and Rescue Robotics (SSRR)* (pp. 108-115). IEEE. 10.1109/SSRR50563.2020.9292589

Shreider, Y. A. (Ed.). (2014). *The Monte Carlo method: the method of statistical trials* (Vol. 87). Elsevier.

Shrestha, N. (2020). Detecting multicollinearity in regression analysis. *American Journal of Applied Mathematics and Statistics*, *8*(2), 39–42. doi:10.12691/ajams-8-2-1

Shukla, S. (2022). Developing pragmatic data pipelines using Apache Airflow on Google Cloud Platform. *International Journal on Computer Science and Engineering*, *10*(8), 1–8. doi:10.26438/ijcse/v10i8.18

Shukla, S. (2023a). Unlocking the power of data: An introduction to data analysis in healthcare. *International Journal on Computer Science and Engineering*, *11*(3), 1–9. doi:10.26438/ijcse/v11i3.19

Shukla, S. (2023b). Real-time monitoring and predictive analytics in healthcare: Harnessing the power of data streaming. *International Journal of Computer Applications*, *185*(8), 32–37. doi:10.5120/ijca2023922738

Shukla, S. (2023c). Enhancing healthcare insights, exploring diverse use-cases with K-means clustering. *International Journal of Management IT and Engineering*, *13*(8), 60–68.

Shukla, S. (2023d). Streamlining integration testing with test containers: Addressing limitations and best practices for implementation. [IJLEMR]. *International Journal of Latest Engineering and Management Research*, *8*(3), 19–26. doi:10.56581/IJLEMR.8.3.19-26

Silva, J. R., & Del Foyo, P. M. (2012). Timed Petri nets. In Petri Nets: Manufacturing and Computer Science (pp. 359-378). InTech.

Sinha, K., & Trivedi, Y. N. (2022). Spectrum sensing based on two state discrete time Markov chain in additive Laplacian noise. *Wireless Networks*, *28*(6), 2393–2402. doi:10.100711276-022-02979-x

Sopov, E., & Semenkina, M. (2021). Genetic programming using cooperative coevolution and problem decomposition for solving large-scale symbolic Regression problems. *2021 IEEE International Conference on Information Technologies (InfoTech-2021)*, (pp. 124-127). IEEE. 10.1109/InfoTech52438.2021.9548435

Spanidis, P. M., Pavloudakis, F., & Roumpos, C. (2021). Introducing the IDEF0 methodology in the strategic planning of projects for reclamation and repurposing of surface mines. *Materials Proceedings*, *5*(1), 26.

Spanidis, P. M., Roumpos, C., & Pavloudakis, F. (2022). A Methodology combining IDEF0 and weighted risk factor analysis for the strategic planning of mine reclamation. *Minerals (Basel)*, *12*(6), 713. doi:10.3390/min12060713

SPEC. (2022*). SPEC's Benchmarks and Tools*. SPEC. https://www.spec.org/benchmarks.html

Staffolani, A., Darvariu, V. A., Bellavista, P., & Musolesi, M. (2023). RLQ: Workload allocation with reinforcement learning in distributed queues. *IEEE Transactions on Parallel and Distributed Systems*, *34*(3), 856–868. doi:10.1109/TPDS.2022.3231981

Stahl, I. (2009). *WebGPSS*. Beliber AB. http://www.webgpss.com/

Stankevich, E., Tananko, I., & Pagano, M. (2022). Optimization of open queuing networks with batch services. *Mathematics*, *10*(16), 3027. doi:10.3390/math10163027

Stoffels, G., & Hohmann, S. (2022). Comparison: Stochastics with a focus on Probability Theory. In *Comparison of Mathematics and Physics Education II: Examples of Interdisciplinary Teaching at School* (pp. 277–297). Springer Fachmedien Wiesbaden. doi:10.1007/978-3-658-36415-1_20

Stone, D. (2015). *The Difference between Multiprocessors & Multicomputer Systems*, Techwalla (https://www.techwalla.com/articles/the-difference-between-multiprocessor-multicomputer-systems)

Su, J., & Schön, S. (2022, April). Deterministic approaches for bounding GNSS uncertainty: A comparative analysis. In *2022 10th workshop on satellite navigation technology (NAVITEC)* (pp. 1-8). IEEE. 10.1109/NAVITEC53682.2022.9847545

Sukhorukova, I., & Chistyakova, N. (2020). Methodology for the formation of a special course on applications of Markov processes. *Revista ESPACIOS, 41*(09).

Swarup, S. (2019, April). Adequacy: what makes a simulation good enough? In 2019 Spring Simulation Conference (SpringSim) (pp. 1-12). IEEE. doi:10.23919/SpringSim.2019.8732895

Sweilam, N. H., Al-Mekhlafi, S. M., Hassan, S. M., Alsenaideh, N. R., & Radwan, A. E. (2022). Numerical treatments for some stochastic–deterministic chaotic systems. *Results in Physics*, *38*, 105628. doi:10.1016/j.rinp.2022.105628

Taifa, I. W., Hayes, S. G., & Stalker, I. D. (2020, June). Computer modelling and simulation of an equitable order distribution in manufacturing through the Industry 4.0 framework. *2020 IEEE International Conference on Electrical, Communication, and Computer Engineering,* (pp. 1-6). IEEE. 10.1109/ICECCE49384.2020.9179275

Tangirala, A. K. (2018). *Principles of system identification: theory and practice*. Crc Press. doi:10.1201/9781315222509

Tan, Z., Chai, M., Chen, D., Liao, J., Chu, Q., Liu, B., Hua, G., & Yu, N. (2021). Diverse semantic image synthesis via probability distribution modeling. In *Proceedings of the IEEE/CVF Conference on Computer Vision and Pattern Recognition* (pp. 7962-7971). IEEE. 10.1109/CVPR46437.2021.00787

Tatarnikova, T., Sikarev, I., Karetnikov, V., & Butsanets, A. (2021). Statistical research and modeling network traffic. *E3S Web of Conferences, 244*, 07002.

Tatashev, A. G., Seleznjev, O. V., & Yashina, M. V. (2022). Approximate Formulas for Characteristics of Multichannel LIFO Preemptive-Resume Priority Queueing System. arXiv preprint arXiv:2206.09263 (5 p.).

Teren, V., Cortadella, J., & Villa, T. (2022, August). Decomposition of transition systems into sets of synchronizing Free-choice Petri Nets. In *2022 25th Euromicro Conference on Digital System Design (DSD)* (pp. 165-173). IEEE. 10.1109/DSD57027.2022.00031

Tian, P., Shukla, A., Nie, L., Zhan, G., & Liu, S. (2018). Characteristics' relation model of asphalt pavement performance based on factor analysis. *International Journal of Pavement Research and Technology, 11*(1), 1–12. doi:10.1016/j.ijprt.2017.07.007

To, H., Asghari, M., Deng, D., & Shahabi, C. (2016). SCAWG: A toolbox for generating synthetic workload for spatial crowdsourcing. In *2016 IEEE International Conference on Pervasive Computing and Communication Workshops (PerCom Workshops)*. IEEE. 10.1109/PERCOMW.2016.7457121

Tomov, Z., Krawczak, M., Andonov, V., Atanassov, K., & Simeonov, S. (2019). Generalized net models of queueing disciplines in finite buffer queueing systems with intuitionistic fuzzy evaluations of the tasks. *Notes on Intuitionistic Fuzzy Sets, 25*(2), 115–122. doi:10.7546/nifs.2019.25.2.115-122

Tsaregorodtsev, A. V., Kravets, O. J., Choporov, O. N., & Zelenina, A. N. (2018). Information security risk estimation for cloud infrastructure. *International Journal on Information Technologies & Security, 10*(4), 67–76.

Tserng, H. P., Cho, I. C., Chen, C. H., & Liu, Y. F. (2021). Developing a risk management process for infrastructure projects using IDEF0. *Sustainability (Basel), 13*(12), 6958. doi:10.3390u13126958

Tsitsiashvili, G. (2021). Alternative designs of high load queuing systems with small queue. In *Informatics and Cybernetics in Intelligent Systems: Proceedings of 10th Computer Science On-line Conference 2021,* Vol. 3 (pp. 69-76). Springer International Publishing. 10.1007/978-3-030-77448-6_8

Turner, C. J., & Garn, W. (2022). Next generation DES simulation: A research agenda for human centric manufacturing systems. *Journal of Industrial Information Integration, 28*, 100354. doi:10.1016/j.jii.2022.100354

UKEssays. (2018). *Shared Memory MIMD Architectures*. UK Essays. (https://www.ukessays.com/essays/architecture/shared-memory-mimd-architecture.php)

Uryvsky, L., & Martynova, K. (2019). Complex analytical model of priority requires service on cloud server. In *2019 International Conference on Information and Telecommunication Technologies and Radio Electronics (UkrMiCo)*, Odessa, Ukraine. 10.1109/UkrMiCo47782.2019.9165323

Van der Linden, P. (1994). *Expert C programming: deep C secrets*. Prentice Hall Professional.

Varenne, F., & Turnbull, K. (2018). *From models to simulations* (1st ed.). Routledge. doi:10.4324/9781315159904

Vassilev, Tz. (2020). Comparison of several paradigms for accelerating physical simulation on a PC. *International Journal on Information Technologies and Security, 12*(3), 13–24.

Vendrell-Herrero, F., Bustinza, O. F., & Opazo-Basaez, M. (2021). Information technologies and product-service innovation: The moderating role of service R&D team structure. *Journal of Business Research*, *128*, 673–687. doi:10.1016/j.jbusres.2020.01.047

Ventisei, A., Yakovlev, A., & Pacheco-Peña, V. (2022). Exploiting Petri nets for graphical modelling of electromagnetic pulse switching operations. *Advanced Theory and Simulations*, *5*(3), 2100429. doi:10.1002/adts.202100429

Wanduku, D. (2022). The multilevel hierarchical data EM-algorithm. Applications to discrete-time Markov chain epidemic models. *Heliyon*, *8*(12), e12622. doi:10.1016/j.heliyon.2022.e12622 PMID:36643325

Wang, G. (1994). *Treading different paths: informatization in Asian nations.*

Wang, S. P. (2021). Advanced computer architecture. In *Computer Architecture and Organization: Fundamentals and Architecture Security* (pp. 163–212). Springer., doi:10.1007/978-981-16-5662-0_7

Wang, X., Zhang, P., Du, Y., & Qi, M. (2020). Trust routing protocol based on cloud-based fuzzy Petri net and trust entropy for mobile ad hoc networks. *IEEE Access : Practical Innovations, Open Solutions*, *8*, 47675–47693. doi:10.1109/ACCESS.2020.2978143

Weder, B., Breitenbücher, U., Leymann, F., & Wild, K. (2020). Integrating quantum computing into workflow modeling and execution. *2020 IEEE/ACM 13th International Conference on Utility and Cloud Computing (UCC)*, (pp. 279-291). IEEE. 10.1109/UCC48980.2020.00046

Wickens, C. D., & Carswell, C. M. (2021). Information processing. Handbook of human factors and ergonomics, 114-158.

Williams, A. E. (2020, April 16). A human-centric functional modeling framework for defining and comparing models of consciousness and cognition, PsyArXiv. doi:10.31234/osf.io/94gw3osf.io/94gw3

WilliamsA. E. (2020). A Human-Centric Functional Modeling Framework for Defining and Comparing Models of Consciousness and Cognition. PsyArXiv, April 16. doi:10.31234/osf.io/94gw3

Wright, J., & Ma, Y. (2022). *High-dimensional data analysis with low-dimensional models: Principles, computation, and applications*. Cambridge University Press., doi:10.1017/9781108779302

Wright, S. A. (2019). Performance modeling, benchmarking and simulation of high-performance computing systems. *Future Generation Computer Systems*, *92*(March), 900–902. doi:10.1016/j.future.2018.11.020

Xie, Q., & Jin, L. (2022). Stabilizing queuing networks with model data-independent control. *IEEE Transactions on Control of Network Systems*, *9*(3), 1317–1326. doi:10.1109/TCNS.2022.3145752

Xu, J., & Gautam, N. (2020). Peak age of information in priority queuing systems. *IEEE Transactions on Information Theory*, *67*(1), 373–390. doi:10.1109/TIT.2020.3033501

Xu, X., He, X., Ai, Q., & Qiu, R. C. (2015). A correlation analysis method for power systems based on random matrix theory. *IEEE Transactions on Smart Grid*, *8*(4), 1811–1820. doi:10.1109/TSG.2015.2508506

Yadavalli, V. S. S., Adetunji, O., & Alrikabi, R. (2021). Optimization of the berth allocation problem to the vessels using priority queuing systems. In *Soft Computing in Inventory Management* (pp. 41–68). Springer., doi:10.1007/978-981-16-2156-7_3

Yoshikawa, H., Lind, M., Matsuoka, T., Hashim, M., Yang, M., & Zhang, Z. (2013). A new functional modeling framework of risk monitor system. *International Electronic Journal of Nuclear Safety and Simulation*, *4*(3), 192–202.

Yu, F., Li, L., Tang, Q., Cai, S., Song, Y., & Xu, Q. (2019). A survey on true random number generators based on chaos. *Discrete Dynamics in Nature and Society,* 1-10. doi:10.1155/2019/2545123

Yu, J., Xiao, B., & Liang, H. (2022, September). Decomposition modeling of uncertain combat missions for air defense and anti-missile based on IDEF. In *2022 6th International Conference on Automation, Control and Robots (ICACR)* (pp. 176-179). IEEE. 10.1109/ICACR55854.2022.9935553

Yu, K., Hua, Q., Wang, S., Li, N., & Zhang, Y. (2015, September). An user interface dialog control model based on UI patterns. In *2015 6th IEEE International Conference on Software Engineering and Service Science (ICSESS)* (pp. 702-705). IEEE. 10.1109/ICSESS.2015.7339154

Yu, X., Jiang, F., Du, J., & Gong, D. (2019). A cross-domain collaborative filtering algorithm with expanding user and item features via the latent factor space of auxiliary domains. *Pattern Recognition, 94*, 96–109. doi:10.1016/j.patcog.2019.05.030

Zanni, M., Sharpe, T., Lammers, P., Arnold, L., & Pickard, J. (2021). Towards a BIM-based decision support system for integrating whole life cost estimation into design development. In *Proceedings of the 18th International Conference on Computing in Civil and Building Engineering: ICCCBE 2020* (pp. 197-206). Springer International Publishing. 10.1007/978-3-030-51295-8_16

Zeifman, A., Korolev, V., & Satin, Y. (2020). Two approaches to the construction of perturbation bounds for continuous-time Markov chains. *Mathematics, 8*(2), 253. doi:10.3390/math8020253

Zhang, L., Ye, F., Laili, Y., Xie, K., Gu, P., Wang, X., Zhao, C., Zhang, X., & Chen, M. (2021, July). X language: an integrated intelligent modeling and simulation language for complex products. In *2021 Annual Modeling and Simulation Conference (ANNSIM)* (pp. 1-11). IEEE. 10.23919/ANNSIM52504.2021.9552057

Zhang, X., & Furnstahl, R. J. (2022). Fast emulation of quantum three-body scattering. *Physical Review. C, 105*(6), 064004. doi:10.1103/PhysRevC.105.064004

Zhao, J., Li, Z., Gao, Q., Zhao, H., Chen, S., Huang, L., Wang, W., & Wang, T. (2021). A review of statistical methods for dietary pattern analysis. *Nutrition journal, 20*(1), 1-18. doi:10.1186/s12937-021-00692-7

Zhao, X., Wang, Y., Li, L., & Delahaye, D. (2022). A queuing network model of a multi-airport system based on point-wise stationary approximation. *Aerospace (Basel, Switzerland), 9*(7), 390. doi:10.3390/aerospace9070390

Zhernovyi, Y. V. (2018). Calculating steady-state characteristics of single-channel queuing systems using phase-type distributions. *Cybernetics and Systems Analysis, 54*(5), 824–832. doi:10.100710559-018-0084-2

Zhou, K. Q., & Zain, A. M. (2016). Fuzzy Petri nets and industrial applications: A review. *Artificial Intelligence Review, 45*(4), 405–446. doi:10.100710462-015-9451-9

Zoppo, G., Korkmaz, A., Marrone, F., Palermo, S., Corinto, F., & Williams, R. S. (2021). Analog solutions of discrete Markov chains via memristor crossbars. *IEEE Transactions on Circuits and Systems. I, Regular Papers, 68*(12), 4910–4923. doi:10.1109/TCSI.2021.3126477

Zoteva, D., & Angelova, N. (2021). An overview of the main results and applications. Research in Computer Science in the Bulgarian Academy of Sciences, 177-226.

Zozulya, M.M., & Kravets, O.Ja., Atlasov, I.V., Aksenov, I.A., Bozhko, L.M. & Rahman, P.A. (2022). Algorithmization of the software testing system based on finite automata. *International Journal on Information Technologies and Security, 14*(1), 77–86.

Zyulkov, A., Kutoyants, Y., Perelevskiy, S., & Korableva, L. (2022, December). Single channel queuing system utilization factor model. [). IOP Publishing.]. *Journal of Physics: Conference Series, 2388*(1), 012043. doi:10.1088/1742-6596/2388/1/012043

About the Authors

Radi Romansky is originally from Botevgrad, Bulgaria. Full professor at Technical University of Sofia in "Computer Systems, Complexes and Networks", Doctor (PhD) in "Computer Engineering" and Doctor of Science (D.Sc.) in "Informatics and Computer Science". He has been Head of Department (2000-2004 & 2010-2015), Vice Chairmen of General Assembly of Technical University of Sofia (2011-2015), and Vice-Rector of Technical University of Sofia (2015-2019). Full member of European Network of Excellence on High Performance and Embedded Architectures and Compilation (HiPEAC) and Member of Advisory Board of "The Swiss Innovation Lab", Swiss Innovation Valley. Member of Management Council of Union of Electronics, Electrical Engineering and Communication; Member of Consultative Council of "Center for Research and Analysis". He is organizer and publisher of International Journal on Information Technologies and Security and International Conference on Information Technologies. Member of Editorial Board of journals Mathematics (Switzerland), Electro-technique and Computer Systems (Ukraine), Electronics and Electro-technique (Bulgaria), etc. He has over 225 scientific publications and over 25 books. Areas of scientific interests: ICT, informatics, computer architectures, computer modelling, privacy, and data protection, etc.

Nikolay Hinov is originally from Lukovit, Bulgaria. Professor at Technical University of Sofia in "Industrial electronics", Doctor (PhD) in "Power electronic converters". He has been Vice Dean of the Faculty of Electronic Engineering and Technology of Technical University of Sofia (2011-2019). Currently, he is the Head of the Department of Power Electronics at the Technical University of Sofia. Member of IEEE Computational Intelligence Society, IEEE Power Electronics Society and IEEE Industrial Electronics Society. Member of Management Council of Union of Electronics, Electrical Engineering and Communication. He is organizer of International Scientific Conference COMPUTER SCIENCE, National Conference with International Participation Electronica, International Conference on High Technology for Sustainable Development, and International Conference on Information Technologies. Member of Editorial Board of journals Electronics (Switzerland) and International Journal on Information Technologies and Security (Bulgaria). He has over 265 scientific publications, 13 patents and utility models and 2 books. Areas of scientific interests: ICT, artificial intelligence, computer architectures, computer modelling, electric and autonomous cars, smart cities, and networks, etc.

Index

D

E

N

O

P

R

S

T

V

www.ingramcontent.com/pod-product-compliance
Lightning Source LLC
LaVergne TN
LVHW080115011026
849160LV00013B/1110